An Introduction To The Theory Of INFINITE SERIES

T. J. I'A Bromwich, M.A., F.R.S.

1908

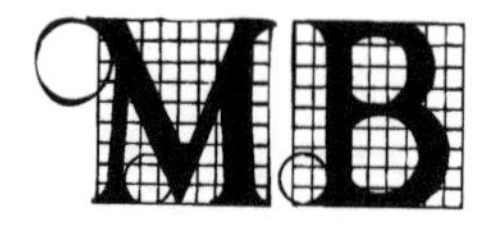

ISBN 1-60386-122-X

PREFACE.

THIS book is based on courses of lectures on Elementary Analysis given at Queen's College, Galway, during each of the sessions 1902-1907. But additions have naturally been made in preparing the manuscript for press: in particular the whole of Chapter XI. and the greater part of the Appendices have been added. In selecting the subject-matter, I have attempted to include proofs of all theorems stated in Pringsheim's article, *Irrationalzahlen und Konvergenz unendlicher Prozesse,** with the exception of theorems relating to continued fractions.

In Chapter I. a preliminary account is given of the notions of a limit and of convergence. I have not in this chapter attempted to supply arithmetic proofs of the fundamental theorems concerning the existence of limits, but have allowed their truth to rest on an appeal to the reader's intuition, in the hope that the discussion may thus be made more attractive to beginners. An arithmetic treatment will be found in Appendix I., where Dedekind's definition of irrational numbers is adopted as fundamental; this method leads at once to the monotonic principle of convergence (Art. 149), from which the existence of extreme limits † is deduced (Arts. 5, 150); it is then easy to establish the general principle of convergence (Art. 151).

In the remainder of the book free use is made of the notation and principles of the Differential and Integral Calculus; I have for some time been convinced that beginners should not attempt to study Infinite Series in any detail until after they have

---

* *Encyklopädie der Mathematischen Wissenschaften*, Bd. I., A, 3 and G, 3 (pp. 47 and 1121).

† Not only here, but in many other places, the proofs and theorems have been made more concise by a systematic use of these maximum and minimum limits.

mastered the differentiation and integration of the simpler functions, and the geometrical meaning of these operations.

The use of the Calculus has enabled me to shorten and simplify the discussion of various theorems (for instance, Arts. 11, 61, 62), and to include other theorems which must have been omitted otherwise (for instance, Arts. 45, 46, and the latter part of 83).

It will be noticed that from Art. 11 onwards, free use is made of the equation

$$\frac{d}{dx}(\log x)=\frac{1}{x},$$

although the limit of $(1+1/\nu)^\nu$ (from which this equation is commonly deduced) is not obtained until Art. 57. To avoid the appearance of reasoning in a circle, I have given in Appendix II. a treatment of the theory of the logarithm of a real number, starting from the equation

$$\log a=\int_1^a \frac{dx}{x}.$$

The use of this definition of a logarithm goes back to Napier, but in modern teaching its advantages have been overlooked until comparatively recently. An arithmetic proof that the integral represents a definite number will be found in Art. 163, although this fact would naturally be treated as axiomatic when the subject is approached for the first time.

In Chapter V. will be found an account of Pringsheim's theory of double series, which has not been easily accessible to English readers hitherto.

The notion of uniform convergence usually presents difficulties to beginners; for this reason it has been explained at some length, and the definition has been illustrated by Osgood's graphical method. The use of Abel's and Dirichlet's names for the tests given in Art. 44 is not strictly historical, but is intended to emphasise the similarity between the tests for uniform convergence and for simple convergence (Arts. 19, 20).

In obtaining the fundamental power-series and products constant reference is made to the principle of uniform convergence, and particularly to Tannery's theorems (Art. 49); the proofs are thus simplified and made more uniform than is otherwise possible. Considerable use is also made of Abel's theorem (Arts. 50, 51, 83) on the continuity of power-series, a theorem which, in spite of its importance, has usually not been adequately discussed in text-books.

Chapter XI. contains a tolerably complete account of the recently developed theories of non-convergent and asymptotic series; the treatment has been confined to the arithmetic side, the applications to function-theory being outside the scope of the book. As might be expected, a systematic examination of the known results has led to some extensions of the theory (see, for instance, Arts. 118-121, 123, and parts of 133).

The investigations of Chapter XI. imply an acquaintance with the convergence of infinite integrals, but when the manuscript was being prepared for printing no English book was available from which the necessary theorems could be quoted.* I was therefore led to write out Appendix III., giving an introduction to the theory of integrals; here special attention is directed to the points of similarity and of difference between this theory and that of series. To emphasise the similarity, the tests of convergence and of uniform convergence (Arts. 169, 171, 172) are called by the same names as in the case of series; and the traditional form of the Second Theorem of Mean Value is replaced by inequalities (Art. 166) which are more obviously connected with Abel's Lemma (Arts. 23, 80). To illustrate the general theory, a short discussion of Dirichlet's integrals and of the Gamma integrals is given; it is hoped that these proofs will be found both simple and rigorous.

The examples (of which there are over 600) include a number of theorems which could not be inserted in the text, and in such cases references are given to sources of further information.

Throughout the book I have made it my aim to keep in view the practical applications of the theorems to every-day work in analysis. I hope that most double-limit problems, which present themselves *naturally*, in connexion with integration of series, differentiation of integrals, and so forth, can be settled without difficulty by using the results given here.

---

* While my book has been in the press, three books have appeared, each of which contains some account of this theory: Gibson's *Calculus* (ch. xxi., 2nd ed.), Carslaw's *Fourier Series and Integrals* (ch. iv.), and Pierpont's *Theory of Functions of a Real Variable* (chs. xiv., xv.).

Mr. G. H. Hardy, M.A., Fellow and Lecturer of Trinity College, has given me great help during the preparation of the book; he has read all the proofs, and also the manuscript of Chapter XI. and the Appendices. I am deeply conscious that the value of the book has been much increased by Mr. Hardy's valuable suggestions and by his assistance in the selection and manufacture of examples.

The proofs have also been read by Mr. J. E. Bowen, B.A., Senior Scholar of Queen's College, Galway, 1906-1907; and in part by Mr. J. E. Wright, M.A., Fellow of Trinity College, and Professor at Bryn Mawr College, Pennsylvania. The examples have been verified by Mr. G. N. Watson, B.A., Scholar of Trinity College, who also read the proofs of Chapter XI. and Appendix III. To these three gentlemen my best thanks are due for their careful work.

T. J. I'A. BROMWICH.

CAMBRIDGE, *December*, 1907.

The following list comprises those books of which I have made most use in arranging the material;

Chrystal, *Algebra*, vol. 2.
Hobson, *Trigonometry*.
Osgood, *Infinite Series*.
De la Vallée Poussin, *Cours d'Analyse Infinitésimale*.
Goursat, *Cours d'Analyse Mathématique*.
Tannery, *Théorie des Fonctions d'une Variable*, t. 1.
Cesàro, *Lehrbuch der Algebraischen Analysis*.
Pringsheim, *Mathematische Annalen*, Bd. 35, pp. 297-394.

Reference has also been made to works on Analysis and Theory of Functions by Baire, Borel, Dini, Harkness and Morley, Hobson, Jordan, Lebesgue, Nielsen, Osgood, Picard, Runge, Schlömilch, Stolz, Vivanti, and various other authors, in addition to the sources mentioned above and in Chapter XI.

CONTENTS.

CHAPTER I.

PAGES

SEQUENCES AND LIMITS, - - - - - - 1-16

Convergence of Sequences. Monotonic Sequences. General Principle of Convergence. Upper and Lower Limits. Maximum and Minimum Limits. Sum of an Infinite Series.

EXAMPLES, - - - - - - - 17

CHAPTER II.

SERIES OF POSITIVE TERMS, - - - - - - 22-42

Cauchy's Condensation Test. Comparison Test. Integral Test. Logarithmic Scale. Ratio Tests. Ermakoff's Tests. Another sequence of Tests. Notes on Tests of Convergence.

EXAMPLES, - - - - - - - 42

CHAPTER III.

SERIES IN GENERAL, - - - - - - - 46-59

Absolute and Non-Absolute Convergence. Tests of Abel and Dirichlet. Alternate Series; Ratio Test. Abel's Lemma. Euler's Transformation.

EXAMPLES, - - - - - - - 60

CHAPTER IV.

ABSOLUTE CONVERGENCE, - - - - - 63-70

Derangement. Applications. Riemann's and Pringsheim's Theorems.

EXAMPLES, - - - - - - - 70

CHAPTER V.

PAGES

DOUBLE SERIES, - - - - - - - - - - 72-90

Sum of a Double Series. Repeated Summation. Series of Positive Terms. Tests for Convergence. Absolute Convergence. Multiplication of Series. Mertens' and Pringsheim's Theorems. Substitution of a Power-Series in another Power-Series. Non-Absolute Convergence.

EXAMPLES, - - - - - - - - - 90

Theta-Series (**16**); other Elliptic Function-Series (**19-24**).

CHAPTER VI.

INFINITE PRODUCTS, - - - - - - - - 95-103

Weierstrass's Inequalities. Positive Terms. Terms of either Sign. Absolute Convergence. Gamma-Product.

EXAMPLES, - - - - - - - - - 103

Theta-Products (**14-20**).

CHAPTER VII.

SERIES OF VARIABLE TERMS, - - - - - - 108-125

Uniform Convergence of Sequences. Uniform Convergence of Series. Weierstrass's, Abel's, and Dirichlet's Tests. Continuity, Integration, and Differentiation of Series. Products. Tannery's Theorems.

EXAMPLES, - - - - - - - - 125

Bendixson's Test of Uniform Convergence (**14**).

CHAPTER VIII.

POWER SERIES, - - - - - - - - - 128-142

Intervals of Convergence. Abel's Theorem. Continuity, Differentiation, Integration. Theorem of Identical Equality between two Power-Series. Multiplication and Division. Reversion of Series. Lagrange's Series.

SPECIAL POWER SERIES, - - - - - - - - 143-160

Exponential Series. Sine and Cosine Series. Binomial Series. Logarithmic Series. arc sin and arc tan Series. $\Sigma r^n \cos n\theta$, $\Sigma r^n \sin n\theta$, $\Sigma \frac{1}{n} r^n \cos n\theta$, $\Sigma \frac{1}{n} r^n \sin n\theta$.

EXAMPLES, - - - - - - - - - 161

Series for π (A. **48**); Abel's Theorem (B. **13-21**); Lagrange's Series (B. **30, 31**); Differential Equations (B. **34-36**).

CHAPTER IX.

PAGES

TRIGONOMETRICAL INVESTIGATIONS, - - - - - 177-187

Sines and Cosines of Multiple Angles. The Sine and Cosine Products. The Cotangent Series.

EXAMPLES, - - - - - - - 188

CHAPTER X.

COMPLEX SERIES AND PRODUCTS, - - - - - 192-240

Complex Numbers. De Moivre's Theorem. Convergence of Complex Sequences. Absolute Convergence of Series and Products. Pringsheim's Ratio Tests for Absolute Convergence. Weierstrass's Test for Power-Series. Abel's and Dirichlet's Tests for Convergence. Uniform Convergence. Circle of Convergence. Abel's Theorem. Poisson's Integral. Taylor's Theorem. Exponential Series. Sine and Cosine Series. Logarithmic Series; arc sin and arc tan Series. Binomial Series. Differentiation of Trigonometrical Series. Sine and Cosine Products. The Cotangent Series. Bernoulli's Numbers. Bernoullian Functions. Euler's Summation Formula.

EXAMPLES, - - - - - - - - 241

Trisection of an Angle (A. 12); Abel's Theorem (B. 27, C. 9-11); Weierstrass's Double Series Theorem (B. 32).

CHAPTER XI.

NON-CONVERGENT AND ASYMPTOTIC SERIES, - - 261-267

Bibliography. Historical Introduction. General Considerations.

BOREL'S METHOD OF SUMMATION, - - - - - 267-296

Borel's Integral. Condition of Consistency. Addition of Terms to a Summable Series. Examples of Summation. Absolute Summability. Multiplication. Continuity, Differentiation, and Integration. Analogue of Abel's Theorem. Summable Power-Series.

OTHER METHODS OF SUMMATION, - - - - - 297-322

Borel's and Le Roy's other Definitions. An Extension of Borel's Definition. Euler's Series and Borel's Integral. Examples of Euler's Series. Cesàro's Mean. Extension of Frobenius's Theorem. Multiplication of Series. Examples of Cesàro's Method. Borel's Sum and Cesàro's Mean.

PAGES

ASYMPTOTIC SERIES, - - - - - - - 322-346

Euler's use of Asymptotic Series. Remainder in Euler's Formula. Logarithmic Integral. Fresnel's Integrals. Stirling's Series. Poincaré's Theory of Asymptotic Series. Stokes' Asymptotic Formula. Summation of Asymptotic Series. Applications to Differential Equations.

EXAMPLES, - - - - - - - 347

Fejér's and de la Vallée Poussin's Theorems (5-7).

APPENDIX I.

ARITHMETIC THEORY OF IRRATIONAL NUMBERS AND LIMITS, 357-389

Infinite Decimals. Dedekind's Definition. Algebraic Operations with Irrational Numbers. Monotonic Sequences. Extreme Limits of a Sequence. General Principle of Convergence. Limits of Quotients. Extension of Abel's Lemma. Theorems on Limits.

EXAMPLES, - - - - - - - - 390

Infinite Sets (15-17); Goursat's Lemma (18-20); Continuous Functions (21-23).

APPENDIX II.

DEFINITIONS OF THE LOGARITHMIC AND EXPONENTIAL FUNCTIONS, - - - - - - - 396-410

Definition and Fundamental Properties of the Logarithm. Exponential Function. Logarithmic Scale of Infinity. Exponential Series. Arithmetic Definition of an Integral (Single and Double).

EXAMPLES, - - - - - - - - 410

APPENDIX III.

SOME THEOREMS ON INFINITE INTEGRALS AND GAMMA-FUNCTIONS, - - - - - - - - 414-466

Convergence, Divergence, and Oscillation. Definition as a Limit of a Sum. Tests of Convergence. Analogue of Abel's Lemma (or Second Theorem of the Mean). Absolute

PAGES

Convergence. Abel's and Dirichlet's Tests. Frullani's Integrals. Uniform Convergence. Weierstrass's, Abel's and Dirichlet's Tests for Uniform Convergence. Continuity, Differentiation, and Integration. Special Integrals. Limiting Values of Integrals. Dirichlet's Integrals. Jordan's Integral. Integration of Series. Inversion of Repeated Integrals. The Gamma Integral. Stirling's Asymptotic Formula. Gamma-Function Formulae.

EXAMPLES, - - - - - - - - 467

Gamma-Functions (35-47, 55-57).

EASY MISCELLANEOUS EXAMPLES.

EXAMPLES, - - - - - - - - 479-483

HARDER MISCELLANEOUS EXAMPLES.

EXAMPLES, - - - - - - - - - 484-506

Riemann's Discontinuous Series (18); Lagrange's Series (19, 20); Weierstrass's Non-Differentiable Function (32-36); Riemann's ζ-Function (44-49); Riemann's Theorems on Trigonometrical Series (64-66); Jacobi's Theta-Functions (70-77); Functions without Analytical Continuations (78-85); Kummer's Series (86); Double Power-Series (87-100).

INDEX OF SPECIAL INTEGRALS, PRODUCTS AND SERIES, - - - - - - - - 507-508

GENERAL INDEX, - - - - - - 509-511

ADDENDA AND CORRIGENDA.

p. 16, l. 17. The series is here supposed to oscillate *finitely*; such a series as $1-2+3-4+\ldots$ is excluded.

p. 29, Art. 11, and p. 80, l. 2. The integral test is commonly attributed to Cauchy: it occurs in Maclaurin's *Fluxions*, 1742, Art. 350.

p. 97. Ex. 1. The value of the second product is $(\sinh\pi)/2\pi = 1\cdot845\ldots$ (Art. 91.)

p. 101. Art. 41. A proof of the first part can also be given on the lines of Art. 77.

p. 122. The discussion can be somewhat shortened by the use of extreme limits; thus, if m is chosen so that

$$1-\epsilon < Q_m(x)/Q_m(c) < 1+\epsilon,$$

we get at once

$$1-\epsilon \leqq \overline{\lim}\, P(x)/P(c) \leqq 1+\epsilon.$$

p. 141. For methods of determining the region of convergence of Lagrange's series, see Goursat, *Cours d'Analyse Math.*, t. 2, p. 131, and Schlömilch, *Kompendium der höheren Analysis*, Bd. 2, p. 100; the relation between these methods will be seen from theorems due to Macdonald, *Proc. Lond. Math. Soc.*, vol. 29, p. 576.

p. 146, Ex. The numerical results should be 4·8105... instead of 4·80, and 23·14... instead of 23·00.

p. 164. Ex. 25, l. 4. The index $^{-1}$ is omitted from $\{(1-xy)(1-x/y)\}^{-1}$.

p. 190. Ex. 18. This is taken from the *Mathematical Tripos Papers*, 1890.

pp. 212, 213. It is assumed that v is a function of ω with a period 2π, so that $v(\omega)=v(\omega-2\pi)$.

p. 226. Professor Dixon's own version of his proof has just been published in the *Quarterly Journal of Mathematics* (vol. 39, p. 94, Oct. 1907).

p. 227. l. 8. The reference should be to Arts. 44, 45 (1), instead of Art. 49.

p. 323. l. 3. The last figure in Euler's constant should be 8 instead of 5, and the following four figures are 6060, according to Gauss.

p. 410, l. 4. The proof that $\Sigma > s_n$ and $\sigma < S_n$ can be made purely arithmetical by using Σ', σ', the sums obtained by superposing the two modes of division. We have then $\Sigma' > \sigma'$, while Σ' is less than both S_n and Σ, and σ' is greater than both s_n and σ. Thus $\Sigma > \Sigma' > \sigma' > s_n$, and similarly $\sigma < \sigma' < \Sigma' < S_n$.

p. 471. Exs. 20, 21. Similar integrals occur in Electron-Theory (compare Sommerfeld, *Göttingen Nachrichten*, 1904, p. 117).

p. 490. Non-differentiable functions. Other examples of a simple character have been given recently by H. von Koch, *Acta Mathematica*, Bd. 30, 1907, p. 145; and by Faber, *Jahresbericht der Deutschen Math. Verein*, Bd. 16, 1907, p. 538.

p. 495, Ex. 51. The function $f(x, n)$ is supposed monotonic with respect to n.

AN INTRODUCTION TO THE THEORY OF INFINITE SERIES

CHAPTER I.

SEQUENCES AND LIMITS.

1. Infinite sequences: convergence and divergence.

Suppose that we have agreed upon some rule, or rules, by which we can associate a definite number a_n with any assigned positive integer n; then the set of numbers

$$a_1, a_2, a_3, a_4, \ldots, a_n, \ldots,$$

arranged so as to correspond to the set of positive integers $1, 2, 3, 4, \ldots, n, \ldots$, will be called an *infinite sequence,* or simply a *sequence.* We shall frequently find it convenient to use the notation (a_n) to represent this sequence. The use of the word *infinite* simply means that *every* term in the sequence is followed by another term.

The rule defining the sequence may either be expressed by some formula (or formulae) giving a_n as an explicit function of n; or by some verbal statement which indicates how each term can be determined, either directly or from the preceding terms.

Ex. 1. If $a_n = 2n - 1$, we have the sequence of odd numbers $1, 3, 5, 7, \ldots$.

Ex. 2. If $a_n = 1/n$, we have the harmonic sequence $1, \frac{1}{2}, \frac{1}{3}, \frac{1}{4}, \ldots$.

Ex. 3. The set consisting of the rational positive proper fractions, *arranged in order of magnitude*, is not a sequence. For if a is any fraction of the set, $\frac{1}{2}a$ also belongs to the set; and since $\frac{1}{2}a$ is less than a, we must place $\frac{1}{2}a$ before a. Thus there can be no *first* number of the set; and so this mode of arrangement does not lead to any correspondence between the set and the positive integers. It is, however, possible to arrange these fractions as a sequence, by adopting a different mode of arrangement; for example $\frac{1}{2}, \frac{1}{3}, \frac{2}{3}, \frac{1}{4}, \frac{3}{4}, \frac{1}{5}, \frac{2}{5}, \frac{3}{5}, \frac{4}{5}$, etc., in which the fractions are arranged, first, according to the magnitude of their denominators, and, secondly, according to the magnitude of their numerators.

The most important sequences in the applications of analysis are those which tend to a limit.

The **limit of a sequence** (a_n) *is said to be* l, *if an index* m *can be found to correspond to every positive number* ϵ, *however small, such that*

$$l-\epsilon < a_n < l+\epsilon,$$

provided only that $n > m$.

It is generally more convenient to contract these two inequalities into the single one

$$|l-a_n| < \epsilon,$$

where the symbol $|x|$ is used to denote the numerical value of x.

The following notations will be convenient abbreviations for the above property:

$$l = \lim_{n\to\infty} a_n; \text{ or } l = \lim a_n; \text{ or } a_n \to l;$$

the two latter being only used when there is no doubt as to what variable tends to infinity.

Amongst sequences having no limit it is useful to distinguish those with an infinite limit.

A sequence (a_n) *has* **an infinite limit**, *if, no matter how large the number* N *may be, an index* m *can be found such that*

$$a_n > N,$$

provided only that $n > m$.

This property is expressed by the equations

$$\lim_{n\to\infty} a_n = \infty; \text{ or } \lim a_n = \infty; \text{ or } a_n \to \infty.$$

In like manner, we interpret the equations

$$\lim_{n\to\infty} a_n = -\infty, \ \lim a_n = -\infty, \ a_n \to -\infty.$$

In case the sequence (a_n) has a finite limit l, it is called *convergent* and is said to *converge to* l *as a limit*; if the sequence has an infinite limit, it is called *divergent.**

Ex. 1 *bis.* With $a_n = 2n-1$ (the sequence of odd integers) we have $a_n \to \infty$; a *divergent* sequence.

Ex. 2 *bis.* With $a_n = 1/n$ (the harmonic sequence) we have $a_n \to 0$; a *convergent* sequence.

---

* Some writers regard *divergent* as equivalent to *non-convergent*; but it seems convenient to distinguish between sequences which tend to infinity as a limit and those which oscillate. We shall call the latter sequences *oscillatory* (Art 5).

Ex. 3 *bis.* If the sequence consists of the rational proper fractions arranged in any definite order no limit (finite or infinite) can exist. For, no matter how far we go in the sequence, there will always remain an unlimited number of terms as close to 0 as we please; and also an unlimited number as close to 1 as we please.

We shall find it convenient sometimes to represent a sequence graphically, indicating a term a_n by an ordinate (y) equal to a_n and an abscissa (x) equal to n; the sequence may then be pictured by joining the successive points with a broken straight line. In the case of a convergent sequence, the representative points lie wholly within a horizontal strip of width 2ϵ, after x exceeds a certain value; if the sequence is divergent, the points lie wholly above (or below) a certain level, after x has passed a certain value.

The graphical representation of the initial terms in the three sequences already considered is given below.

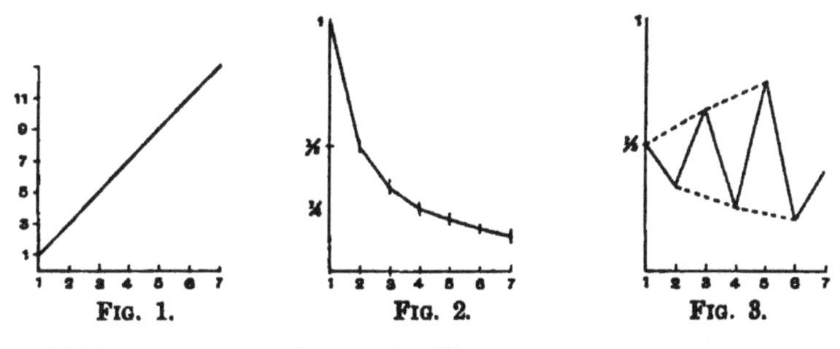

FIG. 1. FIG. 2. FIG. 3.

It will be seen at a glance that the few terms represented in the diagram shew that the first sequence is likely to diverge, the second to converge, and the third to oscillate (see Art. 5).

NOTES.

(1) The definition of a limit is often loosely stated as follows: *The sequence* (a_n) *approaches the limit* l, *if, by taking* n *large enough, we can make* $|l-a_n|$ *as small as we please.*

Such a definition does not exclude the possibility of *oscillation,* as may be seen from the sequence

$$\tfrac{1}{2},\ \tfrac{1}{3},\ \tfrac{2}{3},\ \tfrac{1}{4},\ \tfrac{3}{4},\ \tfrac{1}{5},\ \tfrac{4}{5},\ \tfrac{1}{6},\ \tfrac{5}{6},\ \tfrac{1}{7},\ \tfrac{6}{7},\ \ldots,$$

in which $a_n=2/(n+4)$ if n is even.

Here, by taking n large enough, we can find a term a_n which is as small as we please; but the sequence oscillates between 0 and 1, because $a_n=(n+1)/(n+3)$ when n is *odd.*

(2) *Infinity.*

It is to be remembered that the symbol ∞ and the terms *infinite, infinity, infinitely great,* etc., have purely conventional meanings in the present theory; in fact, anticipating the definitions of Art. 4, we may say that *infinity must be regarded as an upper limit which cannot be attained.* The statement that a set contains an infinite number of objects may be understood as implying that no number suffices to count the set.

Similarly, an equation such as $\lim a_n = \infty$ is merely a conventional abbreviation for the definition on p. 2.

In speaking of a divergent sequence (a_n), some writers use phrases such as: *The numbers a_n* **become** *infinitely great, when n increases without limit.* Of course this phrase is used as an equivalent for **tend** *to infinity*; but we shall avoid the practice in the sequel.

(3) It is evident that *the alteration of a* **finite** *number of terms in a sequence will not alter the limit.*

Ex. 4. The two sequences

$$1,\ 2,\ 3,\ 4,\ \tfrac{1}{5},\ \tfrac{1}{6},\ \tfrac{1}{7},\ \tfrac{1}{8},\ \ldots$$
$$1,\ \tfrac{1}{2},\ \tfrac{1}{3},\ \tfrac{1}{4},\ \tfrac{1}{5},\ \tfrac{1}{6},\ \tfrac{1}{7},\ \tfrac{1}{8},\ \ldots$$

have the same limit zero.

Further, it is evident that *the omission of* **any** *number of terms from a convergent or divergent sequence does not affect the limit; but such omission may change the character of an oscillatory sequence.*

Ex. 5. Thus the sequences $1, 5, 9, 13, \ldots$ and $1, \tfrac{1}{3}, \tfrac{1}{5}, \tfrac{1}{7}, \ldots$ have the same limits as those considered in Exs. 1, 2. But the omission of the alternate terms in the sequence

$$\tfrac{1}{2},\ \tfrac{1}{3},\ \tfrac{2}{3},\ \tfrac{1}{4},\ \tfrac{3}{4},\ \tfrac{1}{5},\ \tfrac{4}{5},\ \ldots$$

changes it into a convergent sequence.

(4) In a convergent sequence, *all, an infinity, a finite number* or *none* of the terms may be equal to the limit.

Examples of these four possibilities (in order) are given by:

$$a_n = 1\,;\quad a_n = \frac{1}{n}\sin(\tfrac{1}{2}n\pi)\,;\quad a_n = \frac{1}{n} - \frac{3}{n^2} + \frac{2}{n^3}\,;\quad a_n = \frac{1}{n}\,;$$

the limits of which are, in order, 1, 0, 0, 0.

(5) We shall usually employ ϵ to denote an arbitrarily small positive number; strictly speaking, the words *arbitrarily small*, or *no matter how small*, or *however small* (which are frequently added to ϵ) are redundant, but serve to emphasise the statement that the variable is *less* than ϵ.

We shall also use N (or G) to denote an arbitrarily great positive number; here again the adjectives *great* or *large* are unnecessary, but are usually added to emphasise the statement that the variable is *greater* than N.

By using ϵ to denote **any** positive number, we could dispense with N; but it avoids confusion to use two distinct symbols. However it is sometimes convenient to use $1/\epsilon$ for N.

(6) It often happens that a certain limit l can be proved to be less than $a+\epsilon_m$, where ϵ_m can be made arbitrarily small by choice of an index *m which does not appear in l or a*; we can then infer that $l \leqq a$. For if l were greater than a, we could not make ϵ_m less than $l-a$, which contradicts the hypothesis.

As a special example of this, suppose that (a_n), (b_n) are two convergent sequences such that $|b_n - a_n|$ can be made less than ϵ by taking $n > m$; then we can choose m, so that each of the differences $|b_n - a_n|$, $|b_n - \lim b_n|$, $|a_n - \lim a_n|$ is less than ϵ, provided that $n > m$. Thus $|\lim b_n - \lim a_n| < 3\epsilon$, by choice of m only; hence we must have $\lim b_n = \lim a_n$.

(7) It should be observed that if (a_n), (b_n) are convergent sequences such that $a_n < b_n$, it may easily happen that

$$\lim a_n = \lim b_n.$$

For the difference $b_n - a_n$, although constantly positive, may converge to 0 as a limit. Thus the correct conclusion from the inequality $a_n < b_n$ is

$$\lim a_n \leqq \lim b_n.$$

2. Monotonic sequences; and conditions for their convergence.

A sequence in which $a_{n+1} \geqq a_n$ for all values of n is called an *increasing sequence*; and similarly if $a_{n+1} \leqq a_n$ for all values of n, the sequence is called *decreasing*. Both increasing and decreasing sequences are included in the term *monotonic sequences*.

The first general theorem on convergence may now be stated:

A monotonic sequence has always a limit, either finite or infinite; the sequence is convergent provided that $|a_n|$ *is less than a number A independent of n; otherwise the sequence diverges.*

For the sake of definiteness, suppose that $a_{n+1} \geqq a_n$, and that a_n is constantly less than the fixed number A. Then, however small the positive fraction ϵ may be, it will be possible to find an index m such that $a_n < a_m + \epsilon$, if $n > m$; for, if not, it would be possible to select an *unlimited* sequence of indices $p, q, r, s, \ldots$, such that $a_p > a_1 + \epsilon$, $a_q > a_p + \epsilon$, $a_r > a_q + \epsilon$, $a_s > a_r + \epsilon$, etc.; and consequently, after going far enough* in the sequence $p, q, r, s, \ldots$, we should arrive at an index v such that $a_v > A$, contrary to hypothesis.

Thus, if we employ the graphical representation described in the last article, we see that all the points to the right of the line $x = m$ will be within a strip of breadth ϵ; and that the breadth of the strip can be made as small as we please by going far enough to the right. From the graphical representation it appears intuitively obvious that the sequence approaches some limit, which cannot exceed A (but may be equal to this value). But inasmuch as intuition has occasionally led to serious blunders in mathematical reasoning, it is desirable to give a proof depending entirely on arithmetical grounds; such a proof will be found in the Appendix, Art. 149.

Ex. 1. As an example consider the increasing sequence

$$\tfrac{1}{2}, \tfrac{2}{3}, \tfrac{3}{4}, \tfrac{4}{5}, \tfrac{5}{6}, \ldots,$$

which is represented by the diagram below.

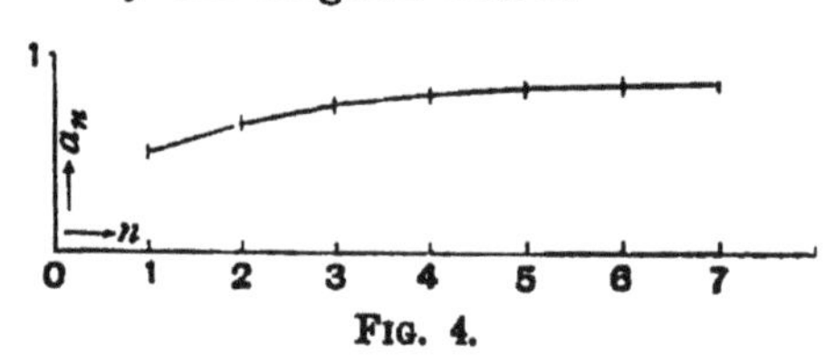

FIG. 4.

In this case we may take $A = 1$, and there is no difficulty in seeing that the limit of the sequence is equal to A; but of course we might have taken $A = 2$, in which case the limit would be less than A.

---

* The number of terms to be taken in the sequence $p, q, r, s, \ldots$ would be equal to the integer next greater than $(A - a_1)/\epsilon$.

Ex. 2. A second example is given by the sequence $(1+1/n)^n$, which has for its first six terms the approximate values 2, 2·25, 2·37, 2·44, 2·49, 2·52.

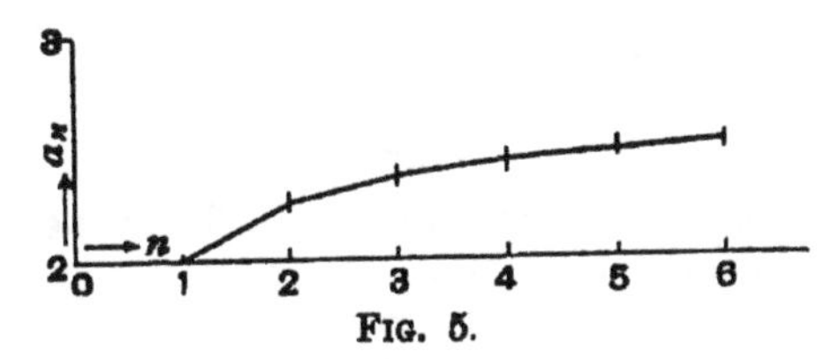

Fig. 5.

The reader is probably aware that this sequence always increases, but that its terms are always less than 3; the limit obtained is the number $e=2{\cdot}71828\ldots$. A formal proof of the monotonic property is given in the Appendix, Art. 158; and the limit is evaluated in Art. 57.

But in case no number such as A can be found, so that, however great A may be, there is always an index m, such that $a_m > A$, then it is plain that the sequence diverges to $+\infty$. For we have $a_n \geqq a_m > A$, if $n \geqq m$.

The reader will have no difficulty in modifying the foregoing work so as to apply to the case of a sequence which never increases, so that $a_{n+1} \leqq a_n$.

Ex. 3. Consider the sequence $a_n=r^n$.

If $0<r<1$, the sequence (a_n) steadily *decreases* but the terms are always positive; and consequently a_n approaches a definite limit l such that $1>l\geqq 0$. Thus we can find m to correspond to ϵ, so that

$$l<r^n<l+\epsilon, \quad \text{if } n>m.$$

Hence
$$r^{n+1}<r(l+\epsilon);$$
and consequently
$$l<r^{n+1}<r(l+\epsilon)$$
or
$$l(1-r)<r\epsilon.$$

Since this inequality is true, however small ϵ may be, we have $l=0$.

When $r>1$, it follows from the last result that $1/r^n\to 0$, and hence we can determine m so that $1/r^n<\epsilon$, if $n>m$.

Thus we find $r^n>1/\epsilon$, if $n>m$, and consequently $r^n\to\infty$. This result can also be established from the monotonic property of the sequence; or by direct reasoning, as in Ex. 1, Art. 6.

If r is *negative*, we have

$$r^n=(-1)^n \cdot |r|^n$$

and so the behaviour of the sequence can be determined from our results already obtained.

Summing up, we conclude that:

If
$$-1<r<1, \quad r^n\to 0;$$
$$r=1, \quad r^n=1;$$
$$r>1, \quad r^n\to\infty.$$

In all other cases the sequence *oscillates*, and we find:

If
$$r<-1,\quad r^{2n}\to\infty,\quad r^{2n+1}\to-\infty;$$
$$r=-1,\quad r^{2n}=1,\quad r^{2n+1}=-1.$$

Ex. 4. Take next $$a_n=r^n/n!.$$

If r is positive, let ρ be its integral part. Then the sequence (a_n) decreases steadily, after n exceeds the value ρ; and since (a_n) is positive it follows that $a_n\to l\geqq 0$.

Now $$\frac{a_{2n}}{a_n}=\frac{r}{n+1}\cdot\frac{r}{n+2}\cdots\frac{r}{2n}<\frac{r}{2n}<\frac{1}{2},\text{ if } n>\rho.$$

Thus we can find m so that

$$\left.\begin{aligned} l<a_n<l+\epsilon \\ a_{2n}<\tfrac{1}{2}a_n\end{aligned}\right\}\text{ if } n>m.$$
and

Hence, as in Ex. 3, we obtain

$$l<\tfrac{1}{2}(l+\epsilon)\text{ or } l<\epsilon.$$

It follows that $l=0$.

When r is negative, we obtain the same result by writing

$$a_n=(-1)^n\,.\,|r|^n/n!.$$

Thus for all values of r we have

$$\lim\frac{r^n}{n!}=0.$$

3. General principle of convergence.

If a sequence is not monotonic, the condition that $|a_n|$ remains constantly less than a fixed number is by no means sufficient to ensure convergence; this may be seen at once from the sequence given in Example 3, Art. 1, for which $0<a_n<1$.

The necessary and sufficient condition for convergence is that it may be possible to find an index m, corresponding to any positive number ϵ, *such that*

$$|a_n-a_m|<\epsilon$$

for all values of n greater than m.

Interpreted graphically, this implies that all points of the sequence which are to the right of $x=m$, lie within a strip of breadth 2ϵ. The statement is then almost intuitive, since the breadth of the strip can be made as small as we please, by going far enough to the right; an arithmetical proof will be found in the Appendix, Art. 151.

Ex. Consider the sequence

$$\tfrac{1}{2},\ 2,\ \tfrac{2}{3},\ \tfrac{3}{2},\ \tfrac{3}{4},\ \tfrac{4}{3},\ \tfrac{4}{5},\ \tfrac{5}{4},\ \tfrac{5}{6},\ \tfrac{6}{5},\ \ldots,$$

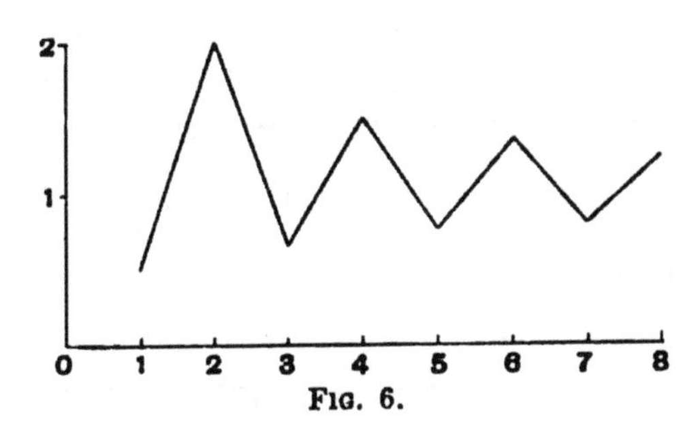

FIG. 6.

for which it is easily seen that the limit is 1. The diagram is as indicated; and it will be seen that m may be taken greater than or equal to $2/\epsilon$.

CAUTION. The reader must be warned not to regard the above test of convergence as equivalent to a condition sometimes given (even in books which are generally accurate), namely: *The necessary and sufficient condition for convergence is that*

$$\lim_{n\to\infty}(a_{n+p}-a_n)=0.$$

This condition is certainly necessary, but is NOT sufficient, *unless p is supposed to be an arbitrary function of n, which may tend towards infinity with n, in an arbitrary way.*

For example, suppose that $a_n=\log n$; then

$$\lim_{n\to\infty}(a_{n+p}-a_n)=\lim_{n\to\infty}\log(1+p/n)=0,$$

if p is any fixed number. But the sequence (a_n) is divergent, as may be seen from the Appendix, Art. 157.

The reader will have no difficulty in proving that *the elementary rules for calculating with limits are as follows:*

$$\lim(a_n \pm b_n)=\lim a_n \pm \lim b_n,$$

$$\lim(a_n \,.\, b_n)=\lim a_n \,.\, \lim b_n,$$

provided that the sequences (a_n), (b_n) are convergent.

$$\lim(a_n/b_n)=\lim a_n/\lim b_n,$$

provided that (a_n), (b_n) are convergent and that $\lim b_n$ is not zero. And generally that

$$\lim f(a_n,\ b_n,\ c_n,\ \ldots)=f(\lim a_n,\ \lim b_n,\ \lim c_n,\ \ldots),$$

where f denotes any combination of the four elementary operations, subject to conditions similar to those already specified.

If the functional symbol contains other operations (such as extraction of roots), the equation above may be taken as a *definition* of the right-hand side, assuming that the left-hand side is found to converge. On this basis the theory of irrational indices and logarithms can be satisfactorily constructed.*

It is to be remembered that the limits on the left may be perfectly definite without implying the existence of $\lim a_n$ and $\lim b_n$. To illustrate this possibility, take $a_n = (-1)^n$, $b_n = (-1)^{n-1}(1+1/n)$.

Then
$$a_n + b_n = (-1)^{n-1}/n \text{ and } (a_n + b_n) \to 0,$$
$$a_n \cdot b_n = -(1+1/n) \quad ,, \quad (a_n \cdot b_n) \to -1,$$
$$a_n/b_n = -n/(n+1) \quad ,, \quad (a_n/b_n) \to -1,$$
so that these three limits are quite definite, in spite of the non-existence of $\lim a_n$ and $\lim b_n$.

If a_n is convergent and $b_n \to 0$, *we cannot infer that* $a_n/b_n \to \infty$ without first proving that a_n/b_n has a *fixed sign.*

If $a_n \to 0$, and $b_n \to 0$, the quotient a_n/b_n may or may not have a limit (see Appendix, Art. 152).

Thus with $a_n = 1$, $b_n = (-1)^n/n$, we see that $b_n \to 0$, but $a_n/b_n = (-1)^n n$ and so a_n/b_n *oscillates* between $-\infty$ and $+\infty$.

Again, with $a_n = 1/n$, $b_n = (-1)^n/n$, the value of a_n/b_n oscillates between -1 and $+1$.

When one of the sequences diverges (say $a_n \to \infty$) and the other converges (say to a *positive* limit) it is easy to see that
$$(a_n \pm b_n) \to \infty\ ; \quad a_n \cdot b_n \to \infty\ ; \quad a_n/b_n \to \infty\ ; \quad b_n/a_n \to 0\ ;$$
the only case of exception arising when $b_n \to 0$, and then the sequences $(a_n \cdot b_n)$ and (a_n/b_n) need special discussion.

Again, if both a_n and b_n diverge to ∞, we have
$$(a_n + b_n) \to \infty\ ; \quad a_n b_n \to \infty\ ;$$
but both $(a_n - b_n)$ and (a_n/b_n) have to be examined specially (see Appendix, Art. 152).

If $a_n \to \infty$ and $b_n \to \infty$, there are three distinct alternatives with respect to the sequence (a_n/b_n), *assuming that it is convergent*†
$$\text{(i) } a_n/b_n \to 0; \quad \text{(ii) } a_n/b_n \to k > 0; \quad \text{(iii) } a_n/b_n \to \infty.$$

---

* Thus we can define $x^{\sqrt{2}}$ as $\lim x^{a_n}$, where $(a_n) = 1, \frac{3}{2}, \frac{7}{5}, \frac{17}{12}, \ldots \to \sqrt{2}$.

† Even when a_n and b_n are both monotonic, the sequence a_n/b_n need not converge (Appendix, Art. 152, Ex. 4).

In case (i), a_n *diverges more slowly* than b_n; in case (iii) a_n *diverges more rapidly* than b_n. In case (ii) it is sometimes convenient to use the notation

$$a_n \sim kb_n,$$

where a_n, b_n are complicated expressions.

Rules are given in the Appendix (Art. 152) for the determination of $\lim (a_n/b_n)$ in a number of cases which are important in practical work.

4. Upper and lower limits of a sequence.

If a sequence (a_n) has a *greatest term* H, this term is called the *upper limit* of the sequence; and similarly, when there is a least term h, it is called the *lower limit*.

But if a sequence has no greatest term, it follows that no matter how large n may be, there is always a larger index p, such that $a_p \geqq a_n$. Further, there is an *infinite* number of such indices p; otherwise there would be a greatest term in the sequence; thus to make p definite we suppose p to be the *least* index satisfying the required condition. Hence the terms of the sequence which fall between a_n and a_p are all less than a_n and a_p.

Choose now a succession of values of p such that

$$a_{p_1} \geqq a_1, \quad a_{p_2} \geqq a_{p_1}, \quad a_{p_3} \geqq a_{p_2}, \quad \text{etc.}$$

$$p_1 > 1, \quad p_2 > p_1, \quad p_3 > p_2,$$

and for simplicity denote a_{p_r} by b_r. Then we have constructed a monotonic sequence $b_1, b_2, b_3, \ldots$; and this sequence has a limit (Art. 2), either a finite number H or ∞. If $\lim b_n = H$, we can find m so that b_m lies between $H - \epsilon$ and H, no matter how small ϵ may be; and consequently we have

$$H - \epsilon < a_{p_r} < H, \quad \text{provided that } r \geqq m.$$

H is then called the upper limit of the sequence; and clearly H is not actually attained by any number belonging to the sequence.

Similarly, if $\lim b_n = \infty$, we can find m so that

$$a_{p_r} > N, \quad \text{provided that } r \geqq m,$$

no matter how large N may be.

If the upper limit of the sequence is H, whether attained or not, the sequence has the two following properties:

(i) *No term of the sequence is greater than H.*

(ii) *At least one term of the sequence is greater than $H-\epsilon$, however small ϵ may be.**

But if the upper limit of the sequence is ∞, it has the property:

An infinity of terms of the sequence exceed N, no matter how great N may be.

It is easy to modify these definitions and results so as to refer to the *lower limit* (h or $-\infty$).

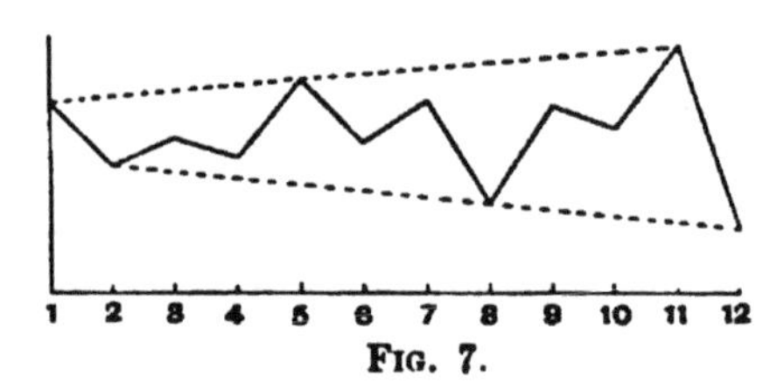

FIG. 7.

The diagram gives an indication of the mode of selecting the sub-sequences for H and h; these are represented by dotted lines.

Ex. 1. (Art. 1) $a_n = 2n - 1$.

Here we have $b_n = a_n$, and so the upper limit is ∞; $h=1$, because 1 is the least number in the sequence.

Ex. 2. (Art. 1) $a_n = 1/n$.

Here $H=1$, because this is the greatest number in the sequence; and h is seen to be 0, which is not actually attained by any number of the sequence.

Ex. 3. (Art. 1) $\frac{1}{2}, \frac{1}{3}, \frac{2}{3}, \frac{1}{4}, \frac{3}{4}, \frac{1}{5}, \frac{2}{5}, \frac{3}{5}, \frac{4}{5}, \ldots$.

Here the sequence (b_n) is $\frac{1}{2}, \frac{2}{3}, \frac{3}{4}, \frac{4}{5}, \ldots$ and gives $H=1$; and similarly h is found from the sequence $\frac{1}{2}, \frac{1}{3}, \frac{1}{4}, \frac{1}{5}, \ldots$ to be 0. These sequences are indicated in Fig. 3 of Art. 1 by the dotted lines.

5. Maximum and minimum limiting values of a sequence.

We have seen in the last article that any infinite sequence has upper and lower limits. Consider successively the sequences

$$\begin{array}{r} a_1, a_2, a_3, a_4, a_5, \ldots, \\ a_2, a_3, a_4, a_5, \ldots, \\ a_3, a_4, a_5, \ldots, \\ a_4, a_5, \ldots, \end{array}$$

and so on.

* If H is not attained, there will be an infinite number of such terms.

Let the corresponding upper and lower limits be denoted by H_1, h_1; H_2, h_2; H_3, h_3; H_4, h_4; and so on.

Then we may have $H_1 = a_1$, in which case a_1 must be the greatest term in the sequence (a_n); otherwise we shall have $H_1 = H_2$. Hence in all cases $H_1 \geqq H_2$. Thus $H_1 \geqq H_2 \geqq H_3 \geqq H_4 \geqq \dots$, and so the sequence (H_n) is monotonic and gives a limit G or $-\infty$ (Art. 2). Similarly $h_1 \leqq h_2 \leqq h_3 \leqq h_4 \leqq \dots$, and therefore (h_n) has a limit g or $+\infty$. It may be noticed that G can only be $+\infty$ in case $H_1 = H_2 = H_3 = \dots = +\infty$; and g can only be $-\infty$, if $h_1 = h_2 = h_3 = \dots = -\infty$.

It is important to notice that G, g *can be obtained as the limits of two sub-sequences properly selected from* (a_n). For, either $H_1, H_2, H_3, \dots$ *all* belong to the sequence (a_n), in which case the sub-sequence for G is coincident with (H_n); or else, after a certain stage, we have $H_m = H_{m+1} = H_{m+2} = \dots = G$, and then H_m is itself the limit of a certain sub-sequence selected from (a_n), so that this same sub-sequence defines G. An exactly similar argument applies to g.

Again *no convergent sub-sequence selected from* (a_n) *can have a limit which is greater than* G *or less than* g. For since $\lim H_n = G$, we can find m so that $H_m \leqq G + \epsilon$, no matter how small ϵ is; but, by the definition of H_m, we have

$$a_n \leqq H_m, \quad \text{if } n \geqq m,$$

so that

$$a_n \leqq G + \epsilon, \quad \text{if } n \geqq m.$$

Thus, if l is the limit of any convergent sub-sequence selected from (a_n), we must have $l \leqq G + \epsilon$; and, as ϵ is arbitrarily small, this requires $l \leqq G$. In like manner we prove that m' can be found to make $a_n \geqq g - \epsilon$, if $n \geqq m'$, and deduce that $l \geqq g$.

The two properties just established justify us in calling G *the maximum limit and* g *the minimum limit of the sequence* (a_n); in symbols we write

$$G = \overline{\lim_{n\to\infty}}\, a_n = \overline{\lim}\, a_n, \quad g = \underline{\lim_{n\to\infty}}\, a_n = \underline{\lim}\, a_n.$$

The symbol $\overline{\underline{\lim}}\, a_n$ is used to denote either the maximum or minimum limit; thus an inequality $f < \overline{\underline{\lim}}\, a_n < F$ implies that $f < g$ and $G < F$.

If it happens that $G = \infty$, we have $H_n = \infty$, and consequently there must be an infinity of terms a_n greater than any assigned number N, however great; similarly when $g=-\infty$, there must be an infinity of terms less than $-N$. On the other hand, if $\lim H_n=-\infty$, it is easy to see that $\lim a_n=-\infty$; and similarly if $\lim h_n=+\infty$ we must have $\lim a_n=+\infty$.

From what has been explained it is clear that *every sequence has a maximum and a minimum limit; and these limits are equal, if, and only if, the sequence converges.*

It is convenient to call sequences *oscillatory* when the maximum and minimum limits are unequal. We shall call these limits the *extreme limits* of the sequence, in case we wish to refer to both maximum and minimum limits.

It will be evident that *the maximum limit coincides with the upper limit, except when the latter is actually attained* by one or more terms of the sequence; and similarly for the minimum and lower limits.

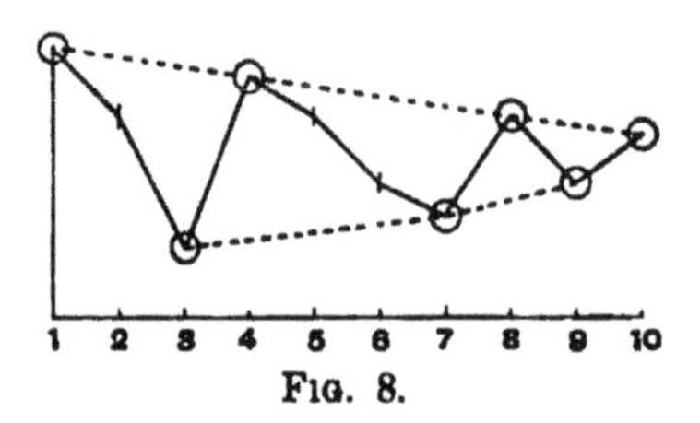

FIG. 8.

The diagram gives an indication of the process. The points H_n and h_n are marked with ⊙ and are joined by dotted lines.

Ex. 1. In the sequence (Ex. 3, Art. 1)

$$\tfrac{1}{2},\ \tfrac{1}{3},\ \tfrac{2}{3},\ \tfrac{1}{4},\ \tfrac{3}{4},\ \tfrac{1}{5},\ \tfrac{2}{5},\ \tfrac{3}{5},\ \tfrac{4}{5},\ \dots$$

we have $H_n=1$, $h_n=0$; so that $G=1$, $g=0$.

Here it is plain that convergent sub-sequences can be selected to give *any* limit between the extreme limits. Thus

$$\tfrac{1}{3},\ \tfrac{2}{5},\ \tfrac{3}{7},\ \tfrac{4}{9},\ \dots \text{ gives the limit } \tfrac{1}{2}$$

and

$$\tfrac{2}{3},\ \tfrac{5}{7},\ \tfrac{12}{17},\ \tfrac{29}{41},\ \dots \text{ gives the limit } \frac{1}{\sqrt{2}}.$$

Ex. 2. With $2,\ -\tfrac{3}{2},\ \tfrac{4}{3},\ -\tfrac{5}{4},\ \tfrac{6}{5},\ \dots \quad a_n=(-1)^{n-1}\left(1+\dfrac{1}{n}\right)$

we get $\quad H_1=2,\ H_2=H_3=\tfrac{4}{3},\ H_4=H_5=\tfrac{6}{5},\ \dots$

and $\quad h_1=h_2=-\tfrac{3}{2},\ h_3=h_4=-\tfrac{5}{4},\ \dots,$

so that $\quad G=1,\ g=-1.$

Ex. 3. With $1,\ -2,\ 3,\ -4,\ 5,\ -6,\ \dots \quad a_n=(-1)^{n-1}n$

we find $\quad H_n=\infty,\ h_n=-\infty$

and so $\quad G=\infty,\ g=-\infty.$

In Exs. 2, 3 it will be seen that no sub-sequences can be found to converge to limits other than the extreme limits.

The reader will find some rules for calculating with extreme limits in Exs. 14-17 at the end of this chapter.

6. Sum of an infinite series; addition of two series.

Suppose that a sequence $a_1, a_2, a_3, \ldots, a_n, \ldots$ is given and that we deduce from this sequence a second $s_1, s_2, s_3, \ldots$ by addition, so that

$$s_1 = a_1, \quad s_2 = a_1 + a_2, \quad s_3 = a_1 + a_2 + a_3,$$
$$s_n = a_1 + a_2 + a_3 + \ldots + a_n.$$

Then *if the sequence (s_n) is convergent and has the limit S, the infinite series*

$$a_1 + a_2 + a_3 + \ldots = \sum_1^\infty a_n = \Sigma a_n$$

is called convergent; and S is called the sum of the series.

It is, however, of fundamental importance to bear in mind that *S is a limit*; and accordingly care must be taken not to assume without proof that familiar properties of finite sums are necessarily true for limits such as S.

Similarly, *if the sequence (s_n) is divergent or oscillatory, the infinite series is said to diverge or to oscillate, respectively.*

Ex. 1. *The geometric series $1 + r + r^2 + r^3 + \ldots$ converges if r is numerically less than 1; it diverges if r is not less than 1; it oscillates if r is not greater than -1.*

For, except when $r = 1$, $s_n = (1 - r^n)/(1 - r)$; and when $r = 1$, $s_n = n$.

Now if
$$-1 < r < 1,$$
we can write
$$|r| = 1/(1 + a),$$
where a is positive. Then
$$|r|^n < 1/(1 + na)$$
by the binomial theorem.

Thus
$$\lim_{n\to\infty} r^n = 0,$$
a result which was obtained independently in Ex. 3, Art. 2.

Hence
$$S = \lim_{n\to\infty} s_n = 1/(1 - r), \qquad \text{if } -1 < r < 1.$$

If $r \geqq 1$, it is obvious that $s_n \geqq n$, and accordingly
$$\lim_{n\to\infty} s_n = \infty,$$
so that the series diverges.

When r is less than -1, we have $r = -(1 + b)$, where b is positive, and so
$$(-r)^n > 1 + nb.$$

Hence
$$s_n > (2 + nb)/(2 + b), \qquad \text{if } n \text{ is odd},$$
or
$$s_n < -nb/(2 + b), \qquad \text{if } n \text{ is even}.$$

Thus $\qquad \underline{\lim}\, s_n = -\infty, \quad \overline{\lim}\, s_n = +\infty,$

and so the series oscillates between $-\infty$ and $+\infty$.

If $r = -1$, $\qquad s_n = 1, \quad$ if n is odd,

and $\qquad s_n = 0, \quad$ if n is even.

Thus the series again oscillates.

We have now justified all the statements of the enunciation.

It follows at once from the results stated in Art. 3 that if

$$S = a_1 + a_2 + a_3 + \dots \text{ to } \infty$$

and
$$T = b_1 + b_2 + b_3 + \dots \text{ to } \infty,$$

then
$$S \pm T = (a_1 \pm b_1) + (a_2 \pm b_2) + (a_3 \pm b_3) + \dots .$$

The rule for multiplication of S, T does not follow quite so readily (see Art. 34).

It should be observed that the insertion of brackets in a series is equivalent to the selection of a sub-sequence from the sequence (s_n); and since an oscillatory sequence always contains at least two convergent sub-sequences (those giving the extreme limits), it is evident that *an oscillatory series can always be made to converge by grouping the terms in brackets; and, conversely, the removal of brackets may cause a convergent series to oscillate.*

Ex. 2. The series $1 - \frac{1}{2} + \frac{2}{3} - \frac{3}{4} + \frac{4}{5} - \frac{5}{6} + \dots$ oscillates between the values $\cdot 306\dots$ and $1\cdot 306\dots$; but the series $(1 - \frac{1}{2}) + (\frac{2}{3} - \frac{3}{4}) + (\frac{4}{5} - \frac{5}{6}) + \dots$ converges to the sum $\cdot 306\dots$, while $1 - (\frac{1}{2} - \frac{2}{3}) - (\frac{3}{4} - \frac{4}{5}) - \dots$ converges to the sum $1\cdot 306\dots$. [$\cdot 306\dots = 1 - \log 2 = \frac{1}{2} - \frac{1}{3} + \frac{1}{4} - \frac{1}{5} + \dots$, see Arts. 21, 24, 63.]

It is evident that when we are only concerned with determining whether a series is convergent or not, we may neglect any finite number of terms of the series; this is often convenient in order to avoid some irregularity of the terms, at the beginning of the series.

In particular, it is clear that the series

$$a_1 + a_2 + a_3 + \dots, \quad a_{m+1} + a_{m+2} + a_{m+3} + \dots,$$

are simultaneously convergent. The sum of the latter is often called *the remainder after m terms of the former.*

EXAMPLES.

Arts. 1-3.

1. If $a_{n+1}=\sqrt{(k+a_n)}$, where $k>0$, $a_1>0$, the sequence (a_n) is monotonic and converges to the positive root of the equation $x^2=x+k$.

2. If $a_{n+1}=k/(1+a_n)$, where $k>0$, $a_1>0$, the sequence (a_n) converges to the positive root of the equation $x^2+x=k$.

3. If $f(x)$ is a continuous function of x such that $|f'(x)|\leqq k<1$, and if $a_{n+1}=f(a_n)$, the sequence (a_n) converges to a root of the equation $x=f(x)$; it should be observed that this equation has only one real root.

Ex. $f(x)=\frac{1}{2}\cos x$.

4. If $a_n\to 0$, prove that $b^{a_n}\to 1$, where b is any positive number; and deduce that if $a_n\to a$, $b^{a_n}\to b^a$. If $a_n\to\infty$, $b^{a_n}\to 0$ or ∞.

5. If $a_{n+1}=ka_n+la_{n-1}$, where k, l are positive, prove that a_n/a^n converges to the limit $(a_2-a_1\beta)/a(a-\beta)$, where a is the positive and β is the negative root of $x^2=kx+l$.

6. If in (5), $k+l=1$, prove that $a_n\to(a_2+la_1)/(1+l)$. In particular, if each term of a sequence is the arithmetic mean of the two preceding terms, the two sequences $a_1, a_3, a_5, \ldots$ and $a_2, a_4, a_6, \ldots$ are separately monotonic and converge to the common value $\frac{1}{3}(2a_2+a_1)$.

Examine similarly the cases in which the geometric and harmonic means are taken.

7. If $a_{n+1}=\frac{1}{2}(a_n+b_n)$, $b_{n+1}=\sqrt{(a_{n+1}b_n)}$, where a_1, b_1 are positive, the sequences (a_n), (b_n) are monotonic and converge to a common limit. If $a_1=\cos\theta$, $b_1=1$, the common limit is $(\sin\theta)/\theta$; and if $a_1=\cosh u$, $b_1=1$, the common limit is $(\sinh u)/u$. [BORCHARDT.]

8. If $a_{n+1}=\frac{1}{2}(a_n+b_n)$, $a_{n+1}b_{n+1}=a_nb_n$, so that a_{n+1}, b_{n+1} are respectively the arithmetic and harmonic means of a_n, b_n, then the sequences (a_n), (b_n) are monotonic and converge to the common limit $\sqrt{(a_1b_1)}$, where a_1, b_1 are positive.

9. If $a_{n+1}=\frac{1}{2}(a_n+b_n)$, $b_{n+1}=\sqrt{(a_nb_n)}$, so that a_{n+1}, b_{n+1} are respectively the arithmetic and geometric means of a_n and b_n, then the sequences (a_n), (b_n) are monotonic and converge to a common limit l.

This limit was called by Gauss the arithmetrico-geometric mean of a_1, b_1, and can be applied to calculate certain elliptic integrals by means of the formula

$$\frac{\pi}{2l}=\int_0^{\frac{1}{2}\pi}\frac{d\theta}{(a_1^2\cos^2\theta+b_1^2\sin^2\theta)^{\frac{1}{2}}}=\int_0^{\frac{1}{2}\pi}\frac{d\theta}{(a_n^2\cos^2\theta+b_n^2\sin^2\theta)^{\frac{1}{2}}}.$$

10. In general, if $a_{n+1}=f(a_n)$, and if (a_n) converges to a limit l, then l must be a root of the equation $x=f(x)$.

But there are two limits which can be derived from (a_n) by means of sub-sequences, they must satisfy the equations $x=f(y)$, $y=f(x)$.

For illustrations, see Ex. 11.

11. If $a_{n+1}=k^{a_n}$, $k>0$, a number of alternatives arise; we can write the condition in the form $\log a_{n+1}=\lambda a_n$, if $\lambda=\log k$. By means of the curve $y=\log x$ and the line $y=\lambda x$ we can easily prove that

(i) if $\lambda>1/e$, (a_n) is a divergent monotonic sequence;

(ii) if $0<\lambda<1/e$, the equation $\lambda x=\log x$ has two real roots α, β (say that $\alpha<\beta$); then the sequence (a_n) is monotonic, and $a_n\to\alpha$ if $a_1<\beta$; but if $a_1>\beta$, $a_n\to\infty$.

When λ is negative, the equation $\log x=\lambda x$ has *one* real root (α); but the sequence (a_n) will be seen to be no longer monotonic. To meet this difficulty we may write $\log\left(\log\frac{1}{a_{n+1}}\right)=\log(-\lambda)+\lambda a_{n-1}$ and use the curve $y=\log\left(\log\frac{1}{x}\right)$ and the line $y=\log(-\lambda)+\lambda x$. It can then be proved that the sequences (a_{2n}), (a_{2n+1}) are separately monotonic, and

(iii) if $-e<\lambda<0$, $a_n\to\alpha$.

(iv) if $\lambda<-e$, $a_{2n+1}\to u$, $a_{2n}\to v$, if $a_1<\alpha$; but $a_{2n}\to u$, $a_{2n+1}\to v$, if $a_1>\alpha$; and $a_n=\alpha$, if $a_1=\alpha$.

Here u, v are such that $u<\alpha<v$ and $k^u=v$, $k^v=u$.

This problem was discussed in the special case $a_1=k$ by Seidel (*Abhandlungen der k. Akad. der Wissensch. zü München*, Bd. 11, 1870), who was the first to point out the possibility of oscillation, in case (iv). Previously, Eisenstein (*Crelle's Journal für Math.*, Bd. 28, 1844, p. 49) had obtained the root α as a series proceeding in powers of λ; this series is the same as the one given in Art. 56, Ex. 4, below.

Arts. 4, 5.

12. The reader may find it instructive to determine the upper and lower limits, and also the extreme limiting values of the following sequences. The relations of the terms a_n to the limits should also be considered.

(1) $a_n=(-1)^n n/(2n+1)$. (2) $a_n=(-1)^n(n+1)/(2n+1)$.

(3) $a_n=n+(-1)^n(2n+1)$. (4) $a_n=2n+1+(-1)^n n$.

13. In an oscillatory sequence there may be a finite number of limits derived from sub-sequences, all, some, or none of the limits being attained, as may be seen by considering:

(1) $a_n=\sin(\frac{1}{3}n\pi)$, which consists of the seven numbers 0, $\pm\frac{1}{2}$, $\pm\frac{1}{2}\sqrt{3}$, ± 1 all repeated infinitely often.

(2) $a_n=\left(1+\frac{1}{n}\right)\sin(\frac{1}{3}n\pi)$ has the same seven limits as in case (1), but only the value 0 is attained.

(3) $a_n=\left(1+\frac{1}{2n}\right)\cos(\frac{1}{3}n\pi)$ has the four limits $\pm\frac{1}{2}$, ± 1, but no term a_n is equal to any of these values.

There may also be a whole interval of limits (see Ex. 1, Art. 5); and an infinity of these limits may be attained. But it is then not possible for a_n to attain *all* the limits, for the set of points forming an interval are not countable (*i.e.* cannot be put into one-to-one correspondence with the set of positive integers), and therefore cannot form a sequence (a_n).

14. *Addition, subtraction and multiplication of oscillatory sequences.*

If
$$\underline{\lim}\, a_n=k, \quad \overline{\lim}\, a_n=K$$
and
$$\underline{\lim}\, b_n=l, \quad \overline{\lim}\, b_n=L,$$
then
$$k+l\leqq\overline{\underline{\lim}}\,(a_n+b_n)\leqq K+L$$
and
$$k-L\leqq\overline{\underline{\lim}}\,(a_n-b_n)\leqq K-l.$$

For multiplication the results are not so simple; we may suppose that by changes of sign, if necessary, we have made K and L positive. Then

(1) if $0<k<K$ and $0<l<L$, we have
$$kl\leqq\overline{\underline{\lim}}\,(a_nb_n)\leqq KL;$$

(2) if $0<k<K$ and $l<0<L$, we have
$$Kl\leqq\overline{\underline{\lim}}\,(a_nb_n)\leqq KL;$$

(3) if $k<0<K$ and $l<0<L$, we have
$$\lambda\leqq\overline{\underline{\lim}}\,(a_nb_n)\leqq\mu,$$

where λ is the numerically greater of kL and Kl, μ of kl and KL.

These rules apply also in the limiting cases $k=K$, $k=0$, $K=\infty$, etc., on the understanding that the rules can give no information in cases when a product $0\times\infty$ is present.

It is also to be noticed that convergence of (a_nb_n) is quite possible, and that these inequalities may easily give very much wider limits for (a_nb_n) than are necessary in particular cases.

15. As regards *division of oscillatory sequences*, it is usually best to reduce the question to one of multiplication by the reciprocal, and we find that

(1) If $0<l<L$, we have
$$1/L=\underline{\lim}\,(1/b_n), \quad 1/l=\overline{\lim}\,(1/b_n),$$

and so $k/L\leqq\overline{\underline{\lim}}\,(a_n/b_n)\leqq K/l$, if $0<k<K$,

but $k/l\leqq\overline{\underline{\lim}}\,(a_n/b_n)\leqq K/l$, if $k<0<K$.

(2) If $l<0<L$, *no general rules can be laid down at all*, as the terms in $(1/b_n)$ which are numerically greatest arise from those terms of (b_n) which are nearest to 0; and the values of l, L give no information whatever as to these terms.

16. To illustrate the results of Exs. 14, 15, the reader may consider the following cases:

(1) $(a_n) = 1, \frac{1}{2}, \frac{1}{3}, \frac{1}{4}, \dots$ $\quad a_n b_n \to 0, \; a_n/b_n \to 0.$

$(b_n) = -\frac{1}{2}, +\frac{2}{3}, -\frac{3}{4}, +\frac{4}{5}, \dots$ $\quad \underline{\lim}\,(b_n/a_n) = -\infty, \; \overline{\lim}\,(b_n/a_n) = \infty.$

(2) $(a_n) = 1, \frac{1}{2}, \frac{1}{3}, \frac{1}{4}, \dots$ $\quad \underline{\lim}\,(a_n b_n) = -1.$

$(b_n) = -2, +3, -4, +5, \dots$ $\quad \overline{\lim}\,(a_n b_n) = +1.$

(3) $(a_n) = 1, \frac{1}{2}, 2, \frac{1}{3}, 3, \frac{1}{4}, 4, \dots$ $\quad a_n b_n \to 1.$

$(b_n) = 1, 1, \frac{1}{2}, 2, \frac{1}{3}, 3, \frac{1}{4}, \dots$ $\quad \underline{\lim}\,(a_n/b_n) = 0, \; \overline{\lim}\,(a_n/b_n) = \infty.$

There is no difficulty in constructing further examples.

17. Verify the following table, and construct examples of each possibility, where (1) denotes convergence to a limit not zero, (2) to zero, (3) divergence to $+\infty$, (3′) divergence to $-\infty$, (4) finite oscillation, (5) infinite oscillation.

a_n	b_n	$a_n b_n$	a_n	b_n	$a_n b_n$
1	1	1	2	5	any way
1	2	2	3	3	3
1	3	3 or 3′	3	4	3, 3′ or 5
1	4	4	3	5	3, 3′ or 5
1	5	5	4	4	1, 2, 4
2	2	2	4	5	any way
2	3	any way	5	5	any way
2	4	2			

[HARDY.]

Art. 6.

18. If the series Σa_n converges to a sum s, and if $a_n = b_n$, then Σb_n is also convergent and has the sum s. [STOLZ.]

19. The series
$$\frac{x}{1-x^2} + \frac{x^2}{1-x^4} + \frac{x^4}{1-x^8} + \dots$$
has the sum to n terms $\dfrac{1}{1-x} - \dfrac{1}{1-x^N}$, if $N = 2^n$.

Thus the series to ∞ converges except when $x = \pm 1$, and the sum is $x/(1-x)$ when $|x| < 1$, or $-1/(x-1)$ when $|x| > 1$. [DE MORGAN.]

20. The series
$$\frac{a_1}{1+a_1} + \frac{a_2}{(1+a_1)(1+a_2)} + \frac{a_3}{(1+a_1)(1+a_2)(1+a_3)} + \dots$$
can be summed to n terms. The series converges if the terms a_n are all positive after a certain stage.

In particular (see Art. 38) the sum to ∞ is 1 if the series Σa_n diverges; examples of which are given by $1/a_n = a+n, \; a+2n$, etc.

21. The series
$$\frac{x}{1+x} + \frac{2x^2}{1+x^2} + \frac{4x^4}{1+x^4} + \frac{8x^8}{1+x^8} + \dots$$
converges to the sum $x/(1-x)$, if $|x| < 1$.

22. Discuss the series

$$1+2r+3r^2+4r^3+\dots,\quad 1+3r+6r^2+10r^3+15r^4+\dots$$

on the same lines as the geometrical progression of Art. 6.

Miscellaneous.

23. If the sequence (a_n) is monotonic, prove that the same is true of the sequence whose nth term is

$$(a_1+a_2+\dots+a_n)/n,$$

and that these sequences vary in the same sense.

Compare similarly the sequences

$$(a_n/b_n) \text{ and } (a_1+a_2+\dots+a_n)/(b_1+b_2+\dots+b_n),$$

where b_n is positive.

24. By taking $a_n=a^{n-1}(1-a)$ in Ex. 23, shew that the sequence $(1-a^n)/n$ is a decreasing sequence when a is positive. Deduce that

$$na^{n-1}(1-a)<1-a^n<n(1-a),\quad \text{if } 0<a<1,$$
$$na^{n-1}(a-1)>a^n-1>n(a-1),\quad \text{if } a>1,$$

and examine the extension of these inequalities to fractional values of n.

25. Deduce from the inequalities of Ex. 24 that $n(a^{\frac{1}{n}}-1)$ decreases as n increases, but remains positive $(a>1)$; deduce that it converges to a limit $f(a)$, and that $\qquad f(b)-f(a)=f(b/a)$.

26. If $a_{n+1}=\dfrac{pa_n+q}{ra_n+s}$, the behaviour of the sequence (a_n) depends on the nature of the roots of the quadratic $rx^2+(s-p)x-q=0$.

If these roots are denoted by α, β, we can prove that if β is not equal to α, the equation reduces to

$$\frac{a_{n+1}-\alpha}{a_{n+1}-\beta}=\frac{p-r\alpha}{p-r\beta}\left(\frac{a_n-\alpha}{a_n-\beta}\right).$$

Consequently, *if* α, β *are real and unequal*, $a_n\to\alpha$, provided that

$$|p-r\alpha|<|p-r\beta|;$$

in the exceptional case when $(p-r\alpha)+(p-r\beta)=0$, we have $p+s=0$, and a_n is alternately equal to the two fixed values a_1 and $(pa_1+q)/(ra_1-p)$.

If α, β *are complex* $|p-r\alpha|=|p-r\beta|$, assuming that p, q, r, s are real, so that convergence is impossible; it is then possible to reduce a_{n+1} to the form

$$\frac{a_1+(k+a_1l)\tan n\theta}{1+(l+a_1m)\tan n\theta},$$

where $(p+s)^2\sec^2\theta=4(ps-qr)$ and k, l, m are easily expressed in terms of p, q, r, s. It is evident that if θ/π is *rational*, the sequence (a_n) will repeat itself in certain periods.

If $\beta=\alpha$, *both being real* (in consequence of the reality of p, q, r, s), we find that

$$\frac{1}{a_{n+1}-\alpha}=\frac{1}{a_n-\alpha}+\frac{r}{r\alpha+s},$$

and so (except when $r=0$, a case requiring no special discussion), we have

$$a_n\to\alpha.$$

CHAPTER II.

SERIES OF POSITIVE TERMS.

7. If all the terms ($a_1, a_2, a_3, \ldots$) of the series are *positive*, the sequence (s_n) steadily increases; and so (by Art. 2) the series Σa_n must be either convergent or divergent; that is, oscillation is impossible. It is therefore clear (from the same article) that:

(1) *The series converges if s_n is less than some fixed number for all values of n.*

(2) *The series diverges if a value of n can be found so that s_n is greater than N, no matter how large N is.*

Ex. 1. Consider the series given by $a_n = 1/n!$, so that

$$s_n = 1 + \frac{1}{2!} + \frac{1}{3!} + \frac{1}{4!} + \ldots + \frac{1}{n!}.$$

Compare s_n with the sum

$$\sigma_n = 1 + \frac{1}{2} + \frac{1}{2^2} + \frac{1}{2^3} + \ldots + \frac{1}{2^{n-1}}.$$

It is clear that $\quad 3! = 3.2 > 2^2; \ 4! = 4.3.2 > 2^3;$

and so on, $\quad n! = n.(n-1)\ldots 3.2 > 2^{n-1}.$

Thus, from the third term onwards, every term in σ_n is greater than the corresponding term in s_n; and the first and second terms in the sums are equal. Thus $\quad \sigma_n > s_n.$

But $$\sigma_n = \left(1 - \frac{1}{2^n}\right) \Big/ \left(1 - \frac{1}{2}\right) = 2 - \frac{1}{2^{n-1}} < 2,$$

so that $$s_n < \sigma_n < 2.$$

Consequently the series Σa_n is convergent and its sum cannot exceed 2.

If the sum is denoted by $e - 1$, as usual, we can prove similarly that

$$e - 1 - s_m < 1/\{m(m!)\}.$$

By direct calculation to 5 decimals we find that $1 + s_7$ lies between 2·71822 and 2·71828 and that $1/\{7(7!)\}$ is less than ·00003, so that e lies between 2·7182 and 2·7183. Further calculations have shewn that

$$e = 2{\cdot}7182818285\ldots.$$

Ex. 2. Consider the harmonic series ($a_n = 1/n$), for which

$$s_n = 1 + \frac{1}{2} + \frac{1}{3} + \frac{1}{4} + \dots + \frac{1}{n}.$$

Then arrange the sum s_n into groups thus:

$$s_n = \left(1 + \frac{1}{2}\right) + \left(\frac{1}{3} + \frac{1}{4}\right) + \left(\frac{1}{5} + \frac{1}{6} + \frac{1}{7} + \frac{1}{8}\right) + \left(\frac{1}{9} + \dots + \frac{1}{16}\right)$$
$$+ \left(\frac{1}{17} + \dots + \frac{1}{32}\right) + \dots + \left(\frac{1}{2^{m-1}+1} + \dots + \frac{1}{2^m}\right),$$

where the last term in each group is a power of 2, and $n = 2^m$. Now compare s_n with the sum

$$\sigma_n = \left(\frac{1}{2} + \frac{1}{2}\right) + \left(\frac{1}{4} + \frac{1}{4}\right) + \left(\frac{1}{8} + \frac{1}{8} + \frac{1}{8} + \frac{1}{8}\right) + \left(\frac{1}{16} + \dots + \frac{1}{16}\right)$$
$$+ \left(\frac{1}{32} + \dots + \frac{1}{32}\right) + \dots + \left(\frac{1}{2^m} + \dots + \frac{1}{2^m}\right),$$

where the number of terms in each group is the same as in the corresponding group of s_n; but all the terms in any group of σ_n are equal to the last term of the group in s_n.

Then $s_n > \sigma_n$, by inspection.

But each group in σ_n (after the first) is equal to $\frac{1}{2}$; for the rth group contains 2^{r-1} terms each equal to $1/2^r$.

Hence $$\sigma_n = 1 + \tfrac{1}{2}(m-1) = \tfrac{1}{2}(m+1),$$
and so $$s_n > \tfrac{1}{2}(m+1).$$

Thus $s_n > N$, if $m \geqq 2N - 1$; and consequently *the series diverges.*

Since all the terms a_n are positive, we need not stop to discuss s_n for cases when n is not a power of 2; of course if some terms in the series were negative, this would be necessary in order to make sure that the series would not oscillate.

If we take similarly

$$\Sigma_n = (1+1) + \left(\frac{1}{2} + \frac{1}{2}\right) + \left(\frac{1}{4} + \frac{1}{4} + \frac{1}{4} + \frac{1}{4}\right)$$
$$+ \left(\frac{1}{8} + \dots + \frac{1}{8}\right) + \dots$$
$$+ \left(\frac{1}{2^{m-1}} + \dots + \frac{1}{2^{m-1}}\right) = 2\sigma_n$$

we can prove that $\Sigma_n > s_n$.

This gives $s_n < m + 1$; and so the divergence is comparatively slow. For instance, the sum to a million terms is less than 21, because

$$2^{20} = (1024)^2 > 10^6.$$

We note that since

$$1 + \frac{1}{3} + \frac{1}{5} + \frac{1}{7} + \dots + \frac{1}{2n-1} > \frac{1}{2}\left(1 + \frac{1}{2} + \frac{1}{3} + \frac{1}{4} + \dots + \frac{1}{n}\right),$$

the series $1 + \frac{1}{3} + \frac{1}{5} + \frac{1}{7} + \dots$ is also divergent.

The method used here can easily be applied to discuss the two series

$$1+\frac{1}{2^p}+\frac{1}{3^p}+\frac{1}{4^p}+\ldots \text{ and } \frac{1}{2\log 2}+\frac{1}{3\log 3}+\frac{1}{4\log 4}+\ldots.$$

But the discussions in Art. 11 are as easy and have the advantage of being more easily remembered.

The method given in Ex. 2 can be put in the following rule (often called *Cauchy's test of condensation*):

The series Σa_n converges or diverges with $\Sigma N a_N$, if $N=2^n$ and $a_n \geqq a_{n+1}$; and it is easy to extend the proof given above so as to shew that we may take N as the integral part of k^n, where k is any number greater than 1.

(3) It is clear also, from the results of Art. 2, that *if we can find n_1, so that $s_{n_1}-s_n>h$ (where h is a fixed positive constant), no matter how large n may be, then the series must be divergent.*

For we can then select a succession of values n_0, n_1, n_2, n_3, n_4, ..., such that

$$s_{n_1}-s_{n_0}>h, \quad s_{n_2}-s_{n_1}>h, \quad s_{n_3}-s_{n_2}>h, \quad s_{n_4}-s_{n_3}>h, \quad \text{etc.}$$

Thus, on adding, we find that

$$s_{n_r}>s_{n_0}+rh,$$

and therefore s_{n_r} can be made arbitrarily large by taking r sufficiently great; and so the series diverges in virtue of (2) above.

As an example, consider Ex. 2 above; we have then

$$s_{n_1}-s_n>(n_1-n)/n_1,$$

because $s_{n_1}-s_n$ contains (n_1-n) terms ranging from $1/(n+1)$ to $1/n_1$; and so, by taking $n_1=2n$, we get

$$s_{2n}-s_n>\tfrac{1}{2}.$$

(4) If S is the sum of a convergent series of *positive* terms, the sequence (s_n) *increases* to the limit S; the value of s_n cannot reach, and *a fortiori*, cannot exceed S. Thus S must be greater than the sum of any number of terms, taken arbitrarily, in the series; for n can be chosen large enough to ensure that s_n includes all these terms.

On the other hand, any number smaller than S, (say $S-\epsilon$), has the property that we can find terms in the series whose sum exceeds $S-\epsilon$.

It is now clear that a series of positive terms remains convergent even if an *infinite* number of its terms are removed.

Also if a series can be proved to converge when its terms are grouped in brackets, it will still converge when the brackets are removed, *provided that all the terms are positive.*

8. Comparison test for convergence (of positive series).

If the series $c_1+c_2+c_3+\ldots$ contains only positive terms and is convergent, and if another series $a_1+a_2+a_3+\ldots$ has the property
$$0 \leqq a_n \leqq c_n$$
(at any rate for values of n greater than some fixed value), then Σa_n is also convergent.

For, if $a_n \leqq c_n$, when $n>m$, we have
$$a_{m+1}+a_{m+2}+\ldots+a_n \leqq c_{m+1}+c_{m+2}+\ldots+c_n < T,$$
if T is the sum $\sum\limits_1^\infty c_n$.

Thus
$$s_n < s_m + T;$$
so that s_n is less than a fixed number (independent of n), which establishes the convergence of Σa_n.

In case the inequality holds for all values of n, we have $s_n<T$; so that the sum cannot exceed T.

The condition that *all* the terms must be positive in Σa_n and Σc_n may be broken *if there are no negative terms after a certain stage.* For the convergence of the series will not be affected by the omission of a finite number of terms at the beginning of the series.

But *if there are negative terms left, however far we go in the series Σc_n, the test is not sufficient.* For instance, take the series
$$1-\frac{1}{1.2}+\frac{1}{2}-\frac{1}{2.3}+\frac{1}{3}-\frac{1}{3.4}+\frac{1}{4}-\frac{1}{4.5}+\ldots$$
and compare it with
$$1-1+\frac{1}{2}-\frac{1}{2}+\frac{1}{3}-\frac{1}{3}+\frac{1}{4}-\frac{1}{4}+\ldots.$$

Every term in the second series is numerically greater than, or equal to, the corresponding term in the first series; and the second series converges to the sum 0. But the first series diverges; for in this series we find
$$1-\frac{1}{1.2}=\frac{1}{2},\quad \frac{1}{2}-\frac{1}{2.3}=\frac{1}{3},\quad \frac{1}{3}-\frac{1}{3.4}=\frac{1}{4},\ \text{etc.},$$
so that
$$s_{2n}=\frac{1}{2}+\frac{1}{3}+\frac{1}{4}+\ldots+\frac{1}{n+1} \text{ and } s_{2n+1}=s_{2n}+\frac{1}{n+1}$$
giving $\lim s_{2n+1}=\lim s_{2n}=\infty$. (Ex. 2, last article.)

9. The comparison test leads at once to the following form, which is often easier to work with:

Let the series $\Sigma(1/C_n)$ be convergent; then Σa_n will converge, provided that

$$\overline{\lim}(a_n C_n)$$

is not infinite, both series containing only positive terms.

For, when this condition is satisfied, we can find a constant G independent of n, such that

$$0 < a_n C_n < G.$$

Hence a_n is less than G/C_n, which is the general term of a convergent series.

It is useful to remark that *there is no need to assume the existence of the limit* $\lim(a_n C_n)$; this is seen by considering the convergent series

$$\frac{1}{C_1}+\frac{2}{C_2}+\frac{1}{C_3}+\frac{2}{C_4}+\frac{1}{C_5}+\frac{2}{C_6}+\dots,$$

for which $a_n C_n$ is alternately equal to 1 and 2.

Further, the test is *sufficient* only and is not *necessary*; as we may see by taking $C_n = n!$ and $a_n = 1/2^{n-1}$; then $a_n C_n > n/2$, so that $\lim(a_n C_n) = \infty$. But Σa_n converges (see Ex. 1, Art. 6).

The corresponding test for divergence runs:

Let the series $\Sigma(1/D_n)$ be divergent, then Σa_n will diverge, provided that

$$\underline{\lim}(a_n D_n) > 0,$$

both series containing only positive terms.

The proof is practically identical with the previous investigation, when the signs of inequality are reversed. We note also that the limit $\lim(a_n D_n)$ need not exist; and that the test is not *necessary*.

It follows immediately that *the following conditions are necessary but not sufficient*:

For convergence, $\underline{\lim}(a_n D_n) = 0$;

for divergence, $\overline{\lim}(a_n C_n) = \infty$.

But, in general there is no need for the limits of $(a_n D_n)$ or of $(a_n C_n)$ to exist; and *the condition*, $\lim(a_n D_n) = 0$, *sometimes given as necessary for convergence, is incorrect.*

Ex. Let $a_n = 1/n^2$, except when n is a squared integer, and let $a_n = 1/n^{\frac{3}{2}}$ when n is a square.

Thus the series is

$$1 + \frac{1}{2^2} + \frac{1}{3^2} + \frac{1}{4^{\frac{3}{2}}} + \frac{1}{5^2} + \frac{1}{6^2} + \frac{1}{7^2} + \frac{1}{8^2} + \frac{1}{9^{\frac{3}{2}}} + \dots.$$

If we take $D_n = n$, we find

$$\underline{\lim}\,(a_n D_n) = 0, \quad \overline{\lim}\,(a_n D_n) = \infty,$$

so that $\lim (a_n D_n)$ does not exist. But yet the series Σa_n converges, as will be seen in Art. 11.

It is easy to see that *if the terms* a_n *steadily decrease, the condition* $\lim (na_n) = 0$ *is necessary for convergence*; but even so, the general condition $\lim (a_n D_n) = 0$ is not necessary.

For if Σa_n is convergent, we can choose m, so that

$$a_{m+1} + a_{m+2} + \dots + a_n < \epsilon.$$

Now each of these terms is not less than a_n, so that

$$(n-m)a_n < \epsilon, \quad \text{if } n > m.$$

But, since $a_n \to 0$, we can choose $\nu (> m)$, so that $ma_n < \epsilon$, if $n > \nu$.

Thus $na_n < 2\epsilon$, if $n > \nu$, and consequently

$$\lim (na_n) = 0.$$

That this condition is not sufficient follows from Abel's example (Art. 11) $a_n = (n \log n)^{-1}$, which gives a divergent series, although $\lim (na_n) = 0$.

No condition such as $\lim (a_n D_n) = 0$ is necessary for convergence if D_n tends to ∞ more rapidly than n; and examples of convergent series for which $(a_n D_n)$ has no definite limit will be found in Pringsheim's article (*Math. Annalen*, Bd. 35, p. 343). Of course, *if the limit exists*, its value must be zero for convergence; but convergence does not imply the existence of a limit for $(a_n D_n)$.

10. If the series Σa_n is compared with the geometric series Σr^n, we can infer Cauchy's test, which is theoretically of fundamental importance:

$$\textit{If } \overline{\lim}\, a_n^{\frac{1}{n}} < 1, \textit{ the series converges;}$$

$$\textit{if } \overline{\lim}\, a_n^{\frac{1}{n}} > 1, \textit{ the series diverges.}$$

It is of great importance to remember that, in contrast with the ratio-tests of Art. 12, these conditions *both* relate to the *maximum* limiting value; and that the condition $\underline{\lim}\, a_n^{\frac{1}{n}} > 1$ is *not* necessary for divergence.

Further, to ensure divergence, it is *not* necessary that $a_n^{\frac{1}{n}}$ should be ultimately greater than unity, in spite of what is sometimes stated in text-books; and if $a_n^{\frac{1}{n}}$ oscillates between limits which include unity, the series *diverges.**

To prove these rules, suppose first that

$$\overline{\lim}\, a_n^{\frac{1}{n}} = G < 1.$$

Take any number ρ between G and 1; then we can find m so that

$$a_n^{\frac{1}{n}} < \rho < 1, \quad \text{if} \quad n > m.$$

Hence, after the mth term, the terms of Σa_n are less than those of the convergent series $\Sigma \rho^n$; that is, Σa_n is convergent. And the remainder after p terms is less than $\rho^p/(1-\rho)$ provided that $p > m$.

But if $\overline{\lim}\, a_n^{\frac{1}{n}} > 1$, there will be an infinite sequence of values of n, (say $n_1, n_2, n_3, \ldots$), such that

$$a_n^{\frac{1}{n}} > 1, \quad \text{if} \quad n = n_p;$$

and therefore

$$a_n > 1, \quad \text{if} \quad n = n_p.$$

Thus the sum Σa_n, taken from 1 to n_p, must be greater than p; and p may be taken as large as we please, so that Σa_n diverges.

We know from Art. 154 that $\overline{\lim}\, a_n^{\frac{1}{n}}$ lies between the extreme limits of (a_{n+1}/a_n); thus the series converges if $\overline{\lim}\,(a_{n+1}/a_n) < 1$, and diverges if $\underline{\lim}\,(a_{n+1}/a_n) > 1$. This shews that d'Alembert's test (Art. 12) is a deduction from Cauchy's.

But on the other hand, since we only know that $\overline{\lim}\, a_n^{\frac{1}{n}}$ falls between the extreme limits of (a_{n+1}/a_n), it is clear that *we cannot deduce Cauchy's test in its full generality from d'Alembert's.*

If we consider a power-series $\Sigma b_n x^n$ (in which b_n and x are supposed positive), Cauchy's test will give

$$xl < 1, \text{ for convergence, and } xl > 1, \text{ for divergence,}$$

where

$$l = \overline{\lim}\, b_n^{\frac{1}{n}}.$$

Thus $x = 1/l$ gives an exact boundary between convergent and divergent series, supposing l to be different from zero and

---

* This might seem to be at variance with a statement made by Chrystal (*Algebra*, 2nd edition, ch. XXVI. § 5, 1); but his remark must be understood as implying the insufficiency of the method used in that article. The complete form of the test is given by Chrystal in § 11 of the same chapter.

finite. If $l=0$, the condition for convergence is satisfied for all positive values of x; but if $l=\infty$, the series will diverge for all values of x, except zero.

But if we apply d'Alembert's test to the power-series, we can only infer that

$x<g$ gives convergence, and $x>G$ gives divergence,

where $$g=\underline{\lim}\,(b_n/b_{n+1}) \text{ and } G=\overline{\lim}\,(b_n/b_{n+1});$$

so that when g, G are unequal (as they may easily be), we can obtain no information as to the behaviour of the series if $g<x<G$.

In spite of this theoretical objection, d'Alembert's test is sufficient to establish the region of convergence of the most useful power-series; and, on account of its simple character, this test (with its extensions in Art. 12) is the most widely used in ordinary work.

11. Second test for convergence; the logarithmic scale.

Suppose that the terms of a positive series are arranged in order of magnitude, so that $a_n \geqq a_{n+1} > 0$.

If we write $f(n)=a_n$, it may happen that the function $f(x)$ is also definite for values of x which are not integers, and that $f(x)$ never increases with x. Then, if x lies between n and $n+1$, it is plain that

$$a_n \geqq f(x) \geqq a_{n+1} > 0.$$

Thus, from the definition of an integral, we have

$$\int_n^{n+1} a_n\,dx \geqq \int_n^{n+1} f(x)\,dx \geqq \int_n^{n+1} a_{n+1}\,dx$$

or $$a_n \geqq \int_n^{n+1} f(x)\,dx \geqq a_{n+1}.$$

Write now $I_n=\int_1^n f(x)\,dx$, and we find successively

$$a_1 \geqq I_2 \geqq a_2,$$
$$a_2 \geqq I_3 - I_2 \geqq a_3,$$
$$\cdots\cdots \quad \cdots\cdots\cdots\cdots$$
$$a_{n-1} \geqq I_n - I_{n-1} \geqq a_n.$$

Adding these inequalities we have

$$s_n - a_n \geqq I_n \geqq s_n - a_1.$$

Hence $$a_1 \geqq s_n - I_n \geqq a_n > 0.$$

Further $(s_{n+1}-I_{n+1})-(s_n-I_n)=a_{n+1}-\int_n^{n+1} f(x)dx \leqq 0,$

and therefore the sequence whose nth term is s_n-I_n *never increases*; and since its terms are contained between 0 and a_1, the sequence must have a limit (Art. 2) and

$$a_1 \geqq \lim (s_n - I_n) \geqq 0.$$

Thus, *the series* Σa_n *converges or diverges with the integral** $\int_1^\infty f(x)dx$; *if convergent, the sum of the series differs from the integral by less than* a_1; *if divergent, the limit of* (s_n-I_n) *nevertheless exists and lies between* 0 *and* a_1.

Ex. If $a_n=1/n(n+1)$, $f(x)=1/x(x+1)$, and $\int_1^\infty f(x)dx=\log 2$.

And $\sum_1^\infty a_n=1$, which is contained between the values $\log 2$ and $\frac{1}{2}+\log 2$, in agreement with the general result.

A large number of very important special series are easily tested by this rule:

(1) Consider $\frac{1}{1^p}+\frac{1}{2^p}+\frac{1}{3^p}+\dots$, where $a_n=n^{-p}$.

Here, if p is positive, the rule applies at once, and gives

$$f(x)=x^{-p}, \quad \int_1^x f(x)dx=\frac{1}{1-p}(x^{1-p}-1);$$

thus the integral to ∞ is convergent only if $p>1$. Thus *the given series converges only if* $p>1$; *and the sum is then contained between the values* $1/(p-1)$ *and* $p/(p-1)$.

If $p=1$, the integral is equal to $\log x$, and shews that the harmonic series is divergent (see Art. 7); we infer also that

$$\lim_{n\to\infty}\left(1+\frac{1}{2}+\frac{1}{3}+\dots+\frac{1}{n}-\log n\right)$$

exists and lies between 0 and 1. This is *Euler's or Mascheroni's constant.*

The convergence of the series used in Art. 9

$$1+\frac{1}{2^2}+\frac{1}{3^2}+\frac{1}{4^{\frac{3}{2}}}+\frac{1}{5^2}+\frac{1}{6^2}+\frac{1}{7^2}+\frac{1}{8^2}+\frac{1}{9^{\frac{3}{2}}}+\dots$$

can now be inferred.

---

* The integral converges or diverges with the sequence (I_n); for further details see Appendix III.

For the first n terms are included in S_n+T_n, where

$$S_n=1+\frac{1}{2^2}+\frac{1}{3^2}+\frac{1}{4^2}+\ldots+\frac{1}{n^2},$$

$$T_n=1+\frac{1}{4^{\frac{2}{3}}}+\frac{1}{9^{\frac{2}{3}}}+\frac{1}{16^{\frac{2}{3}}}+\ldots+\frac{1}{(n^2)^{\frac{2}{3}}};$$

and so the sum of these n terms is less than S_n+T_n.

Now by (1) $$S_n<2, \quad T_n<4,$$

and so $$S_n+T_n<6.$$

Hence the given series converges to a sum not greater than 6 (Arts. 2, 7).

(2) Consider $\frac{1}{2}(\log 2)^{-p}+\frac{1}{3}(\log 3)^{-p}+\frac{1}{4}(\log 4)^{-p}+\ldots,$

for which $$a_1=0 \text{ and } a_n=n^{-1}(\log n)^{-p}.$$

Here $$f(x)=x^{-1}(\log x)^{-p},$$

and so $$\int_2^x f(x)dx=[(\log x)^{1-p}-(\log 2)^{1-p}]/(1-p).$$

or $$=\log(\log x/\log 2), \quad \text{if } p=1.$$

Thus the given series converges if $p>1$ and diverges if $p\leqq 1$; it should be noted that if $p=1$, the divergence is very slow, the sum of a billion terms being less than 5.

(3) It can be proved similarly that if we omit a sufficient number of the early terms to ensure that all the logarithms are positive, and if

$$a_n=(n\log n)^{-1}(\log\log n)^{-p},$$

or $$(n\,.\log n\,.\log\log n)^{-1}[\log(\log\log n)]^{-p},$$

the series converges if $p>1$, diverges if $p\leqq 1$.

(4) Since $\int [F'(x)/F(x)]dx=\log[F(x)]$, it is clear that the two integrals $\int^\infty [F'(x)/F(x)]dx$, and $\int^\infty F'(x)dx$ converge or diverge together; now if we suppose that $F'(x)$ is positive but decreases to zero as a limit, the same will be true of $F'(x)/F(x)$, provided that $F(x)$ is positive and so we can deduce the result:

The series $\Sigma F'(n)/F(n)$ converges or diverges according as the series $\Sigma F'(n)$ does. Similarly, *if $\Sigma F'(n)$ is divergent, the series $\Sigma F'(n)/[F(n)]^p$ converges if $p>1$, but diverges if $p\leqq 1$.*

This result shews that the succession of series begun in 1, 2, 3 can be continued without stopping; but for ordinary work, the two types 1, 2 are sufficient.

The following results, which are independent of the Calculus, have a field of application substantially equivalent to (4):

Let (M_n) denote an increasing sequence such that $\lim M_n = \infty$; then

$$\Sigma(M_{n+1} - M_n)/M_n \text{ and } \Sigma(M_{n+1} - M_n)/M_{n+1}$$

are divergent series, while $\Sigma(M_{n+1} - M_n)/M_n^{p-1}M_{n+1}$ is convergent if $p > 1$.

For, if we take the sum of $(M_{n+1} - M_n)/M_n$ as n ranges from q to r, we see that its value is greater than $\sum_q^r (M_{n+1} - M_n)/M_{r+1} = (M_{r+1} - M_q)/M_{r+1}$ because in the summation $M_n < M_{r+1}$. We can choose r large enough to make $M_{r+1} \geqq 2M_q$; and so this sum is greater than $\frac{1}{2}$, no matter how large q may be. Thus the series diverges. (Art. 7 (3).)

Similarly, $\Sigma(M_{n+1} - M_n)/M_{n+1}$ is divergent.

If $p = 2$, the third series reduces to $\Sigma\left(\frac{1}{M_n} - \frac{1}{M_{n+1}}\right) = \frac{1}{M_1}$, and so is convergent; thus if $p > 2$, the terms are less than those of a convergent series, and so the only case left for discussion is given by $1 < p < 2$.

From Ex. 24, Ch. I., we have the inequality

$$(1 - a^m)/m > (1 - a^n)/n, \quad \text{if } m < n.$$

Write now $$a^n = c, \quad m/n = k,$$

and then we have $$1 - c^k > k(1 - c)$$

where k is a proper fraction.

To apply this lemma, write

$$c = M_n/M_{n+1}, \quad k = p - 1$$

and then we get $$1 - \frac{M_n}{M_{n+1}} < \frac{1}{p-1}\left[1 - \left(\frac{M_n}{M_{n+1}}\right)^{p-1}\right],$$

or $$\frac{M_{n+1} - M_n}{M_n^{p-1} M_{n+1}} < \frac{1}{p-1}\left(\frac{1}{M_n^{p-1}} - \frac{1}{M_{n+1}^{p-1}}\right).$$

From this it is plain that the given series has its terms less than those of a convergent series.

12. Ratio-tests for convergence.

Kummer's test for the series Σa_n runs thus:

If ΣD_n^{-1} is a divergent series, then Σa_n is

$$\text{(C) } \textit{convergent}, \text{ if } \varliminf \left(D_n \frac{a_n}{a_{n+1}} - D_{n+1}\right) > 0,$$

$$\text{(D) } \textit{divergent}, \text{ if } \varlimsup \left(D_n \frac{a_n}{a_{n+1}} - D_{n+1}\right) < 0.$$

For in the first case, if g is the minimum limit and h is any positive number less than g, an integer m can be found such that

$$D_n \frac{a_n}{a_{n+1}} - D_{n+1} > h, \quad \text{if } n \geqq m.$$

Thus $$a_n D_n - a_{n+1} D_{n+1} > h a_{n+1}, \quad \text{if } n \geqq m,$$

or adding, we have

$$a_m D_m - a_n D_n > h(a_{m+1} + a_{m+2} + \dots + a_n).$$

Hence $$a_{m+1} + a_{m+2} + \dots + a_n < a_m D_m / h,$$

and the last expression on the right *does not involve* n; so that $\sum_1^n a_n$ remains always less than a fixed number, and therefore Σa_n is convergent.

In the second case we can find m, so that

$$D_n \frac{a_n}{a_{n+1}} - D_{n+1} < 0, \quad \text{if } n \geqq m,$$

or $$a_n D_n < a_{n+1} D_{n+1}, \quad \text{if } n \geqq m.$$

Hence $$a_n D_n > a_m D_m, \quad \text{if } n > m,$$

and so the terms of Σa_n are, after the mth, greater than those of the divergent series $(a_m D_m) \Sigma D_n^{-1}$. Thus Σa_n is also divergent.

Special cases of importance are:

(1) *d'Alembert's test.*

Let $D_n = D_{n+1} = 1$; then the conditions are

$$\text{(C)} \ \underline{\lim}\, (a_n / a_{n+1}) > 1; \quad \text{(D)} \ \overline{\lim}\, (a_n / a_{n+1}) < 1.$$

This should be compared with Cauchy's test of Art. 10.

Ex. 1. If this test is applied to the series $1 + 2x + 3x^2 + 4x^3 + \dots$ we see that it converges if $x < 1$, diverges if $x > 1$; but the test gives no result if $x = 1$ although the series is then obviously divergent.

(2) *Raabe's test;* to be tried when $\lim (a_n / a_{n+1}) = 1$.

Let $D_n = n$, then the conditions are

$$\text{(C)} \ \underline{\lim}\, [n(a_n / a_{n+1} - 1)] > 1; \quad \text{(D)} \ \overline{\lim}\, [n(a_n / a_{n+1} - 1)] < 1.$$

Ex. 2. If we take

$$1 + \frac{1+\alpha}{1+\beta} + \frac{(1+\alpha)(2+\alpha)}{(1+\beta)(2+\beta)} + \dots,$$

we find $$n\left(\frac{a_n}{a_{n+1}} - 1\right) = \frac{\beta - \alpha}{1 + \alpha/n},$$

and so the series converges if $\beta > \alpha + 1$, diverges if $\beta < \alpha + 1$. If $\beta = \alpha + 1$ the test fails, although the series may then be seen to diverge by comparison with $\Sigma 1/n$.

(3) If the limits used in (2) are both equal to 1, we must use more delicate tests, found by writing

$$D_n = n \log n, \quad n \log n \log(\log n), \quad \text{and so on.}$$

These functions are of the form $f(n)$, where $f(x)$ is continuous and $f''(x)$ tends to zero as x tends to infinity. Then Kummer's test becomes

$$\text{(C)} \ \underline{\lim}\, \kappa_n > 0; \quad \text{(D)} \ \overline{\lim}\, \kappa_n < 0,$$

where
$$\frac{a_n}{a_{n+1}} = 1 + \frac{f'(n)}{f(n)} + \frac{\kappa_n}{f(n)}.$$

For

$$f(n+1) - f(n) - f'(n) = \int_0^1 [f'(n+x) - f'(n)]\,dx = \int_0^1 dx \int_0^x f''(n+t)\,dt.$$

Now we can find ν so that $|f''(x)| < \epsilon$, if $x > \nu$, and so the last integral is easily seen to be less than $\frac{1}{2}\epsilon$, if $n > \nu$. Thus

$$f(n+1) - f(n) - f'(n) \to 0, \quad \text{as } n \to \infty.$$

Writing $f(n+1)$ and $f(n)$ for D_{n+1} and D_n in Kummer's test, we are led at once to the form given above.

In particular if $f(x) = x \log x$, we find $f'(x) = \log x + 1$, $f''(x) = 1/x$; thus we find *de Morgan's and Bertrand's first test,*

$$\text{(C)} \ \underline{\lim}\, \rho_n > 1; \quad \text{(D)} \ \overline{\lim}\, \rho_n < 1,$$

where
$$\frac{a_n}{a_{n+1}} = 1 + \frac{1}{n} + \frac{\rho_n}{n \log n}.$$

Their further tests, given by $f(x) = x \log x \log(\log x)$, etc., are of little practical importance.

(4) It is sometimes more convenient to replace the last test by the following:

$$\text{(C)} \ \underline{\lim}\, \sigma_n > 1; \quad \text{(D)} \ \overline{\lim}\, \sigma_n < 1,$$

where
$$\log \frac{a_n}{a_{n+1}} = \frac{1}{n} + \frac{\sigma_n}{n \log n}.$$

After a certain stage, we have $1 < a_n/a_{n+1} < 1 + (2/n)$;

also
$$0 < \xi - \log(1+\xi) = \int_0^\xi \frac{t}{1+t}\,dt < \tfrac{1}{2}\xi^2, \quad \text{if } \xi > 0;$$

thus we see that $0 < \rho_n - \sigma_n < 2(\log n)/n$, and so $\rho_n - \sigma_n \to 0$.

(5) The most important cases in practical work admit of the quotient a_n/a_{n+1} being expressed in the form

$$\frac{a_n}{a_{n+1}} = 1 + \frac{\mu}{n} + \frac{\omega_n}{n^\lambda},$$

where μ is a constant, λ an index greater than 1, and $|\omega_n|$ remains less than a fixed number A for all values of n.

If this expansion is possible, it is easily seen that d'Alembert's test fails: Raabe's test gives *convergence if* $\mu > 1$, *divergence if* $\mu < 1$. To discuss the case $\mu = 1$, apply the test (3); then we have to consider the limit of $\omega_n . \log n . n^{1-\lambda}$.

But, since $|\omega_n| < A$, and $\lim (\log n . n^{1-\lambda}) = 0$,

we have $$\lim (\omega_n . \log n . n^{1-\lambda}) = 0.$$

Thus $\mu = 1$ also gives divergence. We may therefore sum up these results in the working rule (essentially due to Gauss in his investigations on the Hypergeometric series):

If it is possible to express the quotient a_n/a_{n+1} *in the form*

$$\frac{a_n}{a_{n+1}} = 1 + \frac{\mu}{n} + \frac{\omega_n}{n^\lambda}, \quad \begin{cases} \lambda > 1 \\ |\omega_n| < A \end{cases}$$

the series Σa_n *is divergent if* $\mu \leqq 1$, *convergent if* $\mu > 1$.

If we apply the results of Art. 39, Ex. 3, to the quotient $na_n/(n+1)a_{n+1}$, it is not difficult to prove that, *when* a_n/a_{n+1} *can be expressed in the form above, the condition* $\lim (na_n) = 0$ *is necessary and sufficient for convergence* (in contrast with the results for series in general, Art. 9).

Ex. 3. Consider the Hypergeometric Series

$$1 + \frac{\alpha . \beta}{1 . \gamma} x + \frac{\alpha(\alpha+1)\beta(\beta+1)}{1 . 2 . \gamma(\gamma+1)} x^2 + \frac{\alpha(\alpha+1)(\alpha+2)\beta(\beta+1)(\beta+2)}{1 . 2 . 3 . \gamma(\gamma+1)(\gamma+2)} x^3 + \ldots.$$

By using d'Alembert's test this series is easily seen to converge if $0 < x < 1$, and to diverge if $x > 1$. If $x = 1$, consider

$$\frac{a_n}{a_{n+1}} = \frac{(n+1)(\gamma+n)}{(\alpha+n)(\beta+n)} = 1 + \frac{n(\gamma+1-\alpha-\beta)-\alpha\beta}{n^2+n(\alpha+\beta)+\alpha\beta},$$

which gives $\mu = \gamma + 1 - \alpha - \beta$, so that the series converges if $\gamma > \alpha + \beta$, and diverges if $\gamma \leqq \alpha + \beta$.

It will appear from Art. 50 that the series converges if $-1 < x < 0$; and from Art. 21 that it converges also for $x = -1$, if $\gamma + 1 > \alpha + \beta$.

13. Notes on the ratio-tests.

It is to be noted that *d'Alembert's test does not ensure the convergence of a series if we only know that* $a_n/a_{n+1} > 1$ *for all values of* n.

For, if $\lim (a_n/a_{n+1}) = 1$, it will not be possible to find a number k such that $$\frac{a_n}{a_{n+1}} - 1 > k > 0, \quad \text{for } n \geqq m.$$

In particular, if $a_n = 1/n$, $a_n/a_{n+1} = 1 + 1/n > 1$; and yet the series Σa_n is divergent.

Secondly, *it is not necessary for the convergence of the series* Σa_n *that* a_n/a_{n+1} *should have a definite limit.*

For it will be seen in Art. 26 that the *order* of the terms does not affect the convergence of a series of positive terms; but of course a change in the order may affect the value of $\lim a_n/a_{n+1}$.

Ex. 1. The series $a+1+a^3+a^2+a^5+a^4+\ldots$ is a rearrangement of the geometric series $1+a+a^2+a^3+\ldots$, and so is convergent if $0<a<1$. But in this series the quotient a_n/a_{n+1} is alternately a and $1/a^3$.

Ex. 2. The series $1+\alpha+\beta^2+\alpha^3+\beta^4+\alpha^5+\beta^6+\ldots$
is convergent if $0<\alpha<\beta<1$; as is plain by comparison with

$$1+\beta+\beta^2+\beta^3+\beta^4+\ldots.$$

In this series we have

$$\lim \alpha^n/\beta^{n+1}=0, \quad \lim \beta^n/\alpha^{n+1}=\infty.$$

But even when the terms are arranged in order of magnitude, the convergence of the series does not imply the existence of the limit.

Ex. 3. The series $1+\frac{1}{2}a+\frac{1}{3}a+\frac{1}{4}a^2+\frac{1}{5}a^2+\frac{1}{6}a^3+\frac{1}{7}a^3+\ldots$
has its terms arranged in order of magnitude, if $0<a<1$; and it is then convergent, by comparison with $1+a+a+a^2+a^2+a^3+a^3+\ldots$.

But yet $\quad \overline{\lim}\,(a_n/a_{n+1})=1/a, \quad \underline{\lim}\,(a_n/a_{n+1})=1.$

Thirdly, *if the quotient* a_n/a_{n+1} *has maximum and minimum limits which include unity, the whole scale of ratio-tests will fail.*

For, if $\overline{\lim}\,(a_n/a_{n+1})=G>1>g=\underline{\lim}\,(a_n/a_{n+1})$, we can take K, k such that

$$G>K>1>k>g,$$

and then a_n/a_{n+1} is greater than K for an infinite sequence of values of n, while it is less than k for a second infinite sequence of values.

If n belongs to the first set of values, we shall have

$$n(a_n/a_{n+1}-1)>n(K-1);$$

but if it belongs to the second set,

$$n(a_n/a_{n+1}-1)<-n(1-k).$$

Hence $\quad \overline{\lim}\, n(a_n/a_{n+1}-1)=+\infty, \quad \underline{\lim}\, n(a_n/a_{n+1}-1)=-\infty,$

and therefore Raabe's test fails entirely. It is easy to see that the failure extends to all the following tests.

If we apply Raabe's test to Ex. 3 above, we find

$$\overline{\lim}\, n(a_n/a_{n+1}-1)=+\infty, \quad \underline{\lim}\, n(a_n/a_{n+1}-1)=1;$$

and passing to the next stage we get

$$\overline{\lim}\,(\log n)[n(a_n/a_{n+1}-1)-1]=+\infty, \quad \underline{\lim}\,(\log n)[n(a_n/a_{n+1}-1)-1]=0,$$

so that the ratio-tests can give no information.

It will be seen from the foregoing remarks that the ratio-tests have a comparatively limited range of usefulness; and it may reasonably be asked, why should we trouble to introduce them at all, and not be content with the more general comparison-tests? The answer to this is that, in practice, the quotient a_n/a_{n+1} is often much simpler than a_n, and then it is easier to use the ratio-tests (if they apply) than any others.

Historical Note. Kummer himself gave his test in the form

$$\lim\left[\phi(n)-\phi(n+1)\frac{a_{n+1}}{a_n}\right]>0$$

for convergence, where $\phi(n)$ is an arbitrary sequence of *positive* numbers, subject to the restriction $\lim \phi(n)a_n=0$, which was proved to be superfluous by Dini. Dini also was the first to obtain the condition in the form given above, where the *same* expressions are used to test both for convergence and for divergence.* Further extensions have been given by Pringsheim.

14. Ermakoff's tests.†

The series $\Sigma f(n)$, in which $f(n)$ is subject to the conditions of Art. 11, is

(i) convergent if $\overline{\lim\limits_{x\to\infty}}\dfrac{e^x f(e^x)}{f(x)}<1$,

(ii) divergent if $\underline{\lim\limits_{x\to\infty}}\dfrac{e^x f(e^x)}{f(x)}>1$.

For, in the first case, if ρ is any number between the maximum limit and unity, we can find ξ so that

$$e^x f(e^x)<\rho f(x), \qquad \text{if} \quad x>\xi.$$

Thus

$$\int_\xi^X e^x f(e^x)dx<\rho\int_\xi^X f(x)dx, \quad \text{if} \quad X>\xi,$$

or, changing the independent variable to e^x in the left-hand integral, we have‡

$$\int_{e^\xi}^{e^X} f(x)dx<\rho\int_\xi^X f(x)dx.$$

That is,

$$(1-\rho)\int_{e^\xi}^{e^X} f(x)dx<\rho\left[\int_\xi^X f(x)dx-\int_{e^\xi}^{e^X} f(x)dx\right]$$

$$\text{or} \quad <\rho\left[\int_\xi^{e^\xi} f(x)dx-\int_X^{e^X} f(x)dx\right].$$

* Some variations of the tests have been given by different writers; but Dini's are undoubtedly the most convenient in practice.

† *Bulletin des Sciences Mathématiques*, 1871, t. 2, p. 250.

‡ The reader is advised to use the geometrical representation of $\int f(x)\,dx$ as the area of the curve $y=f(x)$ when following out the argument.

Or, again, since the last term in the bracket is positive, because e^X is greater than X, we have

$$(1-\rho)\int_{e^\xi}^{e^X} f(x)\,dx < \rho\int_\xi^{e^\xi} f(x)\,dx.$$

As this inequality is true for any value of X greater than ξ, it is clear that the integral $\int^\infty f(x)dx$ must converge; and, therefore, so also does the series $\Sigma f(n)$.

In the second case, ξ can be found so that

$$e^x f(e^x) \geqq f(x), \qquad \text{if} \quad x \geqq \xi.$$

As above, this gives

$$\int_{e^\xi}^{e^X} f(x)\,dx \geqq \int_\xi^X f(x)\,dx, \quad \text{if} \quad X > \xi,$$

or

$$\int_X^{e^X} f(x)\,dx \geqq \int_\xi^{e^\xi} f(x)\,dx, \quad \text{if} \quad X > \xi.$$

This indicates that the integral $\int^\infty f(x)dx$ is divergent, because, no matter how great X may be, a number $X' = e^X$ can be found such that $\int_X^{X'} f(x)dx$ is greater than a certain constant K; compare the argument of Art. 7 (3). Thus the series $\Sigma f(n)$ is divergent.

These tests include the whole of the logarithmic scale.

For example, consider

$$f(x) = 1/\{x \,.\, \log x \,.\, [\log(\log x)]^p\},$$

then

$$e^x f(e^x) = e^x/\{e^x \,.\, x \,.\, [\log x]^p\}.$$

Thus

$$e^x f(e^x)/f(x) = [\log(\log x)]^p/[\log x]^{p-1},$$

and so

$$\lim_{x\to\infty} e^x f(e^x)/f(x) = 0, \quad \text{if} \quad p > 1,$$

or

$$= \infty, \quad \text{if} \quad p \leqq 1.$$

That is, the series $\Sigma f(n)$ converges if $p > 1$ and diverges if $p \leqq 1$.

It is easy to see that if $\phi(x)$ is a function which steadily increases with x, in such a way that $\phi(x) > x$, the proof above may be generalised to give Ermakoff's tests:

(i) convergence, if $\overline{\lim}_{x\to\infty} \dfrac{\phi'(x)f(\phi(x))}{f(x)} < 1$,

(ii) divergence, if $\underline{\lim}_{x\to\infty} \dfrac{\phi'(x)f(\phi(x))}{f(x)} > 1$.

15. Another sequence of tests.

Although the following sequence is of less importance than the ratio-tests in ordinary work, it is of theoretical interest, giving a continuation of Cauchy's test in Art. 10.

We have seen in Art. 11, that if $\Sigma F'(n)$ is divergent, $\Sigma F'(n)/[F(n)]^p$ converges only if $p>1$. This gives the following test:

$$\left.\begin{array}{ll}\Sigma a_n \text{ converges if } & \underline{\lim}\,\dfrac{\log[F'(n)/a_n]}{\log[F(n)]}>1\\ \text{and diverges if } & \overline{\lim}\,\dfrac{\log[F'(n)/a_n]}{\log[F(n)]}<1\end{array}\right\}\begin{array}{l}\text{where } F'(n) \text{ is positive}\\ \text{but tends } \textit{steadily} \text{ to } 0.\end{array}$$

For, in the first case, as on previous occasions, we can find $p>1$ and an index m such that

$$\frac{\log[F'(n)/a_n]}{\log[F(n)]}>p, \qquad \text{if } n>m,$$

or

$$a_n<F'(n)/[F(n)]^p, \qquad \text{if } n>m.$$

This shews that Σa_n converges, by the principle of comparison.

But, in the second case, we have an index m such that

$$F'(n)/a_n \leqq F(n), \qquad \text{if } n>m,$$

or

$$a_n \geqq F'(n)/F(n), \qquad \text{if } n>m,$$

this shews that Σa_n diverges.

Special examples of this test are given by

(1) $F(n)=n$; and the function to examine is

$$\frac{\log(1/a_n)}{\log n}.$$

(2) $F(n)=\log n$; and the function is

$$\frac{\log(1/na_n)}{\log(\log n)}.$$

(3) $F(n)=\log(\log n)$; then the function is

$$\frac{\log\{1/(n\,.\log n\,.\,a_n)\}}{\log[\log(\log n)]}$$

and so on.

The test (1) can be transformed into another shape, first given by Jamet, in which the relation to Cauchy's test is easily recognised.

If we write $\lambda=\log(1/a_n)$, it is easy to see that

$$1-\lambda/n<a_n^{\frac{1}{n}}<1/(1+\lambda/n),$$

so that

$$\lambda>n(1-a_n^{\frac{1}{n}})>\lambda/(1+\lambda/n).$$

Thus we have

$$\frac{\overline{\lim}}{\underline{}}\,\frac{\lambda}{\log n}=\frac{\overline{\lim}}{\underline{}}\,\frac{n}{\log n}(1-a_n^{\frac{1}{n}}),$$

provided that $\lim(\lambda/n)=0$; and, if this condition is not satisfied, Cauchy's test will settle the question. So in all cases of practical interest, the test will be

$$\underline{\lim}\,\frac{n}{\log n}(1-a_n^{\frac{1}{n}})>1, \quad \overline{\lim}\,\frac{n}{\log n}(1-a_n^{\frac{1}{n}})<1.$$

Similarly it can be proved that test (2) can be replaced by

$$\overline{\lim} \frac{1}{\log(\log n)}[n(1-a_n^{\frac{1}{n}})-\log n] \lesseqgtr 1.$$

This form proves for example that the series $\sum\left(1-\frac{x}{n}\log n\right)^n$ diverges if $0 \leqq x \leqq 1$, but converges for $x>1$.

16. General notes on series of positive terms.

Although the rules which we have established are sufficient to test the convergence of all series which present themselves naturally in elementary analysis, yet it is impossible to frame any rule which will give a decisive test for an artificially constructed series. In other words, *whatever rule is given, a series can be invented for which the rule fails to give a decisive result.*

The following notes (1)–(3) and (8) show how certain rules which appear plausible at first sight have been proved to be either incorrect or insufficient. Notes (4)–(7) shew that however slowly a series may diverge (or converge) we can always construct series which diverge (or converge) still more slowly; and thus no test of comparison can be sufficient for all series.

Other interesting questions in this connexion have been considered by Hadamard (*Acta Mathematica*, t. 18, 1894, p. 319, and t. 27, 1903, p. 177).

(1) Abel has pointed out that there *cannot* be a positive function $\phi(n)$ such that the two conditions

$$\text{(i) } \lim \phi(n).a_n = 0, \qquad \text{(ii) } \lim \phi(n).a_n > 0$$

are *sufficient*, the first for the convergence, the second for the divergence of any series Σa_n.

For, if so, $\Sigma[\phi(n)]^{-1}$ would diverge; and therefore, if

$$M_n = [\phi(1)]^{-1} + [\phi(2)]^{-1} + \dots + [\phi(n)]^{-1},$$

the sequence M_n would be an increasing sequence tending to ∞.

Hence the series $\Sigma(M_n - M_{n-1})/M_n$ would diverge also (Art. 11); but

$$\phi(n)(M_n - M_{n-1})/M_n = 1/M_n,$$

so that
$$\lim \phi(n)(M_n - M_{n-1})/M_n = 0,$$

contradicting the first condition.

(2) Pringsheim has proved that there *cannot* be a positive function $\phi(n)$ tending to ∞, such that the condition

$$\lim \phi(n).a_n \leqq G \qquad (G \geqq 0)$$

is *necessary* for the convergence of Σa_n. In fact, for *any* such function $\phi(n)$ and for *any* convergent series, the terms of the series can be so rearranged that

$$\overline{\lim}\, \phi(n) . a_n = \infty .$$

See *Math. Annalen*, Bd. 35, p. 344.

(3) Pringsheim has proved that there cannot be a positive function $\phi(n)$ such that the condition

$$\lim \phi(n) . a_n > 0$$

is *necessary* for the divergence of Σa_n. In fact, for *any* function $\phi(n)$ and *any* divergent series, the terms of the series can be so arranged that

$$\underline{\lim}\, \phi(n) . a_n = 0,$$

provided that the terms of the series tend to zero.

See *Math. Annalen*, Bd. 35, p. 358.

(4) Abel remarked that if Σa_n is divergent, a second series Σb_n can be found which is also divergent, but such that

$$\lim (b_n/a_n) = 0.$$

For, write $M_n = a_1 + a_2 + \ldots + a_n$, $b_n = a_n/M_n = (M_n - M_{n-1})/M_n$. The series Σb_n diverges by Art. 11; and

$$\lim (b_n/a_n) = \lim (1/M_n) = 0.$$

(5) du Bois Reymond shewed that if Σa_n is convergent, a second convergent series Σb_n can be found which has the property $\lim (b_n/a_n) = \infty$.

For, write
$$s_n = a_1 + a_2 + \ldots + a_n, \quad s = \lim s_n,$$
$$1/M_1 = s, \quad 1/M_{n+1} = s - s_n = a_{n+1} + a_{n+2} + \ldots \text{ to } \infty .$$

Then $M_n \to \infty$; and consequently the series Σb_n converges if

$$b_n = (M_{n+1} - M_n)/M_n^q M_{n+1} = a_n M_n^{1-q},$$

provided that q is positive (see Art. 11).

But if $q < 1$, it is evident that $b_n/a_n \to \infty$.

(6) Stieltjes shewed that if $u_1, u_2, u_3, \ldots$ is a *decreasing* sequence, tending to zero as a limit, a divergent series Σd_n can be found so that $\Sigma u_n d_n$ is convergent.

For, write $M_n = 1/u_n$; then if $d_n = (M_{n+1} - M_n)/M_{n+1}$ the series Σd_n is divergent (Art. 11). But

$$u_n d_n = \frac{M_{n+1} - M_n}{M_n M_{n+1}} = \frac{1}{M_n} - \frac{1}{M_{n+1}},$$

so that $\Sigma u_n d_n$ converges to the sum $1/M_1 = u_1$.

(7) Stieltjes also proved that if $v_1, v_2, v_3, \ldots$ is an *increasing* sequence, tending to infinity as a limit, a convergent series Σc_n can be found so that $\Sigma v_n c_n$ is divergent.

For, write $c_n = 1/v_n - 1/v_{n+1}$, which makes Σc_n a convergent series; then $v_n c_n = (v_{n+1} - v_n)/v_{n+1}$, so that $\Sigma v_n c_n$ is divergent.

(8) Finally, *even when the terms of the series Σa_n steadily decrease*, the following results have been found by Pringsheim:

However fast the series Σc_n^{-1} may *converge*, yet there are always *divergent* series Σa_n such that $\underline{\lim}\, c_n a_n = 0$.

However slowly $\phi(n)$ may increase to ∞ with n, there are always *convergent* series Σa_n, for which $\overline{\lim}\, n \,.\, \phi(n) \,.\, a_n = \infty$ (although $\lim n \,.\, a_n = 0$ by Art. 9).

See *Math. Annalen*, Bd. 35, pp. 347, 356.

EXAMPLES.

1. Test the convergence of the series Σa_n, where a_n is given by the following expressions:

$$\frac{1}{1+n^2}, \quad \frac{1+n}{1+n^2}, \quad \frac{1}{1+x^n}, \quad \frac{1}{(\log n)^a}, \qquad \text{(Arts. 8, 9)}$$

$$\frac{(n!)^2}{(2n)!}x^n, \quad \frac{n^4}{n!}, \quad \frac{m(m+1)\ldots(m+n-1)}{n!}, \qquad \text{(Art. 12)}$$

$$\frac{1}{n^{a+b/n}}, \quad \frac{n^q}{(n+1)^{p+q}}, \quad \frac{1}{(\log n)^n}, \quad \frac{1}{(\log n)^{\log n}}, \quad \frac{1}{[\log(\log n)]^{\log n}}, \quad \frac{1}{(\log n)^{\log(\log n)}}, \qquad \text{(Art. 15)}$$

$$a^{1/n} - 1. \quad \text{(Ex. 25, Ch. I., and Art. 11)}$$

2. Prove that if $b-1>a>0$, the series

$$1+\frac{a}{b}+\frac{a(a+1)}{b(b+1)}+\frac{a(a+1)(a+2)}{b(b+1)(b+2)}+\ldots$$

converges to the sum $(b-1)/(b-a-1)$.

Shew also that the sum of

$$\frac{a}{b}+2\frac{a(a+1)}{b(b+1)}+3\frac{a(a+1)(a+2)}{b(b+1)(b+2)}+\ldots$$

is $a(b-1)/(b-a-1)(b-a-2)$, if $b-2>a>0$.

[If the first series is denoted by $u_0+u_1+u_2+\ldots$, we get

$$(b+n)u_{n+1} = (a+n)u_n,$$

which gives $\qquad (b-a-1)u_{n+1} = (a+n)u_n - (a+n+1)u_{n+1}.$

Hence $(b-a-1)(s_n - u_0) = au_0 - (a+n)u_n$ by addition. But $\lim (nu_n) = 0$ by Art. 9, since the terms steadily decrease. Hence $\lim s_n$ can be found.

The second series can be expressed as the difference between two series of the first type.]

3. Prove that the series

$$1+\frac{a+1}{b+1}+\frac{(a+1)(2a+1)}{(b+1)(2b+1)}+\frac{(a+1)(2a+1)(3a+1)}{(b+1)(2b+1)(3b+1)}+\dots$$

converges if $b>a>0$ and diverges if $a \geqq b>0$.

4. Prove that the series

$$1+\frac{1}{2}\cdot\frac{1}{3}+\frac{1.3}{2.4}\cdot\frac{1}{5}+\frac{1.3.5}{2.4.6}\cdot\frac{1}{7}+\dots$$

converges.

Shew also that

$$1+\left(\frac{1}{2}\right)^p+\left(\frac{1.3}{2.4}\right)^p+\left(\frac{1.3.5}{2.4.6}\right)^p+\dots$$

converges if $p>2$, and otherwise diverges (Art. 12).

5. Prove that, if $0<q<1$, $\Sigma q^{n^2}x^n$ converges for any positive value of x.

Prove that $1+\frac{1}{2^\beta}+\frac{1}{3^\alpha}+\frac{1}{4^\beta}+\frac{1}{5^\alpha}$ converges if $\alpha>\beta>1$, but that the ratio of two consecutive terms oscillates between 0 and ∞ (Art. 10).

6. Prove that $\Sigma(1/A_n)$ converges if the sequence (B_n) never decreases, where $B_n=A_{n-1}-2A_n+A_{n+1}$.

7. Shew that in Kummer's test for convergence we can write $\phi(n)$ in place of $1/D_n$, where $\phi(n)$ is an arbitrary positive function; but prove also that there is no advantage in making this change.

Deduce that Σa_n is convergent if, after a certain stage,

$$\frac{a_n}{a_{n+1}} \geqq \phi(n)\left[1+\frac{1}{\phi(n+1)}\right]. \qquad \text{[GIUDICE.]}$$

8. If a_{n+1}/a_n can be expressed as the quotient of two polynomials in n, $P(n)/Q(n)$, of the same degree k, whose highest term is n^k, and if the highest term in $Q(n)-P(n)$ is An^{k-l}, prove that Σa_n diverges if $l>1$, converges if $l=1$ and $A>1$. [DE SAINT-GERMAIN.]

9. Test the convergence of the series Σa_n, where

$$a_n=(2-e)(2-e^{\frac{1}{2}})(2-e^{\frac{1}{3}})\dots(2-e^{\frac{1}{n}}).$$

10. Find limits for the sum

$$s_n=\frac{1}{\sqrt{n^2}}+\frac{1}{\sqrt{(n^2+1)}}+\dots+\frac{1}{\sqrt{(n^2+2n)}}$$

in terms of the integral

$$\int_0^{2n}\frac{dx}{\sqrt{(n^2+x)}}=\int_0^2\frac{dt}{\sqrt{(1+t/n)}},$$

and deduce that $s_n \to 2$ as n increases to ∞.

Similarly investigate

$$\sigma_n=\frac{n}{n^2}+\frac{n}{1+n^2}+\frac{n}{2^2+n^2}+\dots+\frac{n}{(n-1)^2+n^2}$$

by means of the integral

$$\int_0^n \frac{n\,dx}{x^2+n^2} = \int_0^1 \frac{dt}{1+t^2},$$

and deduce that $\sigma_n \to \frac{1}{4}\pi$.

11. Prove that if p approaches zero through positive values,

$$\lim_{p\to 0} p \sum_0^\infty n^{-(1+p)} = 1 ;$$

and that

$$\lim_{p\to 0}\left(\sum_1^\infty \frac{1}{n^{1+p}} - \frac{1}{p}\right) = C,$$

where C is Euler's constant. [DIRICHLET.]

[To prove the latter part, note that if

$$f(\nu) = \sum_\nu^\infty \frac{1}{n^{1+p}} - \int_\nu^\infty \frac{dx}{x^{1+p}},$$

$f(\nu)$ is positive but less than $1/\nu$. The desired limit is that of $f(1)$, which can be put in the form

$$1 + \frac{1}{2^{1+p}} + \frac{1}{3^{1+p}} + \dots + \frac{1}{(\nu-1)^{1+p}} - \int_1^\nu \frac{dx}{x^{1+p}} + f(\nu).$$

From this expression we can easily see that the desired limit must be equal to that of

$$\lim_{\nu\to\infty}\left(1 + \frac{1}{2} + \frac{1}{3} + \dots + \frac{1}{\nu-1} - \log\nu\right) = C.]$$

12. More generally, if $M_n = an + b_n$ (where $|b_n|$ is less than a fixed value, and M_n is never zero),

$$\lim_{p\to 0} ap \sum_1^\infty M_n^{-(1+p)} = 1,$$

and $\lim_{p\to 0}\left(\sum_1^\infty \frac{a^{1+p}}{M_n^{1+p}} - \frac{1}{p}\right)$ exists and is finite. [DIRICHLET.]

If M_n tends *steadily* to infinity with n, and

$$d_n = (M_{n+1} - M_n)/M_{n+1},$$

then

$$\lim_{p\to 0}\left(p\sum_1^\infty d_n M_n^{-p}\right) = 1, \qquad \text{if } M_{n+1}/M_n \to 1,$$

$$\text{or } = (1-1/c)/\log c, \qquad \text{if } M_{n+1}/M_n \to c > 1,$$

$$\text{or } = 0, \qquad \text{if } M_{n+1}/M_n \to \infty.$$

[PRINGSHEIM, *Math. Annalen*, Bd. 37.]

Ex. $M_n = n^2,\ 2^n,\ n!$

13. Utilise Theorem II. of Art. 152 (Appendix) to shew that if (u_n) decreases steadily, the condition $\lim(nu_n) = 0$ is necessary (Art. 9) for the convergence of Σu_n, by writing

$$a_n = (S_n/u_n) - n, \quad b_n = 1/u_n,$$

so that

$$\frac{a_n}{b_n} = S_n - nu_n, \quad \frac{a_n - a_{n-1}}{b_n - b_{n-1}} = S_{n-1}.$$ [CESÀRO.]

If $u_n = \left(1 - \frac{1}{n}\log n\right)^n$, prove that $\lim(nu_n) = 1$, and deduce the divergence of Σu_n (compare Art. 15).

14. If Σa_n, Σb_n are both convergent, so also is $\Sigma(a_n b_n)^{\frac{1}{2}}$. But Σa_n, Σb_n may *both* diverge and yet $\Sigma(a_n b_n)^{\frac{1}{2}}$ may converge; a fact illustrated by

$$1+\frac{1}{2^3}+\frac{1}{3}+\frac{1}{4^3}+\frac{1}{5}+\frac{1}{6^3}+\dots \quad \text{and} \quad 1+\frac{1}{2}+\frac{1}{3^3}+\frac{1}{4}+\frac{1}{5^3}+\frac{1}{6}+\dots.$$

If Σa_n converges, so also does $\Sigma(a_n a_{n+1})^{\frac{1}{2}}$; but the converse is not true, as may be seen from either of the two series just written down.

On the other hand, if (a_n) is *monotonic*, the convergence of $\Sigma(a_n a_{n+1})^{\frac{1}{2}}$ implies that of Σa_n. [PRINGSHEIM.]

15. Use the preceding example to prove that if Σa_n^2 is convergent, so also is $\Sigma a_n/n$.

16. If
$$\Sigma_n = 1+\frac{1}{3}+\frac{1}{5}+\dots+\frac{1}{2n-1},$$
shew that
$$\lim(\Sigma_n - \tfrac{1}{2}\log n) = \log 2 + \tfrac{1}{2}C,$$
where C is Euler's constant (see Art. 11, 1).

17. By the method of Ex. 16 or otherwise, prove that

$$\sum_1^\infty [n(4n^2-1)]^{-1} = 2\log 2 - 1, \quad \sum_1^\infty [n(9n^2-1)]^{-1} = \tfrac{3}{2}(\log 3 - 1).$$

18. Shew that, with the notation of Ex. 16,

$$\sum_1^\nu \frac{1}{n(36n^2-1)} = -3 + 3\Sigma_{3\nu+1} - \Sigma_\nu - \left(1+\frac{1}{2}+\dots+\frac{1}{\nu}\right).$$

Deduce that
$$\sum_1^\infty \frac{1}{n(36n^2-1)} = -3 + \tfrac{3}{2}\log 3 + 2\log 2. \quad [\textit{Math. Trip.}\ 1905.]$$

19. Prove similarly that

$$\sum_1^\nu \frac{12n^2-1}{n(4n^2-1)^2} = \frac{1}{2\nu+1} - \frac{1}{(2\nu+1)^2} + 2\Sigma_\nu - \left(1+\frac{1}{2}+\dots+\frac{1}{\nu}\right),$$

and that
$$\sum_1^\infty \frac{12n^2-1}{n(4n^2-1)^2} = 2\log 2. \quad [\textit{Math. Trip.}\ 1896.]$$

20. Shew that

$$\sum_1^\infty \frac{n}{(4n^2-1)^2} = \frac{1}{8}, \quad \sum_1^\infty \frac{1}{n(4n^2-1)^2} = \frac{3}{2} - 2\log 2.$$

21. Examine the convergence of $\Sigma x^{\phi(n)}$, where x is positive; in particular, if $\phi(n) = 1+\frac{1}{2}+\frac{1}{3}+\dots+\frac{1}{n}$, or if $\phi(n)=\log n$, prove that the series converges if $x<1/e$. [Art. 15.]

22. Shew that

$$\sum_1^\infty \frac{1}{4n^2-1} = \frac{1}{2}, \quad \sum_2^\infty \frac{n-1}{n!} = 1, \quad \sum_1^\infty \frac{2}{(t+n-1)(t+n)(t+n+1)} = \frac{1}{t(t+1)},$$

and
$$\sum_1^\infty \frac{3}{(t+n-1)(t+n)(t+n+1)(t+n+2)} = \frac{1}{t(t+1)(t+2)}.$$

CHAPTER III.

SERIES IN GENERAL.

17. The only general test of convergence is simply a transformation of the condition for convergence of the sequence s_n (Art. 3); namely, that we must be able to find m, so that $|s_n - s_m| < \epsilon$, provided only that $n > m$. If we express this condition in terms of the series Σa_n, we get:

It must be possible to find m, *corresponding to the arbitrary positive fraction* ϵ, *so that*

$$|a_{m+1} + a_{m+2} + \dots + a_{m+p}| < \epsilon,$$

no matter how large p *may be.*

It is an obvious consequence that in every convergent series*

$$\lim_{n \to \infty} a_n = 0, \quad \lim_{n \to \infty} (a_{n+1} + a_{n+2} + \dots + a_{n+p}) = 0.$$

But these conditions are not sufficient unless p is allowed to take all possible forms of variation with n; and so they are not practically useful. However, it is sometimes possible to infer non-convergence by using a special form for p and shewing that then the limit is not zero (as in Art. 7 (3)).

We are therefore obliged to employ special tests, which suffice to shew that a large number of interesting series are convergent.

* It is clear from the examples in Chapter II. that the condition $\lim a_n = 0$ does not exclude divergent series; but it does not even exclude *oscillatory* series, as perhaps might be expected. For consider

$$1 - \tfrac{1}{2} - \tfrac{1}{2} + \tfrac{1}{3} + \tfrac{1}{3} + \tfrac{1}{3} - \tfrac{1}{4} - \tfrac{1}{4} - \tfrac{1}{4} - \tfrac{1}{4} + \tfrac{1}{5} + \tfrac{1}{5} + \tfrac{1}{5} + \tfrac{1}{5} + \tfrac{1}{5} - \dots,$$

where $\underline{\lim}\, s_n = 0$, $\overline{\lim}\, s_n = 1$, and yet the terms tend to zero.

18. A convergent series of positive terms remains convergent when each term a_n is multiplied by a factor v_n whose numerical value does not exceed a constant k.

For since Σa_n is convergent, the index m can be chosen so that $\sum_{m+1}^{m+p} a_n < \epsilon/k$, however small ϵ may be.

But
$$\left|\sum_{m+1}^{m+p} a_n v_n\right| \leqq \sum_{m+1}^{m+p} |a_n v_n|$$

and
$$|a_n v_n| = a_n |v_n| \leqq a_n k.$$

Thus
$$\left|\sum_{m+1}^{m+p} a_n v_n\right| \leqq k \sum_{m+1}^{m+p} a_n < \epsilon,$$

and therefore the series $\Sigma a_n v_n$ is convergent.

Two special cases of this theorem deserve mention:

(1) *A series Σa_n is convergent, if the series of its absolute values $\Sigma|a_n|$ is convergent.*

For here $a_n = |a_n|$ and $v_n = a_n/a_n = \pm 1$.

Such series are called *absolutely convergent.*

(2) *A series is convergent if its terms are numerically not greater than the corresponding terms of a convergent series of* **positive** *terms.*

The reader should observe that we cannot apply this method if an infinity of terms are negative in the series which is known to converge. An example is afforded by Ex. 2 below.

Ex. 1. If we take $a_n = n^{-p}$, we know from Art. 11 that Σa_n converges if $p > 1$. The present theorem enables us to deduce the convergence of the two series

$$\left.\begin{aligned} &1 - \frac{1}{2^p} + \frac{1}{3^p} - \frac{1}{4^p} + \frac{1}{5^p} - \frac{1}{6^p} + \dots \\ &1 + \frac{1}{2^p} - \frac{1}{3^p} + \frac{1}{4^p} + \frac{1}{5^p} - \frac{1}{6^p} + \dots \end{aligned}\right\} \qquad p > 1.$$

It will appear from Arts. 21, 22 below that the first of these series converges, but the second diverges if $0 < p \leqq 1$.

Ex. 2. The series $1 - 1 + \frac{1}{2} - \frac{1}{2} + \frac{1}{3} - \frac{1}{3} + \frac{1}{4} - \frac{1}{4} + \dots$ obviously converges to the sum 0. Now take the factors (v_n) to be $1, \frac{1}{2}, 1, \frac{1}{3}, 1, \frac{1}{4}, 1, \frac{1}{5}, \dots$, so that $|v_n| \leqq 1$. The new series is

$$1 - \frac{1}{1.2} + \frac{1}{2} - \frac{1}{2.3} + \frac{1}{3} - \frac{1}{3.4} + \frac{1}{4} - \frac{1}{4.5} + \dots.$$

The sum of the first $2n$ terms is

$$s_{2n}=\frac{2-1}{1.2}+\frac{3-1}{2.3}+\frac{4-1}{3.4}+\frac{5-1}{4.5}+\dots+\frac{(n+1)-1}{n(n+1)}$$

$$=\frac{1}{2}+\frac{1}{3}+\frac{1}{4}+\frac{1}{5}+\dots+\frac{1}{n}.$$

Thus, by Art. 11, $\lim s_{2n}=\infty$.*

But $s_{2n-1}>s_{2n}$, and so also $\lim s_{2n-1}=\infty$.

Thus *the new series is divergent.*

The reason for the failure of the theorem is that the original series contains an infinity of negative terms; and that the series ceases to converge when these terms are made positive (Art. 11).

It is easy to see that the foregoing theorem can also be stated in the form:

An **absolutely** *convergent series remains convergent if each term is multiplied by a factor whose numerical value does not exceed a constant k.*

19. It is often convenient, however, to infer the convergence of a series from one which is not absolutely convergent. For this purpose the following theorem may be used:

A convergent series Σa_n (which need not converge **absolutely**) *remains convergent if its terms are each multiplied by a factor u_n, provided that the sequence (u_n) is monotonic, and that $|u_n|$ is less than a constant k.* **(Abel's test.)**

Under these conditions u_n converges to a limit u; and write $v_n=u-u_n$ when (u_n) is an increasing sequence, but $v_n=u_n-u$ when (u_n) is decreasing. Then it is clear that *the sequence (v_n) never increases and converges to zero as a limit.* Now

$$a_nu_n=a_nu-a_nv_n, \text{ or } a_nu+a_nv_n,$$

so that it will suffice to prove the convergence of Σa_nv_n in order to infer the convergence of Σa_nu_n. But by Abel's lemma (see Art. 23 below)

$$\left|\sum_{m+1}^{m+p} a_nv_n\right|<\rho v_{m+1}<\rho v_1,$$

where ρ is greater than any of the sums

$$|a_{m+1}|, \quad |a_{m+1}+a_{m+2}|, \quad |a_{m+1}+a_{m+2}+a_{m+3}|, \ \dots$$
$$|a_{m+1}+a_{m+2}+a_{m+3}+\dots+a_{m+p}|.$$

Now, since Σa_n is convergent, m can be chosen so that $\rho = \epsilon/v_1$, no matter how small ϵ is; thus $\left|\sum_{m+1}^{m+p} a_n v_n\right|$ is less than ϵ, and consequently $\Sigma a_n v_n$ is convergent.

The reader will observe that the series Σa_n is not subject to as stringent conditions as in Art. 18; but to counterbalance this, the factors are subject to *more* stringent conditions.

Ex. 1. If we take the series $1-1+\frac{1}{2}-\frac{1}{2}+\frac{1}{3}-\frac{1}{3}+\frac{1}{4}-\frac{1}{4}+\dots$ (already used in Ex. 2 of Art. 18) and employ the *monotonic* sequence of factors

$$0, \tfrac{1}{2}, \tfrac{1}{2}, \tfrac{2}{3}, \tfrac{2}{3}, \tfrac{3}{4}, \tfrac{3}{4}, \dots,$$

we obtain the series

$$0-\frac{1}{2}+\frac{1}{2^2}-\frac{1}{3}+\frac{2}{3^2}-\frac{1}{4}+\frac{3}{4^2}-\dots,$$

which must therefore be convergent. To verify that this is the case, we observe that

$$s_{2n-1} = -\left(\frac{1}{2}-\frac{1}{2^2}\right)-\left(\frac{1}{3}-\frac{2}{3^2}\right)-\dots-\left(\frac{1}{n}-\frac{n-1}{n^2}\right)$$

$$= -\frac{1}{2^2}-\frac{1}{3^2}-\dots-\frac{1}{n^2}.$$

Thus $\lim s_{2n-1}$ exists (by Art. 11 (1)), and since $s_{2n}=s_{2n-1}-1/(n+1)$, we have also $\lim s_{2n-1}=\lim s_{2n}$. That is, the series converges.

Ex. 2. It is easy to see why, from our present point of view, the series in Ex. 2, Art. 18, does not converge; *the sequence of factors employed is not monotonic.*

Another important inference is that if the factors u_n depend in any way on a variable x (subject to the condition of forming a monotonic sequence), the remainder after m terms in the series $\Sigma a_n u_n$ is numerically less than $\rho(v_1+|u|)$; and consequently the choice of m, which makes this remainder less than ϵ, is *independent of x*, so long as $v_1+|u|$ is finite.

This property may be expressed by saying that *the convergence of $\Sigma a_n u_n$ is uniform with respect to x.* (See Art. 44, below.)

A special case of this, which was the original object of Abel's lemma, is given by taking $u_n = x^n$, $0 < x \leqq 1$. Then $u=0$, $v_1 = x \leqq 1$.

20. *If an oscillating series Σa_n has finite maximum and minimum limiting values, it will become convergent if its* *terms are multiplied by a decreasing sequence* (v_n) *which tends to zero as a limit.* **(Dirichlet's test.)** *

Abel's lemma gives the inequality

$$\left|\sum_{m+1}^{m+p} a_n v_n\right| < \rho v_{m+1},$$

where ρ is any number not less than the greatest of the sums

$$|a_{m+1}|, \quad |a_{m+1}+a_{m+2}|, \quad |a_{m+1}+a_{m+2}+a_{m+3}|, \ldots, \quad |a_{m+1}+a_{m+2}+\ldots+a_{m+p}|.$$

It is sufficient to suppose ρ not less than each of the differences

$$|s_{m+1}-s_m|, \quad |s_{m+2}-s_m|, \ldots, \quad |s_{m+p}-s_m|.$$

Now, if the extreme limits of s_n are both finite, we can find some constant† l, such that $|s_n|$ is not greater than l, for any value of n. Thus $|s_n - s_m| \leqq 2l$, and we may take $\rho = 2l$.

We can now choose m so that $v_{m+1} < \epsilon/l$, and then

$$\left|\sum_{m+1}^{m+p} a_n v_n\right| < 2\epsilon,$$

proving that the series $\Sigma a_n v_n$ converges.

Ex. The series $\Sigma v_n \cos n\theta$, $\Sigma v_n \sin n\theta$ converge if θ is not 0 or a multiple of 2π.

For $$\sum_{m+1}^{m+p} \cos n\theta = \sin(\tfrac{1}{2}p\theta) \,.\, \cos[m+\tfrac{1}{2}(p+1)]\theta \,.\, \operatorname{cosec}\tfrac{1}{2}\theta$$

and $$\sum_{m+1}^{m+p} \sin n\theta = \sin(\tfrac{1}{2}p\theta) \,.\, \sin[m+\tfrac{1}{2}(p+1)]\theta \,.\, \operatorname{cosec}\tfrac{1}{2}\theta,$$

so that we could here take $\rho = |\operatorname{cosec}\tfrac{1}{2}\theta|$.

When $\theta = 0$, the first series may be convergent or divergent according to the form of v_n; but the second series, being $0+0+0+\ldots$, converges to the sum 0.

21. A special case of the last article is the result:

If the terms of a series $\Sigma(-1)^{n-1}v_n$ *are alternately positive and negative, and never increase in numerical value, the series will converge, provided that the terms tend to zero as a limit.*

---

* It is practically certain that Abel knew of this test: the history is sketched briefly by Pringsheim (*Math. Annalen*, Bd. 25, p. 423, footnote). But to distinguish it clearly from the test of Art. 19, it seems better to use Dirichlet's name, following Jordan (*Cours d'Analyse*, t. 1, § 299).

† This constant l will be either the greatest value of $|s_n|$, or (if there is no greatest value) the greater of $|\overline{\lim}\, s_n|$ and $|\underline{\lim}\, s_n|$.

For if we take $1-1+1-1+1-1+\dots$ as the oscillatory series of the last article, we need only suppose that ρ is 1; and consequently the series

$$v_1-v_2+v_3-v_4+v_5-v_6+\dots$$

is convergent.

On account of the frequent use of this theorem, we shall now give another proof of it. It is plain that

$$s_{2n}=(v_1-v_2)+(v_3-v_4)+\dots+(v_{2n-1}-v_{2n}),$$

and each of these brackets is positive (or at least not negative), so that, as n increases, the sequence of terms (s_{2n}) never decreases.

Also $s_{2n+1}=v_1-(v_2-v_3)-(v_4-v_5)-\dots-(v_{2n}-v_{2n+1}),$

and so the sequence (s_{2n+1}) never increases.

Further $$s_{2n}=s_{2n+1}-v_{2n+1}<v_1$$

and $$s_{2n+1}=s_{2n}+v_{2n+1}>0.$$

Hence, by Art. 2, the sequence (s_{2n}) has a limit not greater than v_1 and (s_{2n+1}) has a limit not less than 0. But these two limits must be equal since $\lim v_{2n+1}=0$, so that

$$\lim s_{2n}=\lim s_{2n+1}.$$

Hence the series converges to a sum lying between 0 and v_1.

Ex. 1. The series already mentioned in Art. 18, Ex. 1,

$$1-\frac{1}{2^p}+\frac{1}{3^p}-\frac{1}{4^p}+\frac{1}{5^p}-\frac{1}{6^p}+\dots$$

is now seen to converge, provided that $0<p\leqq 1$.

In the special case $p=1$ we get the series

$$1-\tfrac{1}{2}+\tfrac{1}{3}-\tfrac{1}{4}+\tfrac{1}{5}-\tfrac{1}{6}+\dots,$$

which is easily seen to be equal to $\log 2$. For

$$\log 2=\int_1^2\frac{dx}{x}=\int_0^1\frac{dx}{1+x}=\int_0^1\left(1-x+x^2-\dots-x^{2n-1}+\frac{x^{2n}}{1+x}\right)dx$$

$$=1-\frac{1}{2}+\frac{1}{3}-\frac{1}{4}+\dots-\frac{1}{2n}+\int_0^1\frac{x^{2n}}{1+x}dx.$$

But $$\int_0^1\frac{x^{2n}}{1+x}dx<\int_0^1 x^{2n}dx \text{ or } <\frac{1}{2n+1},$$

and so $$\log 2=\lim_{n\to\infty}\left(1-\frac{1}{2}+\frac{1}{3}-\frac{1}{4}+\dots-\frac{1}{2n}\right)$$

$$=1-\frac{1}{2}+\frac{1}{3}-\frac{1}{4}+\dots.$$

The diagram indicates the first eight terms in the sequence (s_n) obtained from this series by addition; the dotted lines indicate the monotonic convergence of the two sequences (s_{2n}), (s_{2n+1}).

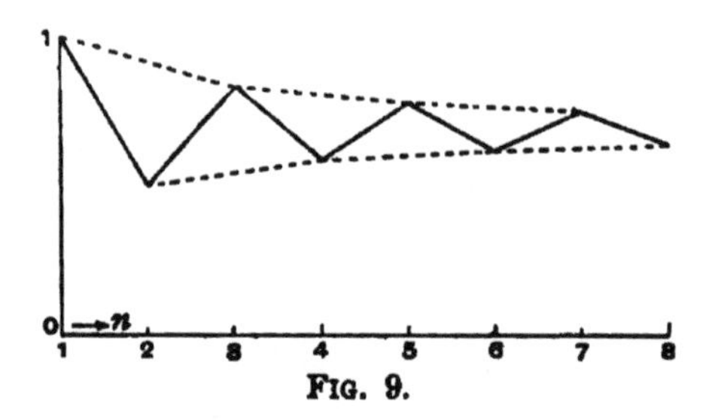

FIG. 9.

It is obvious that *if the sequence (v_n) never increases, but approaches a limit l, not equal to zero, the series $\Sigma(-1)^{n-1}v_n$ will oscillate between two values whose difference is equal to l;* in fact we have by the previous argument $\lim s_{2n+1} = \lim s_{2n} + l$.

A special case of interest is given by the following test which is similar to that of Art. 12:

If v_n/v_{n+1} *can be expressed in the form*

$$\frac{v_n}{v_{n+1}} = 1 + \frac{\mu}{n} + \frac{\omega_n}{n^\lambda}, \quad \begin{cases} \lambda > 1 \\ |\omega_n| < A \end{cases}$$

the series $\Sigma(-1)^{n-1}v_n$ is convergent if $\mu > 0$, oscillatory if $\mu \leqq 0$.

For if $\mu > 0$, after a certain stage we shall have

$$\frac{\mu}{n} > \frac{|\omega_n|}{n^\lambda},$$

so that $v_n > v_{n+1}$; and further (by Art. 39, Ex. 3) $\lim v_n = 0$. But, on the other hand, if $\mu = 0$, it is clear (from Art. 39) that $\lim v_n$ is not zero, and so the series must oscillate. And, if $\mu < 0$, after a certain stage we shall have $v_n < v_{n+1}$, so that $\lim v_n$ cannot be zero, leading to oscillation again.

Ex. 2. Take the series

$$1 - \frac{\alpha.\beta}{1.\gamma} + \frac{\alpha(\alpha+1)\beta(\beta+1)}{1.2.\gamma(\gamma+1)} - \frac{\alpha(\alpha+1)(\alpha+2)\beta(\beta+1)(\beta+2)}{1.2.3.\gamma(\gamma+1)(\gamma+2)} + \dots$$

Here $$\frac{v_n}{v_{n+1}} = \frac{n(\gamma+n-1)}{(\alpha+n-1)(\beta+n-1)}; \quad \mu = \gamma - \alpha - \beta + 1.$$

So the series converges if $\gamma + 1 > \alpha + \beta$.

The same condition applies to the series (Art. 20)

$$1 + \frac{\alpha.\beta}{1.\gamma}\cos\theta + \frac{\alpha(\alpha+1)\beta(\beta+1)}{1.2.\gamma(\gamma+1)}\cos 2\theta + \frac{\alpha(\alpha+1)(\alpha+2)\beta(\beta+1)(\beta+2)}{1.2.3.\gamma(\gamma+1)(\gamma+2)}\cos 3\theta + \dots$$

and to the corresponding series of sines.

It should be observed that if the positive and negative terms in the series form two *separately* decreasing sequences there is no reason to suppose that the theorem is still necessarily true; and in fact it is easy to construct examples of the failure, such as

$$1-\frac{1}{2}+\frac{1}{3^2}-\frac{1}{4}+\frac{1}{5^2}-\frac{1}{6}+\frac{1}{7^2}-\frac{1}{8}+\dots.$$

This is easily recognised as divergent; for the sum of the first n positive terms is less than

$$1+\frac{1}{2^2}+\frac{1}{3^2}+\frac{1}{4^2}+\frac{1}{5^2}+\dots+\frac{1}{n^2},$$

and is therefore less than 2 (Art. 11). But the sum of the first n negative terms is

$$-\frac{1}{2}\left(1+\frac{1}{2}+\frac{1}{3}+\dots+\frac{1}{n}\right);$$

and consequently the sum of the first $2n$ terms of the given series tends to $-\infty$ as its limit.

22. A curious theorem, to some extent a kind of converse of the last, is due to Cesàro:

If a series $(\Sigma \pm v_n)$ is convergent, but not absolutely convergent, and if its terms are arranged in descending order of magnitude, the value of p_n/q_n cannot approach any other limit than unity; where p_n is the number of positive terms and q_n the number of negative terms in the first n terms of the series.

For, the sum of the p_n positive terms is

$$p_1 v_1 + \sum_{r=1}^{n-1} (p_{r+1}-p_r)v_{r+1}$$
$$= p_1(v_1-v_2)+p_2(v_2-v_3)+\dots+p_{n-1}(v_{n-1}-v_n)+p_n v_n.$$

Hence the sum of the first n terms is

$$s_n=(p_1-q_1)(v_1-v_2)+\dots+(p_{n-1}-q_{n-1})(v_{n-1}-v_n)+(p_n-q_n)v_n.$$

Suppose now, if possible, that $(p_n-q_n)/n$ tends to a positive limit l; then, if $l_1<l$, we can find an index m such that

$$(p_n-q_n)/n>l_1, \quad \text{if } n>m.$$

Hence
$$\sum_{m+1}^{n-1}(p_r-q_r)(v_r-v_{r+1})+(p_n-q_n)v_n$$
$$>l_1\left[\sum_{m+1}^{n-1} r(v_r-v_{r+1})+nv_n\right]>l_1(v_m+v_{m+1}+\dots+v_n).$$

But, since the given series is not alsolutely convergent, the series Σv_n is divergent; and consequently $(v_m+v_{m+1}+\dots+v_n)$

can be made greater that N_1/l_1 by taking n greater than (say) n_0. Hence, no matter how large N_1 is, a value n_0 can be found so that

$$s_n > \sum_1^m (p_r - q_r)(v_r - v_{r+1}) + N_1, \quad \text{if } n > n_0;$$

hence s_n must tend to ∞ with n, contrary to hypothesis.

It follows similarly that $(p_n - q_n)/n$ cannot approach a negative limit; so that if $\lim (p_n - q_n)/n$ exists its value must be 0. Now $n = p_n + q_n$, and so if $\lim (p_n/q_n)$ exists, its value must be 1.

This proof is substantially the same as one given by Bagnera.*

Ex. The series $1 + \frac{1}{2} - \frac{1}{3} + \frac{1}{4} + \frac{1}{5} - \frac{1}{6} + \frac{1}{7} + \frac{1}{8} - \frac{1}{9} + \dots$ cannot converge.

As a verification, we note that the sum of $3n$ terms is certainly greater than

$$\frac{1}{3} + \frac{1}{3} - \frac{1}{3} + \frac{1}{6} + \frac{1}{6} - \frac{1}{6} + \frac{1}{9} + \frac{1}{9} - \frac{1}{9} + \dots + \frac{1}{3n} + \frac{1}{3n} - \frac{1}{3n}$$

$$= \frac{1}{3}\left(1 + \frac{1}{2} + \frac{1}{3} + \dots + \frac{1}{n}\right),$$

so that the series is divergent.

23. Abel's Lemma.

If the sequence (v_n) of positive terms never increases, the sum $\sum_1^p a_n v_n$ lies between Hv_1 and hv_1, where H, h are the upper and lower limits of the sums

$$a_1, \; a_1 + a_2, \; a_1 + a_2 + a_3, \dots, \; a_1 + a_2 + \dots + a_p.$$

For, with the usual notation,

$$a_1 = s_1, \; a_2 = s_2 - s_1, \dots, \; a_p = s_p - s_{p-1}.$$

Thus

$$\sum_1^p a_n v_n = s_1 v_1 + (s_2 - s_1) v_2 + \dots + (s_p - s_{p-1}) v_p$$

$$= s_1(v_1 - v_2) + s_2(v_2 - v_3) + \dots + s_{p-1}(v_{p-1} - v_p) + s_p v_p.$$

Now the factors $(v_1 - v_2), (v_2 - v_3), \dots, (v_{p-1} - v_p), v_p$ are *never negative*, and consequently

$$s_1(v_1 - v_2) < H(v_1 - v_2), \; s_2(v_2 - v_3) < H(v_2 - v_3), \dots$$

$$s_{p-1}(v_{p-1} - v_p) < H(v_{p-1} - v_p), \; s_p v_p < Hv_p.$$

* Bagnera, *Bull. Sci. Math.* (2), t. 12, p. 227: Cesàro, *Rom. Acc. Lincei, Rend.* (4), t. 4, p. 133.

Hence $\sum_1^p a_n v_n < H[(v_1 - v_2) + (v_2 - v_3) + \ldots + (v_{p-1} - v_p) + v_p]$

or
$$\sum_1^p a_n v_n < H v_1.$$

In like manner the sum is greater than hv_1.

It follows that
$$\left|\sum_1^p a_n v_n\right| < H' v_1,$$

where H' is the greater of $|H|$ and $|h|$; that is, H' is the upper limit of $|s_1|, |s_2|, \ldots, |s_p|$.

It is sometimes desirable to obtain closer limits for $\Sigma a_n v_n$; suppose that H_m, h_m denote the upper and lower limits of s_m, $s_{m+1}, \ldots, s_p$ while H, h are those of $s_1, s_2, \ldots, s_{m-1}$. Then exactly the same argument gives

$$h(v_1 - v_m) + h_m v_m < \sum_1^p a_n v_n < H(v_1 - v_m) + H_m v_m.$$

On the other hand, when (M_n) is an increasing sequence, we find

$$-(H-h)M_p + HM_1 < \sum_1^p a_n M_n < (H-h)M_p + hM_1$$

$$-(H_m - h_m)M_p + (H_m - H)M_m + HM_1 < \sum_1^p a_n M_n$$
$$< (H_m - h_m)M_p + (h_m - h)M_m + hM_1.$$

24. Transformation of slowly convergent series.

Let us write
$$a_n + a_{n+1} = Ea_n$$

and
$$a_n + 2a_{n+1} + a_{n+2} = Ea_n + Ea_{n+1} = E^2 a_n, \text{ etc.}$$

Then we have

$$(1+x)(a_1 x + a_2 x^2 + a_3 x^3 + \ldots) = a_1 x + (Ea_1)x^2 + (Ea_2)x^3 + \ldots,$$

and consequently $\sum_1^\infty a_n x^n = a_1 y + y[(Ea_1)x + (Ea_2)x^2 + \ldots]$,

where
$$y = x/(1+x).$$

Repeating this operation, we find

$$\sum_1^\infty a_n x^n = a_1 y + (Ea_1)y^2 + \ldots + (E^{p-1}a_1)y^p$$
$$+ y^p[(E^p a_1)x + (E^p a_2)x^2 + \ldots].$$

It can be proved* that in all cases when the original series converges, the remainder term

$$y^p[(E^p a_1)x+(E^p a_2)x^2+\ldots]$$

tends to zero as p increases to infinity, at least when x is positive. Consequently,

$$\sum_1^\infty a_n x^n = a_1 y+(Ea_1)y^2+(E^2 a_1)y^3+\ldots \text{ to } \infty.$$

The cases of chief interest arise when $x=1$, and the terms a_n are alternately positive and negative. Write then

$$a_1=+v_1,\quad a_2=-v_2,\quad a_3=+v_3,\quad a_4=-v_4,\quad \ldots,$$

and we have

$$\sum_1^\infty a_n=\sum_1^\infty(-1)^{n-1}v_n=\frac{1}{2}v_1+\frac{1}{4}(Dv_1)+\frac{1}{8}(D^2v_1)+\frac{1}{16}(D^3v_1)+\ldots$$
$$+\frac{1}{2^p}(D^{p-1}v_1)+\frac{1}{2^p}[(D^pv_1)-(D^pv_2)+(D^pv_3)-\ldots],$$

where we write

$$Dv_1=v_1-v_2,\quad D^2v_1=Dv_1-Dv_2=v_1-2v_2+v_3,\quad \text{etc.}$$

We can write down a simple expression for the remainder, if $v_n=f(n)$, where $f(x)$ is a function such that $f^p(x)$ has a fixed sign for all positive values of x, and steadily decreases in numerical value as x increases.

For we have $$Dv_n=-\int_0^1 f'(x_1+n)\,dx_1,$$

so $$D^2v_n=+\int_0^1 dx_1\int_0^1 f''(x_1+x_2+n)\,dx_2,$$

and generally

$$D^pv_n=(-1)^p\int_0^1 dx_1\int_0^1 dx_2\ldots\int_0^1 f^p(x_1+x_2+\ldots+x_p+n)\,dx_p.$$

Thus the series $D^pv_1-D^pv_2+D^pv_3-\ldots$ consists of a succession of *decreasing* terms, of alternate signs. Its sum is therefore less than D^pv_1 in numerical value; and consequently

$$\sum_1^\infty(-1)^{n-1}v_n=\frac{1}{2}v_1+\frac{1}{4}(Dv_1)+\frac{1}{8}(D^2v_1)+\ldots+\frac{1}{2^p}(D^{p-1}v_1)+R_p,$$

where $$|R_p|<\frac{1}{2^p}|D^pv_1|.$$

---

* For the case $x=1$, see L. D. Ames, *Annals of Mathematics*, series 2, vol. 3, p. 188.

This result applies to any series of the type

$$1-\frac{1}{2^r}+\frac{1}{3^r}-\frac{1}{4^r}+\frac{1}{5^r}-\ldots,\quad \text{where } r>0.$$

Here it is easy to see that $D^p v_n$ is always positive and decreases as n increases; it is a useful test of the accuracy of the work, in arithmetical calculations, to apply the transformation twice, starting first say at $\frac{1}{n^r}$ and secondly at $\frac{1}{(n+1)^r}$; if the results are substantially the same we may be satisfied that the work is correct.

Ex. 1. Take $r=\frac{1}{2}$; if we work to five decimals we get

$$s=1-\cdot 70711+\cdot 57735-\cdot 50000+\cdot 44721-s',$$

and we shall apply the transformation to s', whose first seven terms appear in the table below:

v.	Dv.	D^2v.	D^3v.	D^4v.	D^5v.
$6^{-\frac{1}{2}}=\cdot 40825$					
	3029				
$7^{-\frac{1}{2}}=\cdot 37796$		588			
	2441		169		
$8^{-\frac{1}{2}}=\cdot 35355$		419		62	
	2022		107		29
$9^{-\frac{1}{2}}=\cdot 33333$		312		33	
	1710		74		8
$10^{-\frac{1}{2}}=\cdot 31623$		238		25	
	1472		49		
$11^{-\frac{1}{2}}=\cdot 30151$		189			
	1283				
$12^{-\frac{1}{2}}=\cdot 28868$					

If we apply the transformation at the beginning of s' we get

$$\begin{aligned}
\cdot 20413 &= \tfrac{1}{2}\,(\cdot 40825)\\
757 &= \tfrac{1}{4}\,(\ 3029)\\
73 &= \tfrac{1}{8}\,(\ \ 588)\\
11 &= \tfrac{1}{16}(\ \ 169)\\
2 &= \tfrac{1}{32}(\ \ \ 62)\\
\hline
\cdot 21256 &
\end{aligned}$$

If we start from the second term of s' we get

$$\begin{aligned}
\cdot 18898 &= \tfrac{1}{2}\,(\cdot 37796)\\
610 &= \tfrac{1}{4}\,(\ 2441)\\
52 &= \tfrac{1}{8}\,(\ \ 419)\\
7 &= \tfrac{1}{16}(\ \ 107)\\
1 &= \tfrac{1}{32}(\ \ \ 33)\\
\hline
\cdot 19568 &
\end{aligned}$$

Now $\cdot 40825 - \cdot 19568 = \cdot 21257$, so that s' certainly is contained between $0\cdot 21256$ and $0\cdot 21258$.

But $s = 0\cdot 81746 - s'$, so that $s = 0\cdot 6049$ to four decimal places. If we used the original series, it would need over a hundred million terms to get this result.

Ex. 2. Similarly we may sum the series $1 - \frac{1}{2} + \frac{1}{3} - \frac{1}{4} + \ldots$.

To 6 decimals, the first 8 terms give $0\cdot 634524$ and from the next 7 terms we get the table:

v.	Dv.	D^2v.	D^3v.	D^4v.	D^5v.	D^6v.
$9^{-1} = \cdot 111111$						
	11111					
$10^{-1} = \cdot 100000$		2020				
	9091		505			
$11^{-1} = \cdot 090909$		1515		156		
	7576		349		57	
$12^{-1} = \cdot 083333$		1166		99		24
	6410		250		33	
$13^{-1} = \cdot 076923$		916		66		
	5494		184			
$14^{-1} = \cdot 071429$		732				
	4762					
$15^{-1} = \cdot 066667$						

Thus the sum from the 9th term onwards is given by

(i)
$$\begin{aligned} \cdot 055556 &= \tfrac{1}{2}(\cdot 111111) \\ 2778 &= \tfrac{1}{4}(11111) \\ 252 &= \tfrac{1}{8}(2020) \\ 32 &= \tfrac{1}{16}(505) \\ 5 &= \tfrac{1}{32}(156) \\ 1 &= \tfrac{1}{64}(57) \\ \hline \cdot 058624 & \end{aligned}$$

or by (ii)
$$\begin{aligned} \cdot 050000 &= \tfrac{1}{2}(\cdot 100000) \\ 2273 &= \tfrac{1}{4}(9091) \\ 189 &= \tfrac{1}{8}(1515) \\ 22 &= \tfrac{1}{16}(349) \\ 3 &= \tfrac{1}{32}(99) \\ \hline \cdot 052487 & \\ 111111 & \\ \hline \cdot 058624 & \end{aligned}$$

Thus the sum of the series is $0\cdot 634524 + 0\cdot 058624 = 0\cdot 693148$, that is $0\cdot 69315$ to five decimals.

To reach this degree of accuracy we should have to use over a hundred thousand terms of the original series.*

A number of other numerical examples will be found in the paper by Ames, just quoted.

Ex. 3. A physical application may be found in the theory of Huygens' zones in Physical Optics.†

---

* Of course the actual sum of the series is $\log 2 = \cdot 69314718$. (Art. 21, Ex. 1; and Art. 63.)

† Schuster's *Optics*, § 46; Drude's *Optics*, ch. III. § 2; Schuster, *Phil. Mag.* (5th series), vol. 31, 1891, p. 85.

A reference to either of the authorities quoted will shew that we have there to sum a series of terms $v_1 - v_2 + v_3 - \ldots$, for which Dv_n is very small and D^2v_n has always the same sign. We have then

$$s = \Sigma(-1)^{n-1}v_n = \tfrac{1}{2}v_1 + \tfrac{1}{2}(Dv_1 - Dv_2 + \ldots).$$

Now if D^2v_n is positive we have $Dv_1 > Dv_2 > Dv_3 > \ldots$, and $\lim Dv_n = 0$, because the series in the bracket must converge if s does. Then we get

$$\tfrac{1}{2}v_1 < s < \tfrac{1}{2}(v_1 + Dv_1).$$

Similarly if D^2v_n is negative, we have $\tfrac{1}{2}v_1 > s > \tfrac{1}{2}(v_1 + Dv_1)$.

Thus the series can be represented by $\tfrac{1}{2}v_1$ to a very high degree of approximation.

The transformation described above was first given by Euler, and the first proof of its accuracy is due to Poncelet. Kummer and Markoff have found other transformations for the same purpose; the latter's method includes Euler's as a special case. As an example of Markoff's, we may quote

$$\Sigma \frac{1}{n^3} = \Sigma(-1)^{n-1} \frac{[(n-1)!]^2}{(3n-2)!} \left[\frac{1}{(2n-1)^2} + \frac{5}{12n(3n-1)}\right],$$

13 terms of which give the sum correctly to 20 decimals.*

To apply Euler's method to this example the reader may note that

$$\Sigma \frac{1}{n^3} = \frac{4}{3}\left(1 - \frac{1}{2^3} + \frac{1}{3^3} - \frac{1}{4^3} + \ldots\right).$$

The first ten terms of the series in the bracket give ·9011165, and if we apply Euler's method to the next six, we get ·0004262 for the value of the remainder: thus $\Sigma \frac{1}{n^3} = \frac{4}{3}(0\cdot901427) = 1\cdot202057$ to six places.

Similarly $\Sigma \frac{1}{n^2} = 2\left(1 - \frac{1}{2^2} + \frac{1}{3^2} - \frac{1}{4^2} + \ldots\right)$.

The sum of the first ten terms in the bracket is ·8179622, and Euler's method gives ·0045048 for the remainder.

Thus $\Sigma \frac{1}{n^2} = 2(0\cdot822469) = 1\cdot644934$. See also Art. 130, Exs. 4, 5.

* *Comptes Rendus*, t. 109, 1889, p. 934; *Differenzenrechnung* (Leipzig, 1896), p. 178. For other references, consult Pringsheim (*Encyklopädie*, Bd. I. A. 3, § 37).

EXAMPLES.

1. Prove that the series

$$\frac{1}{x}-\frac{1}{x+1}+\frac{1}{x+2}-\frac{1}{x+3}+\dots$$

converges for any value of x which makes none of the denominators zero; but that both the series

$$\frac{1}{x}+\frac{1}{x+1}-\frac{1}{x+2}+\frac{1}{x+3}+\frac{1}{x+4}-\frac{1}{x+5}+\dots$$

and

$$\frac{1}{x}-\frac{1}{x+1}-\frac{1}{x+2}+\frac{1}{x+3}-\frac{1}{x+4}-\frac{1}{x+5}+\dots$$

are divergent.

2. Prove that if $\Sigma n a_n$ is convergent, so also is Σa_n.

3. If the series Σa_n is convergent and the sequence (M_n) steadily increases to ∞ with n, then (see Art. 23)

$$\lim\,(a_1 M_1 + a_2 M_2 + \dots + a_n M_n)/M_n = 0. \qquad \text{[Kronecker.]}$$

4. Prove that the series

$$a - a^{\frac{1}{2}} + a^{\frac{1}{3}} - a^{\frac{1}{4}} + a^{\frac{1}{5}} - a^{\frac{1}{6}} + \dots$$

oscillates, but can be made to converge to either of its two extreme limits by inserting brackets. On the other hand the series

$$(1-a)-(1-a^{\frac{1}{2}})+(1-a^{\frac{1}{3}})-(1-a^{\frac{1}{4}})+\dots$$

is convergent.

5. Shew that if a series converges, it is still convergent when any number of brackets are inserted, grouping the terms. And shew also that the converse is true, *if all the terms in the brackets are positive.*

6. Calculate, correctly to 20 decimals, the sum of the series

$$1+2x+2x^4+2x^9+2x^{16}+\dots$$

for $x=\pm\frac{1}{10}, \pm\frac{2}{10}$. How many terms would have to be taken, to calculate the sum for $x=\pm\frac{9}{10}$ to 3 decimals?

7. Shew that the series $a_1-a_2+a_3-a_4+\dots$ diverges if $a_n=\dfrac{1}{\sqrt{n}}+\dfrac{(-1)^{n-1}}{n}$ or if $a_n=1/[\sqrt{n}+(-1)^{n-1}]$; although the terms are alternately positive and negative and tend to zero as a limit.

8. If $|x|>1$, prove that

$$\frac{1}{x+1}+\frac{2}{x^2+1}+\frac{4}{x^4+1}+\dots$$

converges to the sum $\dfrac{1}{x-1}$.

9. If $a_n \to a$ and $b_n \to b$, verify that if the series

$$a_1 b_1 + a_2(b_2-b_1) + a_3(b_3-b_2)+\dots$$

converges to the sum S, then the series

$$b_1(a_1-a_2)+b_2(a_2-a_3)+b_3(a_3-a_4)+\dots$$

converges to the sum $S-ab$.

10. Prove that if the series

$$a_1+a_2+a_3+\dots$$

is convergent, so also is

$$\tfrac{1}{2}(a_1+a_2)+\tfrac{1}{2}(a_2+a_3)+\tfrac{1}{2}(a_3+a_4)+\dots,$$

and their sums differ by $\frac{1}{2}a_1$. Is the converse always true? Prove that the converse is certainly true when a_n is *positive*.

11. Discuss the series

$$\frac{x}{c_1}+\frac{x^2}{c_2^{\,2}}+\frac{x^3}{c_3^{\,3}}+\dots+\frac{x^n}{c_n^{\,n}}+\dots$$

$$\frac{1}{x-c_1}+\frac{1}{x-c_2}\frac{x}{c_2}+\dots+\frac{1}{x-c_n}\frac{x^{n-1}}{c_n^{\,n-1}}+\dots,$$

where $c_1, c_2, c_3, \dots$ is an increasing sequence tending to ∞.

12. Verify that

$$\sum_1^\infty\left(\frac{1}{x-c_n}+\frac{1}{c_n}+\frac{x}{c_n^{\,2}}+\dots+\frac{x^{n-1}}{c_n^{\,n}}\right)$$

is absolutely convergent if $|c_n|$ *steadily increases* to ∞, and x is not equal to any of the values $c_1, c_2, c_3, \dots$.

If $c_n=n^{\frac{1}{k}}$, where k is fixed, verify that

$$\sum_1^\infty\left(\frac{1}{x-c_n}+\frac{1}{c_n}+\frac{x}{c_n^{\,2}}+\dots+\frac{x^{r-1}}{c_n^{\,r}}\right)$$

is absolutely convergent if r is the integral part of k.

13. Shew that
$$\sum_1^{\infty}{}'\frac{1}{m^2-n^2}=-\frac{3}{4m^2},$$

where m is an integer and the accented Σ means that $n=m$ is to be omitted from the summation.

[In fact the sum can be written

$$\frac{1}{2m}\left\{\left(\frac{1}{m-1}+\frac{1}{m+1}\right)+\left(\frac{1}{m-2}+\frac{1}{m+2}\right)+\dots+\left(1+\frac{1}{2m-1}\right)\right\}$$

$$-\frac{1}{2m}\left\{\left(1-\frac{1}{2m+1}\right)+\left(\frac{1}{2}-\frac{1}{2m+2}\right)+\left(\frac{1}{3}-\frac{1}{2m+3}\right)+\dots\text{ to }\infty\right\}$$

$$=\frac{1}{2m}\left\{\left(1+\frac{1}{2}+\dots+\frac{1}{2m-1}\right)-\frac{1}{m}\right\}-\frac{1}{2m}\left\{\left(1+\frac{1}{2}+\dots+\frac{1}{2m}\right)\right\}$$

$$=-\frac{3}{4m^2}.\Big]$$

14. With the same notation as in Ex. 13, shew that

$$\sum_1^{\infty}{}'\frac{(-1)^{n-1}}{m^2-n^2}=+\frac{3}{4m^2},$$

if m is even.

Find an expression for the sum when m is odd.

15. Discuss the convergence of the series whose general term is

$$\left(1+\frac{1}{2}+\frac{1}{3}+\ldots+\frac{1}{n-1}\right)\frac{\sin n\theta}{n},$$

and also that of the series with $\cos n\theta$ in place of $\sin n\theta$. [Art. 20.]
[*Math. Trip.*, 1899.]

16. Apply Art. 20 to the series whose sums to n terms are $\sin(n+\frac{1}{2})^2\theta$, $\cos(n+\frac{1}{2})^2\theta$, and deduce that

$$\Sigma v_n \sin n\theta \,.\, \cos n^2\theta, \quad \Sigma v_n \sin n\theta \,.\, \sin n^2\theta$$

are convergent if v_n steadily $\to 0$. [HARDY.]

17. Shew that if v_n tends steadily to zero, in such a way that Σv_n is not convergent, then the series

$$\sum_{r=0}^{\infty}(a_1 v_{kr+1}+a_2 v_{kr+2}+\ldots+a_k v_{kr+k})$$

converges if (and only if) $a_1+a_2+\ldots+a_k=0$.

18. If the sequence (a_n) is convergent, prove that $\lim n(a_{n+1}-a_n)$ must either oscillate or converge to zero.

19. If Σa_n converges, and a_n steadily decreases to 0, $\Sigma n(a_n-a_{n+1})$ is convergent. If, in addition, $a_n-2a_{n+1}+a_{n+2}>0$, prove that

$$n^2(a_n-a_{n+1})\to 0.$$ [HARDY.]

20. Apply Euler's transformation to shew that

$$1+2^2x+3^2x^2+4^2x^3+5^2x^4+\ldots=\frac{1+x}{(1-x)^3}.$$

21. Utilise the result of Ex. 3, p. 58, to shew that the sum of the series

$$\frac{1}{2}-\frac{x}{1+x}+\frac{x^2}{1+x^2}-\frac{x^3}{1+x^3}+\ldots$$

tends to the limit $\frac{1}{4}$ as $x\to 1$.

[It is easy to see that (if $0<x<1$), Dv_n is positive and *decreases*; thus the sum lies between $\frac{1}{2}v_1=\frac{1}{4}$ and $\frac{1}{2}(v_1+Dv_1)=\frac{1}{2(1+x)}$.]

22. By taking $v_n=\log(a+n)$, shew, as on p. 56, that $D^p v_n$ is negative and steadily decreases; deduce that

$$\log a - p\log(a+1)+\frac{p(p-1)}{2!}\log(a+2)\ldots+(-1)^p\log(a+p)<0.$$

CHAPTER IV.

ABSOLUTE CONVERGENCE.

25. It is a familiar fact that a finite sum has the same value, no matter how the terms of the sum are arranged. This property, however, is by no means universally true for infinite series; as an illustration, consider the series

$$s=1-\tfrac{1}{2}+\tfrac{1}{3}-\tfrac{1}{4}+\tfrac{1}{5}-\tfrac{1}{6}+\dots,$$

which we know is convergent (Art. 21, Ex. 1), and has a positive value S greater than $\frac{1}{2}$. Let us arrange the terms of this series so that each positive term is followed by two negative terms: the series then becomes

$$t=1-\frac{1}{2}-\frac{1}{4}+\frac{1}{3}-\frac{1}{6}-\frac{1}{8}+\frac{1}{5}-\frac{1}{10}-\frac{1}{12}+\dots.$$

Now we have

$$\begin{aligned} t_{3n} &= \left(1-\frac{1}{2}\right)-\frac{1}{4}+\left(\frac{1}{3}-\frac{1}{6}\right)-\frac{1}{8}+\dots+\left(\frac{1}{2n-1}-\frac{1}{4n-2}\right)-\frac{1}{4n} \\ &= \frac{1}{2}-\frac{1}{4}+\frac{1}{6}-\frac{1}{8}+\dots+\frac{1}{4n-2}-\frac{1}{4n} \\ &= \frac{1}{2}\left(1-\frac{1}{2}+\frac{1}{3}-\frac{1}{4}+\dots+\frac{1}{2n-1}-\frac{1}{2n}\right) \\ &= \frac{1}{2}s_{2n}. \end{aligned}$$

Thus
$$\lim_{n\to\infty} t_{3n}=\frac{1}{2}S,$$

and it is easily seen that $\lim t_{3n+1}=\lim t_{3n+2}=\lim t_{3n}$, so that the sum of the series t is $\frac{1}{2}S$.

Consequently, *this derangement of the terms in the series alters the sum of the series.*

In view of the foregoing example we naturally ask *under what conditions may we derange the terms of a series without altering its value?* It is to be observed that in the derangement we make a one-to-one correspondence between the terms of two series; so that every term in the first series occupies a perfectly definite place in the second series, and conversely. Thus, corresponding to any number (n) of terms in the first series, we can find a number (n') in the second series, such that the n' terms contain *all* the n terms (and some others); and conversely.

For instance, in the derangement considered above, the first $(2n+2)$ terms of s are all contained in the first $(3n+1)$ terms of t; and the first $3p$ terms of t are all contained in the first $4p$ terms of s.

26. A series of positive terms, if convergent, has a sum independent of the order of its terms; but if divergent it remains divergent, however its terms are deranged.

As above, denote the original series by s and the deranged series by t; and suppose first that s converges to the sum S. Then we can choose n, so that the sum s_n exceeds $S-\epsilon$, however small ϵ may be. Now, t contains *all* the terms of s (and if any term happens to be repeated in s, t contains it equally often); we can therefore find an index p such that t_p contains *all* the terms s_n. Thus we have found p so that t_p exceeds $S-\epsilon$, because all the terms in $t_p - s_n$ are *positive or zero.* Now t contains no terms which are not present in s, so that however great r may be, t_r cannot exceed S; and, combining these two conclusions, we get

$$S \geqq t_r > S-\epsilon, \quad \text{if } r \geqq p.$$

Consequently the series t converges to the sum S.

Secondly, if s is divergent, t cannot converge; for the foregoing argument shews that if t converges, s must also converge. Consequently t is divergent.

If we attempt to apply this argument to the two series considered in Art. 25,

$$s = 1 - \tfrac{1}{2} + \tfrac{1}{3} - \tfrac{1}{4} + \ldots, \quad t = 1 - \tfrac{1}{2} - \tfrac{1}{4} + \tfrac{1}{3} - \tfrac{1}{6} - \tfrac{1}{8} + \ldots,$$

we find that the terms in $t_p - s_n$ are partly *negative.* Thus we cannot prove that $t_r > S - \epsilon$; and as a matter of fact we see from Art. 25 that this inequality is inaccurate. Similarly, the argument used above fails to prove that $S \geqq t_r$, although this happens to be true here if $r > 1$.

It is now easy to prove that **if a series Σa_n is absolutely convergent, its sum S is not altered by derangement.**

For, write $s = \Sigma a_n$, $a = \Sigma |a_n|$, and then introduce the new series $\beta = \Sigma [a_n + |a_n|]$. The series β contains no negative terms, and no term in β is greater than twice the corresponding term in a; so, since a is convergent, β converges to a sum B not greater than $2A$, where A is the sum of a; and so $B = S + A$.

Suppose now that S', A', B' are the sums of the three series after any derangement (supposed the same for each series).

Then $B' = S' + A'$; but by what has been proved $A' = A$ and $B' = B$, because they contain only positive terms.

Hence $$S' = B' - A' = B - A = S,$$
proving the theorem.

Ex. 1. As an example, consider the series

$$1 - \frac{1}{2^2} + \frac{1}{3^2} - \frac{1}{4^2} + \frac{1}{5^2} - \frac{1}{6^2} + \dots.$$

This is absolutely convergent by Art. 11; and therefore the series remains convergent, and has the same sum after any derangement. It is accordingly equal to

$$1 + \frac{1}{3^2} - \frac{1}{2^2} + \frac{1}{5^2} + \frac{1}{7^2} - \frac{1}{4^2} + \frac{1}{9^2} + \frac{1}{11^2} - \frac{1}{6^2} + \dots,$$

or to

$$1 - \frac{1}{2^2} - \frac{1}{4^2} + \frac{1}{3^2} - \frac{1}{6^2} - \frac{1}{8^2} + \frac{1}{5^2} - \frac{1}{10^2} - \frac{1}{12^2} + \dots.$$

Ex. 2. From our present point of view, we observe that the inequality between $1 - \frac{1}{2} + \frac{1}{3} - \frac{1}{4} + \dots$ and $1 - \frac{1}{2} - \frac{1}{4} + \frac{1}{3} - \frac{1}{6} - \frac{1}{8} + \dots$ is explained by the fact that these series are not absolutely convergent (Art. 11). The series a, β are here divergent, and of course we have no right to infer that $S' = S$.

The last result should be contrasted with the state of affairs explained at the beginning of Art. 28; using the notation of that article, we find here

$$\lim (x_p - y_n) = S, \quad \lim (x_p + y_n) = A,$$

so that $$\lim x_p = \tfrac{1}{2}(S + A), \quad \lim y_n = \tfrac{1}{2}(-S + A);$$

whereas there the sequences (x_p), (y_n) are divergent, taken separately, although their difference $(x_p - y_n)$ converges to a sum, whose value depends on the relation between p and n.

27. Applications of absolute convergence.

Consider first *the multiplication of two absolutely convergent series* $A=\Sigma a_n$, $B=\Sigma b_n$. Write the terms of the product so as to form a table of double entry

$$\begin{array}{ccccccccc}
a_1b_1 & & a_1b_2 & & a_1b_3 & & a_1b_4 & \ldots \\
& \swarrow & \uparrow & \swarrow & \uparrow & \swarrow & \uparrow & \swarrow \\
a_2b_1 & \rightarrow & a_2b_2 & & a_2b_3 & & a_2b_4 & \ldots \\
& \swarrow & & \swarrow & \uparrow & \swarrow & \uparrow & \\
a_3b_1 & \rightarrow & a_3b_2 & \rightarrow & a_3b_3 & & a_3b_4 & \ldots \\
& \swarrow & & \swarrow & & & \uparrow & \\
a_4b_1 & \rightarrow & a_4b_2 & \rightarrow & a_4b_3 & \rightarrow & a_4b_4 & \ldots \\
\ldots & \swarrow & \ldots & \ldots & \ldots & \ldots & \ldots & \ldots
\end{array}$$

It is easy to prove that AB is the sum of the series

$$(1)\quad a_1b_1+(a_2b_1+a_2b_2+a_1b_2)+(a_3b_1+a_3b_2+a_3b_3+a_2b_3+a_1b_3)\\ +\ldots,$$

where the order of the terms is the same as is indicated by the arrows in the table. For the sum to n terms of this series (1) is A_nB_n, if

$$A_n=a_1+a_2+\ldots+a_n,\quad B_n=b_1+b_2+\ldots+b_n.$$

Now $A'=\Sigma|a_n|$ and $B'=\Sigma|b_n|$ are convergent by hypothesis. Thus the series

$$(2)\qquad a_1b_1+a_2b_1+a_2b_2+a_1b_2+a_3b_1+\ldots,$$

obtained by removing the brackets from (1), is absolutely convergent, because the sum of the absolute values of any number of terms in (2) cannot exceed $A'B'$. Accordingly, (2) has the same sum AB as the series (1). Since (2) is absolutely convergent, we can arrange it in any order (by Art. 26) without changing the sum. Thus we may replace (2) by

$$(3)\qquad a_1b_1+a_2b_1+a_1b_2+a_3b_1+a_2b_2+a_1b_3+a_4b_1+\ldots,$$

following the order of the diagonals indicated in the diagram. Hence we find, on inserting brackets in (3),

$$AB=c_1+c_2+c_3+\ldots \text{ to } \infty,$$

where $\quad c_1=a_1b_1,\ c_2=a_2b_1+a_1b_2,\ c_3=a_3b_1+a_2b_2+a_1b_3$

and $$c_n=a_nb_1+a_{n-1}b_2+\ldots+a_1b_n.$$

For other results on the multiplication of series the reader should refer to Arts. 34, 35.

A second useful application of the theorem of Art. 26 is to justify the step of arranging a series $\Sigma a_n y^n$ in powers of x, where y is a polynomial in x; say $y = b_0 + b_1 x + \ldots + b_k x^k$.

It is here sufficient to have $\Sigma \alpha_n \eta^n$ convergent where

$$\alpha_n = |a_n|, \quad \eta = \beta_0 + \beta_1 \xi + \ldots + \beta_k \xi^k, \quad \beta_r = |b_r|, \quad \xi = |x|;$$

and from Art. 10, we see that this requires

$$\eta < \lambda, \quad \text{if } \lambda^{-1} = \overline{\lim}\, \alpha_n^{\frac{1}{n}}.$$

The last condition requires that $\beta_0 < \lambda$, and that ξ shall be less than some fixed value; and then the necessary derangement will certainly not alter the sum of the series.

In most of the ordinary cases $\lambda = 1$, and y is of the form $bx \pm x^2$; the condition is then

$$\xi^2 + \beta\xi < 1 \quad \text{or} \quad \xi < \tfrac{1}{2}[(4+\beta^2)^{\frac{1}{2}} - \beta].$$

In particular, if $\beta \leqq 2$, it is enough to take $\xi < \sqrt{2} - 1$, which is certainly satisfied when $\xi < \frac{2}{5}$.

The beginner may be tempted to think that the condition $|y| < \lambda$ would be sufficient; but this is not the case. For we have to ensure the convergence of the series when $a_n y^n$ is written out at length, and every term is made positive in the *expanded* form.

As an illustration of this point, consider the series $1 + \Sigma(2x - x^2)^n$ which has the sum $[1 - (2x - x^2)]^{-1} = (1 - x)^{-2}$, when $|2x - x^2| < 1$. This condition is satisfied by any value of x (except 1) lying between $1 \pm \sqrt{2}$; and in particular by $x = \frac{2}{3}$, because $2x - x^2$ is then $\frac{8}{9}$. But if the series is arranged in powers of x, we get

$$\begin{array}{l|l|l|l|l|l|l|l}
1+2x & -x^2 & & & & & & \\
 & +4x^2 & -4x^3 & +x^4 & & & & \\
 & & +8x^3 & -12x^4 & +6x^5 & -x^6 & & \\
 & & & +16x^4 & -32x^5 & +24x^6 & -8x^7 & +x^8 \\
 & & & & +\ldots\ldots & \ldots\ldots & \ldots\ldots & \ldots\ldots
\end{array}$$

$$= 1 + 2x + 3x^2 + 4x^3 + 5x^4 + \ldots,$$

which diverges if $x = \frac{2}{3}$.

Thus the condition $|2x - x^2| < 1$ is not sufficient to allow the arrangement in powers of x. The condition found in the text above would be $|x| < \sqrt{2} - 1$; and would be obtained from the expansion given at length, by making the negative coefficients positive; this leads to the series

$$1 + 2|x| + 5|x|^2 + 12|x|^3 + 29|x|^4 + \ldots$$

(in which $a_n = 2a_{n-1} + a_{n-2}$).

As a matter of fact, the condition $|x| < \sqrt{2} - 1$ is narrower than is necessary for the truth of the equation

$$1 + \Sigma(2x - x^2)^n = 1 + \Sigma(n+1)x^n.$$

This equation is true if both series converge; although the proof does not follow from our present line of argument. It may be guessed that, in general, the condition found for ξ in the text is unnecessarily narrow; and this is certainly the case in a number of special applications. However, we are not here concerned with finding the widest limits for x; what we wish to shew is that the transformation is certainly legitimate when x is properly restricted.

In view of Riemann's theorem (Art. 28) it may seem surprising that the condition of absolute convergence gives an unnecessarily small value for ξ. However, a little consideration will shew that Riemann's theorem does not imply that *any* derangement of a non-absolutely convergent series will alter its sum; but that such a series can be made to have any value by means of a *special* derangement, which may easily be of a far more sweeping character than the derangement implied in arranging $\Sigma a_n y^n$ according to powers of x.

28. Riemann's Theorem.

If a series converges, but not absolutely, its sum can be made to have any arbitrary value by a suitable derangement of the series; it can also be made divergent or oscillatory.

Let x_p denote the sum of the first p positive terms and $-y_n$ the sum of the first n negative terms; then we are given that

$$\lim (x_p - y_n) = s, \quad \lim (x_p + y_n) = \infty,$$

where p, n tend to ∞ according to some definite relation. Hence

$$\lim_{p \to \infty} x_p = \infty, \quad \lim_{n \to \infty} y_n = \infty.$$

Suppose now that the sum of the series is to be made equal to σ; since $x_p \to \infty$ we can choose p_1 so that $x_{p_1} > \sigma$, and so that p_1 is the *smallest* index which satisfies this condition. Similarly we can find n_1 so that $y_{n_1} > x_{p_1} - \sigma$, and again suppose that n_1 is the *least* index consistent with the inequality.

Then, in the deranged series, we place first a group of p_1 positive terms, second a group of n_1 negative terms, keeping the terms in each group in their original order. Thus, if S_ν is the sum of ν terms, it is plain that

$$S < \sigma, \quad \text{if } \nu < p_1, \quad \text{but } S_\nu > \sigma, \quad \text{if } p_1 \leqq \nu < p_1 + n_1.$$

We now continue the process, placing third a group of $(p_2 - p_1)$ positive terms, where p_2 is the least index such that $x_{p_2} > y_{n_1} + \sigma$; and fourth, a group of $(n_2 - n_1)$ negative terms, where n_2 is the least index such that $y_{n_2} > x_{p_2} - \sigma$.

The method of construction can evidently be carried on indefinitely, and it is clear that if $p_r+n_r>\nu\geqq p_r+n_{r-1}$, $S_\nu-\sigma$ is positive, but cannot exceed the (p_r+n_{r-1})th term of the series; while if $p_{r+1}+n_r>\nu\geqq p_r+n_r$, $\sigma-S_\nu$ is positive, but does not exceed the (p_r+n_r)th term: for $S_\nu-\sigma$ changes sign at these terms.

Thus, since the terms of the series must tend to zero as ν increases, we have
$$\lim S_\nu=\sigma.$$

It is easy to modify the foregoing method so as to get a divergent or oscillatory series, by starting from a sequence (σ_r) which is either divergent or oscillatory and taking p_1, n_1,... in turn to be the first indices which satisfy the inequalities

$$x_{p_1}>\sigma_1,\quad y_{n_1}>x_{p_1}-\sigma_1,\quad x_{p_2}>y_{n_1}+\sigma_2,\quad y_{n_2}>x_{p_2}-\sigma_2,$$

and so on.

As a matter of fact, however, Riemann's process is quite out of the question for practical work; and we have to adopt an entirely different method due to Pringsheim.*

Let $f(x)$ be a positive function, steadily decreasing to zero as x increases; and consider the series $\Sigma(-1)^{n-1}f(n)$, which converges, by Art. 21.

Here every positive term is followed by a negative term; and suppose that, in the deranged series, the first r terms contain $n+\nu$ positive to n negative terms (so that $2n+\nu=r$). Then the sum of these r terms is

$$[f(1)-f(2)+\ldots-f(2n)]$$
$$+[f(2n+1)+f(2n+3)+\ldots+f(2n+2\nu-1)];$$

where the second bracket contains ν terms, and so lies between

$$\nu f(2n) \text{ and } \nu f(2n+2\nu).$$

Suppose first that $nf(n)$ tends *steadily* to infinity with n, then $f(2n+2\nu)/f(2n)$ lies between 1 and $n/(n+\nu)$. Thus if we choose ν to be such a function of n that
$$\lim \nu f(2n)=l,$$
the change in the sum of the series is l, because then $\nu/n\to 0$.

We have thus the first result:

If $nf(n)$ tends steadily to infinity the value of ν requisite for an alteration l in the sum of the series is subject to the condition $\lim \nu f(2n)=l$.

For instance, taking the series $\Sigma(-1)^{n-1}n^{-\frac{1}{2}}$, we see that ν may be the integer nearest† to $l\sqrt{(2n)}$; or again with $\Sigma(-1)^{n-1}\dfrac{\log n}{n}$, ν may be the integer nearest† to $2ln/\log n$.

---

* *Math. Annalen*, Bd. 22, p. 455.

† It is not, of course, essential to take always the *nearest* integer, in order to satisfy the condition. But this is the simplest statement.

Next, if $\lim nf(n)$ is finite, say equal to g, it follows that, for any positive value of ϵ, however small, a value n_0 can be found such that

$$\frac{g-\epsilon}{x} < f(x) < \frac{g+\epsilon}{x}, \quad (x > n_0).$$

Let ν be chosen so that $k = \lim (n+\nu)/n$.

It is easy to see, by an argument similar to that of Art. 11, that

$$\lim_{n\to\infty}\left[\frac{1}{2n+1}+\frac{1}{2n+3}+\ldots+\frac{1}{2n+2\nu-1}\right] = \lim_{n\to\infty}\int_n^{n+\nu}\frac{dx}{2x} = \frac{1}{2}\log k.$$

Hence the alteration l is contained between the two values

$$\tfrac{1}{2}(g\pm\epsilon)\log k.$$

Thus, since ϵ is arbitrarily small, we must have

$$l = \tfrac{1}{2}g\log k.$$

Hence, *if $\lim nf(n) = g$, and if k is the limit of the ratio of the number of positive to the number of negative terms, the alteration l is given by $l = \frac{1}{2}g\log k$.*

In particular, since $1-\frac{1}{2}+\frac{1}{3}-\frac{1}{4}+\ldots = \log 2$ (Art. 21), we see that when this series is arranged so that kn positive terms correspond to n negative terms its sum is $\log 2 + \frac{1}{2}\log k = \frac{1}{2}\log 4k$; and so, if there are two positive terms to each negative term, we get

$$1+\tfrac{1}{3}-\tfrac{1}{2}+\tfrac{1}{5}+\tfrac{1}{7}-\tfrac{1}{4}+\ldots = \tfrac{1}{2}\log 8 = \tfrac{3}{2}\log 2.$$

While, if there are two negative terms to each positive term, we have

$$1-\tfrac{1}{2}-\tfrac{1}{4}+\tfrac{1}{3}-\tfrac{1}{6}-\tfrac{1}{8}+\ldots = \tfrac{1}{2}\log 2,$$

a result which has been proved already (Art. 25).

Finally, if there are four negative terms to each positive term, we find

$$1-\tfrac{1}{2}-\tfrac{1}{4}-\tfrac{1}{6}-\tfrac{1}{8}+\tfrac{1}{3}-\tfrac{1}{10}-\tfrac{1}{12}-\tfrac{1}{14}-\tfrac{1}{16}+\ldots = 0.$$

To save space, we refer to § V. of Pringsheim's paper for the discussion of the rather more difficult case when $\lim nf(n) = 0$.

EXAMPLES.

1. Criticise the following paradox:

$$\begin{aligned} & 1-\tfrac{1}{2}+\tfrac{1}{3}-\tfrac{1}{4}+\tfrac{1}{5}-\tfrac{1}{6}+\ldots \\ &= 1+\tfrac{1}{2}+\tfrac{1}{3}+\tfrac{1}{4}+\tfrac{1}{5}+\tfrac{1}{6}+\ldots \\ &\quad -\ 2(\tfrac{1}{2}\quad +\tfrac{1}{4}\quad +\tfrac{1}{6}+\ldots) \\ &= 1+\tfrac{1}{2}+\tfrac{1}{3}+\tfrac{1}{4}+\ldots\quad -(1+\tfrac{1}{2}+\tfrac{1}{3}+\tfrac{1}{4}+\ldots) \\ &= 0. \end{aligned}$$

2. If a transformation similar to that of Ex. 1 is applied to the series

$$1-\frac{1}{2^p}+\frac{1}{3^p}-\frac{1}{4^p}+\frac{1}{5^p}-\ldots,$$

shew that (if $p<1$) we obtain the paradoxical result that the sum of the series is negative. But, if $p>1$, the result obtained is correct and expresses the sum of the given series in terms of the corresponding series of positive terms.

3. If we write $f(x) = 1/x^2$, and $s = \sum_1^\infty f(n)$, shew that

$$f(1)+f(3)+f(5)+f(7)+f(9)+\ldots = \tfrac{3}{4}s = \tfrac{1}{8}\pi^2,$$
$$f(1)+f(5)+f(7)+f(11)+f(13)+\ldots = \tfrac{2}{3}s = \tfrac{1}{9}\pi^2,$$
$$f(1)-f(2)-f(4)+f(5)+f(7)-f(8)-f(10)+\ldots = \tfrac{4}{9}s = \tfrac{2}{27}\pi^2.$$

[It is proved in Art. 71 that $s = \frac{1}{6}\pi^2$.]

4. Prove that
$$1 - \tfrac{1}{2} - \tfrac{1}{4} + \tfrac{1}{5} + \tfrac{1}{7} - \tfrac{1}{8} - \tfrac{1}{10} + \ldots = \tfrac{2}{3}\log 2.$$

5. Apply a transformation similar to that of Ex. 1 to the series
$$1 - \tfrac{1}{3} + \tfrac{1}{5} - \tfrac{1}{7} + \ldots,$$
and prove that the resulting series is
$$(1 - \tfrac{2}{3}) + (\tfrac{1}{3} - \tfrac{2}{7}) + (\tfrac{1}{5} - \tfrac{2}{11}) + \ldots,$$
which converges to a sum *less* than that of the given series.

6. Any non-absolutely convergent series may be converted into an absolutely convergent series by the insertion of brackets. [See Art. 5.]

Any oscillating series may be converted into a convergent series by the insertion of brackets; and the brackets may be arranged so that the series has a sum equal to any of the limits of s_n.

7. In order that the value of a non-absolutely convergent series may remain unaltered after a certain change in the order of the terms, it is sufficient that the product of the displacement of the nth term by the greatest subsequent term may tend to zero as n increases to ∞.

[Borel, *Bulletin des Sci. Math.* (2), t. 14, 1890, p. 97.]

8. Prove that $\sum_{-\infty}^{\infty} \dfrac{1}{x-n}$ is not a determinate number, but that
$$X = \frac{1}{x} + \sum_{-\infty}^{\infty}{}' \left(\frac{1}{n} + \frac{1}{x-n}\right)$$
is perfectly definite. Here x is supposed not to be an integer, and the accent implies that $n=0$ is to be omitted.

Shew that
$$\lim \left[\sum_{-p}^{q}\left(\frac{1}{x-n}\right)\right] - X = -\log k,$$
where p and q tend to ∞ in such a way that $\lim (q/p) = k$.

9. Find the product of the two series
$$1 + x + \frac{x^2}{2!} + \ldots + \frac{x^n}{n!} + \ldots \quad \text{and} \quad 1 - x + \frac{x^2}{2!} - \ldots + (-1)^n \frac{x^n}{n!} + \ldots.$$

10. Shew that if $s_n = a_0 + a_1 + a_2 + \ldots + a_n$, then
$$(\Sigma a_n x^n)/(1-x) = (\Sigma a_n x^n)(1 + x + x^2 + \ldots) = \Sigma s_n x^n.$$

CHAPTER V.

DOUBLE SERIES.

29. Suppose an infinite number of terms arranged so as to form a network (or lattice) which is bounded on the left and above, but extends to infinity to the right and below, as indicated in the diagram:

$$\begin{array}{l} a_{1,1}+a_{1,2}+a_{1,3}+a_{1,4}+\dots \\ +a_{2,1}+a_{2,2}+a_{2,3}+a_{2,4}+\dots \\ +a_{3,1}+a_{3,2}+a_{3,3}+a_{3,4}+\dots \\ +a_{4,1}+a_{4,2}+a_{4,3}+a_{4,4}+\dots \\ +\dots\dots\dots\dots\dots\dots\dots\dots \end{array}$$

The first suffix refers to the row, the second suffix to the column in which the term stands.

Suppose next that a rectangle is drawn across the network so as to include the first m rows and the first n columns of the array of terms; and denote the sum of the terms contained within this rectangle by the symbol $s_{m,n}$. If $s_{m,n}$ *approaches a definite limit* s *as* m *and* n *tend to infinity at the same time (but independently), then* s *is called the sum of the double series represented by the array.**

In more precise form, this statement requires that it shall be possible to find an index μ, corresponding to an arbitrary positive fraction ϵ, such that

$$|s_{m,n}-s|<\epsilon, \quad \text{if } m, n>\mu.$$

By the last inequality is implied that m, n are subject to no other restriction than the condition of being greater than μ.

---

* This definition is framed in accordance with the one adopted by Pringsheim (*Münchener Sitzungsberichte*, Bd. 27, 1897, p. 101; see particularly pp. 103, 140).

This property is also expressed by the equations

$$\lim_{m, n \to \infty} s_{m,n} = s, \quad \text{or} \quad \lim_{(m, n)} s_{m,n} = s.$$

But the symbol $s_{m,n} \to s$ is not sufficient, unless some indication is added as to the mode of summation adopted; for it is often convenient to use other methods (see Arts. 30, 31) which may give values different from the above.

Since $$a_{m,n} = s_{m,n} - s_{m-1,n} - s_{m,n-1} + s_{m-1,n-1},$$
it follows that when $(s_{m,n})$ converges, we can find μ so that $|a_{m,n}| < \epsilon$, provided that *both* m and n are greater than μ: this of course does not necessarily imply that $a_{m,n}$ will tend to zero, when m, n tend to ∞ *separately*.

The equations

$$\lim_{m, n \to \infty} s_{m,n} = \infty \quad \text{or} \quad \lim_{(m, n)} s_{m,n} = \infty$$

imply that, given any positive number G, however large, we can find μ, such that

$$s_{m,n} > G, \quad \text{if } m, n > \mu;$$

and the double series is then said to *diverge* to ∞. We define similarly divergence to $-\infty$.

It is also possible that the double series may *oscillate*; and there is little difficulty in modifying the method of Art. 5 so as to establish the existence of *extreme limits** for any double sequence $(s_{m,n})$; these may be denoted by

$$\underline{\lim}_{(m, n)} s_{m,n} \quad \text{and} \quad \overline{\lim}_{(m, n)} s_{m,n}.$$

The general condition for convergence is simply that the sum of the terms between two rectangles m, n and p, q must be numerically less than ϵ, if m, n are greater than μ; or in symbols

$$|s_{p,q} - s_{m,n}| < \epsilon, \quad \text{if } p > m > \mu \text{ and } q > n > \mu,$$

where of course the value of μ will depend on ϵ. This condition is obviously *necessary*; and to see that it is *sufficient*, denote by σ_n the value of $s_{m,n}$ when $m = n$ (so that the rectangle is replaced by a *square*). Then our condition yields

$$|\sigma_q - \sigma_n| < \epsilon, \quad \text{if } q > n > \mu.$$

Hence σ_n approaches a limit s (Art. 3), and so we can find μ_1, such that $$|s - \sigma_n| < \tfrac{1}{2}\epsilon, \quad \text{if } n > \mu_1.$$

---

* The proof is given in full by Pringsheim, *Math. Annalen*, Bd. 53, 1900, pp. 294-301.

Now the general condition gives also

$$|s_{p,q}-\sigma_n|<\tfrac{1}{2}\epsilon, \quad \text{if } p, q>n>\mu_2;$$

and so, if μ_3 is the greater of μ_1 and μ_2, and $n>\mu_3$, we find

$$|s_{p,q}-s|<\epsilon, \quad \text{if } p, q>n.$$

Ex. 1. *Convergence*: If $s_{m,n}=1/m+1/n$, $s=0$ and $\mu \geqq 2/\epsilon$.

Ex. 2. *Divergence*: If $s_{m,n}=m+n$, the condition of divergence is satisfied.

Ex. 3. *Oscillation*: If $s_{m,n}=(-1)^{m+n}$, the extreme limits are -1 and $+1$.

30. Repeated series.

In addition to the mode of summation just defined it is often necessary to use the method of *repeated summation*; then we first form the sum of a *row* of terms in the diagram, and obtain $b_m=\sum_{n=1}^{\infty} a_{m,n}$, after which we sum $\sum_{1}^{\infty} b_m$.

This process gives a value which we denote by

$$\sum_{m=1}^{\infty}\left(\sum_{n=1}^{\infty} a_{m,n}\right) \text{ or } \sum_{(m)} \sum_{(n)} a_{m,n};$$

this is called *the sum by rows* of the double series.

In like manner we define the repeated sum

$$\sum_{n=1}^{\infty}\left(\sum_{m=1}^{\infty} a_{m,n}\right) \text{ or } \sum_{(n)} \sum_{(m)} a_{m,n},$$

which is called *the sum by columns* of the double series.

Each of these sums may also be defined as a *repeated limit*, thus:

$$\sum_{(m)} \sum_{(n)} a_{m,n}= \lim_{m\to\infty} (\lim_{n\to\infty} s_{m,n}) \text{ or } \lim_{(m)(n)} s_{m,n},$$

with a similar interpretation for the second repeated sum.

In dealing with a *finite* number of terms it is obvious that

$$s_{M,N}=\sum_{m=1}^{M}\left(\sum_{n=1}^{N} a_{m,n}\right)=\sum_{n=1}^{N}\left(\sum_{m=1}^{M} a_{m,n}\right).$$

But it is by no means necessarily true that if a double series has the sum s in the sense of Art. 29 then also

$$(1) \qquad s=\sum_{m=1}^{\infty}\left(\sum_{n=1}^{\infty} a_{m,n}\right)=\sum_{n=1}^{\infty}\left(\sum_{m=1}^{\infty} a_{m,n}\right);$$

indeed the single series formed by the rows and columns of the double series *need not converge at all, but may oscillate.*

That the rows and columns need not converge is shewn by the example $s_{m,n}=(-1)^{m+n}(1/m+1/n)$ for which $s=0$; but neither of the limits

$$\lim_{m\to\infty} s_{m,n}, \quad \lim_{n\to\infty} s_{m,n}$$

exists at all.

Pringsheim has proved, however, that *if the rows and columns converge, and if the double series is convergent, then the equation* (1) *above is true.*

In fact we have

$$|s_{m,n}-s|<\epsilon, \qquad \text{if } m, n>\mu,$$

so that

$$|\lim_{n\to\infty} s_{m,n}-s| \leqq \epsilon, \qquad \text{if } m>\mu;$$

since, by hypothesis, this limit *exists.*

Hence

$$\lim_{m\to\infty}(\lim_{n\to\infty} s_{m,n})=s.$$

In like manner we can prove the other half of equation (1).

When the double series is not convergent, the equation

$$(2) \qquad \sum_{m=1}^{\infty}\left(\sum_{n=1}^{\infty} a_{m,n}\right)=\sum_{n=1}^{\infty}\left(\sum_{m=1}^{\infty} a_{m,n}\right)$$

is not necessarily valid whenever the two repeated series are convergent.

There is in fact no reason whatever for assuming that the equation

$$\lim_{m\to\infty}(\lim_{n\to\infty} s_{m,n})=\lim_{n\to\infty}(\lim_{m\to\infty} s_{m,n})$$

is true whenever the repeated limits exist.

For instance, with $s_{m,n}=m/(m+n)$, we find

$$\lim_{m\to\infty}(\lim_{n\to\infty} s_{m,n})=0, \quad \lim_{n\to\infty}(\lim_{m\to\infty} s_{m,n})=1.$$

From Pringsheim's theorem it is clear that the double series cannot converge (the rows and columns being supposed convergent) unless equation (2) is valid; but the truth of (2) is no reason for assuming the convergence of the double series.

For instance, with $s_{m,n}=mn/(m+n)^2$, we find

$$\lim_{m\to\infty}(\lim_{n\to\infty} s_{m,n})=0=\lim_{n\to\infty}(\lim_{m\to\infty} s_{m,n}).$$

But yet the double series cannot converge, since if $m=2n$, $s_{m,n}=\frac{2}{9}$; while if $m=n$, $s_{m,n}=\frac{1}{4}$.

For some purposes it is useful to know that equation (2) is true, without troubling to consider the general question of convergence of the double series. In such cases, conditions may be used which will be found in the *Proceedings of the London Mathematical Society*;* the discussion of them here would go somewhat beyond our limits.

A further example, due to Arndt, of the possible failure of equation (2) may be added:

If we write $$a_{m,n}=\frac{1}{m+1}\left(\frac{m}{m+1}\right)^n-\frac{1}{m+2}\left(\frac{m+1}{m+2}\right)^n,$$

we find that $$s_{m,n}=\left(\frac{1}{2}-\frac{1}{2^{n+1}}\right)-\left[\frac{m+1}{m+2}-\left(\frac{m+1}{m+2}\right)^{n+1}\right].$$

Thus $$\lim_{m\to\infty}(\lim_{n\to\infty} s_{m,n})=-\tfrac{1}{2},$$

but $$\lim_{n\to\infty}(\lim_{m\to\infty} s_{m,n})=+\tfrac{1}{2}.$$

Other examples of points in the general theory will be found at the end of the chapter. (See Exs. 1-6 and 9.)

31. Double series of positive terms.

In view of what has been proved in Art. 26, we may anticipate that if a series of positive terms converges to the sum s in any way, it will have the same sum if summed in any other way which includes all the terms. For, however many terms are taken, we cannot get a larger sum than s, but we can get as near to s as we please, by taking a sufficient number of terms. We shall now apply this general principle to the most useful special cases.

(1) *It is sufficient to consider squares only in testing a double series of positive terms for convergence.*

Write for brevity $s_{m,n}=\sigma_n$ when $m=n$; then plainly σ_n must converge to the limit s, if $s_{m,n}$ does so. Further, if σ_n converges to a limit s, so also will $s_{m,n}$. For then we can find μ so that σ_μ lies between s and $s-\epsilon$; but if m and n are greater than μ, we have

$$\sigma_{m+n}\geqq s_{m,n}\geqq\sigma_\mu,$$

so that $$s\geqq s_{m,n}>s-\epsilon.$$

Hence $s_{m,n}$ converges to the limit s.

The reader will find no difficulty in extending the argument to cases of *divergence.*

---

* Bromwich, *Proc. Lond. Math. Soc.*, series 2, vol. 1, 1904, p. 176.

(2) *If more convenient for purposes of summation, we may replace the rectangles by any succession of curves* which tend to infinity in all directions; the terms being positive.*

For, plainly, when the rectangles (and therefore the squares) give a sum s we can suppose any particular curve C_n to be contained between two of the squares and that the sides of these squares are p, q; thus if S_n is the sum for the curve C_n, we have, as in (1) above,

$$\sigma_p \leqq S_n \leqq \sigma_q \leqq s.$$

Further, since C_n is to tend to infinity in all directions, we can make p greater than μ by taking $n > n_0$, say.

Thus, since $\sigma_p > s - \epsilon$,

we have also $s - \epsilon < S_n \leqq s$, if $n > n_0$,

and so $\lim S_n = s$.

In like manner, by enclosing a square between two of the curves, we can shew that if the curves give a sum s, so also do the squares (and therefore the rectangles, too, in virtue of (1) above).

A particular class of the curves used in (2) is formed by drawing diagonals, equally inclined to the horizontal and vertical sides of the network as indicated in the right-hand figure.

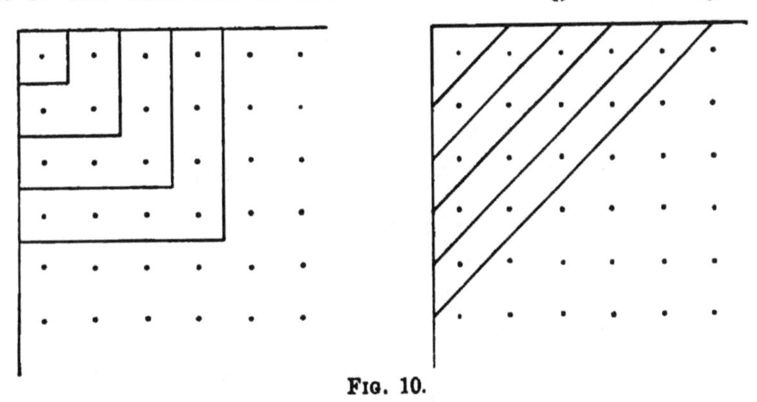

FIG. 10.

The summation by squares is indicated on the left. It should be noticed that *these two modes of summation give two methods of converting a double series into a single series.*

Thus, by squares, we are summing the series

$$a_{11} + (a_{21} + a_{22} + a_{12}) + (a_{31} + a_{32} + a_{33} + a_{23} + a_{13}) + \dots,$$

---

* These "curves" may consist, wholly or in part, of straight lines; and it is supposed that each curve encloses the *whole* of the preceding curve.

and by the diagonals we get

$$a_{11}+(a_{21}+a_{12})+(a_{31}+a_{22}+a_{13})+\dots.$$

Of course the equality between these two series is now seen to be a consequence of Art. 26; but we could not, without further proof, infer theorem (1) from that article since Art. 26 refers only to single and not to double series.

By combining Art. 26 with (1) above, it will be seen that:

(3) *No derangement of the (positive) terms of a double series can alter the sum, nor change divergence into convergence.*

It is also important to note that:

(4) *When the terms of the double series are positive, its convergence implies the convergence of all the rows and columns, and its sum is equal to the sums of the two repeated series.*

For, when the double series has the sum s, it is clear that $s_{m,n}$ cannot exceed s; and consequently the sum of any number of terms in a single row cannot be greater than s. Also, for any fixed value of m, $\lim\limits_{(n)} s_{m,n}$ exists and is not greater than s. Now we can find μ so that $s_{m,n}>s-\epsilon$, if m, n are greater than μ. Consequently

$$s \geqq \lim_{(n)} s_{m,n} > s-\epsilon, \quad \text{if } m>\mu.$$

Hence
$$\lim_{(m)} [\lim_{(n)} s_{m,n}] = s,$$

or
$$\sum_{m=1}^{\infty}\left(\sum_{n=1}^{\infty} a_{m,n}\right)=s.$$

In a similar way, we see that each column converges and that

$$\sum_{n=1}^{\infty}\left(\sum_{m=1}^{\infty} a_{m,n}\right)=s.$$

As a converse to (4), we have:

(5) *The terms being always positive, if either repeated series is convergent, so also is the other and also the double series; and the three sums are the same.*

For, suppose that
$$\lim_{(m)} [\lim_{(n)} s_{m,n}] = s,$$

then
$$\sigma_m \leqq \lim_{(n)} s_{m,n} \leqq s.$$

Hence by Art. 2 the sequence (σ_m) converges to a limit σ; and it then follows from (4) above that $s=\sigma$, and that the other repeated series has the same sum.

The reader will find little difficulty in modifying the proofs in (4) and (5) so as to cover the case of divergence.

32. Tests for convergence of a double series of positive terms.

If we compare Art. 8 with (1) of the last article, we see that:

(1) *If the (positive) terms of a double series are less than those of another double series which is known to converge, the series converges.*

Similarly for divergence, with "greater" in place of "less."

The most important type of convergent series* is given by $a_{m,n}=(C_m C_n)^{-1}$, where ΣC_ν^{-1} is a convergent single series; to see that this double series is convergent, we note that the sum σ_n contained in a square of side n is plainly equal to

$$(C_1^{-1}+C_2^{-1}+\dots+C_n^{-1})^2,$$

and therefore σ_n has a limit. Consequently the double series converges, by (1) of the last article.

On the other hand, the chief type for divergence* is given by $a_{m,n}=(pD_p)^{-1}$, where $p=m+n$ and ΣD_ν^{-1} is a divergent single series; to recognise the divergence, take the sum by diagonals, as in (2) of the last article. In this way we get the series

$$\tfrac{1}{2}D_2^{-1}+\tfrac{2}{3}D_3^{-1}+\tfrac{3}{4}D_4^{-1}+\tfrac{4}{5}D_5^{-1}+\dots,$$

which is seen to be divergent, on comparing it with

$$\tfrac{1}{2}(D_2^{-1}+D_3^{-1}+D_4^{-1}+\dots).$$

Ex. 1. $\Sigma m^{-\alpha}n^{-\beta}$ converges if $\alpha>1$, $\beta>1$.

Ex. 2. $\Sigma(m+n)^{-\alpha}$ diverges if $\alpha\leqq 2$ and converges if $\alpha>2$; for divergence is assured if $p^{\alpha-1}=D_p$, *i.e.* if $\alpha\leqq 2$; but if $\alpha>2$, we have $(m+n)^{-\alpha}<m^{-\frac{1}{2}\alpha}n^{-\frac{1}{2}\alpha}$ and so we may take $C_\nu=\nu^{\frac{1}{2}\alpha}$.

Ex. 3. If a, c are positive, and $ac>b^2$, in case $b<0$, the series

$$\Sigma(am^2+2bmn+cn^2)^{-\lambda},$$

converges if $\lambda>1$, diverges if $\lambda\leqq 1$; for we have

$$A(m+n)^2>am^2+2bmn+cn^2>2(b+\sqrt{ac})mn,$$

where A is the greatest of a, c, and $|b|$.

Thus the conditions of convergence or divergence follow from Exs. 1 and 2.

---

* Pringsheim, *Münchener Sitzungsberichte*, Bd. 27, pp. 146-150.

The reader will have no difficulty in seeing that the following generalisation of Cauchy's test (Art. 11) is correct:

(2) *If the function $f(x, y)$ is positive and steadily decreases to zero as x and y increase to infinity,* then the double series $\Sigma f(m, n)$ converges or diverges with the double integral*

$$\int^{\infty}\int^{\infty} f(x, y)\,dx\,dy.$$

However, nearly all cases of interest which come under the test (2) can be as easily tested by the following method, which depends only on a single integral:

(3) *If the positive function $f(x, y)$ has a lower limit $g(\xi)$ and an upper limit $G(\xi)$ when $y=\xi-x$ and x varies from 0 to ξ, and if $\xi G(\xi)$, $\xi g(\xi)$ tend steadily to zero as $\xi\to\infty$, then the double series $\Sigma\Sigma f(m, n)$ converges if the integral $\int^{\infty} G(\xi)\xi\,d\xi$ converges; but the series diverges if the integral $\int^{\infty} g(\xi)\xi\,d\xi$ diverges.*†

For then the sum of the terms on the diagonal $x+y=n$ lies between $(n-1)g(n)$ and $(n-1)G(n)$; thus the series converges with $\Sigma(n-1)G(n)$, that is, with the integral $\int^{\infty} G(\xi)\xi\,d\xi$; but the series diverges with $\Sigma(n-1)g(n)$, that is, with the integral $\int^{\infty} g(\xi)\xi\,d\xi$.

Ex. 4. A particular case of (3) which has some interest is given by the double series $f(am^2+2bmn+cn^2)$, where $f(x)$ is a function which steadily decreases as its argument increases, and $am^2+2bmn+cn^2$ is subject to the same conditions as in Ex. 3 above.

If A is the greatest of a, $|b|$, c, it is evident that

$$ax^2+2bx(\xi-x)+c(\xi-x)^2$$

is less than $A[x^2+2x(\xi-x)+(\xi-x)^2]=A\xi^2$. When b is positive, we see in the same way that if B is the least of a, b, c, the expression is greater than $B\xi^2$. But if b is negative, we can put the expression in the form

$$[\{(a+c-2b)x+(b-c)\xi\}^2+(ac-b^2)\xi^2]/(a+c-2b)\geqq B\xi^2, \quad \text{if } 4AB=ac-b^2.$$

Hence $g(\xi)=f(A\xi^2)$ and $G(\xi)=f(B\xi^2)$.

---

* That is, we suppose $f(\xi, \eta)\leqq f(x, y)$,
if $\xi\geqq x$ *and* $\eta\geqq y$.

† The use of a single integral for testing multiple series seems to be due to Riemann (*Ges. Werke*, 1876, p. 452); an alternative investigation is given by Hurwitz (*Math. Annalen*, Bd. 44, p. 83). The above form seems to be novel.

Thus the series converges if $\int^{\infty} f(B\xi^2)\xi d\xi$ converges; that is, if $\int^{\infty} f(x)dx$ is convergent. Similarly the series is seen to diverge when $\int^{\infty} f(x)dx$ is divergent.

This result confirms Ex. 3 above; and it shews also that when

$$f(x)=e^{-x} \quad \text{or} \quad 1/x(\log x)^{1+a}, \quad (a>0)$$

the series converges; on the other hand, the series diverges if

$$f(x)=1/x\log x.$$

33. Absolutely convergent double series.

Just as in the theory of single series, we call the series $\Sigma a_{m,n}$ *absolutely convergent* if $\Sigma |a_{m,n}|$ is convergent.

The method used in Art. 26 can be applied at once to shew that all the results proved in Art. 31 for double series of positive terms are still true for any absolutely convergent double series. In this connexion the reader who has advanced beyond the elements of the subject should consult a paper by Hardy (*Proc. Lond. Math. Soc.* (2) vol. 1, 1903, p. 285).

That these results are not necessarily true for non-absolutely convergent series may be seen by taking two simple examples:

(1) Consider first

$$\begin{array}{l} 1+1+1+1+\ldots \\ +1-1-1-1-\ldots \\ +1-1+0+0+\ldots \\ +1-1+0+0+\ldots \\ +\ldots\ldots\ldots\ldots\ldots\ldots, \end{array}$$

where all the terms are 0 except in the first two rows and columns.

Here $s_{m,n}=2$ if $m, n>1$, so that the series has the sum 2, according to Pringsheim's definition. But if we convert the double series into a single series by summing the diagonals (as in (2), Art. 31), we get

$$1+2+1+0+0+\ldots=4.$$

Obviously, too, the convergence of this series does not imply the convergence of the first two rows and columns (compare (4) Art. 31).

(2) Consider next the double series suggested by Cesàro:

$$\begin{array}{l} \frac{1}{2}-\frac{1}{4}+\frac{1}{4}-\frac{1}{8}+\frac{1}{8}-\frac{1}{16}+\frac{1}{16}-\ldots \\ +\frac{1}{2^2}-\frac{3}{4^2}+\frac{3}{4^2}-\frac{7}{8^2}+\frac{7}{8^2}-\frac{15}{16^2}+\frac{15}{16^2}-\ldots \\ +\frac{1}{2^3}-\frac{3^2}{4^3}+\frac{3^2}{4^3}-\frac{7^2}{8^3}+\frac{7^2}{8^3}-\frac{15^2}{16^3}+\frac{15^2}{16^3}-\ldots \\ +\frac{1}{2^4}-\frac{3^3}{4^4}+\frac{3^3}{4^4}-\frac{7^3}{8^4}+\frac{7^3}{8^4}-\frac{15^3}{16^4}+\frac{15^3}{16^4}-\ldots \\ +\ldots\ldots\ldots\ldots\ldots\ldots\ldots\ldots\ldots\ldots\ldots\ldots. \end{array}$$

Here the sums of the rows in order are

$$\frac{1}{2}, \frac{1}{2^2}, \frac{1}{2^3}, \frac{1}{2^4}, \ldots,$$

and so the sum of all the rows is 1.

But the sums of the columns are

$$+1, -1, +1, -1, +1, -1, +1, \ldots,$$

proving that (5) of Art. 31 does not apply.

This result is specially striking because each row converges absolutely (the terms being less than $\frac{1}{2}+\frac{1}{4}+\frac{1}{4}+\frac{1}{8}+\frac{1}{8}+\ldots$), and secondly, the series formed by the sums of the rows is $\frac{1}{2}+\frac{1}{2^2}+\frac{1}{2^3}+\ldots$, which also converges absolutely.

But the justification for applying (5) of Art. 31 is that the double series still converges when all the terms are made positive, which is not the case here; since the sum of the first n columns then becomes equal to n.

The fact that the sum of a non-absolutely convergent double series may have different values according to the mode of summation has led Jordan* to frame a definition which admits only absolute convergence. Such restriction seems, however, unnecessary, provided that, when a non-absolutely convergent series is used, we do not attempt to employ theorems (1) to (5) of Art. 31 without special justification.

For example in Lord Kelvin's discussion of the force between two electrified spheres in contact, the repeated series

$$\sum_{m=1}^{\infty}\left[\sum_{n=1}^{\infty}\frac{(-1)^{m+n}mn}{(m+n)^3}\right]$$

is used.† This series has the sum $\frac{1}{8}(\log 2-\frac{1}{4})$, and it has the same value if we sum first with respect to m. However, Pringsheim's sum does not exist but oscillates between limits‡ $\frac{1}{8}(\log 2-\frac{5}{8})$ and $\frac{1}{8}(\log 2+\frac{1}{8})$; while the diagonal series oscillates between $-\infty$ and $+\infty$.

34. A special example of deranging a double series is given by the *rule for multiplying single series* given in Art. 27 above.

Suppose we take the two single series $A=\Sigma a_n$, $B=\Sigma b_n$, and construct from them the double series $P=\Sigma a_m b_n$.

---

* *Cours d'Analyse*, t. 1, p. 302; compare Goursat's *Analysis* (translation by Hedrick), vol. 1, p. 357.

† Kelvin, *Reprint of Electrical Papers*, § 140.

‡ Bromwich and Hardy, *Proc. Lond. Math. Soc.*, series 2, vol. 2, 1904, p. 161 (see § 9).

It is clear that P converges in Pringsheim's sense, provided that A, B converge; for we have

$$s_{m,n}=(a_1+a_2+\dots+a_m)(b_1+b_2+\dots+b_n),$$

so that $\lim\limits_{m,n\to\infty} s_{m,n}=AB$.

But for practical work in analysis it is generally necessary to convert the double series P into a single series; the one usually chosen being the sum by diagonals (see (2) Art. 31). This single series is Σc_n, where

$$c_n=a_1b_n+a_2b_{n-1}+\dots+a_nb_1.$$

It follows at once from Art. 33 that: *If the two series Σa_n, Σb_n are absolutely convergent, their product is equal to Σc_n, which is also absolutely convergent.*

For under these circumstances the double series is clearly absolutely convergent, because $\sum\limits_{m,n} |a_m|.|b_n|$ converges to the sum $(\Sigma|a_m|).(\Sigma|b_n|)$; and $\Sigma|c_n|$ converges because the sum of any number of terms from $\Sigma|c_n|$ cannot exceed the product

$$(\Sigma|a_m|).(\Sigma|b_n|).$$

If, however, one or both of A, B should not converge absolutely, we have **Abel's theorem**: *Provided that the series Σc_n converges, its sum is equal to the product AB.** For then, if we write $A_n=a_1+a_2+\dots+a_n$, $B_n=b_1+b_2+\dots+b_n$, we find

$$C_n=c_1+c_2+\dots+c_n=a_1B_n+a_2B_{n-1}+\dots+a_nB_1.$$

Hence $C_1+C_2+\dots+C_n=A_1B_n+A_2B_{n-1}+\dots+A_nB_1$,

or $$\frac{1}{n}(C_1+C_2+\dots+C_n)=\frac{1}{n}(A_1B_n+A_2B_{n-1}+\dots+A_nB_1).$$

Now (App., Art. 154), when $\lim C_n=C$, we have also

$$\lim\frac{1}{n}(C_1+C_2+\dots+C_n)=C;$$

and again (App., Art. 154, IV.)

$$\lim\frac{1}{n}(A_1B_n+A_2B_{n-1}+\dots+A_nB_1)=AB.$$

Hence $C=AB$. (For an alternative proof, see Art. 54 below.)

* Pringsheim has proved, by a similar method, that if a double series is convergent, its sum is equal to the sum of the diagonal series, when the latter converges, provided that every row converges and also every column.

It should be observed that *the series Σc_n cannot diverge (if Σa_n, Σb_n are convergent), although it may oscillate.* For, if Σc_n is divergent, we should have $\lim C_n = \infty$, and therefore also

$$\lim \frac{1}{n}(C_1 + C_2 + \ldots + C_n) = \infty,$$

by Art. 154; whereas this limit must be equal to AB. If Σc_n oscillates, it is clear from the article quoted that AB lies between the extreme limits of Σc_n; that in some cases Σc_n does oscillate (and that its extreme limits may be $-\infty$ and $+\infty$) is evident from Ex. 3 below; but in all cases the oscillation is of such a character* that

$$\lim \frac{1}{n}(C_1 + C_2 + \ldots + C_n) = AB.$$

Ex. 1. Undoubtedly the cases of chief interest arise in the multiplication of power-series. Thus, if the two series

$$a_0 + a_1x + a_2x^2 + \ldots + a_nx^n + \ldots, \quad b_0 + b_1x + b_2x^2 + \ldots$$

are both absolutely convergent for $|x| < r$ (see Art. 50), their product is given by

$$c_0 + c_1x + c_2x^2 + \ldots,$$

which is also absolutely convergent for $|x| < r$; where we have written

$$c_0 = a_0b_0, \quad c_n = a_0b_n + a_1b_{n-1} + \ldots + a_nb_0,$$

the notation being slightly changed from that used in the text.

Ex. 2. If we apply the rule to square the series

$$1 - \frac{1}{2} + \frac{1}{3} - \frac{1}{4} + \ldots,$$

we have no reason (so far) to anticipate a convergent series; but we find the series

$$1 - \left(\frac{1}{2} + \frac{1}{2}\right) + \left(\frac{1}{3} + \frac{1}{2^2} + \frac{1}{3}\right) - \ldots,$$

in which the general term is $(-1)^{n-1}w_n$, where

$$w_n = \frac{1}{1 \cdot n} + \frac{1}{2(n-1)} + \frac{1}{3(n-2)} + \ldots + \frac{1}{n \cdot 1},$$

so that

$$(n+1)w_n = \left(1 + \frac{1}{n}\right) + \left(\frac{1}{2} + \frac{1}{n-1}\right) + \ldots + \left(\frac{1}{n} + 1\right)$$

$$= 2\left(1 + \frac{1}{2} + \frac{1}{3} + \ldots + \frac{1}{n}\right).$$

---

* Cesàro (to whom this result is due) calls such series ***simply indeterminate***; the degree of indeterminacy being measured by the number of means which have to be taken before a definite value is obtained. (See Art. 122 below.)

Now $$\lim\left(1+\frac{1}{2}+\ldots+\frac{1}{n}-\log n\right)=C \quad \text{(Art. 11)},$$

so that $$\lim w_n=\lim\frac{2}{n+1}(C+\log n)=0.$$

Further $$w_{n-1}-w_n=\frac{2}{n}\left(1+\frac{1}{2}+\frac{1}{3}+\ldots+\frac{1}{n-1}\right)-\frac{2}{n+1}\left(1+\frac{1}{2}+\frac{1}{3}+\ldots+\frac{1}{n}\right)$$
$$=\frac{2}{n(n+1)}\left[\left(1+\frac{1}{2}+\frac{1}{3}+\ldots+\frac{1}{n-1}\right)-1\right]$$
$$=\frac{2}{n(n+1)}\left(\frac{1}{2}+\frac{1}{3}+\frac{1}{4}+\ldots+\frac{1}{n-1}\right).$$

That is, $w_{n-1}>w_n$, so that $\Sigma(-1)^{n-1}w_n$ is convergent (Art. 21), and therefore, by Abel's theorem,

$$\frac{1}{2}\left(1-\frac{1}{2}+\frac{1}{3}-\frac{1}{4}+\ldots\right)^2=\frac{1}{2}-\frac{1}{3}\left(1+\frac{1}{2}\right)+\frac{1}{4}\left(1+\frac{1}{2}+\frac{1}{3}\right)-\frac{1}{5}\left(1+\frac{1}{2}+\frac{1}{3}+\frac{1}{4}\right)+\ldots.$$

Of course this agrees with Pringsheim's general theorem (Art. 35 below). [*Math. Trip.* 1900.]

Ex. 3. But if we square the series

$$1-\frac{1}{2^p}+\frac{1}{3^p}-\frac{1}{4^p}+\ldots, \quad (0<p<1),$$

we obtain $\Sigma(-1)^{n-1}w_n$, where

$$w_n=(1\,.\,n)^{-p}+[2(n-1)]^{-p}+\ldots+(n\,.\,1)^{-p}.$$

Now $$x(n+1-x)<n^2, \quad \text{if } 0<x<n+1,$$

so that $$[x(n+1-x)]^{-p}>n^{-2p}$$

and $$w_n>n^{1-2p}.$$

Consequently *if* $p\leqq\frac{1}{2}$, *the series* $\Sigma(-1)^{n-1}w_n$ *is oscillatory and the rule for multiplication fails.*

On the other hand, if $p>\frac{1}{2}$, Pringsheim's theorem (Art. 35) shews that the rule is correct.

35. Mertens has proved that *the series* Σc_n *will converge to the sum* AB, *provided that* one *of the series* Σa_n, Σb_n *is absolutely convergent.*

Suppose that $\Sigma|a_n|$ is convergent, and then write for brevity

$$v_n=|a_n|, \quad V_n=v_1+v_2+\ldots+v_n, \quad (V=\lim V_n).$$

Also write $$r_n=b_{n+1}+b_{n+2}+\ldots \text{ to } \infty, \quad \rho_n=|r_n|,$$

so that $$B=r_n+B_n.$$

Consequently $$C_n=a_1B_n+a_2B_{n-1}+\ldots+a_nB_1$$
$$=a_1(B-r_n)+a_2(B-r_{n-1})+\ldots+a_n(B-r_1)$$
$$=A_nB-R_n,$$

where $$R_n=a_1r_n+a_2r_{n-1}+\ldots+a_nr_1,$$

so that $$|R_n|\leqq v_1\rho_n+v_2\rho_{n-1}+\ldots+v_n\rho_1.$$

Now, since the series B is convergent, the sequence $\rho_1, \rho_2, \rho_3, \ldots$ tends to zero as a limit; it has therefore a *finite* upper limit H. Also we can find m so that $\rho_m, \rho_{m+1}, \rho_{m+2}, \ldots$ are all less than ϵ; thus, if $p=n-m$, we have

$$|R_n| < (v_1+v_2+\ldots+v_p)\epsilon+(v_{p+1}+v_{p+2}+\ldots+v_n)H$$
$$< V\epsilon+(V-V_p)H.$$

Take the limit of the last inequality as n tends to ∞, and we find

$$\overline{\lim_{n\to\infty}}\, |R_n| \leqq \epsilon V,$$

because p tends to ∞ with n. Now ϵ is arbitrarily small and V is fixed, so that we must have

$$\overline{\lim}\, |R_n|=0, \quad \text{(see Note (6), p. 5),}$$

that is,
$$\lim R_n=0.$$

Hence
$$\lim (A_nB-C_n)=0;$$

or
$$\lim C_n=AB.$$

It must not, however, be supposed that the condition of Mertens is *necessary* for the convergence of Σc_n; in fact Pringsheim has established a large number of results on the multiplication of two series neither of which converges absolutely. The simplest of these (including most cases of interest) is as follows:

If $U=\Sigma(-1)^{n-1}u_n$, $V=\Sigma(-1)^{n-1}v_n$ *are convergent in virtue of the conditions*

$$u_n \geqq u_{n+1},\ v_n \geqq v_{n+1},\ \lim u_n=0,\ \lim v_n=0,$$

their product is given by $\Sigma(-1)^{n-1}w_n$, *where*

$$w_n=u_1v_n+u_2v_{n-1}+\ldots+u_nv_1,$$

provided that the series $\Sigma(u_nv_n)$ *is convergent.*

Since $\lim u_n=0$, we can write

$$2U=u_1-(u_2-u_1)+(u_3-u_2)-(u_4-u_3)+\ldots$$
$$=\delta_1-\delta_2+\delta_3-\delta_4+\ldots,$$

where
$$\delta_1=u_1,\quad \delta_2=u_2-u_1,\quad \delta_3=u_3-u_2,\ \ldots .$$

Now, by hypothesis, $\delta_2, \delta_3, \delta_4, \ldots$ are all negative, and accordingly

$$|\delta_1|+|\delta_2|+\ldots+|\delta_n|=u_1+(u_1-u_2)+(u_2-u_3)+\ldots+(u_{n-1}-u_n)=2u_1-u_n.$$

Hence, $\Sigma|\delta_n|$ is convergent, and therefore the series $\delta_1-\delta_2+\delta_3-\delta_4+\ldots$ is absolutely convergent; we can therefore apply Mertens' theorem and obtain

$$2UV=\gamma_1-\gamma_2+\gamma_3-\gamma_4+\ldots,$$

where
$$\gamma_1=\delta_1v_1=w_1,\quad \gamma_n=\delta_1v_n+\delta_2v_{n-1}+\ldots+\delta_nv_1=w_n-w_{n-1}.$$

Thus
$$2UV=w_1-(w_2-w_1)+(w_3-w_2)-(w_4-w_3)+\ldots,$$

and therefore we have

$$UV=w_1-w_2+w_3-w_4+\ldots,$$

provided that $\lim w_n=0$.

Now since $\Sigma(-1)^{n-1}\gamma_n$ is convergent (in virtue of Mertens' theorem), it is clear that $\lim \gamma_n=0$, so that we need only prove that $w_n \to 0$ for *even* suffixes; we may therefore write $n=2p$. Then

$$w_n < \sum_{\nu=1}^{p}(u\, v_{n-\nu}) + \sum_{\nu=1}^{p}(u_{n-\nu}v_\nu),$$

and suppose that q is any index less than p; as ν varies from $q+1$ to p, we have $v_{n-\nu} \leqq v_\nu$ and $u_{n-\nu} \leqq u_\nu$, because $n-\nu \geqq \nu$; thus

$$\sum_{q+1}^{p}(u_\nu v_{n-\nu} + v_\nu u_{n-\nu}) \leqq 2\sum_{q+1}^{p} u_\nu v_\nu .$$

And as ν varies from 1 to q, we have $v_{n-\nu} \leqq v_{n-q}$ and $u_{n-\nu} \leqq u_{n-q}$, so that

$$\sum_{1}^{q}(u_\nu v_{n-\nu} + v_\nu u_{n-\nu}) \leqq q(u_1 v_{n-q} + v_1 u_{n-q}).$$

Thus we find

$$w_n \leqq q(u_1 v_{n-q} + v_1 u_{n-q}) + 2\sum_{q+1}^{p} u_\nu v_\nu .$$

If, as is supposed, $\Sigma u_n v_n$ is convergent, we can find q so that $\sum_{q+1}^{p} u_\nu v_\nu$ is less than $\frac{1}{2}\epsilon$; q having been fixed, take the limit of the last inequality as n tends to infinity, and we find $\overline{\lim}\, w_n \leqq \epsilon$. Thus, since ϵ is arbitrarily small, $\overline{\lim}\, w_n=0$, and it follows that $\lim w_n=0$, because w_n is positive.

Other results are due to Voss and Cajori, in addition to those found by Pringsheim. For references, see § 34 of Pringsheim's article in the *Encyklopädie*, Bd. I.; two more recent papers will be found in the *Trans. Amer. Math. Society*, vol. 2, 1901, pp. 25 and 404.

36. Substitution of a power-series in another power-series.

This operation gives another example of deranging a double series. Consider the series $z=f(y)=a_0+a_1y+a_2y^2+\dots$ and $y=b_0+b_1x+b_2x^2+\dots$; if convergent at all, they converge absolutely for $|y|<s$, $|x|<r$, say (see Art. 50). The question then arises whether the result of substituting the second series in the first and arranging in powers of x is ever convergent, and if so, for what values of x. It appears from Ex. 1, Art. 34, that the powers of y can be calculated by using the rule for the multiplication of series, and then z is equal to *the sum by rows* of the double series

$$\left.\begin{array}{l|l|l|l} a_0 & & & \\ +a_1b_0 & +\ a_1b_1x & +\ a_1b_2x^2 & +\dots \\ +a_2b_0^2 & +2a_2b_0b_1x & +\ a_2(b_1^2+2b_0b_2)x^2 & +\dots \\ +a_3b_0^3 & +3a_3b_0^2b_1x & +3a_3(b_0b_1^2+b_0^2b_2)x^2 & +\dots \\ +\dots\dots & \dots\dots\dots & \dots\dots\dots\dots\dots & \dots\dots \end{array}\right\} \quad \dots\dots(1)$$

If this double series is arranged according to powers of x, we are summing it by *columns*; these two sums are certainly equal if the double series still converges, after every term is made positive (Art. 31 (5) and Art. 33).

Write
$$|a_n| = \alpha_n, \quad |b_n| = \beta_n, \quad |x| = \xi,$$
and then the new series is not greater than

$$\left.\begin{array}{l|l|l|l} \alpha_0 & & & \\ +\alpha_1\beta_0 & +\alpha_1\beta_1\xi & +\ \alpha_1\beta_2\xi^2 & +\dots \\ +\alpha_2\beta_0^2 & +2\alpha_2\beta_0\beta_1\xi & +\ \alpha_2(\beta_1^2+2\beta_0\beta_2)\xi^2 & +\dots \\ +\alpha_3\beta_0^3 & +3\alpha_3\beta_0^2\beta_1\xi & +3\alpha_3(\beta_0\beta_1^2+\beta_0^2\beta_2)\xi^2 & +\dots \\ +\dots\dots & \dots\dots\dots\dots & \dots\dots\dots\dots\dots\dots\dots\dots & \dots\dots \end{array}\right\} \quad \dots(2)$$

Now this series, summed by rows, gives
$$\alpha_0 + \Sigma\alpha_n(\beta_0 + \Sigma\beta_m\xi^m)^n,$$
which converges, provided that $\beta_0 + \Sigma\beta_m\xi^m < s$.

Take now any positive number less than r, say ρ, then the series $\Sigma\beta_m\rho^m$ is convergent, and consequently the terms $\beta_m\rho^m$ have a finite upper limit M. Thus our condition is satisfied if
$$\beta_0 + \Sigma M(\xi/\rho)^m < s,$$
or if
$$\beta_0 + M\xi/(\rho - \xi) < s.$$

Hence if $\beta_0 < s$, and $\xi < (s - \beta_0)\rho/(M + s - \beta_0)$, the series (2) of positive terms will converge. Consequently the derangement of the series (1) will not alter its sum. Thus *the transformation is permissible if the two conditions*
$$\text{(i) } |b_0| < s, \qquad \text{(ii) } |x| < (s - |b_0|)\rho/(M + s - |b_0|)$$
are satisfied (where $\rho < r$, $|b_n|\rho^n \leqq M$*). In particular, if* $b_0 = 0$, *the conditions may be replaced by the one*
$$|x| < \rho s/(M + s).$$

If the series $z = \Sigma a_n y^n$ *converges for all values of* y, *it is evident that the condition* $|x| < r$ *is sufficient to justify the derangement.*

The case $b_0 = 0$ is of special interest in practice; and then the coefficient of x^n in the final series is not itself an infinite series, but terminates; a few of the coefficients are
$$c_0 = a_0, \quad c_1 = a_1 b_1, \quad c_2 = a_1 b_2 + a_2 b_1^2, \quad c_3 = a_1 b_3 + 2a_2 b_1 b_2 + a_3 b_1^3,$$
and generally, if $n > 2$, c_n will contain the terms
$$a_1 b_n + 2a_2 b_1 b_{n-1} + \dots + a_n b_1^n.$$

Ex. Take $$z = 1 + y + \frac{y^2}{2!} + \frac{y^3}{3!} + \frac{y^4}{4!} + \dots,$$
$$y = \mu(x - \tfrac{1}{2}x^2 + \tfrac{1}{3}x^3 - \tfrac{1}{4}x^4 + \dots),$$
then the transformation is allowable, provided that $|x| < 1$, since z converges for all values of y, and $r = 1$. The result is obviously $1 + \Sigma c_n x^n$, where c_n is a polynomial in μ, such that the term of highest degree is $\mu^n/n!$.

Assuming that $z = e^y$ and $y = \mu \log(1+x)$ (see Arts. 58, 62), we see that $z = (1+x)^\mu$. Thus c_n will vanish for $\mu = 0, 1, 2, \dots, n-1$, because in these cases the series terminates before reaching x^n.

Hence, $$c_n = \frac{1}{n!}\mu(\mu-1)(\mu-2)\dots(\mu-n+1),$$
and so we obtain the binomial series (Arts. 61 and 89).

37. Non-absolutely convergent double series.

Almost the only general type of such series has been given, comparatively recently, by Hardy;* it corresponds to the type of series discussed in Arts. 19, 20. The theorem is the extension of Dirichlet's test, and runs:

If in a double series $\Sigma a_{m,n}$ the sum $s_{m,n}$ is numerically less than a constant C for all values of m, n, the double series $\Sigma a_{m,n} v_{m,n}$ converges, provided that the expressions
$$v_{m,n} - v_{m+1,n}, \quad v_{m,n} - v_{m,n+1}, \quad v_{m,n} - v_{m+1,n} - v_{m,n+1} + v_{m+1,n+1}$$
are all positive, and that $v_{m,n}$ tends to zero as either m or n tends to ∞.

In fact, just as in the proof of Abel's lemma (Art. 23), we can shew that under the given conditions for $v_{m,n}$, we have
$$\left|\sum_\mu^M \sum_\nu^N a_{m,n} v_{m,n}\right| < H v_{\mu,\nu}, \quad (M > \mu, N > \nu),$$
where H is an upper limit to
$$\left|\sum_\mu^\xi \sum_\nu^\eta a_{m,n}\right|, \quad (\xi = \mu, \mu+1, \dots, M;\ \eta = \nu, \nu+1, \dots, N).$$

But $$\sum_\mu^\xi \sum_\nu^\eta a_{m,n} = s_{\xi,\eta} - s_{\xi,\nu-1} - s_{\mu-1,\eta} + s_{\mu-1,\nu-1},$$
so that if either μ or ν is 1, $H \leqq 2C$, and otherwise $H \leqq 4C$.

Now $$\sum_1^M \sum_1^N - \sum_1^{\mu-1} \sum_1^{\nu-1} = \sum_1^{\mu-1} \sum_\nu^N + \sum_\mu^M \sum_1^{\nu-1} + \sum_\mu^M \sum_\nu^N,$$
so that $\left|\left(\sum_1^M \sum_1^N - \sum_1^{\mu-1} \sum_1^{\nu-1}\right) a_{m,n} v_{m,n}\right| < 2C(v_{1,\nu} + v_{\mu,1} + 2v_{\mu,\nu}) < 4C(v_{1,\nu} + v_{\mu,1}),$
which can be made as small as we please by proper choice of μ, ν, because $v_{1,\nu}$ and $v_{\mu,1}$ both tend to 0; and so the double series $\Sigma a_{m,n} v_{m,n}$ converges.

An application is given by the series
$$a_{m,n} = \cos(m\theta + n\phi), \quad v_{m,n} = (m\alpha + n\beta)^{-p},$$
where α, β, p are positive.

---

* *Proc. Lond. Math. Soc.*, series 2, vol. 1, 1903, p. 124; vol. 2, 1904, p. 190.

For then $$|s_{m,n}| < 4\,|\operatorname{cosec}\tfrac{1}{2}\theta \operatorname{cosec}\tfrac{1}{2}\phi|$$

and $$v_{m,n} - v_{m+1,n} = \int_m^{m+1} \frac{p\alpha\, dx}{(\alpha x+\beta n)^{p+1}} > 0,$$

$$v_{m,n} - v_{m+1,n} - v_{m,n+1} + v_{m+1,n+1} = \int_m^{m+1} dx \int_n^{n+1} \frac{p(p+1)\alpha\beta\, dy}{(\alpha x+\beta y)^{p+2}} > 0,$$

while $\lim v_{m,n} = 0$.

EXAMPLES.

1. As examples of *double sequences* we take the following:

(1) $s_{m,n} = \dfrac{1}{m} + \dfrac{1}{n}$; here the double and repeated limits exist and are all equal to 0.

(2) $s_{m,n} = (-1)^{m+n}\left(\dfrac{1}{m} + \dfrac{1}{n}\right)$; here the double limit is again 0, but the single and repeated limits do not exist, although we have

$$\lim_{m\to\infty}\left(\overline{\lim_{n\to\infty}}\, s_{m,n}\right) = 0 = \lim_{n\to\infty}\left(\overline{\lim_{m\to\infty}}\, s_{m,n}\right).$$

(3) $s_{m,n} = mn/(m^3+n^3)$; here the double and repeated limits are again all 0.

(4) $s_{m,n} = m/(m+n)$; here the double limit does not exist, but we have

$$0 < s_{m,n} < 1, \text{ and } \underline{\lim}\, s_{m,n} = 0,\ \overline{\lim}\, s_{m,n} = 1,$$

because, however large μ may be, we can find values of m, n greater than μ, such that $s_{m,n} < \epsilon$; and other values of m, n for which $s_{m,n} > 1 - \epsilon$. But the repeated limits exist and are such that

$$\lim_{m\to\infty}\left(\lim_{n\to\infty} s_{m,n}\right) = 0, \quad \lim_{n\to\infty}\left(\lim_{m\to\infty} s_{m,n}\right) = 1.$$

Similar features present themselves in the sequences

$$s_{m,n} = mn/(m^2+n^2) \text{ and } s_{m,n} = 1/[1+(m-n)^2].$$

(5) If $s_{m,n} = (-1)^m\, m^2n^3/(m^3+n^6)$, we have

$$\lim_{(m)}\left(\lim_{(n)} s_{m,n}\right) = 0 = \lim_{(n)}\left(\lim_{(m)} s_{m,n}\right);$$

but yet $$\underline{\lim}\, s_{m,n} = -\infty,\ \overline{\lim}\, s_{m,n} = +\infty;$$

as may be seen by taking $m = n^2$. Here it should be noticed that the limit of the single sequence given by putting $m = n$ exists and is equal to 0; although the double limit does not exist. [PRINGSHEIM.]

2. The double series given by

$$\begin{array}{lllll} (a_0+b_0) + & (a_1-b_0) + a_2 + a_3 + a_4 + \dots \\ (-a_0+b_1) + & (-a_1-b_1) - a_2 - a_3 - a_4 - \dots \\ \quad b_2 \quad - & \quad b_2 \quad + 0 + 0 + 0 + \dots \\ \quad b_3 \quad - & \quad b_3 \quad + 0 + 0 + 0 + \dots \\ \dots\dots\dots\dots & \dots\dots\dots\dots\dots\dots \end{array}$$

gives the sum 0 in Pringsheim's sense, whatever may be the values of a_n, b_n. But the sum by rows is only convergent if Σa_n converges; and the sum by columns converges only if Σb_n is convergent. The sum by diagonals is $\lim(a_n + b_n)$, if this limit exists; and is otherwise oscillatory.

3. In the double series

$$\begin{array}{l} 1+2+4+8+\dots \\ -\tfrac{1}{2}-1-2-4-\dots \\ -\tfrac{1}{4}-\tfrac{1}{2}-1-2-\dots \\ -\tfrac{1}{8}-\tfrac{1}{4}-\tfrac{1}{2}-1-\dots \\ \dots\dots\dots\dots\dots \end{array}$$

every column converges to 0, but every row diverges. Of course Pringsheim's sum cannot exist; and the sum by diagonals is divergent.

4. The series given by

$$\begin{array}{r} 0+1+0+0+0+\dots \\ -1+0+1+0+0+\dots \\ 0-1+0+1+0+\dots \\ 0+0-1+0+1+\dots \\ 0+0+0-1+0+\dots \\ \dots\dots\dots\dots\dots \end{array}$$

has the sum 1 by rows; -1 by columns; 0 by diagonals; and naturally the double series cannot converge in Pringsheim's sense. In fact, $s_{m,n}$ is 0 if $m=n$, and is -1 if $m>n$, or $+1$ if $m<n$.

5. The series given by

$$\begin{array}{r} -2+1+0+0+0+\dots \\ +1-2+1+0+0+\dots \\ 0+1-2+1+0+\dots \\ 0+0+1-2+1+\dots \\ 0+0+0+1-2+\dots \\ \dots\dots\dots\dots\dots \end{array}$$

has the sum -1 both by rows and columns; and the diagonal sum oscillates between -2 and 0. There is no sum in Pringsheim's sense, because $s_{m,n}$ is -1 if $m=n$, and is otherwise 0.

6. The double series

$$\begin{array}{r} 2+0-1+0+0+0+\dots \\ 0+2+0-1+0+0+\dots \\ -1+0+2+0-1+0+\dots \\ 0-1+0+2+0-1+\dots \\ 0+0-1+0+2+0+\dots \\ 0+0+0-1+0+2+\dots \\ +\dots\dots\dots\dots\dots \end{array}$$

has the sum by rows $1+1+0+0+\dots=2$, and the same sum by columns; the sum by diagonals is $2+0+0+0+\dots=2$. Thus *these three sums are the same, but the series does not converge in Pringsheim's sense, since* $a_{n,n}=2$.

7. Prove that the multiplication rule for $\Sigma a_n x^n$, $\Sigma b_n x^n$ can be established by summing the double series

$$\begin{array}{r} a_0b_0+a_0b_1x+a_0b_2x^2+\dots \\ +a_1b_0x+a_1b_1x^2+\dots \\ +a_2b_0x^2+\dots \\ +\dots \end{array}$$

first by rows and secondly by columns.

8. Discuss the following paradox:

If we sum the double series of positive terms

$$\begin{aligned}\frac{1}{1.2}+\frac{1}{2.3}+\frac{1}{3.4}+\frac{1}{4.5}+\ldots\\ +\frac{1}{2.3}+\frac{1}{3.4}+\frac{1}{4.5}+\ldots\\ +\frac{1}{3.4}+\frac{1}{4.5}+\ldots\\ +\frac{1}{4.5}+\ldots\\ +\ldots\end{aligned}$$

first by rows and secondly by columns, we obtain $s+1=s$ or $1=0$, where $s=\frac{1}{2}+\frac{1}{3}+\frac{1}{4}+\ldots$. [J. BERNOULLI.]

9. If the double series $\Sigma\Sigma a_{m,n}$ is convergent in Pringsheim's sense, it does not follow (in contrast to the case of single series) that a constant C can be found such that $|s_{m,n}|<C$ for all values of m, n; this is seen by considering the series of Ex. 2, and supposing Σa_n, Σb_n to be divergent.

In like manner we cannot infer $|s_{m,n}|<C$ from the convergence of the sum by columns or by rows (see for instance Ex. 3).

10. The double series in which $a_{m,n}=(-1)^{m+n}/mn$ does *not* converge absolutely; but yet its sums by rows, columns and diagonals are equal to one another and to Pringsheim's sum. The common value is, in fact, $(\log 2)^2$.

Exactly similar results apply to the series in which $a_{m,n}=(-1)^{m+n}u_m v_n$, where the sequences (u_m), (v_n) steadily decrease to zero.

11. Consider the double series in which

$$a_{m,n}=1/(m^2-n^2) \qquad (m \gtrless n)$$

and

$$a_{m,n}=0. \qquad (m=n)$$

Here we find

$$\sum_{(m)}\sum_{(n)} = -\tfrac{3}{4}\Sigma(1/m^2)$$

and

$$\sum_{(n)}\sum_{(m)} = +\tfrac{3}{4}\Sigma(1/n^2).$$

(Ex. 13, Chap. III.) [HARDY.]

12. Prove that

$$\sum_{m=-\nu}^{\nu}\sum_{n=-\nu}^{\nu}\frac{2x+m+n}{(x+m)^2(x+n)^2}$$

tends to zero when ν tends to ∞, provided that all terms for which $m=n$ are omitted from the summation. [*Math. Trip.* 1895.]

13. If

$$a_{m,n}=\frac{m-n}{2^{m+n}}\frac{(m+n-1)!}{m!\,n!} \qquad (m, n>0)$$

and

$$a_{m,0}=2^{-m}, \quad a_{0,n}=-2^{-n}, \quad a_{0,0}=0,$$

then

$$\sum_{m=0}^{\infty}\left(\sum_{n=0}^{\infty}a_{m,n}\right)=-1, \quad \sum_{n=0}^{\infty}\left(\sum_{m=0}^{\infty}a_{m,n}\right)=+1.$$

14. If $a_{m,n}=(-1)^{m+n}mn/(m+n)^2$, we have

$$\sum_{m=0}^{\infty}\left(\sum_{n=0}^{\infty}a_{m,n}\right)=\sum_{n=0}^{\infty}\left(\sum_{m=0}^{\infty}a_{m,n}\right)=\tfrac{1}{8}(\log 2-\tfrac{1}{4})=l,$$

but the sum by diagonals oscillates between $-\infty$ and ∞; and Pringsheim's sum oscillates between $l-\frac{1}{16}$ and $l+\frac{1}{16}$.

[For the details of Exs. 2, 3, 9, 13, 14, see Bromwich and Hardy, *Proc. Lond. Math. Soc.* (2), vol. 2, 1904, p. 175.]

15. Prove that the product of the two series

$$1+\frac{x}{1^2}+\frac{x^2}{(2!)^2}+\frac{x^3}{(3!)^2}+\dots,\quad 1-\frac{x}{1^2}+\frac{x^2}{(2!)^2}-\frac{x^3}{(3!)^2}+\dots$$

is equal to

$$1-\frac{x^2}{1^2.2!}+\frac{x^4}{(2!)^2.4!}-\frac{x^6}{(3!)^2.6!}+\dots.$$

16. If $f(x,q)=1+q(x+1/x)+q^4(x^2+1/x^2)+q^9(x^3+1/x^3)+\dots$

$$=\sum_{-\infty}^{\infty}q^{n^2}x^n,\quad \text{where } |q|<1,$$

and $\qquad g(x,q)=q^{\frac{1}{4}}x^{-\frac{1}{2}}f(x/q,q)=\sum_{-\infty}^{\infty}q^{\nu^2}x^{\nu},\quad \nu=n-\tfrac{1}{2},$

then
$$f(xy,q).f(x/y,q)=f(x^2,q^2).f(y^2,q^2)+g(x^2,q^2).g(y^2,q^2),$$
$$g(xy,q).g(x/y,q)=f(x^2,q^2).g(y^2,q^2)+f(y^2,q^2).g(x^2,q^2).$$

17. Verify that

$$\frac{1}{x(x+1)\dots(x+n)}$$
$$=\frac{1}{n!}\frac{1}{x}-\frac{1}{1!(n-1)!}\frac{1}{x+1}+\frac{1}{2!(n-2)!}\frac{1}{x+2}-\frac{1}{3!(n-3)!}\frac{1}{x+3}+\dots\pm\frac{1}{n!}\frac{1}{x+n},$$

and use Arts. 33, 57 to infer Prym's identity,

$$\frac{1}{x}+\frac{1}{x(x+1)}+\frac{1}{x(x+1)(x+2)}+\dots=e\left[\frac{1}{x}-\frac{1}{1!}\frac{1}{x+1}+\frac{1}{2!}\frac{1}{x+2}-\frac{1}{3!}\frac{1}{x+3}+\dots\right].$$

18. Shew that $\dfrac{1}{t^2}=\dfrac{1}{t(t+1)}+\dfrac{1}{t(t+1)(t+2)}+\dfrac{1.2}{t(t+1)(t+2)(t+3)}+\dots.$

Hence convert $\dfrac{1}{t^2}+\dfrac{1}{(t+1)^2}+\dfrac{1}{(t+2)^2}+\dots$

into a double series, and transform it to

$$\frac{1}{t}+\frac{1}{2t(t+1)}+\frac{1.2}{3t(t+1)(t+2)}+\frac{1.2.3}{4t(t+1)(t+2)(t+3)}+\dots.$$

Take $t=10$ and so calculate $\Sigma 1/n^2$ to 7 decimal places. [STIRLING.]

19. Convert the series $\dfrac{x}{1+x^2}+\dfrac{x^2}{1+x^4}+\dfrac{x^3}{1+x^6}+\dots\quad(|x|<1)$

into a double series, and deduce that it is equal to

$$\frac{x}{1-x}-\frac{x^3}{1-x^3}+\frac{x^5}{1-x^5}+\dots.$$

20. Shew that (if $|x|<1$) Lambert's series

$$\frac{x}{1-x}+\frac{x^2}{1-x^2}+\frac{x^3}{1-x^3}+\ldots=\sum_1^\infty\sum_1^\infty x^{mn},$$

and deduce that this series is equal to Clausen's series,

$$x\frac{1+x}{1-x}+x^4\frac{1+x^2}{1-x^2}+x^9\frac{1+x^3}{1-x^3}+x^{16}\frac{1+x^4}{1-x^4}+\ldots.$$

Hence evaluate Lambert's series to five decimal places, for $x=\frac{1}{10}$.

Shew that each of these series is also equal to

$$x+2x^2+2x^3+3x^4+2x^5+4x^6+\ldots,$$

the coefficient of x^n being the number of divisors of n (1 and n included).

21. From Ex. 19, or directly, prove that, if $|x|<1$,

$$\frac{x}{1+x^2}+\frac{x^3}{1+x^6}+\frac{x^5}{1+x^{10}}+\ldots=\frac{x}{1-x^2}-\frac{x^3}{1-x^6}+\frac{x^5}{1-x^{10}}-\ldots,$$

$$\frac{x}{1+x^2}-\frac{x^2}{1+x^4}+\frac{x^3}{1+x^6}-\ldots=\frac{x}{1+x}-\frac{x^3}{1+x^3}+\frac{x^5}{1-x^5}-\ldots.$$

22. Shew that, if $|x|<1$,

$$\frac{x}{1+x}-\frac{2x^2}{1+x^2}+\frac{3x^3}{1+x^3}-\ldots=\frac{x}{(1+x)^2}-\frac{x^2}{(1+x^2)^2}+\frac{x^3}{(1+x^3)^2}-\ldots,$$

$$\frac{x}{1-x^2}+\frac{3x^3}{1-x^6}+\frac{5x^5}{1-x^{10}}+\ldots=\frac{x(1+x^2)}{(1-x^2)^2}+\frac{x^3(1+x^6)}{(1-x^6)^2}+\frac{x^5(1+x^{10})}{(1-x^{10})^2}+\ldots.$$

[For the connexion between the series in Exs. 19-22 and elliptic functions, see Jacobi, *Fundamenta Nova*, § 40.]

23. If $|x|<1$, shew that

$$\frac{x}{(1-x)^2}+\frac{x^2}{(1-x^2)^2}+\frac{x^3}{(1-x^3)^2}+\ldots=\sum_1^\infty\phi_n x^n,$$

where ϕ_n is the sum of the divisors of n (including 1 and n). Deduce that, if $\phi_{-1}=0=\phi_0$,

$$1+\frac{x}{(1+x)^2}+\frac{x^2}{(1+x+x^2)^2}+\ldots=\sum_0^\infty(\phi_{n+1}+\phi_{n-1}-2\phi_n)x^n.$$

[*Math. Trip.* 1899.]

24. If $|x|<1$, prove that

$$f(1)\frac{x}{1-x}+f(2)\frac{x^2}{1-x^2}+f(3)\frac{x^3}{1-x^3}+\ldots=\sum_1^\infty\theta(n)x^n,$$

where $\theta(n)$ denotes the sum $\Sigma f(d)$ for all the divisors of n (including 1 and n).

In particular $\frac{x}{1-x}-\frac{x^2}{1-x^2}+\frac{x^3}{1-x^3}+\frac{x^4}{1-x^4}-\frac{x^5}{1-x^5}+\frac{x^6}{1-x^6}-\ldots=\sum_1^\infty x^{n^2}.$

[LAGUERRE.]

25. Shew that in the special series of Art. 37, the repeated series also converge to the same sum as the double series; but the diagonal series may oscillate, for instance $\alpha=\beta=1$, $\theta=\phi=\pi$, $p=1$, gives for the diagonal series

$$\tfrac{1}{2}-\tfrac{2}{3}+\tfrac{3}{4}-\tfrac{4}{5}+\ldots.$$

[HARDY.]

CHAPTER VI.

INFINITE PRODUCTS.

38. Weierstrass's inequalities.

In this article the numbers a_1, a_2, a_3, ... are supposed to be positive and less than 1; this being the case, we see that

$$(1+a_1)(1+a_2)=1+(a_1+a_2)+a_1a_2>1+(a_1+a_2).$$

Hence

$$(1+a_1)(1+a_2)(1+a_3)>[1+(a_1+a_2)](1+a_3)>1+(a_1+a_2+a_3),$$

and continuing this process we see that

$$(1)\quad (1+a_1)(1+a_2)(1+a_3)\dots(1+a_n)>1+(a_1+a_2+a_3+\dots+a_n).$$

In like manner we have

$$(1-a_1)(1-a_2)=1-(a_1+a_2)+a_1a_2>1-(a_1+a_2).$$

Thus, since $1-a_3$ is positive, we have

$$(1-a_1)(1-a_2)(1-a_3)>[1-(a_1+a_2)](1-a_3)>1-(a_1+a_2+a_3),$$

and so we have, generally,

$$(2)\quad (1-a_1)(1-a_2)(1-a_3)\dots(1-a_n)>1-(a_1+a_2+a_3+\dots+a_n).$$

Next,
$$1+a_1=\frac{1-a_1^2}{1-a_1}<\frac{1}{1-a_1},$$

so that $(1+a_1)(1+a_2)\dots(1+a_n)<\dfrac{1}{(1-a_1)(1-a_2)\dots(1-a_n)}$,

and thus, *if** $a_1+a_2+\dots+a_n$ *is less than* 1, we have, by the aid of (2), the result

$$(3)\quad (1+a_1)(1+a_2)\dots(1+a_n)<[1-(a_1+a_2+\dots+a_n)]^{-1}.$$

* If $a_1+a_2+\dots+a_n$ were greater than 1, the inequality (3) would be untrue, since it would then make a positive number less than a negative number.

Similarly, we find

(4) $$(1-a_1)(1-a_2)\dots(1-a_n)<[1+(a_1+a_2+\dots+a_n)]^{-1}.$$

By combining these four inequalities, we find the results

(5) $$(1-\Sigma a)^{-1}>\Pi(1+a)>1+\Sigma a,$$

(6) $$(1+\Sigma a)^{-1}>\Pi(1-a)>1-\Sigma a,$$

where all the letters a denote numbers between 0 *and* 1, *such that* Σa *is less than* 1.

39. If $a_1, a_2, a_3, \dots$ are numbers between 0 and 1, the convergence of the series Σa_n is necessary and sufficient for the convergence of the products P_n, Q_n to positive limits P, Q as n increases to ∞, where

$$P_n=(1+a_1)(1+a_2)\dots(1+a_n),\quad Q_n=(1-a_1)(1-a_2)\dots(1-a_n).$$

For clearly P_n increases as n increases, and Q_n decreases.

Now, if Σa_n is convergent, we can find a number m such that

$$\sigma=a_{m+1}+a_{m+2}+a_{m+3}+\dots \text{ to } \infty<1.$$

Then by the inequalities (5), (6) of the last article, we have

$$\frac{1}{1-\sigma}>\frac{1}{1-(a_{m+1}+a_{m+2}+\dots+a_n)}>\frac{P_n}{P_m},$$

and

$$\frac{Q_n}{Q_m}>1-(a_{m+1}+a_{m+2}+\dots+a_n)>1-\sigma.$$

Hence, provided that n is greater than m, we have

$$P_n<P_m/(1-\sigma),$$

and

$$Q_n>Q_m(1-\sigma).$$

Thus, by Art. 2, P_n and Q_n approach definite finite limits P, Q, such that

$$P\leqq P_m/(1-\sigma),\quad Q\geqq Q_m(1-\sigma).$$

But, if Σa_n is divergent, we can find m so that

$$a_1+a_2+\dots+a_n>N,\quad \text{if } n>m,$$

no matter how large N may be.

Hence, by the same inequalities,

$$P_n>1+N,\quad Q_n<1/(1+N),\quad \text{if } n>m,$$

and consequently $\lim P_n=\infty$, $\lim Q_n=0$.

It should be observed that *if a product tends to zero as a limit, without any of its factors being zero, the product is said to diverge.* This of course is merely a convention; but it preserves the parallelism with the theory of series.

Ex. 1. Since $\Sigma 1/n^2$ converges, the product

$$\left(1-\frac{1}{2^2}\right)\left(1-\frac{1}{3^2}\right)\left(1-\frac{1}{4^2}\right)\ldots$$

will approach a limit *different from zero.* This is at once obvious because

$$1-\frac{1}{n^2}=\frac{(n-1)(n+1)}{n^2},$$

and so we find the product

$$Q_{n-1}=\frac{1.3}{2^2}\cdot\frac{2.4}{3^2}\cdot\frac{3.5}{4^2}\cdot\frac{4.6}{5^2}\ldots\frac{(n-1)(n+1)}{n^2}=\frac{1}{2}\frac{n+1}{n},$$

so that $\lim Q_n=\frac{1}{2}$.

Similarly $\left(1+\frac{1}{2^2}\right)\left(1+\frac{1}{3^2}\right)\left(1+\frac{1}{4^2}\right)\ldots$ is convergent, although its value is not calculated so readily.

Ex. 2. Since $\Sigma 1/n$ is divergent, the products

$$(1+\tfrac{1}{2})(1+\tfrac{1}{3})(1+\tfrac{1}{4})\ldots,\quad (1-\tfrac{1}{2})(1-\tfrac{1}{3})(1-\tfrac{1}{4})\ldots$$

will diverge also.

In fact $P_{n-1}=\frac{3}{2}\cdot\frac{4}{3}\cdot\frac{5}{4}\ldots\frac{n+1}{n}=\frac{n+1}{2}$, $Q_{n-1}=\frac{1}{2}\cdot\frac{2}{3}\cdot\frac{3}{4}\ldots\frac{n-1}{n}=\frac{1}{n}$,

so that $\lim P_n=\infty$, $\lim Q_n=0$.

Ex. 3. *If* $a_n/a_{n+1}=1+(b_n/n)$, *where* $\lim b_n=b>0$, *then* $\lim a_n=0$.

For, under the given circumstances, we can find an index m, such that

$$b_n>\tfrac{1}{2}b>0 \quad \text{if } n\geqq m.$$

Thus we have

$$\frac{a_m}{a_{m+1}}>1+\frac{b}{2m},\quad \frac{a_{m+1}}{a_{m+2}}>1+\frac{b}{2(m+1)},\ \ldots,\ \frac{a_n}{a_{n+1}}>1+\frac{b}{2n},$$

and therefore

$$\frac{a_m}{a_{n+1}}>\left(1+\frac{b}{2m}\right)\ldots\left(1+\frac{b}{2n}\right)>1+\frac{b}{2}\left(\frac{1}{m}+\frac{1}{m+1}+\ldots+\frac{1}{n}\right).$$

Hence $\lim\limits_{n\to\infty}(a_m/a_n)=\infty$,

so that $\lim a_n=0$.

It is easy to see that the argument of Art. 26 can be modified to prove that *when* $a_1, a_2, a_3 \ldots$ *are between* 0 *and* 1, *the values of the two infinite products*

$$P=(1+a_1)(1+a_2)(1+a_3)\ldots,\quad Q=(1-a_1)(1-a_2)(1-a_3)\ldots$$

are both independent of the order of the factors.

For if, in any new arrangement (represented by accents), we have to take p factors to include the first n-factors of P and Q, then

$$P \geqq P_r' > P_n, \quad Q \leqq Q_r' < Q_n, \quad \text{if } r \geqq p.$$

Now n can be taken large enough to bring P_n and Q_n as close to P and Q, respectively, as we please.

Consequently $\lim_{r \to \infty} P_r' = P, \quad \lim_{r \to \infty} Q_r' = Q.$

In like manner, if P diverges to ∞, so does P'; and if Q diverges to 0, so does Q'.

By taking logarithms we can see at once that the present theorem is deducible from the theorem of Art. 26.

40. Convergence of infinite products in general.

It is quite possible that in an infinite product

$$(1+u_1)(1+u_2)(1+u_3)\ldots,$$

the numbers $u_1, u_2, u_3, \ldots$ may have both signs. But without loss of generality it may be supposed that they are all numerically less than 1; for there can only be a finite number of them greater than 1 (otherwise the product would certainly diverge or oscillate), and the corresponding factors can be omitted without affecting the convergence.

Now we have*

$$0 < u - \log(1+u) < \tfrac{1}{2}u^2 \qquad \text{if } u \text{ is positive,}$$

$$\text{or } < \tfrac{1}{2}u^2/(1+u) \text{ if } 0 > u > -1.$$

Thus, if λ is the lower limit of the numbers

$$1,\ 1+u_1,\ 1+u_2, \ldots,\ 1+u_n, \ldots,$$

we have

$$0 < (u_{m+1} + u_{m+2} + \ldots + u_n) - \log[(1+u_{m+1})(1+u_{m+2})\ldots(1+u_n)]$$
$$< \tfrac{1}{2}(u_{m+1}^2 + u_{m+2}^2 + \ldots + u_n^2)/\lambda.$$

---

*For $\log(1+u) = \int_0^u \frac{dx}{1+x},$

so that $u - \log(1+u) = \int_0^u \left(1 - \frac{1}{1+x}\right) dx = \int_0^u \frac{x\,dx}{1+x}.$

Hence, if u is positive $\int_0^u \frac{x\,dx}{1+u} < u - \log(1+u) < \int_0^u x\,dx,$

but, if u is negative $\int_0^u x\,dx < u - \log(1+u) < \int_0^u \frac{x\,dx}{1+u}.$

Consequently, *if the series* Σu_n^2 *is convergent*, the difference

$$(u_{m+1}+u_{m+2}+\ldots+u_n)-\log[(1+u_{m+1})(1+u_{m+2})\ldots(1+u_n)]$$

can be made arbitrarily small by properly choosing m, no matter how large n is. Thus we have the theorem:

If the series Σu_n^2 *is convergent, the infinite product*

$$(1+u_1)(1+u_2)(1+u_3)\ldots$$

converges if Σu_n *converges; diverges to* ∞ *if* Σu_n *diverges to* $+\infty$; *diverges to* 0 *if* Σu_n *diverges to* $-\infty$; *oscillates if* Σu_n *oscillates.*

Again* we have

$$u-\log(1+u)>\tfrac{1}{2}u^2/(1+u) \quad \text{if } u \text{ is positive,}$$

$$\text{or } >\tfrac{1}{2}u^2 \quad \text{if } 0>u>-1,$$

so that

$$(u_{m+1}+u_{m+2}+\ldots+u_n)-\log[(1+u_{m+1})(1+u_{m+2})\ldots(1+u_n)]$$
$$>\tfrac{1}{2}(u_{m+1}^2+u_{m+2}^2+\ldots+u_n^2)/L,$$

where L is the upper limit of 1, $(1+u_1)$, $(1+u_2)$, ..., $(1+u_n)$, ...

Hence, *if* Σu_n *converges*† (*or oscillates so that its maximum limit is not* $+\infty$) *while* Σu_n^2 *diverges, the infinite product*

$$(1+u_1)(1+u_2)(1+u_3)\ldots$$

diverges to the value 0.

The only cases not covered by the foregoing method are those in which Σu_n^2 diverges and Σu_n either diverges to $+\infty$, or has $+\infty$ as its maximum limit (in case of oscillation).

It is, perhaps, a little perplexing at first sight that when Σu_n, Σu_n^2 both diverge to ∞, the product may nevertheless converge; but it is quite easy to construct a product of this type. For, let Σc_n be a convergent, Σd_n a divergent, series of positive terms, and form the product of which the $(2n-1)$th and $2n$th terms are given by

$$1+u_{2n-1}=1+d_n, \quad 1+u_{2n}=\frac{1+c_n}{1+d_n}.$$

Then, provided that $\lim d_n=0$, the product $\Pi(1+u_n)=\Pi(1+c_n)$ and so obviously converges (by Art. 39). Further Σu_n will diverge if $\Sigma\left(d_n-1+\frac{1+c_n}{1+d_n}\right)=\Sigma\left(\frac{d_n^2+c_n}{1+d_n}\right)$ is divergent; and then

* See footnote on previous page.

† Of course not absolutely (see Art. 41).

Σu_n^2 must also diverge.* This condition can be satisfied in many ways; one simple method is to take $c_n = d_n^3$, and then we must suppose Σd_n, Σd_n^2 both divergent, and Σd_n^3 convergent; for instance, we may take $d_n = n^{-p}$, where $\frac{1}{2} \geqq p > \frac{1}{3}$. The product is then given by $u_{2n-1} = n^{-p}$, $u_{2n} = -n^{-p} + n^{-2p}$.

Ex. 1. Since the series

$$\frac{1}{2} - \frac{1}{3} + \frac{1}{4} - \frac{1}{5} + \ldots \text{ and } \frac{1}{2^2} + \frac{1}{3^2} + \frac{1}{4^2} + \frac{1}{5^2} + \ldots$$

are both convergent, the two infinite products

$$(1 + \tfrac{1}{2})(1 - \tfrac{1}{3})(1 + \tfrac{1}{4})(1 - \tfrac{1}{5}) \ldots \text{ and } (1 - \tfrac{1}{2})(1 + \tfrac{1}{3})(1 - \tfrac{1}{4})(1 + \tfrac{1}{5}) \ldots$$

converge also. In fact the first is obviously equal to 1 and the second to $\frac{1}{2}$.

Ex. 2. Since the series $\frac{1}{2} + \frac{1}{3} - \frac{1}{4} + \frac{1}{5} + \frac{1}{6} - \frac{1}{7} + \ldots$ diverges in virtue of Cesàro's theorem (Art. 22) it follows that the infinite products

$$(1 + \tfrac{1}{2})(1 + \tfrac{1}{3})(1 - \tfrac{1}{4})(1 + \tfrac{1}{5})(1 + \tfrac{1}{6})(1 - \tfrac{1}{7}) \ldots \text{ and } (1 - \tfrac{1}{2})(1 - \tfrac{1}{3})(1 + \tfrac{1}{4})(1 - \tfrac{1}{5})(1 - \tfrac{1}{6})(1 + \tfrac{1}{7}) \ldots$$

are both divergent. In fact the first diverges to ∞ and the second to 0: for they are equivalent to the products

$$(1 + \tfrac{1}{2})(1 + \tfrac{1}{5})(1 + \tfrac{1}{8})(1 + \tfrac{1}{11}) \ldots \text{ and } (1 - \tfrac{1}{2})(1 - \tfrac{1}{3})(1 - \tfrac{1}{6})(1 - \tfrac{1}{9})(1 - \tfrac{1}{12}) \ldots.$$

Ex. 3. Since the series

$$\frac{1}{\sqrt{2}} - \frac{1}{\sqrt{3}} + \frac{1}{\sqrt{4}} - \frac{1}{\sqrt{5}} + \ldots$$

is convergent, but $\frac{1}{2} + \frac{1}{3} + \frac{1}{4} + \frac{1}{5} + \ldots$ is divergent, it is clear that the two products

$$\left(1 + \frac{1}{\sqrt{2}}\right)\left(1 - \frac{1}{\sqrt{3}}\right)\left(1 + \frac{1}{\sqrt{4}}\right)\left(1 - \frac{1}{\sqrt{5}}\right) \ldots$$

and

$$\left(1 - \frac{1}{\sqrt{2}}\right)\left(1 + \frac{1}{\sqrt{3}}\right)\left(1 - \frac{1}{\sqrt{4}}\right)\left(1 + \frac{1}{\sqrt{5}}\right) \ldots$$

both diverge to the value 0.

In fact

$$\left(1 + \frac{1}{\sqrt{n}}\right)\left[1 - \frac{1}{\sqrt{(n+1)}}\right] = 1 - \frac{1}{\sqrt{[n(n+1)]}}\left[1 - \frac{1}{\sqrt{n} + \sqrt{(n+1)}}\right],$$

so that this product is always less than 1, and can be put in the form $(1 - a_n)$. Further $\lim (na_n) = 1$, so that our two products diverge to 0 by Art. 39.

Ex. 4. If $u_n = (-1)^n \frac{1}{2}$, it is evident that Σu_n oscillates, while Σu_n^2 diverges; thus the product

$$(1 - \tfrac{1}{2})(1 + \tfrac{1}{2})(1 - \tfrac{1}{2})(1 + \tfrac{1}{2}) \ldots$$

must diverge to 0, which may be verified by inspection.

---

* If Σu_n^2 were convergent, the divergence of Σu_n would imply the divergence of the product.

41. Absolute convergence of an infinite product.

If the series Σu_n is absolutely convergent, the infinite product $\Pi(1+u_n)$ converges to a sum which is independent of the order of the factors.

Write $|u_n|=a_n$, then Σa_n converges; and, as explained in the last article, we can suppose $a_n<1$, so that $a_n^2<a_n$. Hence Σa_n^2 is also convergent (Art. 8), that is, Σu_n^2 converges; and Σu_n is convergent, because Σa_n converges: and therefore $\Pi(1+u_n)$ is convergent in virtue of a theorem proved in the last article.

Suppose next that the series Σu_n is deranged so as to become Σv_n while Σa_n becomes Σb_n; and write for brevity

$$U_n=(1+u_1)(1+u_2)\dots(1+u_n),\quad V_n=(1+v_1)(1+v_2)\dots(1+v_n)$$
$$A_n=(1+a_1)(1+a_2)\dots(1+a_n),\quad B_n=(1+b_1)(1+b_2)\dots(1+b_n).$$

Then suppose p chosen $(>n)$ so that U_p contains the whole of V_n (and consequently A_p contains the whole of B_n); on multiplying out it is evident that A_p/B_n-1 contains every term in U_p/V_n-1, but with the signs made positive.

Hence $$|U_p/V_n-1|\leqq A_p/B_n-1,$$
and $$V_n\leqq B_n,$$
so that $$|U_p-V_n|\leqq A_p-B_n.$$

Now, as explained in Art. 39, $\lim B_n=\lim A_n=A$ say. Consequently n_0 can be found so great that

$$A>A_p>B_n>A-\tfrac{1}{2}\epsilon,\quad \text{if } p>n>n_0.$$
Hence $$A_p-B_n<\tfrac{1}{2}\epsilon,\quad \text{if } n>n_0,$$
and so $$|U_p-V_n|<\tfrac{1}{2}\epsilon,\quad \text{if } n>n_0.$$

But U_p approaches a limit U as p tends to ∞: and therefore if $p>n>n_1$ we have $|U-U_p|<\frac{1}{2}\epsilon$.

Thus, if n' is the greater of n_0 and n_1, we have

$$|U-V_n|<\epsilon,\quad \text{if } n>n',$$
that is $$\lim V_n=U.$$

Alteration in the value of a non-absolutely convergent infinite product by deranging the factors.

The argument used to establish Riemann's theorem (Art. 28) requires but little change to shew that a non-absolutely convergent infinite product may be made to converge to any value, or to diverge, or to oscillate, by altering the order of the factors.

Perhaps the case of chief interest is that afforded by the infinite product $\Pi[1+(-1)^{n-1}a_n]$, where a_n is positive and $\lim(na_n)=g$. Suppose that the value of the product is P when the positive and negative terms occur alternately; and let its value be X when the limit of the ratio of the number of positive to the number of negative terms is k.

Then $$X/P=\lim_{n\to\infty}(1+a_{2n+1})(1+a_{2n+3})\dots(1+a_{2\nu-1}),$$
where $$\lim(\nu/n)=k.$$

Now it is plain that Σa_n^2 is convergent, and therefore, by the last article, it is clear that
$$(a_{2n+1}+a_{2n+3}+\dots+a_{2\nu-1})-\log[(1+a_{2n+1})(1+a_{2n+3})\dots(1+a_{2\nu-1})]$$
can be made arbitrarily small by taking n large enough. Further, by Pringsheim's method (Art. 28), it is clear that
$$\lim(a_{2n+1}+a_{2n+3}+\dots+a_{2\nu-1})=\tfrac{1}{2}g\log k,$$
and therefore $$\log(X/P)=\tfrac{1}{2}g\log k,$$
or $$X/P=k^{\frac{1}{2}g}.$$

42. The Gamma-product.

It is evident from the foregoing articles (39, 40) that the product
$$P_n=\left(1+\frac{x}{1}\right)\left(1+\frac{x}{2}\right)\left(1+\frac{x}{3}\right)\dots\left(1+\frac{x}{n}\right),\quad (x>-1),$$
is divergent except for $x=0$. But we have
$$\frac{x}{n}-\log\frac{P_n}{P_{n-1}}=\frac{x}{n}-\log\left(1+\frac{x}{n}\right)>0,$$
so that the expression
$$S_n=x\left(1+\frac{1}{2}+\frac{1}{3}+\dots+\frac{1}{n}\right)-\log P_n$$
increases with n. Also, as in Art. 40, we have
$$S_n<\frac{x^2}{2\lambda}\left(1+\frac{1}{2^2}+\frac{1}{3^2}+\dots+\frac{1}{n^2}\right)<\frac{x^2}{\lambda},\qquad \text{[Art. 11 (1)]},$$
where λ is either 1, if x is positive, or $1+x$, if x is negative. Hence, by Art. 2, S_n approaches a definite limit S as n increases to ∞.

Now (Art. 11) $$1+\frac{1}{2}+\frac{1}{3}+\dots+\frac{1}{n}-\log n$$
approaches a definite limit C, and therefore
$$\lim(x\log n-\log P_n)=\lim\left[S_n-x\left(1+\frac{1}{2}+\dots+\frac{1}{n}-\log n\right)\right]=S-Cx.$$

Now $$x\log n-\log P_n=\log(n^x/P_n),$$
so that n^x/P_n has also a definite limit; this limit is denoted by $\Pi(x)$ in Gauss's notation.

Thus $$\Pi(x)=\lim_{n\to\infty}\frac{n^x\,.\,n!}{(1+x)(2+x)\dots(n+x)},$$
which, again, can be written in Weierstrass's form,
$$1/\Pi(x)=e^{Cx-S}=e^{Cx}\lim_{n\to\infty}e^{-S_n}=e^{Cx}\prod_{r=1}^{\infty}\left(1+\frac{x}{r}\right)e^{-\frac{x}{r}}.$$

When x is a positive integer, Gauss's form gives $\Pi(x)=x!$, because
$$\frac{n^x\,.\,n!}{(1+x)(2+x)\dots(n+x)}=\frac{n^x\,.\,x!}{(1+n)(2+n)\dots(x+n)}=x!\Big/\left(1+\frac{1}{n}\right)\left(1+\frac{2}{n}\right)\dots\left(1+\frac{x}{n}\right).$$

Although we have found it convenient to restrict $(1+x)$ to be positive, yet this is not necessary for convergence; and it is easy to see that the products for $\Pi(x)$ still converge if x has any negative value which is not an integer.*

It is easy to verify by integration by parts that Euler's integral

$$\Gamma(1+x)=\int_0^\infty e^{-t}t^x dt = x\int_0^\infty e^{-t}t^{x-1}dt$$

has the property of being equal to $x!$ when x is an integer; and we may therefore anticipate the equation

$$\Gamma(1+x)=\Pi(x),$$

which will be proved to be correct in Art. 178 of the Appendix.

If we change x to $x+1$ in the definition of $\Pi(x)$, we find

$$\Pi(x+1)=\lim_{n\to\infty}\frac{n^{x+1}.n!}{(2+x)(3+x)\ldots(n+1+x)}.$$

Thus
$$\frac{\Pi(x+1)}{\Pi(x)}=(x+1)\lim_{n\to\infty}\frac{n}{n+1+x}=x+1,$$

or
$$\Pi(x+1)=(x+1)\Pi(x).$$

It follows that $(1+x)(2+x)\ldots(n+x)=\Pi(n+x)/\Pi(x)$, and consequently the definition leads to the equation

$$\Pi(x)=\lim_{n\to\infty}\frac{n^x.\Pi(n)\Pi(x)}{\Pi(n+x)},$$

or
$$1=\lim_{n\to\infty}\frac{n^x\Pi(n)}{\Pi(n+x)}.$$

It is often convenient to write the last equation in the form

$$\Pi(n+x)\sim n^x\Pi(n),$$

using the notation explained in Art. 3.

By reversing the foregoing argument we see that *the function* $\Pi(x)$ *is completely defined by the properties*

$$\Pi(x+1)=(x+1)\Pi(x),\quad \Pi(n+x)\sim n^x\Pi(n)$$

together with the condition $\Pi(0)=1$.

EXAMPLES.

1. Discuss the convergence of

$$\Pi[1+f(n)x^n],\ \Pi\left[\left(1-\frac{1}{n}\right)^x\left(1+\frac{x}{n}\right)\right],\ \Pi\left(\frac{x+x^{2n}}{1+x^{2n}}\right),$$

where $f(n)$ is a polynomial in n.

2. Prove that
$$\Pi\left[\left(1-\frac{x}{c+n}\right)e^{x/n}\right]\quad (n=1, 2, 3, \ldots)$$

converges absolutely for any value of x, provided that c is not a negative integer; and that

$$\Pi\left[1-\left(\frac{nx}{n+1}\right)^n\right]$$

is absolutely convergent if $|x|<1$.

---

* The convergence persists also for complex values of x (see Ex. 7, p. 255).

3. If
$$w_n = \frac{1}{m}\frac{1+a/n+b/n^2+\dots}{1+c/n+d/n^2+\dots},$$
then Πw_n diverges to 0 if $m>1$; to ∞ if $m<1$.

If $m=1$, Πw_n diverges to 0 if $a<c$, and to ∞ if $a>c$; and converges if $a=c$. [STIRLING.]

4. If $u_1=0$, $u_2=0$, $u_{2n-1}=-n^{-p}$, $u_{2n}=n^{-p}+n^{-2p}$, where $n>1$ and $\frac{1}{2}\geqq p>\frac{1}{3}$, then Σu_n, Σu_n^2 are both divergent, but $\Pi(1+u_n)$ is convergent. Verify that the same is true if
$$u_{2n-1}=-n^{-p},\ u_{2n}=n^{-p}+n^{-2p}+n^{-3p},\quad (n>1).$$
[*Math. Trip.* 1906.]

5. Verify the identity
$$\left(1-\frac{x}{1}\right)\left(1-\frac{x}{2}\right)\dots\left(1-\frac{x}{n}\right)$$
$$=1-x+\frac{x(x-1)}{2!}-\frac{x(x-1)(x-2)}{3!}+\dots+(-1)^n\frac{x(x-1)\dots(x-n+1)}{n!}.$$
Shew that as n tends to infinity, the product diverges for all values of x except 0, but the series converges, provided that $x>0$.

6. Prove that $(1+x)(1+x^2)(1+x^4)(1+x^8)\dots=1/(1-x)$, if $|x|<1$.

7. Verify that
$$\cos\frac{x}{2}\cdot\cos\frac{x}{2^2}\cdot\cos\frac{x}{2^3}\dots=\frac{\sin x}{x},$$
and that
$$\frac{1}{2}\tan\frac{x}{2}+\frac{1}{2^2}\tan\frac{x}{2^2}+\frac{1}{2^3}\tan\frac{x}{2^3}+\dots=\frac{1}{x}-\cot x.$$
[EULER.]

8. Determine the value of
$$\Pi\left(1+\frac{x}{c+n}\right)e^{-x/n}$$
in terms of the Gamma functions $\Gamma(1+c)$ and $\Gamma(1+x+c)$.

9. Shew that
$$\lim_{n\to\infty}\frac{x(x+1)(x+2)\dots(x+2n-1)}{1.3.5\dots(2n-1).2x(2x+2)\dots(2x+2n-2)}=2^{x-1}.$$
$\left[\text{The product is } \frac{1}{2}\frac{\Gamma(x+2n)}{\Gamma(x+n)}\frac{\Gamma(n)}{\Gamma(2n)}.\right]$

10. Prove that, if k is an integer,
$$\lim_{n\to\infty}\frac{2.4.6\dots2kn}{1.3.5\dots(2kn-1)}\cdot\frac{1.3.5\dots(2n-1)}{2.4.6\dots2n}=\sqrt{k}.$$

11. If
$$a_n=\left[\frac{\Gamma(n)\Gamma(t)}{\Gamma(n+t)}\right]^2,\quad b_n=\left[\frac{\Gamma(n)\Gamma(1+t)}{\Gamma(n+1+t)}\right]^2$$
verify that
$$na_n-(n+1)a_{n+1}=ta_n+(t-1)a_{n+1}-(n+t)b_n,$$
and that
$$(n-1)^2nb_{n-1}-n^2(n+1)b_n=t^2[(2t-1)a_{n+1}+nb_n].$$
Shew that $\lim na_n=0$, $\lim n^3b_n=0$, if $t>\frac{1}{2}$, and deduce that
$$(2t-1)\sum_1^\infty a_n-t\sum_1^\infty b_n-\sum_1^\infty nb_n=ta_1,$$
$$(2t-1)\sum_1^\infty a_n+\sum_1^\infty nb_n=2ta_1.$$
Hence prove that
$$t\sum_1^\infty b_n=2(2t-1)\sum_1^\infty a_n-3/t.$$

12. Let a_n, b_n, c_n denote the general terms of the three hypergeometric series

$$A=F(\alpha, \beta, \gamma, 1), \quad B=F(\alpha-1, \beta, \gamma, 1), \quad C=F(\alpha, \beta, \gamma+1, 1),$$

in which $\gamma>\alpha+\beta$. Then prove that

$$a_n-a_{n+1}=(1-\beta/\gamma)c_n-b_{n+1},$$
$$(\gamma-\alpha)(a_n-b_n)=\beta a_{n-1}+(n-1)a_{n-1}-na_n,$$
$$\lim_{n\to\infty}(na_n)=0.$$

Deduce that $$\gamma B=(\gamma-\beta)C, \quad (\gamma-\alpha)(A-B)=\beta A,$$

and that $$A=\frac{(\gamma-\alpha)(\gamma-\beta)}{\gamma(\gamma-\alpha-\beta)}C.$$

13. From Ex. 12 prove that

$$F(\alpha, \beta, \gamma, 1)\,.\,\frac{\Gamma(\gamma-\alpha)\Gamma(\gamma-\beta)}{\Gamma(\gamma)\Gamma(\gamma-\alpha-\beta)}=F(\alpha, \beta, \gamma+n, 1)\frac{\Gamma(\gamma+n-\alpha)\Gamma(\gamma+n-\beta)}{\Gamma(\gamma+n)\Gamma(\gamma+n-\alpha-\beta)},$$

and shew that the last expression tends to the limit 1 as $n\to\infty$. Deduce that

$$F(\alpha, \beta, \gamma, 1)=\frac{\Gamma(\gamma)\Gamma(\gamma-\alpha-\beta)}{\Gamma(\gamma-\alpha)\Gamma(\gamma-\beta)}. \qquad \text{[Gauss.]}$$

14. If $$\left.\begin{array}{ll} q_0=\Pi(1-q^{2n}), & q_1=\Pi(1+q^{2n}), \\ q_2=\Pi(1+q^{2n-1}), & q_3=\Pi(1-q^{2n-1}) \end{array}\right\} \quad (n=1, 2, 3, \ldots),$$

the four products are absolutely convergent if $|q|<1$.

Also $$q_0q_3=\Pi(1-q^n), \quad q_1q_2=\Pi(1+q^n),$$

and $$q_1q_2q_3=1.$$

Thus $$1/[(1-q)(1-q^3)(1-q^5)\ldots]=(1+q)(1+q^2)(1+q^3)\ldots. \qquad \text{[Euler.]}$$

15. If Σu_n is absolutely convergent, the product $\Pi(1+xu_n)$ is absolutely convergent for any value of x; and it can be expanded in an absolutely convergent series

$$1+U_1x+U_2x^2+\ldots, \quad \text{where } U_1=\Sigma u_n.$$

Shew also that

$$\Pi(1+xu_n)(1+u_n/x)=V_0+V_1(x+1/x)+V_2(x^2+1/x^2)+\ldots,$$

where $$V_n=U_n+U_1U_{n+1}+U_2U_{n+2}+\ldots.$$

16. If $$f(x)=(1+qx)(1+q^3x)(1+q^5x)\ldots,$$

we have at once $(1+qx)f(q^2x)=f(x)$; and by the last example

$$f(x)=1+U_1x+U_2x^2+\ldots.$$

Thus we find

$$q+q^2U_1=U_1, \quad q^{2n-1}U_{n-1}+q^{2n}U_n=U_n,$$

which give

$$U_1=\frac{q}{1-q^2}, \quad U_2=\frac{q^4}{(1-q^2)(1-q^4)}, \quad U_3=\frac{q^9}{(1-q^2)(1-q^4)(1-q^6)},$$

and generally $$U_n=q^{n^2}/P_n,$$

where $$P_n=(1-q^2)(1-q^4)\ldots(1-q^{2n}).$$

17. If $$F(x)=\Pi(1+q^{2n-1}x)(1+q^{2n-1}/x),$$
we see, from Ex. 15, that we can write
$$F(x)=V_0+V_1(x+1/x)+V_2(x^2+1/x^2)+\ldots.$$
But $qxF(q^2x)=F(x)$, and so we find
$$V_1=V_0q,\quad V_n=V_{n-1}q^{2n-1},$$
yielding $$V_2=V_0q^4,\quad V_3=V_0q^9,\quad V_n=V_0q^{n^2}.$$
Thus $$F(x)=V_0[1+q(x+1/x)+q^4(x^2+1/x^2)+q^9(x^3+1/x^3)+\ldots].$$
To determine V_0 we may use the results of Exs. 15, 16, from which we find
$$V_0q^{n^2}=U_n+U_1U_{n+1}+U_2U_{n+2}+\ldots$$
or
$$V_0=\frac{1}{P_n}+\frac{q^{2n+2}}{P_1P_{n+1}}+\frac{q^{4n+8}}{P_2P_{n+2}}+\ldots.$$
Thus $$P_nV_0-1<\frac{q^{2n}}{q_0^2}+\frac{q^{4n}}{q_0^2}+\ldots,\quad \text{where } q_0=\Pi(1-q^{2n}),$$
because $$|q|<1 \text{ and } P_{n+r}/P_n>q_0,\ P_r>q_0.$$
Hence $$P_nV_0-1<q^{2n}/q_0^2(1-q^{2n}),$$
so that $$\lim(P_nV_0)=1,\ \text{or}\ V_0=1/q_0.$$
Thus, using the notation of Ex. 16, Ch. V., we have
$$f(x,q)=q_0\Pi(1+q^{2n-1}x)(1+q^{2n-1}/x),$$
from which a number of interesting results follow. [JACOBI.]

18. From Ex. 17 we find, with the notation of Ex. 14,
$$f(1,q)=q_0q_2^2,\ f(-1,q)=q_0q_3^2,\ f(q,q)=2q_0q_1^2.$$
Or, writing these equations at length, we have
$$q_0q_2^2=1+2q+2q^4+2q^9+\ldots,$$
$$q_0q_3^2=1-2q+2q^4-2q^9+\ldots,$$
$$q_0q_1^2=1+q^2+q^6+q^{12}+q^{20}+\ldots,$$
where the indices in the third series are of the type $n(n+1)$.

Again, by taking the limit of $f(x,q)/(1+q/x)$ as x approaches $-q$, we have
$$q_0^3=1-3q^2+5q^6-7q^{12}+9q^{20}-\ldots,$$
the indices being the same as in the third series. [Compare Art. 46.]

19. Again, from Ex. 17, we get
$$f(-1,q^2)=\Pi(1-q^{4n}).\Pi(1-q^{4n-2})^2=q_0q_2q_3=q_0/q_1,$$
so that $$q_0/q_1=1-2q^2+2q^8-2q^{18}+\ldots,$$
the indices being of the form $2n^2$.

Also $$f(\sqrt{q},\sqrt{q})=2\Pi(1-q^n).\Pi(1+q^n)^2=q_0q_1q_2=q_0/q_3,$$
so that $$q_0/q_3=1+q+q^3+q^6+q^{10}+\ldots,$$
the indices being of the form $\frac{1}{2}n(n+1)$. [GAUSS.]

Similarly,
$$f(-q^{\frac{1}{2}},q^{\frac{3}{2}})=\Pi(1-q^{3n}).\Pi(1-q^{3n-1})(1-q^{3n-2})=\Pi(1-q^n)$$
or $$q_0q_3=1-(q+q^2)+(q^5+q^7)-(q^{12}+q^{15})+\ldots,$$
the indices being alternately $\frac{1}{2}n(3n\pm1)$. [EULER.]

20. Write $y=1$, and put $\sqrt{x}$, $\sqrt{q}$ in place of x, q, in the first result of Ex. 16, Ch. V. Then we have

$$[f(\pm\sqrt{x}, \sqrt{q})]^2 = f(x, q) \cdot f(1, q) \pm g(x, q) \cdot g(1, q).$$

Now $$f(\sqrt{x}, \sqrt{q}) = \Pi(1-q^n) \cdot \Pi(1+q^{n-\frac{1}{2}}x^{\frac{1}{2}})(1+q^{n-\frac{1}{2}}x^{-\frac{1}{2}}),$$

so that $$f(\sqrt{x}, \sqrt{q}) \cdot f(-\sqrt{x}, \sqrt{q}) = (q_0 q_3)^2 \cdot \Pi(1-q^{2n-1}x)(1-q^{2n-1}/x)$$
$$= q_0 q_3^2 f(-x, q).$$

Thus, on multiplication, we find

$$(q_0 q_3^2)^2 [f(-x, q)]^2 = [f(x, q) \cdot f(1, q)]^2 - [g(x, q) \cdot g(1, q)]^2.$$

But $$f(1, q) = q_0 q_2^2, \quad g(1, q) = q^{\frac{1}{4}} f(q, q) = 2q^{\frac{1}{4}} q_0 q_1^2,$$

and so we have the identity

$$q_3^4 [f(-x, q)]^2 = q_2^4 [f(x, q)]^2 - 4q^{\frac{1}{2}} q_1^4 [g(x, q)]^2.$$

In particular, if we write $x=1$, we find the interesting result

$$q_2^8 = q_3^8 + 16q \cdot q_1^8,$$

which leads again to the identity

$$(1+2q+2q^4+2q^9+\dots)^4 - (1-2q+2q^4-2q^9+\dots)^4$$
$$= 16q(1+q^2+q^6+q^{12}+q^{20}+\dots)^4,$$

where the series are those given in Ex. 18 above. [JACOBI.]

21. Prove that if $u_n = \dfrac{(n+a_1)(n+a_2)\dots(n+a_k)}{(n+b_1)(n+b_2)\dots(n+b_l)}$, the product $\prod\limits_1^\infty u_n$ can only converge if $k=l$ and $\Sigma a = \Sigma b$. When these conditions are satisfied, express the product in the form

$$\frac{\Gamma(1+b_1)\Gamma(1+b_2)\dots\Gamma(1+b_k)}{\Gamma(1+a_1)\Gamma(1+a_2)\dots\Gamma(1+a_k)}.$$

In particular, prove that

$$\prod_1^\infty \frac{n(n+a+b)}{(n+a)(n+b)} = \frac{\Gamma(1+a)\Gamma(1+b)}{\Gamma(1+a+b)}.$$

22. Prove that

$$(1-x)(1+\tfrac{1}{2}x)(1-\tfrac{1}{3}x)(1+\tfrac{1}{4}x)\dots = \frac{\Gamma(\frac{1}{2})}{\Gamma(1+\frac{1}{2}x)\Gamma(\frac{1}{2}-\frac{1}{2}x)}.$$

[Take the terms in pairs and use the last example.]

23. If $\psi(x)$ denotes $\Gamma'(x)/\Gamma(x)$, we can write (see Art. 46)

$$\psi(x) = \lim_{n\to\infty}\left(\log n - \frac{1}{x} - \frac{1}{1+x} - \frac{1}{2+x} - \dots - \frac{1}{n+x}\right).$$

Then we find $$\frac{\Gamma(x)}{\Gamma(x+y)} e^{y\psi(x)} = \prod_0^\infty \left(1+\frac{y}{x+n}\right) e^{-y/(x+n)}.$$ [MELLIN.]

24. It is easy to deduce from the theory of infinite products Abel's result (Arts. 11, 16), that Σa_n and $\Sigma a_n/s_n$ converge or diverge together. In fact consider the product $\Pi(1-a_n/s_n) = \Pi(s_{n-1}/s_n)$, which diverges to 0 if $s_n \to \infty$; so that $\Sigma(a_n/s_n)$ must also diverge (Art. 39). [Here $a_n > 0$.]

Other examples on products will be found at the ends of Chapters IX., X.

CHAPTER VII.

SERIES OF VARIABLE TERMS.

43. Uniform convergence of a sequence.

It may happen that the terms of a sequence depend on some variable x in addition to the index n; and this is indicated by using the notation $S_n(x)$. We assume that the sequence is convergent for all values of x within a certain interval (a, b), and then the limit

$$\lim_{n\to\infty} S_n(x)$$

defines a certain function of x, say $F(x)$, in the interval (a, b).

The condition of convergence (Art. 1) implies that, given an arbitrarily small positive number ϵ, we can determine an integer m such that

$$|S_n(x)-F(x)|<\epsilon, \quad \text{if } n>m.$$

Obviously the definition of m is not yet precise, but we can make it precise by agreeing to always select the *least* integer m which satisfies the prescribed inequality. When this is done, it is natural to expect that the value of m will depend on x, and so we are led to consider a new function $m(x)$, which depends on ϵ and on the nature of the sequence.

We note incidentally that, regarded as a function of ϵ, $m(x)$ is monotonic since (for any assigned value of x) it cannot decrease as ϵ diminishes.

Ex. 1. If $S_n(x)=1/(x+n)$, where $x \geqq 0$, we have

$$F(x)=\lim_{n\to\infty} S_n(x)=0.$$

Then the condition of convergence gives

$$x+n>1/\epsilon,$$

so that $\quad m(x)=$ the integral part of $(1/\epsilon)-x$, when $x<1/\epsilon$,

or $\quad m(x)=0$, when $x \geqq 1/\epsilon$.

Ex. 2. If $S_n(x)=x^n$, where $0 \leqq x \leqq 1$, we have

$$F(x)=\lim_{n\to\infty} S_n(x)=0, \text{ if } x<1\text{ ; and } F(1)=1.$$

Then we are to have

$$(1/x)^n > 1/\epsilon, \quad \text{if } x<1,$$

so that $\quad m(x)=$ the integral part of $\dfrac{\log(1/\epsilon)}{\log(1/x)}$, when $x<1$.

Also, since $S_n(1)=1$ for all values of n, we must take $m(1)=0$.

Ex. 3. If $S_n(x)=\text{arc}\tan(nx)$, where $x \geqq 0$, we have

$$F(x)=\lim_{n\to\infty} S_n(x)=\tfrac{1}{2}\pi, \text{ if } x>0\text{ ; and } F(0)=0.$$

It is easily seen that

$$m(x)=\text{the integral part of } (\cot\epsilon)/x, \text{ when } x>0,$$

and $\quad m(0)=0.$

Ex. 4. If $S_n(x)=nx/(1+n^2x^2)$, x being unrestricted, we have

$$F(x)=\lim_{n\to\infty} S_n(x)=0.$$

Thus $F(x)$ is here *continuous*, in contrast to Exs. 2, 3.

The condition of convergence is

$$n|x|>[1+(1-4\epsilon^2)^{\frac{1}{2}}]/2\epsilon, \quad \text{if } \epsilon<\tfrac{1}{2}.$$

Thus $\quad m(x)=$ the integral part of $[1+(1-4\epsilon^2)^{\frac{1}{2}}]/2|x|\epsilon$, $\quad$ if $|x|>0$,

although $\quad m(0)=0$.

It will be seen that in Ex. 1 the function $m(x)$ is always less than $1/\epsilon$; but in Ex. 2, $m(x)\to\infty$ as $x\to1$; and in Exs. 3, 4, $m(x)\to\infty$ as $x\to0$ (assuming that $\epsilon<\frac{1}{2}$). This consideration suggests a further subdivision of convergent sequences, which will prove of great importance in the sequel, and introduces a more subtle distinction.

We shall say that *the sequence $S_n(x)$ converges* **uniformly** *in the interval (a, b), provided that $m(x)$ is less than $\mu(\epsilon)$ for* **all** *points of the interval*; here of course $\mu(\epsilon)$ may vary with ϵ, but must be *independent of x*. Thus, as x varies from a to b, $m(x)$ has a fixed upper limit and so cannot tend to infinity at any point in the interval (a, b).

Thus in Ex. 1 the convergence is uniform for all positive values of x, since we can take $\mu(\epsilon)=1/\epsilon$. But in Ex. 2, the convergence is not uniform in an interval reaching up to $x=1$; although it *is* uniform in the interval $(0, c)$, if $0<c<1$, because we can then take

$$\mu(\epsilon)=\log(1/\epsilon)/\log(1/c).$$

Hence in Ex. 2, $x=1$ cannot be included in any interval of uniform convergence: such a point will be called *a point of non-uniform convergence*. Similarly in Exs. 3, 4 the point $x=0$ must be excluded to ensure uniform convergence.

This distinction may be made more tangible by means of a graphical method suggested by Osgood.* The curves $y=S_n(x)$ are drawn for a succession of values of n in the same diagram; this is done in Figs. 11–14 for the sequences of Exs. 1–4. Then, if $S_n(x)\to F(x)$ *uniformly* in the interval (a, b), the *whole* of

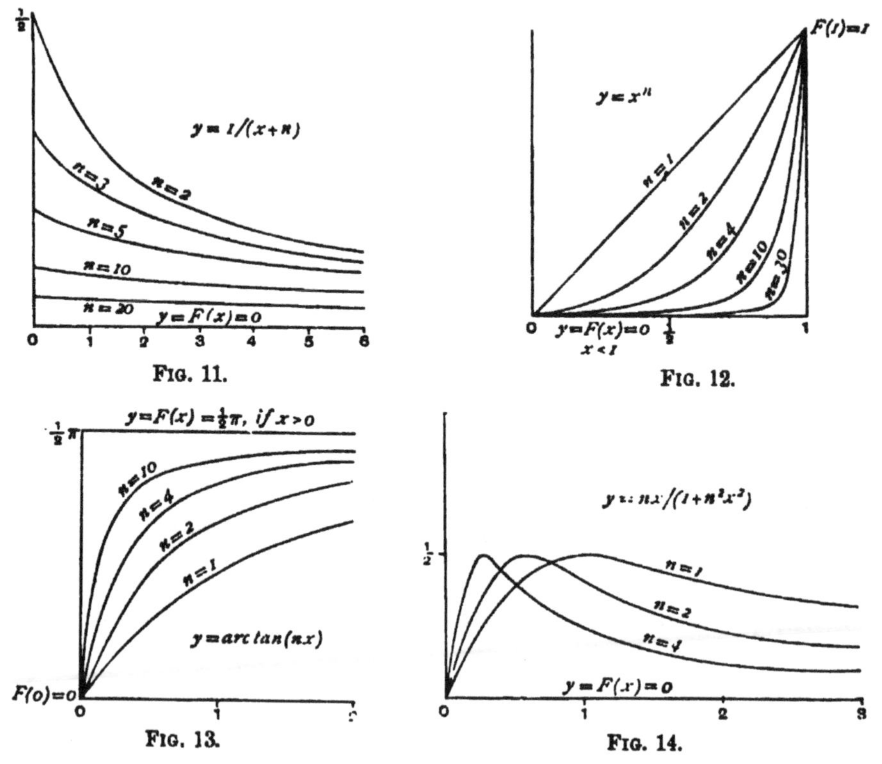

FIG. 11. FIG. 12. FIG. 13. FIG. 14.

the curves for which $n>\mu(\epsilon)$ will lie in the strip bounded by $y=F(x)\pm\epsilon$. A glance at Fig. 11 will shew that this does occur in Ex. 1. But in Ex. 2, as we see from Fig. 12, every curve $y=S_n(x)$ finally rises above $y=\epsilon$; and the larger n is taken, the nearer to $x=1$ is the point of crossing; thus $x=1$ is a point of non-uniform convergence. In the same way, Figs. 13, 14 shew that $x=0$ is a point of non-uniform convergence for each of the sequences in Exs. 3, 4.

* *Bulletin of the American Math. Society* (2), vol. 3, 1897, p. 59.

In order to give a definition of uniform convergence which does not involve the actual determination of $F(x)$, we introduce the following test, corresponding to that of Art. 3 for convergence.

The necessary and sufficient condition for uniform convergence in an interval is that, corresponding to any positive number ϵ, *it may be possible to find an index* m, **which is independent of x**, *and is such that*

$$|S_n(x)-S_m(x)|<\epsilon,$$

for all values of n *greater than* m, *and for* **all** *points of the interval.*

It will be seen on comparison with Art. 3 that the only fresh condition is that m is to be independent of x, whereas the terms of the sequence are functions of x.

That the condition is *necessary* is evident, for if $S_n(x)$ tends to $F(x)$, *uniformly*, we can write $m-1=\mu(\frac{1}{2}\epsilon)$, so that

$$|S_n(x)-F(x)|<\tfrac{1}{2}\epsilon, \quad \text{if } n>m-1.$$

Hence the condition for uniform convergence leads to the inequality
$$|S_n(x)-S_m(x)|<\epsilon, \quad \text{if } n>m.$$

The condition is also *sufficient*; for if it is satisfied, $S_n(x)$ must converge to some limit, $F(x)$ say, in virtue of Art. 3; and since
$$\lim S_n(x)=F(x),$$
we have
$$|F(x)-S_m(x)|\leqq\epsilon, \qquad \text{[Art. 1 (7)].}$$

Hence
$$|F(x)-S_n(x)|<2\epsilon, \quad \text{if } n>m,$$

and so the condition of uniform convergence to $F(x)$ is satisfied.

It should be noticed that *the interval of uniform convergence is always* **closed**; that is, if $S_n(x)$ converges uniformly for $a<x<b$, it will also converge uniformly for $a\leqq x\leqq b$, assuming that $x=a$, b are not discontinuities of $S_n(x)$. For m can be found so that

$$|S_n(x)-S_m(x)|<\tfrac{1}{3}\epsilon, \quad \text{if } a<x<b \text{ and } n>m.$$

Also, since $S_n(x)$ is a continuous function at $x=a$, we can find δ so that

$$\left.\begin{aligned}&|S_n(x)-S_n(a)|<\tfrac{1}{3}\epsilon\\ \text{and}\quad &|S_m(x)-S_m(a)|<\tfrac{1}{3}\epsilon\end{aligned}\right\}, \quad \text{if } x-a<\delta,$$

where δ may depend on m, n, as well as ϵ.

Hence $$|S_n(a)-S_m(a)|<\epsilon,$$
so that $S_n(a)$ converges, and $x=a$ can therefore be included in the interval of uniform convergence. Similarly for $x=b$.

Ex. 5. The sequence $S_n(x)=x^n(1-x)$, $0\leqq x\leqq 1$, converges *uniformly*, because $\lim S_n(x)=0$ and $S_n(x)<1/n$, since the maximum of $S_n(x)$ in the interval is given by $x=n/(n+1)$. The reader should contrast this result with Ex. 2, and should draw the curves $y=S_n(x)$ for a few values of n.

44. Uniform convergence of a series.

If, in Art. 43, we suppose the sequence to be derived from a series of variable terms
$$f_0(x)+f_1(x)+f_2(x)+\dots \text{ to } \infty,$$
by writing $$S_n(x)=f_0(x)+f_1(x)+\dots+f_n(x),$$
we obtain the *test for uniform convergence of a series in an interval* (a, b) in the form:

It must be possible to find a number m **independent of x**, *so as to satisfy the condition*
$$|f_{m+1}(x)+f_{m+2}(x)+\dots+f_{m+p}(x)|<\epsilon, \quad \text{where } p=1,\ 2,\ 3,\ \dots,$$
at **all** *points of the interval* (a, b).

Each of the examples given in Art. 43 can be used to construct a series by writing
$$f_n(x)=S_n(x)-S_{n-1}(x),$$
but a more natural type of non-uniform convergence is the following:
$$x^2+\frac{x^2}{1+x^2}+\frac{x^2}{(1+x^2)^2}+\dots.$$
Here we find $$S_n(x)=(1+x^2)-(1+x^2)^{-n+1},$$
so that $$F(x)=1+x^2 \quad (x\gtrless 0)$$
and $$F(0)=0.$$

There is a point of non-uniform convergence at $x=0$, as the reader will see by considering the condition
$$(1+x^2)^{-n+1}<\epsilon, \quad \text{or } (1+x^2)^{n-1}>1/\epsilon.$$

But, just as the general test for convergence is usually replaced by narrower tests (compare Chap. III.) which are more convenient in ordinary practice, so here we usually replace the test above by one of the three following tests:

(1) Weierstrass's M-test for uniform convergence.

The majority of series met with in elementary analysis can be proved to converge uniformly by means of a test due to Weierstrass and described briefly as the *M*-test:

Suppose that for all values of x in the interval (a, b), the function $f_n(x)$ has the property

$$|f_n(x)| \leqq M_n,$$

where M_n is a positive **constant, independent of x**; *and suppose that the series ΣM_n is convergent. Then the series $\Sigma f_n(x)$ is uniformly and absolutely convergent in the interval (a, b).*

The absolute convergence follows at once from Art. 18; to realise the uniform convergence, it is only necessary to remember that for any integral value of p,

$$\left|\sum_{m+1}^{m+p} f_n(x)\right| \leqq \sum_{m+1}^{m+p} M_n < \sum_{m+1}^{\infty} M_n.$$

Consequently, if we choose m so as to make the remainder in ΣM_n less than ϵ, $\left|\sum_{m+1}^{m+p} f_n(x)\right|$ is also less than ϵ; and this choice of m is obviously independent of x, so that the condition of uniform convergence is satisfied. [Compare Stokes, *Math. and Phys. Papers*, vol. 1, p. 281.]

(2) Abel's test for uniform convergence.

A more delicate test for uniform convergence is due, in substance, to Abel, and has been already mentioned in Art. 19:

The series $\Sigma a_n v_n(x)$ is uniformly convergent in an interval (a, b), provided that Σa_n is convergent; that for any particular value of x in the interval $v_n(x)$ is positive and never increases with n; and that $v_0(x)$ remains less than a fixed number κ for all values of x in the interval.

For, in virtue of the convergence of Σa_n, we can find m, so that, whatever positive integer p may be,

$$a_{m+1}, \quad a_{m+1}+a_{m+2}, \quad \ldots, \quad a_{m+1}+a_{m+2}+\ldots+a_{m+p}$$

are all numerically less than ϵ/κ. Then, in virtue of Abel's lemma (Art. 23), we see that

$$\left|\sum_{m+1}^{m+p} a_n v_n(x)\right| < \epsilon v_m(x)/\kappa < \epsilon,$$

since, by hypothesis, $v_m(x) \leqq v_0(x) < \kappa$.

Further, since a_n and κ are independent of x, the choice of m is independent of x also; and therefore $\Sigma a_n v_n(x)$ converges uniformly in the interval.

It is obvious that the terms a_n may be themselves functions of x, *provided that* Σa_n *is uniformly convergent in the interval.*

(3) Dirichlet's test for uniform convergence.

This is also more delicate than the M-test. (Compare Exs. 1, 4 below.)

The series $\Sigma a_n v_n(x)$ *is uniformly convergent in an interval* (a, b), *provided that* Σa_n *oscillates between finite limits; that for any particular value of* x *in the interval* $v_n(x)$ *is positive and never increases with* n; *and that, as* n *tends to* ∞, $v_n(x)$ *tends uniformly to zero for all values of* x *in the interval.*

For then the expressions

$$|a_{m+1}|, \quad |a_{m+1}+a_{m+2}|, \quad \ldots, \quad |a_{m+1}+a_{m+2}+\ldots+a_{m+p}|$$

are less than a fixed number κ; and we can find an index m such that

$$v_m(x) < \epsilon/\kappa$$

for all values of x in the interval.

Thus, using Abel's lemma as before, we see that

$$\left|\sum_{m+1}^{m+p} a_n v_n(x)\right| < \epsilon$$

for all points in the interval.

Again, the terms a_n may be changed to functions of x, provided that the maximum limit of $|\Sigma a_n|$ remains less than a fixed number throughout the interval.

Ex. 1. *Weierstrass's M-test.*

Consider
$$\sum_1^\infty \frac{\sin nx}{n^p}, \quad \sum_1^\infty \frac{\cos nx}{n^p}, \quad (p>1);$$
these converge uniformly for all real values of x, because then
$$\left|\frac{\sin nx}{n^p}\right| \leqq \frac{1}{n^p}, \quad \left|\frac{\cos nx}{n^p}\right| \leqq \frac{1}{n^p} \text{ and } \sum_1^\infty(1/n^p) \text{ is convergent.}$$

Ex. 2. *Abel's test.*

Consider the case $v_n(x)=1/n^x$, $(0 \leqq x \leqq 1)$; then $\Sigma(a_n/n^x)$ converges uniformly in the interval $(0, 1)$ if Σa_n converges. [DIRICHLET.]

Ex. 3. *Abel's test.*

If Σa_n is convergent,
$$\Sigma a_n \frac{x^n}{1+x^n}, \quad \Sigma a_n \frac{x^n}{1+x^{2n}}, \quad \Sigma a_n \frac{nx^n(1-x)}{1-x^n}, \quad \Sigma \frac{2na_n x^n(1-x)}{1-x^{2n}},$$
converge uniformly in the interval $(0, 1)$. [HARDY.]

Ex. 4. *Dirichlet's test.*

Consider $$\sum_1^\infty \frac{\sin nx}{n^p}, \quad \sum_1^\infty \frac{\cos nx}{n^p}, \quad (p > 0);$$

then writing $v_n(x) = 1/n^p$, $a_n = \sin nx$ (or $\cos nx$),

we see that both series converge uniformly in an interval $(a, 2\pi - a)$, where a is any positive angle.

45. Fundamental properties of uniformly convergent series.

Cauchy and the earlier analysts (with the exception of Abel) assumed that the continuity of $F(x) = \lim_{n\to\infty} S_n(x)$ could be deduced from that of $S_n(x)$; that this assumption is not correct follows immediately from Exs. 2, 3 of Art. 43. Further, these examples suggest that a discontinuity in $F(x)$ implies a point of non-uniform convergence; although Ex. 4, Art. 43, indicates that non-uniform convergence does not necessarily involve the discontinuity of $F(x)$.

Again, if we wish to integrate $F(x)$, the equation

$$\int_{c_1}^{c_2} [\lim_{n\to\infty} S_n(x)]dx = \lim_{n\to\infty} \int_{c_1}^{c_2} S_n(x)dx$$

is not necessarily true either, as will be seen from the example on p. 118 below.

In 1847 Stokes* published his discovery of the distinction between uniform and non-uniform convergence, and gave theorem (1) below, which establishes the continuity of most series required in elementary analysis. Theorem (2) on integration is due to Weierstrass, and seems not to have been published (except in lectures) until 1870.

(1) *If the series $F(x) = \Sigma f_n(x)$ is uniformly convergent in the interval (a, b), and if each of the functions $f_n(x)$ is continuous in the interval, so also is the sum $F(x)$.*

For, in virtue of the definition of uniform convergence, the number m can be chosen **independently of** x (provided only that $a \leqq x \leqq b$), in such a way that

$$|f_m(x) + f_{m+1}(x) + \dots \text{ to } \infty| < \tfrac{1}{3}\epsilon,$$

---

*The discovery was made also by Weierstrass and Seidel, but Stokes's paper was published a year before Seidel's; however, Weierstrass must have been aware of the distinction some five years at least before Stokes's paper was published. This is clear from papers in the first volume of Weierstrass's works (see pp. 67–84), which remained unpublished for about 50 years.

no matter how small the constant ϵ may be. Now write

$$f_0(x)+f_1(x)+f_2(x)+\ldots+f_{m-1}(x)=S_m(x),$$

and it is then clear that

$$|F(x)-S_m(x)|<\tfrac{1}{3}\epsilon, \quad (a\leqq x\leqq b),$$

where it must be remembered that m, ϵ are quite independent of x.

Thus if c is any value of x within the interval, we have

$$|F(c)-S_m(c)|<\tfrac{1}{3}\epsilon,$$

so that $$|F(c)-F(x)|<\tfrac{2}{3}\epsilon+|S_m(c)-S_m(x)|.$$

Now m being fixed, $S_m(x)$ is a continuous function of x, and therefore we can find a value δ, such that

$$|S_m(c)-S_m(x)|<\tfrac{1}{3}\epsilon, \quad \text{if } |c-x|<\delta.$$

Hence $$|F(c)-F(x)|<\epsilon, \quad \text{if } |c-x|<\delta,$$

which proves the continuity of $F(x)$ within the interval (a, b).

It is not unusual for beginners to miss the point of the foregoing proof; and it is therefore advisable to show how the argument fails when applied to such a series as

$$(1-x)+(x-x^2)+(x^2-x^3)+\ldots, \quad \text{(Ex. 2, Art. 43)}$$

when we take $c=1$.

Here $$f_m(x)+f_{m+1}(x)+\ldots \text{ to } \infty = x^m \quad \text{if } 0<x<1,$$

and $$f_m(1)+f_{m+1}(1)+\ldots \text{ to } \infty = 0.$$

Thus, if we wish to make both these remainders less than $\frac{1}{3}\epsilon$, we must choose m, if we can, so that

$$x^m<\tfrac{1}{3}\epsilon, \quad \ldots\ldots\ldots\ldots(A)$$

but to make $$|S_m(1)-S_m(x)|<\tfrac{1}{3}\epsilon$$

we must take $$1-x^m<\tfrac{1}{3}\epsilon$$

or $$x^m>1-\tfrac{1}{3}\epsilon, \quad \ldots\ldots\ldots\ldots(B)$$

and the two inequalities (A) and (B) are mutually contradictory (supposing that $\epsilon<1$).

Consequently the two steps used in the general argument are incompatible here; and the reason for this difference lies in the fact that the inequality (A) does not lead to a determination of m *independent of* x, when x can approach as near to 1 as we please. The assumption that the series converges uniformly enables us to avoid the difficulty involved in such a condition as (A).

(2) *If the series $F(x)=\Sigma f_n(x)$ is uniformly convergent in the interval (a, b), and if each of the functions $f_n(x)$ is continuous in the interval, we may write*

$$\int_{c_1}^{c_2} F(x)\,dx=\Sigma\int_{c_1}^{c_2} f_n(x)\,dx, \quad \text{if } a\leqq c_1<c_2\leqq b.$$

For, in virtue of the uniform convergence of $\Sigma f_n(x)$, we can find m so that

$$|f_m(x)+f_{m+1}(x)+\ldots+f_p(x)|<\epsilon, \quad \text{if } p>m,$$

however small ϵ may be; and the value of m will be *independent* of x, as before.

Hence we have

$$\left|\sum_m^p \int_{c_1}^{c_2} f_n(x)dx\right| < \epsilon(c_2-c_1) \leqq \epsilon(b-a),$$

and since this is true, no matter how large p may be, we see that

$$\sum_0^\infty \int_{c_1}^{c_2} f_n(x)dx$$

converges, and that

$$\left|\sum_m^\infty \int_{c_1}^{c_2} f_n(x)dx\right| \leqq \epsilon(b-a).$$

At the same time we have

$$|F(x)-S_m(x)| \leqq \epsilon,$$

and consequently

$$\left|\int_{c_1}^{c_2} F(x)dx - \int_{c_1}^{c_2} S_m(x)dx\right| \leqq \epsilon(c_2-c_1) \leqq \epsilon(b-a).$$

This last inequality can be written

$$\left|\int_{c_1}^{c_2} F(x)dx - \sum_0^{m-1} \int_{c_1}^{c_2} f_n(x)dx\right| \leqq \epsilon(b-a),$$

so that, combining the two inequalities, we find

$$\left|\int_{c_1}^{c_2} F(x)dx - \sum_0^\infty \int_{c_1}^{c_2} f_n(x)dx\right| \leqq 2\epsilon(b-a),$$

where ϵ may be as small as we please, by proper choice of m. Since m is no longer present on the left-hand side, the argument given in Note (6), p. 5, shews that

$$\int_{c_1}^{c_2} F(x)dx = \sum_0^\infty \int_{c_1}^{c_2} f_n(x)dx.$$

This operation is often described as *term-by-term integration*.

The reader will probably find less difficulty here in realising the importance of the condition that m should be independent of x. It is not, however, easy to give a really simple example of a non-uniformly convergent series in which term-by-term integration leads to erroneous results. The following, although artificial, is perhaps as good as any.

Let
$$S_n(x) = nxe^{-nx^2},$$
so that
$$F(x) = \lim_{n\to\infty} S_n(x) = 0, \quad \text{if } x > 0,$$
and obviously
$$S_n(0) = 0, \quad \text{so that } F(0) = 0.$$
Hence
$$\int_0^1 F(x)dx = 0.$$
But on the other hand
$$\sum_0^m \int_0^1 f_n(x)dx = \int_0^1 S_m(x)dx = \tfrac{1}{2}\Big[-e^{-mx^2}\Big]_0^1 = \tfrac{1}{2}(1-e^{-m}),$$
so that
$$\sum_0^\infty \int_0^1 f_n(x)dx = \lim_{m\to\infty} \tfrac{1}{2}(1-e^{-m}) = \tfrac{1}{2},$$
which is obviously not equal to $\int_0^1 F(x)dx$.

The figure shews the non-uniform convergence at $x=0$ and gives some indication as to the reason why the area under $y=S_n(x)$ does not tend to zero.

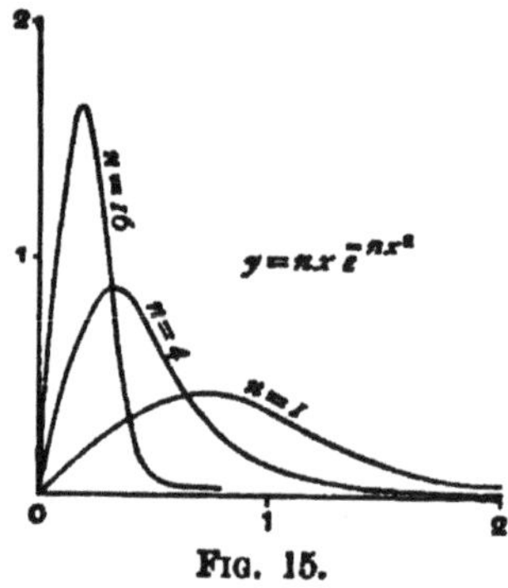

FIG. 15.

Of course the argument above assumes that the range of integration is *finite*; the conditions under which an infinite series can be integrated from 0 to ∞, say, belong more properly to the Integral Calculus; but some special cases are given in Art. 176 of the Appendix.

46. Differentiation of an infinite series.

If we consider Ex. 4 of Art. 43, for which
$$S_n(x) = \frac{nx}{1+n^2x^2}, \quad F(x) = 0,$$
we see that
$$S_n'(0) = \lim_{x\to 0} \frac{S_n(x) - S_n(0)}{x} = \lim_{x\to 0} \frac{n}{1+n^2x^2} = n.$$
Thus $\lim_{n\to\infty} S_n'(0) = \infty$, although $F'(0) = 0$.

It follows that the equation
$$\frac{d}{dx}[\lim_{n\to\infty} S_n(x)] = \lim_{n\to\infty} S_n'(x)$$

is not necessarily true when non-uniform convergence presents itself. But it should be noticed that it is the non-uniform convergence of the *derivate* $S_n'(x)$ which is the cause of the failure, as will be apparent from the general theorem below.

The reader may consider similarly the case

$$S_n(x) = \frac{1}{n}\sin(nx), \quad F(x) = 0;$$

here $S_n(x)$ converges uniformly to zero, but $S_n'(x)$ *oscillates.*

If the series of differential coefficients $\Sigma f_n'(x)$ *is uniformly convergent within the interval* (a, b), *its sum is equal to* $F'(x)$, *the differential coefficient of* $F(x) = \Sigma f_n(x)$, *assuming the latter series to converge in the interval.**

Write $$G(x) = \Sigma f_n'(x),$$
then, by (2) Art. 45, we have

$$\int_{c_1}^{c_2} G(x)dx = \Sigma[f_n(c_2) - f_n(c_1)]$$
$$= F(c_2) - F(c_1).$$

Thus, by the fundamental property of an integral, we have

$$F'(c_2) = G(c_2)$$

or

$$F'(x) = G(x), \quad (a \leqq x \leqq b).$$

A *direct* proof of the foregoing theorem is not easy without some use of the Integral Calculus; but the following method is easier than those proofs which depend *entirely* on the Differential Calculus, and only one result is obtained by integration.

Let m be chosen so as to make

$$\left|\sum_m^{m+p} f_n'(x)\right| < \epsilon, \quad a \leqq x \leqq b.$$

Thus

$$\left|\sum_m^{m+p} \int_{c_1}^{c_2} f_n'(x)\,dx\right| < \epsilon(c_2 - c_1),$$

provided that c_1, c_2 belong to the interval (a, b). This gives

$$\left|\sum_m^{m+p} \{f_n(c_2) - f_n(c_1)\}\right| < \epsilon(c_2 - c_1),$$

and so

$$\left|\sum_m^{\infty} \{f_n(c_2) - f_n(c_1)\}\right| \leqq \epsilon(c_2 - c_1).$$

* We can infer the convergence of $\Sigma f_n(x)$ from that of $\Sigma f_n'(x)$, *if the constants of integration are properly adjusted* (as in Art. 45); but this would require an additional assumption which we do not introduce here.

Or, changing the notation, we have

$$\sum_m^\infty \frac{1}{h}\{f_n(x+h)-f_n(x)\} \leqq \epsilon,$$

provided that both x and $x+h$ belong to the interval (a, b).

Now we have

$$\frac{F(x+h)-F(x)}{h} = \sum_0^{m-1}\left[\frac{f_n(x+h)-f_n(x)}{h}\right]+\sum_m^\infty\left[\frac{f_n(x+h)-f_n(x)}{h}\right]$$
$$=X_m+Y_m, \text{ say}.$$

We have proved that $|Y_m| \leqq \epsilon$, and in virtue of our original choice of m,

$$\left|\sum_m^\infty f_n'(x)\right| \leqq \epsilon.$$

But we have identically

$$\frac{F(x+h)-F(x)}{h} - G(x) = \left\{X_m - \sum_0^{m-1} f_n'(x)\right\} + Y_m - \sum_m^\infty f_n'(x),$$

and so we find

$$\left|\frac{F(x+h)-F(x)}{h} - G(x)\right| \leqq \left|\sum_0^{m-1}\left\{\frac{f_n(x+h)-f_n(x)}{h} - f_n'(x)\right\}\right| + 2\epsilon.$$

Since m has been fixed without reference to h, we can allow h to tend to zero, without changing m; the right-hand side of the last inequality then approaches the limit 2ϵ, because each term under the summation sign tends to zero. Consequently, the maximum limiting value of the left-hand side is not greater than 2ϵ; and since ϵ is arbitrarily small, this maximum limit must be zero (see Note (6), p. 5).

Thus we have

$$\overline{\lim_{h\to 0}}\left|\frac{F(x+h)-F(x)}{h} - G(x)\right| = 0,$$

or

$$\lim_{h\to 0}\frac{F(x+h)-F(x)}{h} = G(x),$$

which is the required theorem.

47. It is important to bear in mind that the condition of uniform convergence is merely *sufficient* for the truth of the theorems in Arts. 45, 46; but it is by no means a *necessary* condition. In other words, this condition is too narrow; but in spite of this, no other condition of equal simplicity has been discovered as yet, and we shall not go further into the subject* here.

That uniform convergence is not necessary may be seen by considering the two following examples:

(1) Ex. 4, Art. 43, shews that non-uniform convergence does not always imply discontinuity.

(2) Consider the series

$$1-x+x^2-x^3+\ldots=1/(1+x), \quad (0<x<1).$$

Then

$$\int_0^1 \frac{dx}{1+x}=\log 2,$$

and log 2 is also equal to the series

$$1-\tfrac{1}{2}+\tfrac{1}{3}-\tfrac{1}{4}+\ldots$$

found by integrating term-by-term.

Nevertheless $x=1$ is a point of non-uniform convergence of the series in x; because the remainder is greater than $\frac{1}{2}x^n$, and the condition $\frac{1}{2}x^n<\epsilon$ leads to a determination of n, which *cannot* be independent of x (when $x=1$ is included in the interval considered).

48. Uniform convergence of an infinite product.

The definition of uniform convergence can be extended at once to an infinite product; but applications of the principle occur less frequently in elementary analysis, and for our present purpose the following theorem will be sufficient:

If for all values of x in the interval (a, b) the function $f_n(x)$ has the property $|f_n(x)|\leqq M_n$, where M_n is a positive constant (independent of x), then if the series ΣM_n is convergent, the product

$$P(x)=[1+f_0(x)][1+f_1(x)][1+f_2(x)]\ldots \text{ to } \infty$$

is a continuous function of x in the interval, provided that all the functions $f_n(x)$ are continuous in the interval.

For, write

$$[1+f_0(x)][1+f_1(x)]\ldots[1+f_{m-1}(x)]=P_m(x),$$
$$[1+f_m(x)][1+f_{m+1}(x)]\ldots \text{ to } \infty = Q_m(x);$$

then

$$\prod_m^\infty\{1-|f_n|\}<Q_m(x)<\prod_m^\infty\{1+|f_n|\},$$

and

$$1-R_m<\prod_m^\infty\{1-|f_n|\}; \quad \prod_m^\infty\{1+|f_n|<1/(1-R_m), \quad \text{(Art. 38.)}$$

where

$$R_m=|f_m(x)|+|f_{m+1}(x)|+\ldots \text{ to } \infty.$$

Thus

$$|Q_m(x)-1|<R_m/(1-R_m).$$

---

* Reference may be made to a paper by the author (*Proc. Lond. Math. Soc.* series 2, vol. 1, 1904, p. 187) for the general question. Many wider tests for term-by-term integration have been given by various writers; some simple ones are given in the Appendix, Arts. 175, 176.

Now $$R_m \leqq M_m + M_{m+1} + M_{m+2} + \dots \text{ to } \infty,$$
where all the M's are *independent* of x. Hence m can be chosen (independently of x) to make $R_m < \frac{1}{4}\epsilon$, for all values of x within the interval (a, b).

Hence, assuming ϵ to be less than 1, we have
$$|Q_m(x) - 1| < \epsilon/(4-\epsilon) < \tfrac{1}{3}\epsilon, \quad \text{if } a \leqq x \leqq b,$$
where (as already explained) m is chosen quite independently of x.

Thus, if c is any value of x within the interval, we have the inequalities
$$1 - \tfrac{1}{3}\epsilon < Q_m(c) < 1 + \tfrac{1}{3}\epsilon$$
and $$1 - \tfrac{1}{3}\epsilon < Q_m(x) < 1 + \tfrac{1}{3}\epsilon.$$
Or $$\frac{1-\frac{1}{3}\epsilon}{1+\frac{1}{3}\epsilon} < \frac{Q_m(x)}{Q_m(c)} < \frac{1+\frac{1}{3}\epsilon}{1-\frac{1}{3}\epsilon}.$$
Hence $$\frac{1-\frac{1}{3}\epsilon}{1+\frac{1}{3}\epsilon} \cdot \frac{P_m(x)}{P_m(c)} < \frac{P(x)}{P(c)} < \frac{1+\frac{1}{3}\epsilon}{1-\frac{1}{3}\epsilon} \frac{P_m(x)}{P_m(c)}.$$

Now, since m is *fixed* and independent of x, and since
$$f_0(x),\ f_1(x),\ \dots,\ f_{m-1}(x)$$
are continuous functions of x, so also is the product $P_m(x)$; thus we can find δ so that
$$1 - \tfrac{1}{3}\epsilon < \frac{P_m(x)}{P_m(c)} < \frac{1}{1-\frac{1}{3}\epsilon},$$
provided that $|x - c| < \delta$.

Thus $$\frac{(1-\frac{1}{3}\epsilon)^2}{1+\frac{1}{3}\epsilon} < \frac{P(x)}{P(c)} < \frac{1+\frac{1}{3}\epsilon}{(1-\frac{1}{3}\epsilon)^2},$$
and so* $$1 - \epsilon < \frac{P(x)}{P(c)} < \frac{1}{1-\epsilon},$$
provided that $|x - c| < \delta$. Hence $P(x)$ is a continuous function of x within the interval (a, b).

If the function $f_n(x)$ has a derivate, such that $|f_n'(x)| < M_n$, and if
$$1 + f_n(x) \geqq a > 0$$

---

* For $$(1-\epsilon)(1+\tfrac{1}{3}\epsilon) = 1 - \tfrac{2}{3}\epsilon - \tfrac{1}{3}\epsilon^2,$$
$$(1-\tfrac{1}{3}\epsilon)^2 = 1 - \tfrac{2}{3}\epsilon + \tfrac{1}{9}\epsilon^2,$$
so that $$(1-\epsilon)(1+\tfrac{1}{3}\epsilon) < (1-\tfrac{1}{3}\epsilon)^2.$$

at all points of the interval; then the infinite product has a derivate $P'(x)$ given by

$$\frac{P'(x)}{P(x)} = \Sigma \frac{f_n'(x)}{1+f_n(x)}.$$

For under these conditions we have

$$\frac{f_n'(x)}{1+f_n(x)} < \frac{M_n}{a},$$

so that Art. 46 can be applied; and we find, accordingly,

$$\Sigma \frac{f_n'(x)}{1+f_n(x)} = \frac{d}{dx} \Sigma \log [1+f_n(x)] = \frac{d}{dx} \log P(x) = \frac{P'(x)}{P(x)}.$$

49. Closely connected with the theory of uniform convergence is the following theorem* which is of frequent use in subsequent investigations:

Suppose that we are given a sum

$$F(n) = v_0(n) + v_1(n) + v_2(n) + \ldots + v_p(n)$$

and that we want to find the limit $\lim_{n\to\infty} F(n)$, *it being understood that p tends steadily to infinity with n. Then, if we have*

$$\lim_{n\to\infty} v_r(n) = w_r \quad (r \text{ being } \textbf{fixed}),$$

the limit is given by

$$\lim_{n\to\infty} F(n) = w_0 + w_1 + w_2 + \ldots \text{ to } \infty = W,$$

provided that $|v_r(n)| \leqq M_r$, *where* M_r *is* **independent of** n, *and the series* ΣM_r *is convergent.*

The reader will note that the test for the theorem is substantially the same† as the M-test due to Weierstrass (Art. 44). The proof, too, is almost the same.

First choose a number q (which of course is independent of n), such that

$$M_q + M_{q+1} + \ldots \text{ to } \infty < \epsilon,$$

and let n be taken large enough to make $p > q$; then we have

$$|v_q + v_{q+1} + \ldots + v_p| \leqq M_q + M_{q+1} + \ldots \text{ to } \infty < \epsilon$$

or

$$|F(n) - (v_0 + v_1 + v_2 + \ldots + v_{q-1})| < \epsilon.$$

* Tannery, *Fonctions d'une variable*, § 183 (in the 2nd edition).

† Here of course n takes the place of x in the test of Art. 44.

Also $|w_q+w_{q+1}+w_{q+2}+\dots \text{ to } \infty| \leqq M_q+M_{q+1}+\dots \text{ to } \infty < \epsilon$

Thus $|F(n)-(w_0+w_1+w_2+\dots \text{ to } \infty)|$

$$<|(v_0+v_1+\dots+v_{q-1})-(w_0+w_1+\dots+w_{q-1})|+2\epsilon,$$

and it is to be remembered that so far n has only been restricted by the condition $p>q$.

Now, since q is *fixed* and *independent* of n, we can allow n to tend to infinity in the last inequality, and then we find

$$\overline{\lim}\, |F(n)-W| \leqq 2\epsilon,$$

because
$$\lim_{n\to\infty} v_r(n) = w_r.$$

Hence, since ϵ is arbitrarily small, we find, as on p. 120,

$$\lim\,[F(n)-W]=0,$$

or
$$\lim_{n\to\infty} F(n) = W = w_0+w_1+w_2+\dots \text{ to } \infty.$$

The following example will serve to shew the danger of trying to use the foregoing theorem when the M-test does not apply.

Consider the sum

$$F(n)=\log\left(1+\frac{1}{n^2}\right)+\log\left(1+\frac{2}{n^2}\right)+\dots+\log\left(1+\frac{n}{n^2}\right),$$

so that
$$v_r(n)=\log\left(1+\frac{r}{n^2}\right) \text{ and } p=n.$$

Then obviously
$$w_r=\lim_{n\to\infty} v_r(n)=0,$$

and so the sum of the series $w_0+w_1+w_2+\dots$ is 0.

But $v_r(n)$ lies between r/n^2 and $r/(n^2+n)$, and $\sum_1^n r=\frac{1}{2}(n^2+n)$,

so that
$$\frac{1}{2}\left(1+\frac{1}{n}\right) > F(n) > \frac{1}{2},$$

and hence
$$\lim_{n\to\infty} F(n)=\tfrac{1}{2}.$$

Another theorem of importance in this connexion is the analogous result for products:

Suppose that

$$P(n)=[1+v_0(n)][1+v_1(n)]\dots[1+v_p(n)]$$

where p tends steadily to infinity with n.

Then if $\lim_{n\to\infty} v_r(n)=w_r$, and if $|v_r(n)|\leqq M_r$ where M_r is independent of n and ΣM_r is convergent, we have the equation

$$\lim_{n\to\infty} P(n)=(1+w_0)(1+w_1)(1+w_2)\dots \text{ to } \infty.$$

The reader should have little difficulty in constructing a proof of this theorem on the lines of the foregoing, employing the results of Arts. 38, 39 to find limits for the products

$$(1+v_q)(1+v_{q+1})\dots(1+v_p)$$

and

$$(1+w_q)(1+w_{q+1})\dots \text{ to } \infty$$

in terms of the remainder $M_q+M_{q+1}+\dots$ to ∞.

To shew the need of some condition such as the M-test, we may consider the example

$$P(n)=\left(1+\frac{1}{n}\right)^p$$

in which

$$v_0=v_1=v_2=\dots=1/n,$$

so that

$$w_0=w_1=w_2=\dots=0.$$

But the equation

$$\lim_{n\to\infty}\left(1+\frac{1}{n}\right)^p=1$$

is not necessarily true. In fact the value of the limit depends on the value of $\lim(p/n)$, because

$$\frac{1}{n+1}<\log\left(1+\frac{1}{n}\right)<\frac{1}{n},$$

so that

$$\frac{p}{n+1}<\log\left(1+\frac{1}{n}\right)^p<\frac{p}{n}.$$

Thus

$$\lim_{n\to\infty}\log\left(1+\frac{1}{n}\right)^p=\lim_{n\to\infty}\frac{p}{n}.$$

EXAMPLES.

1. Shew that if $S_n(x)=x^n/(1+x^{2n})$, $x=1$ is a point of non-uniform convergence of $S_n(x)$ to its limit. Draw graphs of $S_n(x)$ and $\lim S_n(x)$.

2. Shew that the series $f(x)=\sum_1^\infty \frac{1}{n^3+n^4x^2}$ is uniformly convergent for all values of x; and that $f'(x)$ is given by term-by-term differentiation.

3. If $f_n(x)$ is never negative in the interval (a, b), and if $\Sigma f_n(x)$ is a continuous function of x in the same interval, shew that the series converges uniformly in the interval. [Dini.]

[For if $x=c$ is any point of the interval, we can find m such that

$$0\leqq F(c)-S_m(c)<\tfrac{1}{3}\epsilon.$$

Further, since $F(x)$ and $S_m(x)$ are continuous, we can find δ such that

$$|F(x)-F(c)|<\tfrac{1}{3}\epsilon,\quad |S_m(x)-S_m(c)|<\tfrac{1}{3}\epsilon,$$

provided that $|x-c|<\delta$.

Hence we have

$$0\leqq F(x)-S_m(x)<\epsilon$$

at all points of the interval $(c-\delta, c+\delta)$.

Now, since $S_n(x)$ never decreases (as n increases), we have

$$S_n(x) \geqq S_m(x), \quad \text{if } n > m.$$

Thus $$F(x) - S_n(x) < \epsilon, \quad \text{if } n > m,$$

at *all* points of the interval $(c-\delta, c+\delta)$. Consequently there can be no point of non-uniform convergence in the neighbourhood of c; and therefore there is no such point in the whole interval (a, b).

The reader should observe that this argument fails in cases such as those illustrated in Figs. 14, 15, because then $S_n(x)$ may be further from $F(x)$ than $S_m(x)$ is. In the cases considered here, *no two of the approximation curves can intersect.*]

4. Shew that the series $\sum_1^\infty \frac{x}{n(1+nx^2)}$ is continuous for all values of x, and deduce from Ex. 3 that it converges uniformly.

[It is easy to prove that $x=0$ is the only possible point of discontinuity, by means of the M-test. Now if we take ν as equal to the greatest integer contained in $1/x^2$, we have $\sum_1^\nu < x(1+\log \nu)$, $\sum_{\nu+1}^\infty < 1/x\nu$, if $0<x<1$. Hence

since $$\frac{1}{x\nu} = \frac{x}{x^2\nu} < \frac{x}{1-x^2}, \text{ we find } \sum^\infty < x\left(1 + 2\log\frac{1}{x} + \frac{1}{1-x^2}\right),$$

the limit of which as $x \to 0$ is zero, so that the series is continuous at $x=0$.]

5. Shew that $f(x) = \sum_0^\infty \frac{1}{1+n^2+n^4x^2}$ converges uniformly for all values of x; examine whether $f'(0)$ can be found by term-by-term differentiation.

6. Shew that $$\Sigma \frac{1}{n!} \frac{a^n}{1+x^2a^{2n}}, \quad \Sigma \frac{(-1)^n}{n!} \frac{a^n}{1+x^2a^{2n}}$$

converge uniformly for all values of x; and that if $a<1$ and $x<1$, they are respectively equal to the series

$$e^a - x^2e^{a^3} + x^4e^{a^5} - \dots,$$

and $$e^{-a} - x^2e^{-a^3} + x^4e^{-a^5} - \dots,$$

obtained by expanding each fraction in powers of x. [PRINGSHEIM.]

7. If $$f_n(x) = x^n(1-x^n);$$

then we have $$\Sigma f_n(x) = x/(1-x^2), \quad \text{if } |x|<1,$$

but $$\Sigma f_n(1) = 0, \text{ although } \lim_{x\to 1}[\Sigma f_n(x)] = \infty.$$

8. Shew that the series $\sum_1^\infty 1/(n+x)^2$ converges uniformly if $x \geqq 0$; but that it cannot be integrated from 0 to ∞. [OSGOOD.]

9. If $|f_n(x)| < M_n/x^p$, where M_n is a positive constant such that ΣM_n converges and $p>1$, then, if $a>0$,

$$\int_a^\infty [\Sigma f_n(x)]\,dx = \Sigma \int_a^\infty f_n(x)\,dx.$$

Apply this to the series $\sum_1^\infty 1/(n+x)^3$; and shew that it does *not* apply to Ex. 8. [OSGOOD.]

10. If Σa_n oscillates finitely or converges, then the series $\Sigma(a_n/n^x)$ is a continuous function of x, if $x \geqq c > 0$. [DIRICHLET.]

11. Shew that $$2 \lim_{x\to 1} \sum_1^\infty \frac{x^n(1-x)}{n(1-x^{2n})} = \sum_1^\infty \frac{1}{n^2}.$$

[For $1-x^{2n} \geqq 2nx^n(1-x)$ and the M-test can be used.]

12. If Σu_n is an absolutely convergent series of constants, shew that $\Pi(1+u_n x)$ converges absolutely and uniformly in any finite interval.

If Σu_n converges (not absolutely) and $\Sigma u_n{}^2$ converges, $\Pi(1+u_n x)$ converges uniformly (but not absolutely) in any finite interval.

13. Shew that the products
$$\Pi[1+(-1)^n x/n],\quad \Pi[1+(-1)^n \sin(x/n)],\quad \Pi\cos(x/n)$$
converge uniformly in any finite interval, and that the third converges absolutely.

14. If $\left|\sum_{n=0}^{m} f'(x)\right|$ is less than a fixed number G at all points of (a, b) and for all values of m, then if $\sum_0^\infty f_n(x)$ converges at all points of the interval (a, b), it converges uniformly. [BENDIXSON.]

[For, divide the interval into ν sub-intervals each of length $l=\delta/G$, where $\delta < \frac{1}{2}\epsilon$, ϵ being any assigned small positive number. Next find m so that at the ends of each sub-interval
$$\phi(x_r) = \sum_{m+1}^{m+p} f_n(x_r),\quad (p=1, 2, 3, \ldots)$$
is numerically less than δ. This is possible because the series converges at each of these points, *and they are finite in number* $(\nu+1)$. Now if x is any point of the interval the nearest end of a sub-interval (say x_r) is not further distant than $\frac{1}{2}l$; hence
$$|\phi(x)-\phi(x_r)| < (\tfrac{1}{2}l)(2G) = \delta,$$
because $$|\phi'(x)| < 2G.$$
Thus $$|\phi(x)| < |\phi(x_r)| + \delta < 2\delta < \epsilon,$$
and so the test of uniform convergence is satisfied.]

15. Apply Bendixson's test to the series
$$\Sigma(1/n)\cos nx,\quad \Sigma(1/n)\sin nx.$$

16. If Σa_n converges and (μ_n) is a sequence which tends steadily to ∞ with n, the series $\Sigma a_n \mu_n{}^{-x}$ converges uniformly if $x \geqq 0$. Deduce that there is, in general, some number ξ such that $\Sigma c_n \mu_n{}^{-x}$ converges if $x > \xi$, and does not converge if $x < \xi$. Of course it is possible that the latter series may converge for *all* values of x or for *no* values of x; examples are given by $c_n = 1/n!$ or $n!$ and $\mu_n = n$. [CAHEN.]

17. Shew that $$\lim_{x\to 1} \sum_1^\infty (-1)^{n-1} n^{-x} = \log 2.$$

CHAPTER VIII.

POWER SERIES.

50. The power-series $\Sigma a_n x^n$ is one of the most important types of uniformly convergent series.

We recall the result proved in Art. 10, that if

$$\overline{\lim_{n\to\infty}} |a_n|^{\frac{1}{n}} = l$$

the power-series converges absolutely when $|x| < 1/l$; but the series cannot converge if $|x| > 1/l$, for then $\overline{\lim} |a_n x^n| > 1$, and so there will be an infinity of terms in the series whose absolute values are greater than 1.

Thus any power-series has an interval $(-1/l, +1/l)$ within which it converges absolutely, and outside which convergence is impossible. By writing x in place of lx, we can reduce this interval to the special one $(-1, +1)$: and we shall suppose this done in what follows (we exclude for the moment the cases $l=0$ or ∞).

Thus suppose that we have a power-series which is absolutely convergent for values of x between -1 and $+1$: so that if k is any number between 0 and 1 the series $\Sigma |a_n| k^n$ is convergent. Then, by Weierstrass's M-test, it is clear that the series $\Sigma a_n x^n$ converges uniformly in the interval $(-k, +k)$, because in that interval $|a_n x^n| \leqq |a_n| k^n$. Hence we have the result that *a power series converges uniformly in an interval which falls entirely within its interval of absolute convergence.*

It sometimes happens that further tests (such as those given in Art. 11) shew that the series is absolutely convergent for $|x|=1$; and then the interval of uniform convergence extends from -1 to $+1$, because we can compare the series $\Sigma a_n x^n$ with $\Sigma |a_n|$ and apply Weierstrass's rule again.

But it may also happen that Σa_n is convergent although not absolutely convergent: in this case we can apply Abel's test (Art. 44), because the sequence of variable factors x^n never increases with n, and is never greater than 1 (if $0 \leqq x \leqq 1$). Consequently, since Σa_n is supposed to converge, the series $\Sigma a_n x^n$ converges uniformly in an interval which includes $x=1$ (but need not extend as far as $x=-1$). Similarly if $\Sigma(-1)^n a_n$ is convergent the interval of uniform convergence includes $x=-1$.

Ex. 1. The series $1+2x+3x^2+4x^3+\ldots$ converges uniformly in the interval $(-k, +k)$, where k is any number between 0 and 1; but the points -1, $+1$ do not belong to the region of uniform convergence.

Ex. 2. The series $1+\frac{x}{2^2}+\frac{x^2}{3^2}+\frac{x^3}{4^2}+\ldots$

converges uniformly in the interval $(-1, +1)$.

Ex. 3. The series $1-\frac{x}{2}+\frac{x^2}{3}-\frac{x^3}{4}+\ldots$

converges uniformly in the interval $(-k, +1)$, where k is any number between 0 and 1; but the point -1 does not belong to the region of uniform convergence.

We now return to the cases $l=0$ or ∞, which we have hitherto left on one side. If it happens that

$$l=\overline{\lim}\,|a_n|^{\frac{1}{n}}=0$$

the series $\Sigma a_n x^n$ will converge absolutely for any value of x; and the interval of uniform convergence may be taken as $(-A, +A)$, where A can be arbitrarily large.

Thus the series

$$1+x+\frac{x^2}{2!}+\frac{x^3}{3!}+\frac{x^4}{4!}+\ldots$$

converges absolutely for any value of x and is uniformly convergent in the interval $(-A, +A)$.

On the other hand, if $l=\overline{\lim}\,|a_n|^{\frac{1}{n}}=\infty$, the series $\Sigma a_n x^n$ cannot converge for any value of x other than zero.

An example of this is afforded by the series

$$1+x+(2!)x^2+(3!)x^3+(4!)x^4+\ldots.$$

There is one important distinction between the intervals of absolute and of uniform convergence; the interval of uniform convergence *must* include its end-points, but the interval of absolute convergence need not. Or, to use a convenient terminology, the former interval is *closed*; the latter may be *unclosed*.

That the interval of absolute convergence of a power-series need not be closed is evident from Ex. 1 above, in which the series is absolutely convergent for any value of x numerically less than 1, but the series diverges for $x=1$ and oscillates for $x=-1$. On the other hand, Ex. 2 gives an illustration of a closed interval of absolute convergence.

But we proved (at the end of Art. 43) that the interval of uniform convergence *must* be closed, whenever the function $S_n(x)$ is a continuous function of x. Now for a power-series $\Sigma a_n x^n$, we have

$$S_n(x)=a_0+a_1x+a_2x^2+\ldots+a_nx^n,$$

which is obviously continuous for all values of x. Consequently the interval of uniform convergence of a power-series is certainly closed. This fact is not deducible from Abel's theorem (see p. 129, top, or Art. 51), for it does not appear impossible *a priori* that a power-series might diverge at $x=1$ and yet be uniformly convergent for $|x|<1$.

51. Abel's theorem is expressed by the equation

$$\lim_{x\to 1}(\Sigma a_n x^n)=\Sigma a_n,$$

provided that Σa_n is convergent; this of course follows from the fact (pointed out in Art. 50) that $x=1$ belongs to the region of uniform convergence and from the theorem (1) in Art. 45.

Abel also shewed that when Σa_n diverges, say to $+\infty$, then $\Sigma a_n x^n$ also tends to $+\infty$ as x approaches 1. This theorem cannot be proved by any appeal to uniform convergence; but the following method applies to both theorems.

Write $\quad A_0=a_0,\ A_1=a_0+a_1,\ldots,\ A_n=a_0+a_1+a_2+\ldots+a_n.$

Then since $1, x, x^2,\ldots$ is a decreasing sequence, we have by the second form of Abel's Lemma (Art. 23)

$$h(1-x^m)+h_m x^m<\sum_0^p a_n x^n<H(1-x^m)+H_m x^m,$$

where H, h are the upper and lower limits of $A_0, A_1, \ldots, A_{m-1}$, and H_m, h_m are those of $A_m, A_{m+1}, \ldots$ to ∞.

Since these limits are independent of p, we have

$$(1) \qquad h(1-x^m)+h_m x^m \leqq \sum_0^\infty a_n x^n \leqq H(1-x^m)+H_m x^m.$$

Suppose first that Σa_n is convergent and has the sum s, then we can choose m so that

$$h_m \geqq s-\epsilon, \quad H_m \leqq s+\epsilon,$$

however small ϵ may be.

Now $$\lim_{x\to 1}\,[h(1-x^m)+h_m x^m]=h_m \geqq s-\epsilon,$$

and $$\lim_{x\to 1}\,[H(1-x^m)+H_m x^m]=H_m \leqq s+\epsilon,$$

because the index m depends only on the series Σa_n and is therefore independent of x.

Thus we see from (1), that

$$s-\epsilon \leqq \varliminf_{x\to 1} \Sigma a_n x^n \leqq \varlimsup_{x\to 1} \Sigma a_n x^n \leqq s+\epsilon.$$

But ϵ is arbitrarily small, and so (see Note (6), p. 5) both the maximum and minimum limiting values of $\Sigma a_n x^n$ must be equal to s.

Hence $$\lim_{x\to 1}\,(\sum_0^\infty a_n x^n)=s=\Sigma a_n.$$

Secondly, if Σa_n diverges, say to ∞, we can find m so that $h_m \geqq N$, however large N may be. Thus

$$\lim_{x\to 1}\,[h(1-x^m)+h_m x^m]=h_m \geqq N,$$

so that from (1) $$\varliminf_{x\to 1} \sum_0^\infty a_n x^n \geqq N.$$

Thus, by an argument similar to that used above, we prove that

$$\lim_{x\to 1} \Sigma a_n x^n = \infty.$$

Thirdly, if Σa_n oscillates, we can see by the same argument that

$$\varliminf_{n\to\infty} A_n \leqq \varliminf_{x\to 1} \Sigma a_n x^n \leqq \varlimsup_{n\to\infty} A_n.$$

Closely connected with the foregoing results is **the theorem of comparison for two divergent series.**

Suppose that Σa_n, Σb_n are two divergent series, then we have an inequality similar to (1),

$$k(1-x^m)+k_m x^m \leqq \Sigma b_n x^n \leqq K(1-x^m)+K_m x^m,$$

where k, K and k_m, K_m are derived from $B_n=b_0+b_1+\ldots+b_n$ in the same way as h, H and h_m, H_m from A_n.

Thus, *provided that h, h_m are positive,** we find

$$(2) \quad \phi(x)=\frac{k(1-x^m)+k_m x^m}{H(1-x^m)+H_m x^m} \leqq \frac{\Sigma b_n x^n}{\Sigma a_n x^n} \leqq \frac{K(1-x^m)+K_m x^m}{h(1-x^m)+h_m x^m}=\Phi(x).$$

* In applications of the theorem it is generally the case that a_n is always positive.

Obviously $\lim_{x\to 1} \phi(x) = k_m/H_m$, and $\lim_{x\to 1} \Phi(x) = K_m/h_m$;

and now suppose that B_n/A_n approaches a limit l as n increases; it is then possible to choose m, so that

$$k_m/H_m \geqq l-\epsilon, \quad K_m/h_m \leqq l+\epsilon$$

however small ϵ may be. Then we have at once

$$\lim_{x\to 1} \phi(x) \geqq l-\epsilon \text{ and } \lim_{x\to 1} \Phi(x) \leqq l+\epsilon.$$

Hence from (2) we have, if $f(x) = \Sigma b_n x^n / \Sigma a_n x^n$,

$$l-\epsilon \leqq \underline{\lim}_{x\to 1} f(x) \leqq \overline{\lim}_{x\to 1} f(x) \leqq l+\epsilon.$$

Repeating the former argument, we find that

$$\lim_{x\to 1} f(x) = l.$$

In a similar way we can prove that if $B_n/A_n \to \infty$, then $f(x) \to \infty$; and that if B_n/A_n oscillates, $f(x)$ may oscillate between limits which cannot be wider than those of B_n/A_n.

Thus we have the theorem, due to Cesàro:

If $B_n/A_n = (b_0+b_1+\dots+b_n)/(a_0+a_1+\dots+a_n)$ approaches a definite limit, finite or infinite, then

$$\lim_{x\to 1} [(\Sigma b_n x^n)/(\Sigma a_n x^n)] = \lim_{n\to\infty} B_n/A_n;$$

a result which can be obtained also from Art. 153 of the Appendix.

It should be noticed that if b_n/a_n approaches a limit, B_n/A_n approaches the same limit (Appendix, Art. 152); thus a particular case of the theorem is:

If b_n/a_n approaches a definite limit, finite or infinite, then

$$\lim_{x\to 1} [(\Sigma b_n x^n)/(\Sigma a_n x^n)] = \lim_{n\to\infty} (b_n/a_n).$$

Ex. 1. It is possible to obtain the first form of Abel's theorem from the last theorem, by comparing the two series

$$A_0 + A_1 x + A_2 x^2 + \dots, \quad 1 + x + x^2 + \dots. \qquad \text{[Cesàro.]}$$

Ex. 2. Similarly, by comparing the series

$$A_0 + (A_0+A_1)x + (A_0+A_1+A_2)x^2 + \dots \text{ and } 1 + 2x + 3x^2 + \dots,$$

we see that if $\lim_{n\to\infty} \frac{1}{n}(A_0+A_1+A_2+\dots+A_{n-1}) = l$,

then $\lim_{x\to 1} (\Sigma a_n x^n) = l$. [Frobenius.]

Ex. 3. Again, if the limit in Ex. 2 is not definite, we may consider a further mean. Suppose that

$$\lim_{n\to\infty} \frac{nA_0 + (n-1)A_1 + (n-2)A_2 + \dots + A_{n-1}}{\frac{1}{2}n(n+1)} = l,$$

then we can compare the series

$$A_0 + (2A_0+A_1)x + (3A_0+2A_1+A_2)x^2 + \dots \text{ and } 1 + 3x + 6x^2 + \dots,$$

and prove that $\lim_{x\to 1} (\Sigma a_n x^n) = l$. [Compare Art. 123.]

We note that each of the examples 1, 2, 3 includes the preceding one.

Ex. 4. As other applications, the reader may shew that

(i) $\lim\limits_{x\to 1}(1-x)^{\frac{1}{2}}(x+x^4+x^9+x^{16}+\ldots)=\frac{1}{2}\sqrt{\pi}$,

(ii) $\lim\limits_{x\to 1}[(x+x^a+x^{a^2}+x^{a^3}+\ldots)/\log(1-x)]=-1/\log a$,

(iii) $\lim\limits_{x\to 1}(1-x)^p(1^{p-1}x+2^{p-1}x^2+3^{p-1}x^3+\ldots)=\Gamma(p)$,

(iv) $\lim\limits_{x\to 1}(x-x^4+x^9-x^{16}+\ldots)=\frac{1}{2}$.

In case (i), the series $x+x^4+x^9+\ldots$ gives $A_n \sim n^{\frac{1}{2}} \sim \Gamma(n+\frac{3}{2})/\Gamma(n+1)$, while the series for $(1-x)^{-\frac{1}{2}}$ gives $B_n=3.5.7\ldots(2n+1)/2.4.6\ldots 2n$.

In case (ii) we find $A_n \sim \log n/\log a$, while the series for $\log(1-x)$ gives $B_n \sim -\log n$.

In case (iii) we use the fact that $a_n \sim \Gamma(n+p)/\Gamma(n+1)$.

Finally, in case (iv) we have $A_0+A_1+\ldots+A_n \sim \frac{1}{2}n$.

Lasker and Pringsheim* have proved theorems of great generality on series which diverge at $x=1$. As an example we quote the following:

If $\lambda(x)$ is a function of x, steadily increasing to ∞ with x, but more slowly than x, so that $\lim[\lambda(x)/x]=0$, then $\Sigma\lambda'(n)x^n$ is represented approximately by $\lambda[1/(1-x)]$ for values of x near to 1.

52. Properties of a power-series.

The general theorems proved in Arts. 45, 46 of course apply to a power-series, so that we can make the following statements:

(1) *A power-series $\Sigma a_n x^n$ is a continuous function of x in any interval contained within its region of convergence.*

(2) *If (c_1, c_2) is an interval within the region of convergence*

$$\int_{c_1}^{c_2} x^k(\Sigma a_n x^n)dx=\Sigma\frac{a_n}{n+k+1}(c_2^{n+k+1}-c_1^{n+k+1}).$$

(3) *If x is any point within the region of convergence*

$$\frac{d}{dx}(\Sigma a_n x^n)=\Sigma n a_n x^{n-1}.$$

We note that the interval of absolute convergence of a power-series is not altered by differentiation or integration. This follows from the fact that† $\lim n^{\frac{1}{n}}=1$,

so that $$\overline{\lim}\,|na_n|^{\frac{1}{n}}=\overline{\lim}\,|a_n|^{\frac{1}{n}}=\overline{\lim}\left|\frac{a_n}{n+1}\right|^{\frac{1}{n}}.$$

---

* Pringsheim, *Acta Mathematica*, Bd. 28, 1904, p. 1, where full references will be found.

† For $\lim(n+1)/n=1$, so that $\lim n^{\frac{1}{n}}=1$ by Art. 154 in the Appendix.

By applying Abel's theorem (Art. 51) to the integrated series we see that *in* (2) *the point* c_2 *may be taken at the boundary of the interval of absolute convergence, provided that the* **integrated** *series converges there, no matter whether the original series does so or not.*

An example of this has occurred already in Art. 47.

(4) *If a power-series* $f(x)=\sum_0^\infty a_n x^n$ *converges within an interval* $(-k, +k)$, *there is an interval within which* $f(x)=0$ *has no root except, perhaps,* $x=0$.

For suppose that a_m is the first coefficient in $f(x)$ which does not vanish; then the series

$$f_m(x)=a_{m+1}x+a_{m+2}x^2+a_{m+3}x^3+\dots$$

is a continuous function of x, in virtue of (1) above, and $f_m(x)=0$ for $x=0$. Thus we can find c, so that $|f_m(x)|$ is less than $\frac{1}{2}|a_m|$, if $|x|<c<k$; consequently in the interval $(-c, c)$,

$$|f(x)|\geqq|a_m x^m|-|x^m f_m(x)|>\tfrac{1}{2}|a_m|.|x|^m.$$

Hence $f(x)=0$ has no root other than $x=0$ in the interval $(-c, c)$; and *if* a_0 *is different from zero (so that* $m=0$), $f(x)=0$ *has no root in the interval* $(-c, c)$.

(5) It is an immediate deduction from (4) that: *If two power-series* $f(x)=\sum_0^\infty a_n x^n$, $g(x)=\sum_0^\infty b_n x^n$ *are both convergent in the interval* $(-k, +k)$, *and if, however small* δ *may be, we can find a non-zero value* x_1 *in the interval* $(-\delta, \delta)$, *which satisfies*

$$f(x_1)=g(x_1),$$

then $$a_0=b_0,\ a_1=b_1,\ a_2=b_2,\ \dots,\ a_n=b_n,\ \dots,$$

and the two series are identical.

In practice, we hardly ever need this theorem except when $f(x)$ is known to be equal to $g(x)$ for *all* values of x, such that $|x|<\delta$.

53. We have hitherto discussed the continuity of the power-series from the point of view of the variable x; but it sometimes happens that we wish to discuss a series $\Sigma f_n(y).x^n$ regarded as a function of the variable y. The following theorem (due to Pringsheim) throws some light on this question:*

* Further results have been established by Hartogs (*Math. Annalen*, Bd. 62, 1906, p. 9), using more elaborate analysis.

Suppose that a positive value X can be found such that

$$|f_n(y)|X^n < An^p, \quad a \leqq y \leqq b,$$

where A, p are **fixed** *and positive, and n has* **any** *value. Then $\Sigma f_n(y).x^n$ is a continuous function of y in the interval (a, b), provided that $f_n(y)$ is continuous for all finite values of n, and that $|x| < X$.*

To prove the theorem, we need only compare the series with $\Sigma An^p\left(\frac{|x|}{X}\right)^n$, which is independent of y and is convergent when $|x| < X$; thus the series $\Sigma f_n(y).x^n$ (by Weierstrass's test) converges uniformly with regard to y in the interval (a, b), and is therefore a continuous function of y in that interval.

It was erroneously supposed by Abel that the convergence of $\Sigma f_n(y).X^n$ in the interval (a, b) was sufficient to ensure the continuity of $\Sigma f_n(y)x^n$ for $0 < x < X$ (assuming $f_n(y)$ continuous). But Pringsheim has constructed an example shewing that this condition is not sufficient (see Example (5) below).

The following examples are due to Abel, with the exception of (5):

(1) The series $\quad 1^y x + 2^y x^2 + \ldots + n^y x^n + \ldots \quad (|x| < 1)$

represents a continuous function of y.

(2) The series $\quad x \sin y + \frac{1}{2}x^2 \sin 2y + \frac{1}{3}x^3 \sin 3y + \ldots$

is continuous as regards y when $|x| < 1$; but although the series still converges if $x = 1$, it is discontinuous at $y = 0, \pm 2\pi, \pm 4\pi, \ldots$ (see Art. 65).

(3) The series $\quad f(y) = \frac{y}{1+y^2} + \frac{y}{4+y^2}x + \frac{y}{9+y^2}x^2 + \ldots$

is a continuous function of y if $|x| < 1$; and thus $\lim_{y\to\infty} f(y) = 0$. But if $x = 1$, the series $\quad \frac{y}{1+y^2} + \frac{y}{4+y^2} + \frac{y}{9+y^2} + \ldots$ (see Art. 11)

differs from $\quad \int_1^\infty \frac{y\,dx}{x^2+y^2} = \tan^{-1} y$

by less than the first term $y/(1+y^2)$. Thus it is evident that

$$\lim_{y\to\infty}\left(\frac{y}{1+y^2} + \frac{y}{4+y^2} + \frac{y}{9+y^2} + \ldots\right) = \frac{\pi}{2}.$$

(4) The convergence of the series

$$\Sigma[\lim_{y\to 0} f_n(y)]x^n$$

does not follow from that of $\Sigma f_n(y)x^n$ for all values of $y>0$. Thus the series

$$\frac{\sin y}{y}+\frac{\sin 2y}{y}x+\frac{\sin 2^2y}{y}x^2+\ldots+\frac{\sin 2^ny}{y}x^n+\ldots$$

converges if $x<1$, when $y>0$; but the series $1+2x+2^2x^2+\ldots+2^nx^n+\ldots$ diverges if $x>\frac{1}{2}$.

(5) *Pringsheim's Example*:

Let M_n tend steadily to ∞ with n in such a way that $\lim M_{n+1}/M_n=\infty$, and let $M_0=0$. [For example $M_0=0$, $M_n=n^n$.] Then write

$$f_n(y)=\frac{M_{n+1}y^2}{1+M_{n+1}y^2}-\frac{M_ny^2}{1+M_ny^2},$$

and it is evident that the series $\Sigma f_n(y)x^{2n}$ converges for all real values of y and for any value* of x. Further, the functions $f_0, f_1, \ldots$ are continuous for all real values of y. *But if* $x\geqq 1$, *the series* $\Sigma f_n(y)x^{2n}$ *is discontinuous at* $y=0$.

For
$$\Sigma f_n(0)x^{2n}=0.$$

But
$$f_0(y)+f_1(y)+\ldots+f_{n-1}(y)=\frac{M_ny^2}{1+M_ny^2},$$

and so if $|y|>0$,
$$\sum_0^\infty f_n(y)=1.$$

Now the terms $f_n(y)$ are *positive*, so that

$$\sum_0^\infty f_n(y)\,.\,x^{2n}\geqq 1 \text{ if } x\geqq 1 \quad (|y|>0).$$

From these facts it is clear that the series is discontinuous at $y=0$, if $x\geqq 1$.

Of course if $|x|<1$, the series is continuous at $y=0$, because $f_n(y)$ is positive but less than 1, and so

$$|f(y)x^{2n}|<x^{2n};$$

and thus the Weierstrass-test applies.

54. Multiplication and division of power-series.

As regards *multiplication* of two power-series, the results of Art. 34 shew that if both series

$$a_0+a_1x+a_2x^2+\ldots,\quad b_0+b_1x+b_2x^2+\ldots$$

converge absolutely in the interval† $(-k, +k)$, their product is given by

$$c_0+c_1x+c_2x^2+\ldots,$$

---

* Because
$$f_n(y)x^{2n}=\left(\frac{1}{1+M_ny^2}-\frac{1}{1+M_{n+1}y^2}\right)x^{2n}<\frac{x^{2n}}{M_ny^2},$$
and $\Sigma x^{2n}/M_n$ converges for any value of x, since $\lim M_{n+1}/M_n=\infty$. Of course we have taken y not to be zero; if $y=0$, all the terms of the series are zero, and $\Sigma f_n(0)\,.\,x^{2n}=0$.

† If the two series have different regions of convergence, this interval will be the *smaller* of the two.

which converges absolutely in the same interval, where

$$c_0=a_0b_0, \quad c_1=a_0b_1+a_1b_0, \ \ldots,$$
$$c_n=a_0b_n+a_1b_{n-1}+\ldots+a_nb_0, \ \ldots.$$

If we apply Abel's theorem (Art. 51) to the equation

$$\Sigma c_nx^n=(\Sigma a_nx^n).(\Sigma b_nx^n),$$

we can deduce at once his theorem (Art. 34) that $C=AB$, provided that all three series converge.

For *division*, we assume first that the constant term is different from zero; and for simplicity we take it as 1. Thus we consider first

$$\frac{1}{1+b_1x+b_2x^2+\ldots}=\frac{1}{1+y} \text{ say,}$$

where $$y=b_1x+b_2x^2+\ldots.$$

Now $$(1+y)^{-1}=1-y+y^2-y^3+\ldots,$$

and by Art. 36, this series may be arranged in powers of x, provided that

$$|x|<\rho/(M+1),$$

ρ being any number less than the radius of convergence of Σb_nx^n, and M the upper limit of $|b_n|\rho^n$ (of course here $s=1$).

We obtain

$$(1+y)^{-1}=1-b_1x+(b_1^2-b_2)x^2-(b_1^3-2b_1b_2+b_3)x^3+\ldots.$$

This series may then be multiplied by any other power-series in x, and we obtain a power-series for the quotient

$$(a_0+a_1x+a_2x^2+\ldots)/(1+b_1x+b_2x^2+\ldots).$$

Of course if some of the initial terms in the denominator happen to be zero, the quotient may still be found as a power-series *together with a rational fraction.*

Thus suppose $b_0=0$, $b_1=0$, but that b_2 is not zero; then we have

$$\frac{\Sigma a_nx^n}{\Sigma b_nx^n}=\frac{\Sigma a_nx^n}{b_2x^2(1+B_1x+B_2x^2+\ldots)},$$

where $B_1=b_3/b_2$, $B_2=b_4/b_2, \ldots$.

Then, as above, we find

$$\frac{\Sigma a_nx^n}{1+B_1x+B_2x^2+\ldots}=a_0+(a_1-a_0B_1)x+(a_2-a_1B_1+a_0B_1^2-a_0B_2)x^2+\ldots.$$

Thus $$\frac{\Sigma a_nx^n}{\Sigma b_nx^n}=\frac{a_0}{b_2x^2}+\frac{a_1-a_0B_1}{b_2x}+\text{a power series in } x.$$

In practice it is often better to use the method of undetermined coefficients; thus we should write

$$\frac{a_0+a_1x+a_2x^2+\dots}{b_0+b_1x+b_2x^2+\dots}=d_0+d_1x+d_2x^2+\dots,$$

multiply up, and obtain, in virtue of Art. 52 (5),

$$a_0=b_0d_0, \quad a_1=b_0d_1+b_1d_0,$$
$$a_2=b_0d_2+b_1d_1+b_2d_0, \dots,$$

from which we get successively

$$d_0, d_1, d_2, \dots.$$

A more exact determination of the interval of convergence is given in Art. 84 below.

55. Reversion of a power-series.

Suppose that the series

$$y=a_1x+a_2x^2+a_3x^3+\dots$$

converges absolutely in the interval $(-k, +k)$, and that it is required to express x, if possible, as a power-series in y.

Let us try to solve the equation, formally, by inserting a power-series

$$x=b_1y+b_2y^2+b_3y^3+\dots.$$

If $\Sigma b_n y^n$ is convergent for any value of y other than zero (by Art. 36), the resulting series may certainly be re-arranged in powers of y without altering its value, at any rate for some values of y; leaving the question of what these values are for subsequent examination, we have, in a certain interval,

$$y=(a_1b_1)y+(a_1b_2+a_2b_1^2)y^2+(a_1b_3+2a_2b_1b_2+a_3b_1^3)y^3+\dots$$

or $\qquad a_1b_1=1, \quad a_1b_2+a_2b_1^2=0, \quad a_1b_3+2a_2b_1b_2+a_3b_1^3=0,$

and so on, in virtue of Art. 52 (5).

Thus we can determine, step-by-step, the succession of coefficients,

$$b_1=1/a_1, \quad b_2=-a_2/a_1^3, \quad b_3=2a_2^2/a_1^5-a_3/a_1^4, \quad \dots.$$

It is evident from these results that a_1 must be supposed different from 0, or the assumed solution will certainly fail.*

---

* For the case when $a_1=0$ and a_2 is not zero, the reader may refer to Exs. B, 31–33, at the end of the chapter.

We may then take $a_1=1$ without loss of generality, for the given equation can be written

$$y/a_1=x+(a_2/a_1)x^2+(a_3/a_1)x^3+\dots,$$

and so, with a slight change of notation, we can start from

$$y=x+a_2x^2+a_3x^3+\dots.$$

Then the equations for b_1, b_2, b_3, ... give

$$b_1=1,\quad |b_2|=|a_2|,\quad |b_3|\leqq 2|a_2|\,.\,|b_2|+|a_3|,$$

$$|b_4|\leqq|a_2|\{2|b_3|+b_2^2\}+3|a_3|\,.\,|b_2|+|a_4|,\quad \dots.$$

These equations shew that $|b_n|\leqq\beta_n$, where the β's are given by

$$\beta_1=1,\quad \beta_2=\alpha_2,\quad \beta_3=2\alpha_2\beta_2+\alpha_3,$$

$$\beta_4=\alpha_2(2\beta_3+\beta_2^2)+3\alpha_3\beta_2+\alpha_4,\quad \dots,$$

provided that $$|a_n|\leqq\alpha_n.$$

Now the equations for the β's are those which would be obtained by inserting the series

$$\xi=\beta_1\eta+\beta_2\eta^2+\beta_3\eta^3+\dots$$

in the equation

$$\eta=\xi-\alpha_2\xi^2-\alpha_3\xi^3-\dots.$$

But if ρ is any positive number less than k, the series $\Sigma|a_n|\rho^n$ is convergent, and so we can find* a number M such that

$$|a_n|\rho^n\leqq M$$

for all values of n.

Thus we can put $$\alpha_n=M/\rho^n.$$

Then $$\eta=\xi-\frac{M\xi^2}{\rho(\rho-\xi)}.$$

Now this equation gives

$$(M+\rho)\xi^2-\rho(\rho+\eta)\xi+\rho^2\eta=0$$

or $$2(M+\rho)\xi=\rho(\rho+\eta)-\rho[(\rho+\eta)^2-4\eta(M+\rho)]^{\frac{1}{2}},$$

---

* For a more precise determination of M, see Cauchy's inequalities in Art. 82 below.

the negative sign being taken for the square-root, because ξ and η vanish together.

But
$$(\rho+\eta)^2-4\eta(M+\rho)=(\lambda-\eta)(\mu-\eta),$$
where
$$\lambda=2M+\rho+2[M(M+\rho)]^{\frac{1}{2}},$$
$$\mu=2M+\rho-2[M(M+\rho)]^{\frac{1}{2}}.$$

Thus we have
$$2(M+\rho)\xi=\rho(\rho+\eta)-\rho^2\left(1-\frac{\eta}{\lambda}\right)^{\frac{1}{2}}\left(1-\frac{\eta}{\mu}\right)^{\frac{1}{2}},$$
and thus, since $\lambda>\mu$, the value of ξ can be expanded* in a convergent series of powers of η, provided that $0<\eta<\mu$. But this series is clearly the same as $\Sigma\beta_n\eta^n$; which therefore converges if $0<\eta<\mu$. Now $\beta_n\geqq|b_n|$, so that finally $\Sigma b_n y^n$ is absolutely convergent in the interval $(-\mu, +\mu)$. Consequently the *formal* solution proves to be a real one, in the sense that it is certainly convergent for sufficiently small values of y.

It is perhaps advisable to point out that the interval $(-\mu, +\mu)$ has *not* been proved to be the extreme range of convergence of the series $\Sigma b_n y^n$; we only know that the region of convergence is not *less* than the interval $(-\mu, +\mu)$.

For instance, with the series
$$y=x+\frac{x^2}{2!}+\frac{x^3}{3!}+\frac{x^4}{4!}+\dots$$
we could take as the comparison-series
$$\eta=\xi-\left(\frac{\xi^2}{2}+\frac{\xi^3}{2^2}+\frac{\xi^4}{2^3}+\dots\right)$$
or
$$\eta=\xi-\frac{\xi^2}{2-\xi}.$$

This is found to give
$$\lambda=6+4\sqrt{2},\quad \mu=6-4\sqrt{2}=\cdot 343\dots,$$
and so the method above gives an interval only slightly greater than $(-\frac{1}{3}, +\frac{1}{3})$. But actually (see Arts. 58, 62 below)
$$y=e^x-1,\ \text{so that}\ x=\log(1+y),$$
and the series for x converges absolutely in the interval $(-1, +1)$.

56. Lagrange's Series.

In books on the Differential Calculus, an investigation† is commonly given for the expansion of x in powers of y, when an equation holds of the form
$$x=yf(x).$$

---

* We anticipate here the binomial expansion of Art. 61, for the case $\nu=\frac{1}{2}$.

† See for instance Williamson, *Differential Calculus*, chap. 7; Edwards, *Differential Calculus*, chap. 18.

This process gives an analytical expression for the coefficients in the expansion; but it gives no information as to the conditions under which such an expansion is *possible*. As a matter of fact, the expansion is generally not possible unless $f(x)$ can be expanded in a convergent power-series, *the first term not being zero.** It is then possible to write the equation in the form

$$y = x/f(x) = \sum_1^\infty a_n x^n,$$

on carrying out the division. Thus Lagrange's problem is now seen to be, in reality, equivalent to the reversion of the power-series $\sum_1^\infty a_n x^n$, in the form $x = \sum_1^\infty b_n y^n$. Lagrange's investigation shews that nb_n is equal to the coefficient of x^{n-1} in the expansion of $[f(x)]^n$; or, what is the same thing, in the expansion of $(x/y)^n$.

We can easily establish the more general form of Lagrange's series in which $g(x)$ is expanded in powers of y, where $g(x) = \sum_0^\infty c_n x^n$ is another given power-series. We know in fact (from Arts. 36 and 55) that for sufficiently small values of $|x|$ and $|y|$, we can write

$$g(x) = c_0 + \sum_1^\infty d_n y^n,$$

where $d_1 = c_1/a_1$ and the other coefficients have still to be found. The interval of convergence cannot be found, by elementary methods, until the coefficients have been determined.

Now we can differentiate this series term-by-term (Art. 52), and we find

$$g'(x) = \sum_1^\infty \left(n d_n y^{n-1} \frac{dy}{dx} \right).$$

Divide now by y^r, where r is any whole number, and we get

$$\frac{g'(x)}{y^r} = \sum_1^\infty \left(n d_n y^{n-r-1} \frac{dy}{dx} \right).$$

Suppose both sides of this equation to be expanded in ascending powers of x; then, on the right, there is only one term† containing x^{-1}; and this one is the term $r d_r \frac{1}{y} \frac{dy}{dx}$, given by $n = r$.

It is now clear that rd_r is the coefficient of x^{-1} in the expansion of

---

* Compare Exs. B, 31–33, at the end of the chapter.

† Except for $n = r$, we have

$$y^{n-r-1} \frac{dy}{dx} = \frac{1}{(n-r)} \frac{d}{dx}(y^{n-r}) = \frac{1}{(n-r)} \frac{d}{dx}[x^{n-r}(A_0 + A_1 x + A_2 x^2 + \ldots)]$$
$$= A_0 x^{n-r-1} + \ldots;$$

but in this expansion, even if n is less than r, there can be no term in x^{-1}, because x^{-1} is not the differential coefficient of any power of x.

On the other hand, if $n = r$, we have

$$\frac{1}{y} \frac{dy}{dx} = \frac{d}{dx}(\log y) = \frac{d}{dx}[\log x + B_0 + B_1 x + \ldots]$$
$$= \frac{1}{x} + B_1 + 2B_2 x + \ldots.$$

$g'(x)/y^r$ in ascending powers of x; and this is equivalent to Lagrange's formula, although first put in the above form by Jacobi.*

It is to be observed that if the equation in x,

$$y=\sum_1^\infty a_n x^n,$$

is solved by some algebraic (or other) process, there will usually be additional solutions as well as the series found by reversion. This series gives the solution which tends to zero with y.

Ex. 1. If $y=x-ax^2$, and $g(x)=x$, we have to find the coefficient of x^{-1} in the expansion of

$$(x-ax^2)^{-r}=x^{-r}(1-ax)^{-r}.$$

Thus we get $$rd_r=\frac{r(r+1)\ldots(2r-2)}{(r-1)!}a^{r-1},$$

and so $$x=y+ay^2+2a^2y^3+5a^3y^4+\ldots,$$

which converges if $|ay|<\frac{1}{4}$.

Of course this series gives only *one* root of the quadratic in x, namely

$$\{1-\sqrt{(1-4ay)}\}/2a.$$

Ex. 2. In like manner, if $y=x-ax^{m+1}$, we find for one root

$$x=y+ay^{m+1}+\frac{2m+2}{2!}a^2y^{2m+1}+\ldots+\frac{(sm+2)(sm+3)\ldots(sm+s)}{s!}a^s y^{sm+1}+\ldots.$$

Ex. 3. The reader will find similarly that if $y=x(1+x)^n$, then

$$x=y-\frac{2n}{2!}y^2+\frac{3n(3n+1)}{3!}y^3-\frac{4n(4n+1)(4n+2)}{4!}y^4+\ldots.$$

Ex. 4. To illustrate the method of expanding $g(x)$, we take the following example: To expand e^{ax} in powers of $y=xe^{bx}$.

Here $g'(x)/y^r=ax^{-r}e^{(a-rb)x}$, and so the coefficient of x^{-1} is easily seen to be

$$a(a-rb)^{r-1}/(r-1)!.$$

Thus we have

$$e^{ax}=1+ay+\frac{a(a-2b)}{2!}y^2+\frac{a(a-3b)^2}{3!}y^3+\ldots,$$

which converges if $|y|<1/e|b|$.

In particular, with $a=1$, $b=-1$, $\xi=e^x$, we obtain Eisenstein's solution of the equation $\log\xi=y\xi$ (see Ex. 11, p. 18), in the form of the series

$$\xi=1+y+3\frac{y^2}{2!}+4^2\frac{y^3}{3!}+5^3\frac{y^4}{4!}+\ldots.$$

---

* *Ges. Werke*, Bd. 6, p. 37.

CERTAIN SPECIAL POWER SERIES.

57. The exponential limit.*

We proceed to prove that

$$\lim_{\nu\to\infty}(1+\xi)^\nu = 1+x+\frac{x^2}{2!}+\frac{x^3}{3!}+\frac{x^4}{4!}+\dots,$$

where $$x=\lim_{\nu\to\infty}(\nu\xi).$$

Consider first the special case when ν tends to infinity through integral values n, and write $n\xi = X$.

Then we find,† on expanding,

$$(1+\xi)^n = 1+X+\left(1-\frac{1}{n}\right)\frac{X^2}{2!}+\left(1-\frac{1}{n}\right)\left(1-\frac{2}{n}\right)\frac{X^3}{3!}$$
$$+\dots+\left(1-\frac{1}{n}\right)\left(1-\frac{2}{n}\right)\dots\left(1-\frac{n-1}{n}\right)\frac{X^n}{n!}.$$

Now, this expression satisfies the conditions of Art. 49; we can use as the comparison-series,

$$1+X_0+\frac{1}{2!}X_0^2+\frac{1}{3!}X_0^3+\dots,$$

where X_0 is the greatest value‡ of $|X|$ for any value of n. For we have

$$|v_r(n)|=\left(1-\frac{1}{n}\right)\left(1-\frac{2}{n}\right)\dots\left(1-\frac{r-1}{n}\right)\frac{|X|^r}{r!}<\frac{X_0^r}{r!},$$

where X_0 is of course independent of n. Also

$$\lim_{n\to\infty} v_r(n) = x^r/r!,$$

because $\lim_{n\to\infty}\left(1-\frac{r}{n}\right)=1$ and $\lim_{n\to\infty} X = x$. Finally the index p is equal to n, and so of course tends steadily to infinity.

Thus $$\lim(1+\xi)^n = 1+x+\frac{1}{2!}x^2+\frac{1}{3!}x^3+\dots \text{ to } \infty.$$

---

* The reader is recommended to refer to Appendix II. before proceeding further.

† For the general term in the binomial expansion of $(1+\xi)^n$ is

$$\frac{n(n-1)(n-2)\dots(n-r+1)}{r!}\xi^r=\left(1-\frac{1}{n}\right)\left(1-\frac{2}{n}\right)\dots\left(1-\frac{r-1}{n}\right)\frac{(n\xi)^r}{r!}.$$

‡ That there is a greatest value is evident, because it is supposed that X approaches the limit x, as n increases to infinity.

If now ν tends to infinity in any other way, ν will, at any stage, be contained between the two integers n and $(n+1)$ say; and of course n will tend to infinity as ν does. Thus $(1+\xi)^\nu$ will be contained between* $(1+\xi)^n$ and $(1+\xi)^{n+1}$; and $\nu\xi$ will be contained between $n\xi$ and $(n+1)\xi$, so that

$$\lim_{n\to\infty}(n\xi)=\lim_{n\to\infty}(n+1)\xi=x.$$

Thus, from what has already been proved, we see that

$$\lim_{n\to\infty}(1+\xi)^n=1+x+\frac{x^2}{2!}+\frac{x^3}{3!}+\ldots=\lim_{n\to\infty}(1+\xi)^{n+1},$$

and since $(1+\xi)^\nu$ is contained between $(1+\xi)^n$ and $(1+\xi)^{n+1}$, it follows that

$$\lim_{\nu\to\infty}(1+\xi)^\nu=1+x+\frac{x^2}{2!}+\frac{x^3}{3!}+\ldots.$$

If we write for brevity

$$(1+\xi)^n=\sum_{r=0}^{n}f_r(x,\ n),$$

it will be seen that we have used the theorem

$$\lim_{n\to\infty}\sum_{r=0}^{n}f_r(x,\ n)=\sum_{r=0}^{\infty}[\lim_{n\to\infty}f_r(x,\ n)].$$

That is, we have replaced a *single* limit by a *double* (repeated) limit; and of course such a step needs justification (see the examples in Art. 49).

Special cases.

If $\xi=1/\nu$, we have the equation

$$\lim_{\nu\to\infty}\left(1+\frac{1}{\nu}\right)^\nu=1+1+\frac{1}{2!}+\frac{1}{3!}+\ldots \text{ to } \infty=e.$$

The value of e has been calculated in Art. 7 and proved to be

$$2{\cdot}7182\ldots.$$

If $\xi=1/(\nu-1)$, we have

$$(1+\xi)^\nu=[\nu/(\nu-1)]^\nu=(1-1/\nu)^{-\nu},$$

so that

$$\lim_{\nu\to\infty}(1-1/\nu)^{-\nu}=e.$$

These two results may be combined into the single equation

$$\lim_{\lambda\to 0}(1+\lambda)^{1/\lambda}=e,$$

where λ approaches 0 from either side.

---

*It is of course understood that the positive value of $(1+\xi)^\nu$ is taken; and then this value is obviously contained between $(1+\xi)^n$ and $(1+\xi)^{n+1}$ if ν is rational. On the other hand, if ν is irrational, the statement is a consequence of the definition of an irrational power.

58. The exponential function.

We may denote by the symbol $E(x)$ the exponential series

$$1+x+\frac{x^2}{2!}+\frac{x^3}{3!}+\frac{x^4}{4!}+\dots.$$

Then (by Arts. 45, 46) we see that $E(x)$ is a continuous function of x, and that its differential coefficient is given by term-by-term differentiation, so that

$$\frac{d}{dx}E(x)=1+x+\frac{x^2}{2!}+\frac{x^3}{3!}+\dots=E(x),$$

because
$$\frac{d}{dx}\frac{x^r}{r!}=\frac{x^{r-1}}{(r-1)!}.$$

Further, we see from Art. 57 that

$$\begin{aligned}E(x)\times E(y)&=\lim_{n\to\infty}\left(1+\frac{x}{n}\right)^n.\lim_{n\to\infty}\left(1+\frac{y}{n}\right)^n\\&=\lim_{n\to\infty}\left[\left(1+\frac{x}{n}\right)\left(1+\frac{y}{n}\right)\right]^n\\&=\lim_{n\to\infty}\left(1+\frac{x+y}{n}+\frac{xy}{n^2}\right)^n\\&=E(x+y).\end{aligned}$$

This result can also be proved directly from the series for $E(x)$, by applying the rule (Art. 34) for multiplying two absolutely convergent series. The result leads directly to the equations (in which n, n' are positive integers)

$$e^n=[E(1)]^n=E(n)=[E(n/n')]^{n'},\quad E(-x)=[E(x)]^{-1}.$$

Thus we see that $E(x)$ is the *positive* value of e^x for any *rational* value of x. But the equation must also be true for irrational values of x; for if $a_0, a_1, \dots, a_n, \dots$ represents a sequence of rational numbers whose limit is x, we have

$$e^x=\lim_{n\to\infty}e^{a_n}=\lim_{n\to\infty}E(a_n)=E(x),$$

the last step being valid because $E(x)$ is a *continuous* function (Art. 45). In future, we shall generally write e^x instead of $E(x)$; but when the exponent x is a complicated expression it is sometimes clearer to use $\exp x$, as in Ex. A 26, at the end of the chapter.

The reader will find an independent proof (depending on the Integral Calculus) of the equation

$$e^x = 1 + x + \frac{x^2}{2!} + \frac{x^3}{3!} + \ldots$$

in the Appendix (Art. 162).

Ex. As a numerical example, the reader may shew that

$$e^{\frac{1}{2}\pi} = 4{\cdot}800\ldots,$$

and hence that $$e^{\pi} = 23{\cdot}0\ldots.$$

[Reference may be made to the example of Art. 59 for some of the results needed in the calculation.]

59. The sine and cosine power-series.

Write $$\sin x - \left[x - \frac{x^3}{3!} + \frac{x^5}{5!} - \ldots + (-1)^n \frac{x^{2n+1}}{(2n+1)!}\right] = S_n(x)$$

and $$\cos x - \left[1 - \frac{x^2}{2!} + \frac{x^4}{4!} - \ldots + (-1)^n \frac{x^{2n}}{(2n)!}\right] = C_n(x).$$

Then it is plain that $S_n(x)$, $C_n(x)$ are both continuous functions of x and

$$\frac{d}{dx}[S_n(x)] = C_n(x), \quad \frac{d}{dx}[C_n(x)] = -S_{n-1}(x),$$

as may be seen by differentiation.

Now $C_0(x) = \cos x - 1$ is negative, and consequently $\frac{d}{dx}[S_0(x)]$ is negative; but $S_0(x)$ vanishes with x, and consequently $S_0(x)$ is negative when x is positive.*

Thus $\frac{d}{dx}[C_1(x)] = -S_0(x)$ is positive when x is positive; but $C_1(x)$ vanishes with x, and therefore $C_1(x)$ is positive when x is positive.

Hence $\frac{d}{dx}[S_1(x)] = C_1(x)$ is positive when x is positive; and $S_1(x)$ vanishes with x, so that $S_1(x)$ must be positive when x is positive.

That is, $\frac{d}{dx}[C_2(x)] = -S_1(x)$ is negative when x is positive; and therefore, since $C_2(x)$ vanishes with x, $C_2(x)$ is negative when x is positive.

* We make use throughout this article of the obvious fact that if y is increasing with x and is zero for $x=0$, then y (if continuous) must be positive for positive values of x.

We can continue this argument, and by doing so we find that

$$\left.\begin{array}{l} C_0(x),\ C_2(x),\ C_4(x),\ C_6(x),\ldots \\ S_0(x),\ S_2(x),\ S_4(x),\ S_6(x),\ldots \end{array}\right\} \begin{array}{l}\text{are negative when} \\ x \text{ is positive,}\end{array}$$

while the expressions with odd suffixes are positive.

This shews that $\sin x$ lies between the two expressions

$$x-\frac{x^3}{3!}+\frac{x^5}{5!}-\ldots+(-1)^n\frac{x^{2n+1}}{(2n+1)!}$$

and

$$x-\frac{x^3}{3!}+\frac{x^5}{5!}-\ldots+(-1)^n\frac{x^{2n+1}}{(2n+1)!}+(-1)^{n+1}\frac{x^{2n+3}}{(2n+3)!}.$$

Hence, since $\lim\limits_{n\to\infty}\dfrac{x^{2n+3}}{(2n+3)!}=0$ (see Ex. 4, p. 8),

we have

$$\sin x = x-\frac{x^3}{3!}+\frac{x^5}{5!}-\frac{x^7}{7!}+\ldots \text{ to } \infty.$$

In like manner we find

$$\cos x = 1-\frac{x^2}{2!}+\frac{x^4}{4!}-\frac{x^6}{6!}+\ldots \text{ to } \infty.$$

These results have been established only for positive values of x; but it is evident that $\sin x$ and its series both change sign with x, while $\cos x$ and its series do not change sign, so that the results are valid for negative values of x also.

The figure below will serve to shew the relation between $\sin x$ and the first two or three terms in the infinite series.

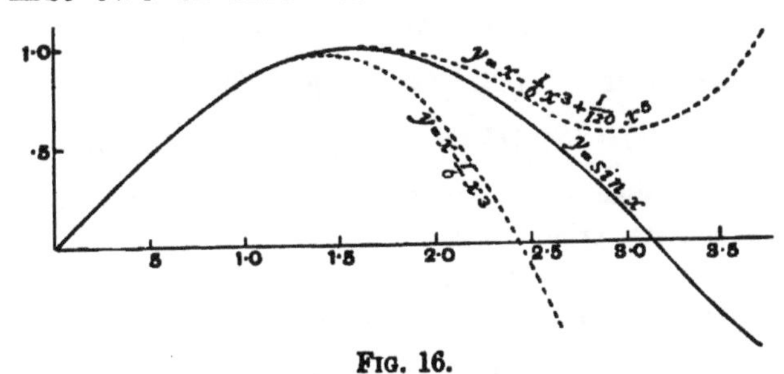

FIG. 16.

Ex. Let us calculate $\cos(\frac{1}{2}\pi)$ and $\sin(\frac{1}{2}\pi)$.

We have $x=\frac{1}{2}\pi=1{\cdot}5708$ very nearly.

This gives

$$\begin{array}{ll} \frac{1}{2}x^2=1{\cdot}2337, & \frac{1}{5040}x^7={\cdot}0047, \\ \frac{1}{6}x^3={\cdot}6460, & \frac{1}{40320}x^8={\cdot}0009, \\ \frac{1}{24}x^4={\cdot}2537, & \frac{1}{362880}x^9={\cdot}0002, \\ \frac{1}{120}x^5={\cdot}0797, & \frac{1}{3628800}x^{10}={\cdot}00003. \\ \frac{1}{720}x^6={\cdot}0209, & \end{array}$$

Hence $\cos(\frac{1}{2}\pi)=1{\cdot}2546-1{\cdot}2546=0$, the error being less than ·00003.

Also $\sin(\frac{1}{2}\pi)=1{\cdot}6507-0{\cdot}6507=1$, the error being less than ·00003.

60. Other methods of establishing the sine and cosine power-series.

(1) Probably the most rapid method of recalling the series to memory is to assume that $\sin x$ and $\cos x$ may be represented by power-series.

Thus if
$$\sin x = a_0 + a_1 x + a_2 x^2 + a_3 x^3 + \ldots$$
we have
$$\cos x = \frac{d}{dx}(\sin x) = a_1 + 2a_2 x + 3a_3 x^2 + \ldots,$$
and so
$$-\sin x = \frac{d}{dx}(\cos x) = 1 \,.\, 2a_2 + 2 \,.\, 3 \,.\, a_3 x + 3 \,.\, 4 \,.\, a_4 x^2 + \ldots .$$

Further, $a_0 = 0$, $a_1 = 1$, because $\sin x$ is 0 and $\cos x$ is 1, for $x = 0$.

Hence we get
$$1 \,.\, 2 \,.\, a_2 = -a_0 = 0,$$
$$2 \,.\, 3 \,.\, a_3 = -a_1 = -1, \text{ or } a_3 = -\frac{1}{3!},$$
$$3 \,.\, 4 \,.\, a_4 = -a_2 = 0,$$
$$4 \,.\, 5 \,.\, a_5 = -a_3, \text{ or } a_5 = \frac{1}{5!},$$
and so on.

But of course we have no *a priori* reason for supposing that $\sin x$ and $\cos x$ can be expressed as power-series; and therefore this method is not logically complete.

(2) We may start from the series, and call them, say, $S(x)$, $C(x)$. Then multiplication of the series gives
$$S(x)\,C(y) + S(y)\,C(x) = S(x+y),$$
$$C(x)\,C(y) - S(x)\,S(y) = C(x+y).$$

Hence in particular
$$[C(x)]^2 + [S(x)]^2 = C(0) = 1$$
and
$$S(2x) = 2S(x)\,C(x), \quad C(2x) = [C(x)]^2 - [S(x)]^2.$$

From these formulae we can shew that $S(x)$ and $C(x)$ satisfy the ordinary results of elementary trigonometry.

Further, it can be seen without much trouble that $C(2)$ is negative,* so that $C(x) = 0$ has at least one root between 0 and 2. But
$$\frac{d}{dx}[C(x)] = -S(x),$$
and $S(x)$ is always positive† for any value of x between 0 and 2. Thus $C(x)$ can have only one root between 0 and 2, because it steadily decreases in that interval.

---

* $C(2) = 1 - 2 + \frac{2}{3} - \frac{2^6}{6!}\left(1 - \frac{4}{7 \,.\, 8}\right) - \frac{2^{10}}{10!}\left(1 - \frac{4}{11 \,.\, 12}\right) - \ldots .$

† Because $S(x) = x\left(1 - \frac{x^2}{2 \,.\, 3}\right) + \frac{x^5}{5!}\left(1 - \frac{x^2}{6 \,.\, 7}\right) + \ldots .$

Call this root $\frac{\pi}{2}$: then we have

$$C\left(\frac{\pi}{2}\right)=0, \quad \left[S\left(\frac{\pi}{2}\right)\right]^2=1,$$

and so
$$S\left(\frac{\pi}{2}\right)=1, \text{ since } S\left(\frac{\pi}{2}\right) \text{ must be positive.}$$

Hence
$$S(\pi)=2S\left(\frac{\pi}{2}\right)C\left(\frac{\pi}{2}\right)=0,$$

$$C(\pi)=\left[C\left(\frac{\pi}{2}\right)\right]^2-\left[S\left(\frac{\pi}{2}\right)\right]^2=-1,$$

and so
$$S(x+\pi)=-S(x), \quad C(x+\pi)=-C(x).$$

Thus
$$S(x+2\pi)=+S(x), \quad C(x+2\pi)=+C(x).$$

On these facts the whole of Analytical Trigonometry can be based.

(3) It is not difficult to prove, by induction or by the methods given in Chap. IX. below, that

$$\sin n\theta=\cos^n\theta\left[nt-\frac{n(n-1)(n-2)}{3!}t^3+\ldots\right],$$

$$\cos n\theta=\cos^n\theta\left[1-\frac{n(n-1)}{2!}t^2+\frac{n(n-1)(n-2)(n-3)}{4!}t^4-\ldots\right],$$

where $t=\tan\theta$, and both series terminate after $\frac{1}{2}n$ or $\frac{1}{2}(n+1)$ or $\frac{1}{2}(n+2)$ terms.

Thus we have, on putting $n\theta=x$,

$$\sin x=\left(\cos^n\frac{x}{n}\right)\left[\xi-\left(1-\frac{1}{n}\right)\left(1-\frac{2}{n}\right)\frac{\xi^3}{3!}+\left(1-\frac{1}{n}\right)\left(1-\frac{2}{n}\right)\left(1-\frac{3}{n}\right)\left(1-\frac{4}{n}\right)\frac{\xi^5}{5!}-\ldots\right],$$

$$\cos x=\left(\cos^n\frac{x}{n}\right)\left[1-\left(1-\frac{1}{n}\right)\frac{\xi^2}{2!}+\left(1-\frac{1}{n}\right)\left(1-\frac{2}{n}\right)\left(1-\frac{3}{n}\right)\frac{\xi^4}{4!}-\ldots\right],$$

where $\xi=n\tan(x/n)$.

To these expressions in brackets we can apply the theorem of Art. 49, using as the comparison-series

$$t_0+\frac{t_0^3}{3!}+\frac{t_0^5}{5!}+\ldots \text{ and } 1+\frac{t_0^2}{2!}+\frac{t_0^4}{4!}+\ldots,$$

where $t_0=|\tan x|\geqq|\xi|$.

The argument is, in fact, almost identical with that employed for the exponential limit in Art. 57; and we get

$$\lim_{n\to\infty}\left[\xi-\left(1-\frac{1}{n}\right)\left(1-\frac{2}{n}\right)\frac{\xi^3}{3!}+\ldots\right]=x-\frac{x^3}{3!}+\frac{x^5}{5!}-\ldots \text{ to } \infty,$$

$$\lim_{n\to\infty}\left[1-\left(1-\frac{1}{n}\right)\frac{\xi^2}{2!}+\ldots\right]=1-\frac{x^2}{2!}+\frac{x^4}{4!}-\ldots \text{ to } \infty.$$

Finally, if we write
$$1+\eta=\cos\frac{x}{n},$$

we have
$$|n\eta|=2n\sin^2\frac{x}{2n}<\frac{x^2}{2n},$$

so that
$$\lim_{n\to\infty} n\eta=0,$$

and thus
$$\lim_{n\to\infty}\cos^n\frac{x}{n}=\lim_{n\to\infty}(1+\eta)^n=E(0)=1 \text{ (see Art. 58).}$$

(4) Another instructive method is to apply the process of integration by parts to the two equations

$$\sin x = \int_0^x \cos(x-t)\,dt, \quad \cos x - 1 = -\int_0^x \sin(x-t)\,dt.$$

If we integrate twice by parts, we obtain

$$\sin x = \left[t\cos(x-t) - \frac{t^2}{2!}\sin(x-t)\right]_0^x - \int_0^x \frac{t^2}{2!}\cos(x-t)\,dt$$

$$= x - \int_0^x \frac{t^2}{2!}\cos(x-t)\,dt$$

and $$\cos x = 1 - \left[t\sin(x-t) + \frac{t^2}{2!}\cos(x-t)\right]_0^x + \int_0^x \frac{t^2}{2!}\sin(x-t)\,dt$$

$$= 1 - \frac{x^2}{2!} + \int_0^x \frac{t^2}{2!}\sin(x-t)\,dt,$$

and so on.

Thus we find that, in the notation of Art. 59,

$$S_{n-1}(x) = (-1)^n \int_0^x \frac{t^{2n}}{(2n)!}\cos(x-t)\,dt,$$

$$C_{n-1}(x) = (-1)^n \int_0^x \frac{t^{2n-1}}{(2n-1)!}\cos(x-t)\,dt.$$

Hence we find

$$|S_{n-1}(x)| \leqq \int_0^{|x|} \frac{t^{2n}}{(2n)!}\,dt \leqq \frac{|x|^{2n+1}}{(2n+1)!}$$

and

$$|C_{n-1}(x)| \leqq \int_0^{|x|} \frac{t^{2n-1}}{(2n-1)!}\,dt \leqq \frac{|x|^{2n}}{(2n)!}.$$

61. The binomial series.

Let us examine the series

$$f(x) = 1 + \nu x + \nu(\nu-1)\frac{x^2}{2!} + \nu(\nu-1)(\nu-2)\frac{x^3}{3!} + \dots \text{ to } \infty.$$

We know from elementary algebra that if ν is a positive integer this series terminates and represents $(1+x)^\nu$. We now proceed to examine the corresponding theorem for other values of ν.

By Art. 12, the region of absolute convergence is given by $|x| < 1$; and so (Art. 50) the series is uniformly convergent in *any* interval $(-k, +k)$, where $0 < k < 1$.

Now $$f'(x) = \nu\left[1 + (\nu-1)x + (\nu-1)(\nu-2)\frac{x^2}{2!} + \dots \text{ to } \infty\right]$$

$$= \nu g(x) \text{ say,}$$

where $g(x)$ differs from $f(x)$ by having $(\nu-1)$ in place of ν.

Also $$(1+x)g(x)=1+(\nu-1)x+(\nu-1)(\nu-2)\frac{x^2}{2!}+\dots$$
$$+x \qquad +2(\nu-1)\frac{x^2}{2!}+\dots$$
$$=1 \qquad +\nu x \qquad +\nu(\nu-1)\frac{x^2}{2!}+\dots$$
$$=f(x),$$
so that $(1+x)f'(x)=\nu f(x)$.

Hence we see that $$\frac{d}{dx}\left[\frac{f(x)}{(1+x)^\nu}\right]=0$$
or $$f(x)=A(1+x)^\nu,$$
where A is a constant independent of x. But $f(0)=1$; and consequently, if we choose the *positive* value for $(1+x)^\nu$, we shall have $A=1$; that is,
$$f(x)=(1+x)^\nu.$$

This result has, of course, been proved only for the interval $(-k, +k)$; let us now see if it can be extended to include the points -1, $+1$. The quotient of the nth term in the series by the $(n+1)$th is
$$-n/(n-\nu-1)x, \quad \text{if } n>\nu+1,$$
and so the series converges for $x=-1$ if ν is positive (Art. 12), and for $x=+1$ if $(\nu+1)$ is positive (Art. 21). Thus by Abel's theorem (Art. 51) the sum of the series for $x=-1$ is 0, if ν is positive; and for $x=+1$ the sum is 2^ν if $\nu+1$ is positive.

Other methods.

(1) The most rapid method for recalling the series to memory is to solve the differential equation
$$(1+x)f'(x)=\nu f(x),$$
by assuming a series $f(x)=1+a_1x+a_2x^2+\dots$.

On substitution, we find that
$$a_1=\nu, \quad 2a_2+a_1=\nu a_1, \quad 3a_3+2a_2=\nu a_2, \quad \text{etc.}$$

Of course this must be supplemented as above in order to complete the proof.

(2) We can multiply together two series with different values of ν (say ν_1 and ν_2) and verify (by Art. 27) that their product is another series in which $\nu=\nu_1+\nu_2$; compare Art. 89, below. Then we can apply the same argument as was used for the exponential series (Art. 58) to prove that $f(x)$ must be the νth power of its value for $\nu=1$.

(3) We have $$(1+x)^\nu - 1 = \nu\int_0^x (1+x-t)^{\nu-1}dt$$
$$= \nu x + \nu(\nu-1)\int_0^x (1+x-t)^{\nu-2}t\,dt,$$
where we obtain the last line by integrating by parts. Continuing thus, we get
$$(1+x)^\nu - \left[1+\nu x+\nu(\nu-1)\frac{x^2}{2!}+\ldots+\nu(\nu-1)\ldots(\nu-r+1)\frac{x^r}{r!}\right]$$
$$= \nu(\nu-1)\ldots(\nu-r)\int_0^x (1+x-t)^{\nu-r-1}\frac{t^r}{r!}dt.$$

Now, if x is positive, $(1+x-t)$ lies between 1 and $(1+x)$, so that
$$\int_0^x (1+x-t)^{\nu-r-1}\frac{t^r}{r!}dt < \frac{x^{r+1}}{(r+1)!},$$
provided that $r>\nu-1$.

On the other hand, if x is negative, we can only say that
$$\left|\int_0^x (1+x-t)^{\nu-r-1}\frac{t^r}{r!}dt\right| < (1+x)^{\nu-r-1}\frac{|x|^{r+1}}{(r+1)!}, \quad \text{if } r>\nu-1.$$

Thus, in either case, the difference
$$(1+x)^\nu - \left[1+\nu x+\ldots+\nu(\nu-1)(\nu-2)\ldots(\nu-r+1)\frac{x^r}{r!}\right]$$
tends to zero as r increases to infinity.

If $x=-1$, it is interesting to note that we can sum the binomial series to a finite number of terms. Thus we have
$$1-\nu+\tfrac{1}{2}\nu(\nu-1)=(1-\nu)(1-\tfrac{1}{2}\nu),$$
$$1-\nu+\tfrac{1}{2}\nu(\nu-1)-\tfrac{1}{6}\nu(\nu-1)(\nu-2)=(1-\nu)(1-\tfrac{1}{2}\nu)(1-\tfrac{1}{3}\nu),$$
and so on.

Hence the sum to $(r+1)$ terms is
$$(1-\nu)\left(1-\frac{1}{2}\nu\right)\left(1-\frac{1}{3}\nu\right)\ldots\left(1-\frac{1}{r}\nu\right).$$

Therefore it is clear from Art. 39 that this sum tends to 0 if ν is positive, to infinity if ν is negative.*

62. The logarithmic series.

We take as our definition (see Appendix II.) of the natural logarithm the equation
$$\log(1+x)=\int_0^x \frac{dt}{1+t}.$$

* Because the series $1+\frac{1}{2}+\frac{1}{3}+\ldots$ to ∞ is divergent.

Now when $|x|<1$, we can write

$$(1+t)^{-1}=1-t+t^2-t^3+\dots \text{ to } \infty,$$

the series converging uniformly from $t=0$ to $t=x$. Hence (by Art. 52 (2)) we have the expansion

$$\log(1+x)=x-\tfrac{1}{2}x^2+\tfrac{1}{3}x^3-\tfrac{1}{4}x^4+\dots \text{ to } \infty.$$

However, it is not necessary to make use of the uniform convergence of the series in order to integrate term-by-term; for we can write

$$\frac{1}{1+t}=1-t+t^2-\dots+(-1)^{n-1}t^{n-1}+(-1)^n\frac{t^n}{1+t}.$$

Thus $$\log(1+x)=x-\tfrac{1}{2}x^2+\tfrac{1}{3}x^3-\dots+(-1)^{n-1}\frac{1}{n}x^n$$
$$+(-1)^n\int_0^x \frac{t^n}{1+t}dt.$$

If x is positive, the last integral is clearly less than

$$\int_0^x t^n dt=\frac{x^{n+1}}{n+1},$$

which tends to zero as n tends to infinity, provided that $0<x\leqq 1$. Thus *the logarithmic series is valid* even for* $x=1$.

On the other hand, when x is negative, we can only say that

$$\left|\int_0^x \frac{t^n}{1+t}dt\right|$$

is less than $$\frac{1}{1+x}\cdot\int_0^{|x|} t^n dt=\frac{|x|^{n+1}}{(n+1)(1+x)},$$

and from this expression it would be expected that $x=-1$ must be excluded from the region of convergence of the logarithmic series; and, as a matter of fact, the series

$$-(1+\tfrac{1}{2}+\tfrac{1}{3}+\tfrac{1}{4}+\dots)$$

has been proved to diverge (Art. 7, Ex. 2).

Another method.

From the identity $(1+x)^y=e^{y\log(1+x)}$,

we see that (Arts. 58, 61)

$$1+yx+y(y-1)\frac{x^2}{2!}+y(y-1)(y-2)\frac{x^3}{3!}+\dots$$
$$=1+y\log(1+x)+\frac{1}{2!}[y\log(1+x)]^2+\frac{1}{3!}[y\log(1+x)]^3+\dots.$$

* That is, the operation of term-by-term integration can here be extended beyond the region of uniform convergence of the integrated series (compare Art. 47 and Art. 52).

It is now necessary to consider whether the first of these series can be re-arranged in powers of y without changing its value. By Art. 26, this derangement will be permissible if the series

$$1+\eta\xi+\eta(\eta+1)\frac{\xi^2}{2!}+\eta(\eta+1)(\eta+2)\frac{\xi^3}{3!}+\dots$$

is convergent, where $\xi=|x|$, $\eta=|y|$.

But the last series is the expanded form of $(1-\xi)^{-\eta}$, which is convergent if $\xi<1$; that is, if $|x|<1$. Thus the derangement will not alter the sum.

Hence we get (from the coefficients of y and y^2) the equations

$$\log(1+x)=x-\tfrac{1}{2}x^2+\tfrac{1}{3}x^3-\tfrac{1}{4}x^4+\dots \text{ to } \infty,$$

$$\tfrac{1}{2}[\log(1+x)]^2=\tfrac{1}{2}x^2-\tfrac{1}{3}x^3(1+\tfrac{1}{2})+\tfrac{1}{4}x^4(1+\tfrac{1}{2}+\tfrac{1}{3})-\tfrac{1}{5}x^5(1+\tfrac{1}{2}+\tfrac{1}{3}+\tfrac{1}{4})+\dots \text{ to } \infty.$$

Similar (but less simple) series may be deduced for higher powers of $\log(1+x)$. [Compare Chrystal's *Algebra*, Ch. XXVIII., § 9.]

63. For purposes of numerical computation of logarithms it is better to use the series

$$\log\frac{1+x}{1-x}=2(x+\tfrac{1}{3}x^3+\tfrac{1}{5}x^5+\tfrac{1}{7}x^7+\dots),$$

which can be found from the previous series by writing $-x$ for x and then subtracting; or directly, by integrating $1/(1-x^2)$. In either way the remainder after n terms is seen to be less than

$$\frac{2x^{2n+1}}{(2n+1)(1-x^2)}.$$

Ex. $\log\frac{3}{2}=\log\frac{1+\frac{1}{5}}{1-\frac{1}{5}}$, $\log\frac{4}{3}=\log\frac{1+\frac{1}{7}}{1-\frac{1}{7}}$, $\log\frac{5}{4}=\log\frac{1+\frac{1}{9}}{1-\frac{1}{9}}$.

The details of calculation for $\log\frac{3}{2}$ may be arranged as follows:

$\frac{1}{5}=$	·20 00 00 00		·20 00 00 00
$\frac{1}{5^3}=$	80 00 00	÷3	26 66 67
$\frac{1}{5^5}=$	3 20 00	÷5	64 00
$\frac{1}{5^7}=$	12 80	÷7	1 83
$\frac{1}{5^9}=$	51	÷9	06
			·20 27 32 56
			2
	$\log\frac{3}{2}=$	·40 54 65	

The error involved in neglecting terms beyond the fifth is less than $\frac{2}{11}\frac{(\frac{1}{5})^{11}}{1-(\frac{1}{5})^2}=\frac{1}{132}\left(\frac{1}{5}\right)^9$, which cannot affect the eighth decimal. Hence the result is correct to the sixth decimal.

Similarly, we get

$$\log \tfrac{4}{3} = 2[\cdot 14285714 + \cdot 00097182 + \cdot 00001190 + \cdot 00000017]$$
$$= \cdot 287682 \text{ to six decimals,}$$

the error involved being again less than a unit in the eighth decimal.

Also $$\log \tfrac{5}{4} = 2[\cdot 11111111 + \cdot 00045725 + \cdot 00000339 + \cdot 00000003]$$
$$= \cdot 223144 \text{ to six decimals.}$$

From these results we find the natural logarithms of all integers from 1 to 10 (with the exception of 7, which can be found similarly from $\log \frac{7}{6}$).

In particular, we have

$$\log 2 = \cdot 693147, \quad \log 3 = 1\cdot 098612,$$
$$\log 5 = 1\cdot 609438, \quad \log 10 = 2\cdot 302585.$$

Of course other series than the above have been found, which converge more rapidly, and so enable the logarithms to be easily calculated to a great number of places. To illustrate, the reader may find formulae for $\log 2$, $\log 3$, $\log 5$ in terms of the three series obtained by writing $x=1/31$, $1/49$, $1/161$ in $$2(x+\tfrac{1}{3}x^3+\tfrac{1}{5}x^5+\dots).$$

64. The power-series for $\text{arc}\sin x$ and $\text{arc}\tan x$.

If $x=\sin\theta$, we have

$$x=\theta-\frac{\theta^3}{3!}+\frac{\theta^5}{5!}-\dots,$$

and so, by the principle of reversion of series (Art. 55), we can express the numerically least value of θ as a convergent power-series in x, the first two terms of which are obviously

$$\theta = x+\tfrac{1}{6}x^3+\dots.$$

However, it is not easy to obtain the general law of the coefficients in this manner; but we can overcome the difficulty by using the Calculus.

We have in fact

$$\frac{d\theta}{dx}=\frac{1}{\cos\theta}=\frac{1}{\sqrt{(1-x^2)}}=1+\frac{1}{2}x^2+\frac{1.3}{2.4}x^4+\frac{1.3.5}{2.4.6}x^6+\dots,$$

where θ is supposed to lie between $-\frac{1}{2}\pi$ and $+\frac{1}{2}\pi$, so that $\cos\theta$ is positive.

The series for $\dfrac{d\theta}{dx}$ is obtained from the binomial series by writing $\nu=-\frac{1}{2}$ and $-x^2$ for $+x$: it will therefore converge uniformly in any interval $(-k, +k)$ if $0<k<1$.

Hence we may integrate term-by-term and so obtain

$$\text{arc}\sin x=\theta=x+\frac{1}{2}\frac{x^3}{3}+\frac{1.3}{2.4}\frac{x^5}{5}+\frac{1.3.5}{2.4.6}\frac{x^7}{7}+\dots,$$

which converges absolutely and uniformly in the interval $(-1, +1)$, as may be seen from the test of Art. 12 (5). Thus, writing $x=1$, we have a series for $\frac{1}{2}\pi$.

Although we have not found* a series for $\tan x$, we can easily find one for $\arctan x$. For, writing $x=\tan\phi$, we have

$$\frac{d\phi}{dx}=\frac{1}{\sec^2\phi}=\frac{1}{1+x^2}.$$

Thus $\dfrac{d\phi}{dx}=1-x^2+x^4-\ldots+(-1)^{n-1}x^{2n-2}+(-1)^n\dfrac{x^{2n}}{1+x^2}$

or
$$\phi=x-\tfrac{1}{3}x^3+\tfrac{1}{5}x^5-\ldots+(-1)^{n-1}\frac{x^{2n-1}}{2n-1}$$
$$+(-1)^n\int_0^x\frac{t^{2n}}{1+t^2}dt,$$

where ϕ is supposed to lie between $-\frac{1}{2}\pi$ and $+\frac{1}{2}\pi$.

The integral last written is less, in numerical value, than

$$\int_0^{|x|}t^{2n}dt=\frac{|x|^{2n+1}}{2n+1};$$

and this tends to 0 as n tends to ∞, provided that $|x|\leqq 1$.

Hence we have†

$$\arctan x=x-\tfrac{1}{3}x^3+\tfrac{1}{5}x^5-\ldots \text{ to } \infty,$$

where $\qquad -1\leqq x\leqq +1, \quad -\frac{1}{4}\pi\leqq \arctan x\leqq +\frac{1}{4}\pi.$

In particular we have

$$\tfrac{1}{4}\pi=1-\tfrac{1}{3}+\tfrac{1}{5}-\tfrac{1}{7}+\ldots.$$

Of course this series converges very slowly, but by the aid of the method given in Art. 24, the reader will find no great difficulty in calculating $\frac{1}{4}\pi$ to five decimals, from the first 13 or 14 terms. The result is $\frac{1}{4}\pi=\cdot 78540$..

* We can, of course, form such a series by dividing $\sin x$ by $\cos x$ (compare Art. 54), but there is no simple general law for the coefficients. The first three terms are

$$\tan x=x+\tfrac{1}{3}x^3+\tfrac{2}{15}x^5+\ldots,$$

and more coefficients are given in Chrystal's *Algebra*, vol. II., p. 344.

The method used in Art. 54 shews that this series will certainly be convergent in any interval for which $\dfrac{x^2}{2!}+\dfrac{x^4}{4!}+\dfrac{x^6}{6!}+\ldots<1$; and a short calculation will shew that this is satisfied in the interval $(-1\cdot 3, +1\cdot 3)$; but by means of a theorem given in Art. 84, we can shew that the region of convergence is $(-\frac{1}{2}\pi, +\frac{1}{2}\pi)$.

† Of course the term-by-term integration could also have been justified by making use of the uniform convergence of the series $1-x^2+x^4-\ldots$.

For the actual calculation of π to a large number of places, it is necessary to use special devices to increase the convergence of the series; a well-known method is to write $a = \text{arc} \tan \frac{1}{5}$.

Then we find $\tan 2a = \frac{5}{12}$, $\tan 4a = \frac{120}{119}$.

Hence
$$\tan(4a - \tfrac{1}{4}\pi) = \tfrac{1}{239}$$
or
$$\tfrac{1}{4}\pi = 4(\text{arc} \tan \tfrac{1}{5}) - (\text{arc} \tan \tfrac{1}{239}).$$

For other series to calculate π see Ex. A 48, p. 168.

65. Various trigonometrical power-series.

It is clear from Art. 27 that the expansion of

$$(1 - 2r\cos\theta + r^2)^{-1} = 1 + (2r\cos\theta - r^2) + (2r\cos\theta - r^2)^2 + \dots \text{ to } \infty$$

may be arranged in powers of r without altering its value, provided that* $0 < r \leqq \frac{2}{5}$.

The sequence of coefficients is, however, more easily determined for the fraction $(1 - r\cos\theta)/(1 - 2r\cos\theta + r^2)$.

Write
$$\frac{1 - r\cos\theta}{1 - 2r\cos\theta + r^2} = 1 + A_1 r + A_2 r^2 + A_3 r^3 + \dots,$$

where A_1, A_2, $A_3, \dots$ are of course functions of θ. Then we have the identity

$$\begin{aligned} 1 - r\cos\theta = 1 + A_1 r \quad &+ A_2 r^2 \qquad\quad + A_3 r^3 + \dots \\ -2r\cos\theta &- 2A_1 r^2 \cos\theta - 2A_2 r^3 \cos\theta + \dots \\ &+ r^2 \qquad\qquad\;\; + A_1 r^3 + \dots, \end{aligned}$$

and hence we get, using Art. 52 (5),

$$\begin{aligned} A_1 &= \cos\theta, \\ A_2 &= 2A_1 \cos\theta - 1 \;\; = \cos 2\theta, \\ A_3 &= 2A_2 \cos\theta - A_1 = \cos 3\theta, \\ A_4 &= 2A_3 \cos\theta - A_2 = \cos 4\theta, \end{aligned}$$

and so on.

Thus we have the series

$$\frac{1 - r\cos\theta}{1 - 2r\cos\theta + r^2} = 1 + r\cos\theta + r^2\cos 2\theta + r^3\cos 3\theta + \dots \text{ to } \infty.$$

If we subtract 1 and divide by r, we get

$$\frac{\cos\theta - r}{1 - 2r\cos\theta + r^2} = \cos\theta + r\cos 2\theta + r^2\cos 3\theta + r^3\cos 4\theta + \dots \text{ to } \infty.$$

---

* For $|2r\cos\theta| + r^2 \leqq 2|r| + r^2$; and when r satisfies the condition above we have $2|r| + r^2 = \frac{24}{25} < 1$. Thus $|2r\cos\theta| + r^2 < 1$, as required by Art. 27.

Combining these, we get also

$$\frac{1-r^2}{1-2r\cos\theta+r^2}=1+2r\cos\theta+2r^2\cos 2\theta+2r^3\cos 3\theta+\dots \text{ to } \infty.$$

An exactly similar argument will give the result

$$\frac{r\sin\theta}{1-2r\cos\theta+r^2}=r\sin\theta+r^2\sin 2\theta+r^3\sin 3\theta+\dots \text{ to } \infty.$$

By inspection we see that all these series converge when $r+r^2+r^3+\dots$ converges, or when $0<r<1$. Thus we are led to enquire whether the equations are not also true for the interval (0, 1). Now we find, identically,

$$\frac{1-r\cos\theta}{1-2r\cos\theta+r^2}=1+r\cos\theta+r^2\cos 2\theta+\dots+r^{n-1}\cos(n-1)\theta+R_n,$$

where $$R_n=\frac{r^n\cos n\theta-r^{n+1}\cos(n-1)\theta}{1-2r\cos\theta+r^2}.$$

Hence, as for the geometrical progression (Art. 6), we see that $\lim R_n=0$, if $0<r<1$, and accordingly the first equation holds for the interval (0, 1). And the other equations can be extended similarly.

Again we have

$$\frac{d}{dr}\log(1-2r\cos\theta+r^2)=-\frac{2(\cos\theta-r)}{1-2r\cos\theta+r^2}$$
$$=-2(\cos\theta+r\cos 2\theta+r^2\cos 3\theta+\dots)$$

by what has been proved.

Hence, integrating,* we have

$$\log(1-2r\cos\theta+r^2)=-2(r\cos\theta+\tfrac{1}{2}r^2\cos 2\theta+\tfrac{1}{3}r^3\cos 3\theta+\dots),$$

no constant being needed because both sides are zero for $r=0$.

Also we have

$$\int_0^r\frac{dt}{1-2t\cos\theta+t^2}=\int_0^r\frac{dt}{(t-\cos\theta)^2+\sin^2\theta}=\frac{1}{\sin\theta}\arctan\left(\frac{r\sin\theta}{1-r\cos\theta}\right).$$

Thus we find

$$\arctan\left(\frac{r\sin\theta}{1-r\cos\theta}\right)=\int_0^r\frac{\sin\theta\,.\,dt}{1-2t\cos\theta+t^2}$$
$$=\int_0^r dt(\sin\theta+t\sin 2\theta+t^2\sin 3\theta+\dots)$$
$$=r\sin\theta+\tfrac{1}{2}r^2\sin 2\theta+\tfrac{1}{3}r^3\sin 3\theta+\dots.$$

---

* Term-by-term integration is permissible because, if $0\leqq r\leqq k<1$, the series may be compared with $1+k+k^2+k^3+\dots$, and Weierstrass's M-test can be applied.

We have only established the equations above on the hypothesis that $0<r<1$; but we know from Art. 20 that the two series

$$\cos\theta+\tfrac{1}{2}\cos 2\theta+\tfrac{1}{3}\cos 3\theta+\ldots$$
$$\sin\theta+\tfrac{1}{2}\sin 2\theta+\tfrac{1}{3}\sin 3\theta+\ldots$$

are convergent, except the first for $\theta=0$ or $2k\pi$.

Thus, by Abel's theorem (Art. 51), we have

$$\cos\theta+\tfrac{1}{2}\cos 2\theta+\tfrac{1}{3}\cos 3\theta+\ldots=\lim_{r\to 1} -\tfrac{1}{2}\log(1-2r\cos\theta+r^2)$$
$$=-\tfrac{1}{2}\log(4\sin^2\tfrac{1}{2}\theta),$$

and similarly,

$$\cos\theta-\tfrac{1}{2}\cos 2\theta+\tfrac{1}{3}\cos 3\theta-\ldots=\lim_{r\to 1}\tfrac{1}{2}\log(1+2r\cos\theta+r^2)$$
$$=\tfrac{1}{2}\log(4\cos^2\tfrac{1}{2}\theta),$$

although a fresh investigation is unnecessary, because the last series can be deduced from the preceding by changing from θ to $\pi+\theta$.

In like manner we have

$$\sin\theta+\tfrac{1}{2}\sin 2\theta+\tfrac{1}{3}\sin 3\theta+\ldots=\lim_{r\to 1}\operatorname{arc\,tan}\left(\frac{r\sin\theta}{1-r\cos\theta}\right).$$

Now from the figure it is evident that the angle in question is the angle ϕ, which (according to the definition of the arc tan function) must lie between $-\tfrac{1}{2}\pi$ and $+\tfrac{1}{2}\pi$; so that $\lim\phi=\tfrac{1}{2}(\pi-\theta)$.

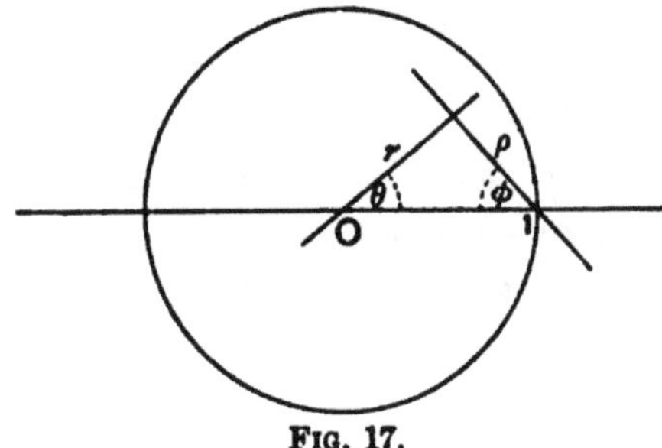

FIG. 17.

Thus $\quad \sin\theta+\tfrac{1}{2}\sin 2\theta+\tfrac{1}{3}\sin 3\theta+\ldots=\tfrac{1}{2}(\pi-\theta),\quad$ if $0<\theta<2\pi$.

But if $\theta=0$, or 2π, the value of the series is 0 because each term in it vanishes: thus the series is discontinuous* at $\theta=0$ and 2π.

If θ lies between $2k\pi$ and $2(k+1)\pi$, where k is an integer, positive or negative, we have

$$\sin\theta+\tfrac{1}{2}\sin 2\theta+\tfrac{1}{3}\sin 3\theta+\ldots=\tfrac{1}{2}[\pi-(\theta-2k\pi)]=\tfrac{1}{2}[(2k+1)\pi-\theta].$$

* Hence these are points of non-uniform convergence for the series (Art. 45).

It is not difficult to discuss the series

$$f(\theta)=\sin\theta+\tfrac{1}{2}\sin 2\theta+\tfrac{1}{3}\sin 3\theta+\dots$$

by a direct method. In fact, if $S_n(\theta)$ is the sum of the first n terms in $f(\theta)$, we have, by differentiating,

$$S_n'(\theta)=\cos\theta+\cos 2\theta+\cos 3\theta+\dots+\cos n\theta=\tfrac{1}{2}\left[\frac{\sin(n+\frac{1}{2})\theta}{\sin\frac{1}{2}\theta}-1\right].$$

Thus

$$S_n(\theta)=\int_0^{\frac{1}{2}\theta}\frac{\sin(2n+1)t}{\sin t}\,dt-\tfrac{1}{2}\theta.$$

Now, by Art. 174, Ex. 2 (Appendix), the limit of this integral is $\frac{1}{2}\pi$, provided that $\frac{1}{2}\theta$ lies between 0 and π; and consequently

$$f(\theta)=\tfrac{1}{2}(\pi-\theta),\quad \text{if } 0<\theta<2\pi.$$

But

$$f(0)=0=f(2\pi).$$

Thus the curve $y=f(\theta)$ consists of a line making an angle $\text{arc}\tan\frac{1}{2}$ with the horizontal and two points on the horizontal axis.

A glance at Figs. 12 and 13 of Art. 43 suggests the conjecture that the limiting form of the curve $y=S_n(\theta)$ consists of the slanting line and two vertical lines, joining the slanting line to the axis (see the figure below). But as a matter of fact this is not quite correct, and the vertical lines really project above and below the slanting line.

For clearly the point $\theta=\lambda/n,\ y=S_n(\lambda/n)$,

belongs to the curve $y=S_n(\theta)$, whatever the positive number λ may be. Now, as $n\to\infty$, this point approaches the limiting position

$$\theta=0,\ \ y=\int_0^{\lambda}\frac{\sin t}{t}\,dt,$$

in virtue of Ex. 3, Art. 174. Similarly the point

$$\theta=2\pi,\ \ y=-\int_0^{\lambda}\frac{\sin t}{t}\,dt,$$

belongs to the limiting form of the curve.

Now the integral $\int_0^{\lambda}(\sin t/t)\,dt$ can have any value from 0 to its maximum $1{\cdot}85194=\frac{1}{2}\pi\times(1{\cdot}1790)$, which occurs for $\lambda=\pi$; so that in the limiting form of the curve $y=S_n(\theta)$, the two vertical lines have lengths $1{\cdot}85194$, instead of $\frac{1}{2}\pi=1{\cdot}57080$, as conjectured.

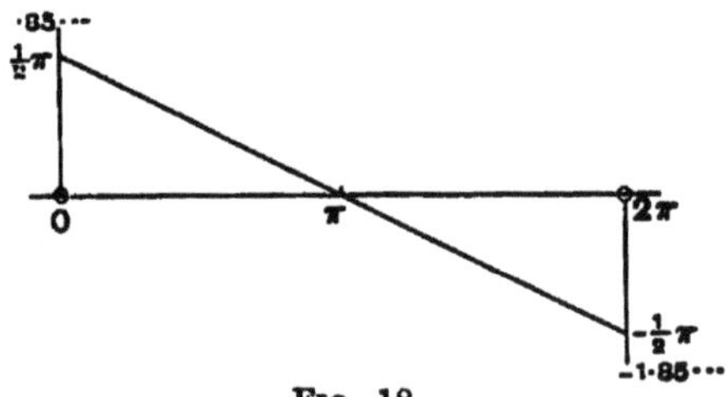

FIG. 18.

Some of the approximation curves $y=S_n(\theta)$ are drawn for various values of n in Byerly's *Fourier Series*, etc., p. 63 (No. II.), and in Carslaw's *Fourier's Series*, etc., p. 49; and the curve $n=80$ is given by Michelson and Stratton, *Phil. Mag.* (5), vol. 45, 1898, Pl. XII., Fig. 5.

EXAMPLES* A.

Differentiation and Integration.

1. Justify the equation

$$\frac{1}{a}-\frac{1}{a+b}+\frac{1}{a+2b}-\frac{1}{a+3b}+\ldots=\int_0^1\frac{t^{a-1}}{1+t^b}dt \quad (a,\ b>0).$$

Thus the series can be found in finite terms if b/a is rational. [GAUSS.]

Deduce that
$$1-\frac{1}{4}+\frac{1}{7}-\frac{1}{10}+\ldots=\frac{1}{3}\left(\frac{\pi}{\sqrt{3}}+\log 2\right),$$
$$\frac{1}{2}-\frac{1}{5}+\frac{1}{8}-\frac{1}{11}+\ldots=\frac{1}{3}\left(\frac{\pi}{\sqrt{3}}-\log 2\right),$$
$$1-\frac{1}{5}+\frac{1}{9}-\frac{1}{13}+\ldots=\frac{1}{4\sqrt{2}}[\pi+2\log(\sqrt{2}+1)].$$

[*Math. Trip.* 1896.]

2. Shew that if k and n are positive integers

$$\frac{k!}{(a+nb)(a+nb+1)\ldots(a+nb+k)}=\int_0^1 t^{a+nb-1}(1-t)^k dt. \quad (a,\ b>0)$$

Deduce that
$$\sum_0^\infty \frac{(k!)x^n}{(a+nb)\ldots(a+nb+k)}=\int_0^1\frac{t^{a-1}(1-t)^k}{1-xt^b}dt.$$

3. Utilise Ex. 2 to prove that

$$\frac{1}{1.2.3}+\frac{x}{2.3.4}+\frac{x^2}{3.4.5}+\ldots=\frac{3}{4x}-\frac{1}{2x^2}+\frac{(1-x)^2}{2x^3}\log\left(\frac{1}{1-x}\right),$$

and find a formula for the sum of the series

$$\frac{1}{1.3.5}+\frac{x}{3.5.7}+\frac{x^2}{5.7.9}+\ldots.$$

Obtain the former result also by integrating the logarithmic series twice.

4. Prove that
$$\frac{1}{1.2.3}+\frac{1}{3.4.5}+\frac{1}{5.6.7}+\ldots=\log 2-\tfrac{1}{2},$$
$$\frac{1}{1.2.3}-\frac{1}{3.4.5}+\frac{1}{5.6.7}-\ldots=\tfrac{1}{2}(1-\log 2),$$
$$\frac{1}{2.3.4}-\frac{1}{4.5.6}+\frac{1}{6.7.8}-\ldots=\tfrac{1}{4}(\pi-3).$$

[Ex. 2.]

5. Shew that with certain restrictions on α, β, γ,

$$F(\alpha,\ \beta,\ \gamma,\ x)=\frac{\Gamma(\gamma)}{\Gamma(\alpha)\Gamma(\gamma-\alpha)}\int_0^1 t^{\alpha-1}(1-t)^{\gamma-\alpha-1}(1-xt)^{-\beta}dt,$$

and deduce the equation (Ex. 13, Ch. VI.),

$$F(\alpha,\ \beta,\ \gamma,\ 1)=\frac{\Gamma(\gamma)\Gamma(\gamma-\alpha-\beta)}{\Gamma(\gamma-\alpha)\Gamma(\gamma-\beta)}.$$

---

*In a number of these examples, the word *expansion* is used as equivalent to *power-series*; and in some cases the words *for sufficiently small values of x* are implied.

6. The complete elliptic integrals are

$$\left.\begin{aligned} K&=\int_0^{\frac{1}{2}\pi}\frac{d\phi}{\sqrt{(1-k^2\sin^2\phi)}}=\frac{\pi}{2}\left[1+\left(\frac{1}{2}\right)^2k^2+\left(\frac{1.3}{2.4}\right)^2k^4+\dots\right],\\ E&=\int_0^{\frac{1}{2}\pi}\sqrt{(1-k^2\sin^2\phi)}\,d\phi=\frac{\pi}{2}\left[1-\left(\frac{1}{2}\right)^2\frac{k^2}{1}-\left(\frac{1.3}{2.4}\right)^2\frac{k^4}{3}-\dots\right],\end{aligned}\right\}\ 0<k<1.$$

7. Prove that

$$\frac{x}{\log[1/(1-x)]}=\int_0^1(1-x)^t\,dt=1-\tfrac{1}{2}x-\tfrac{1}{12}x^2-\tfrac{1}{24}x^3-\dots.$$

8. From Ex. 13, Ch. VI., or Ex. 5 above, prove that

$$1+\left(\frac{1}{2}\right)^2+\left(\frac{1}{2.4}\right)^2+\left(\frac{1.3}{2.4.6}\right)^2+\left(\frac{1.3.5}{2.4.6.8}\right)^2+\dots=\frac{\Gamma(2)}{[\Gamma(\frac{3}{2})]^2}=\frac{4}{\pi}.$$

[Write $\alpha=-\frac{1}{2}$, $\beta=-\frac{1}{2}$, $\gamma=1$.]

9. Multiply the expansions of $(1-x)^{-1}$ and $\log(1-x)$, and deduce by integration that

$$\tfrac{1}{2}[\log(1-x)]^2=\tfrac{1}{2}x^2+\tfrac{1}{3}(1+\tfrac{1}{2})x^3+\tfrac{1}{4}(1+\tfrac{1}{2}+\tfrac{1}{3})x^4+\dots.$$

[Compare Art. 62.]

10. If $y=(1+x)^{-n}\log(1+x)$, shew that

$$(1+x)\frac{dy}{dx}+ny=(1+x)^{-n}.$$

Deduce the expansion

$$y=x-n(n+1)\left(\frac{1}{n}+\frac{1}{n+1}\right)\frac{x^2}{2!}+n(n+1)(n+2)\left(\frac{1}{n}+\frac{1}{n+1}+\frac{1}{n+2}\right)\frac{x^3}{3!}-\dots.$$

11. Prove that

$$-\log(1+x).\log(1-x)=x^2+\left(1-\frac{1}{2}+\frac{1}{3}\right)\frac{x^4}{2}+\left(1-\frac{1}{2}+\frac{1}{3}-\frac{1}{4}+\frac{1}{5}\right)\frac{x^6}{3}+\dots.$$

[By direct multiplication, or by expanding the differential coefficient

$$(1-x)^{-1}\log(1+x)-(1+x)^{-1}\log(1-x).]$$

12. Prove that

$$\sqrt{(1+x^2)}\log[x+\sqrt{(1+x^2)}]=x+\frac{x^3}{3}-\frac{2}{3}\frac{x^5}{5}+\frac{2.4}{3.5}\frac{x^7}{7}-\dots.$$

$\left[\text{Use the equation } (1+x^2)\left(\frac{du}{dx}-1\right)=xu.\right]$

Shew similarly that

$$\sqrt{(1-x^2)}\arcsin x=x-\frac{x^3}{3}-\frac{2}{3}\frac{x^5}{5}-\frac{2.4}{3.5}\frac{x^7}{7}-\dots.$$

13. Prove that, if $|x|<1$,

$$\frac{1}{2}(\arctan x)^2=\frac{x^2}{2}-\left(1+\frac{1}{3}\right)\frac{x^4}{4}+\left(1+\frac{1}{3}+\frac{1}{5}\right)\frac{x^6}{6}-\dots.$$

Is the result true for $x=1$?

14. Shew that, if $|x|<1$,

$$\log\left[\tfrac{1}{2}\{1+\sqrt{(1-x)}\}\right]=-\frac{1}{2}\frac{x}{2}-\frac{1.3}{2.4}\frac{x^2}{4}-\frac{1.3.5}{2.4.6}\frac{x^3}{6}-\dots.$$

15. Prove that

$$\tfrac{1}{2}(\arctan x)\log\frac{1+x}{1-x}=x^2+\left(1-\frac{1}{3}+\frac{1}{5}\right)\frac{x^6}{3}+\left(1-\frac{1}{3}+\frac{1}{5}-\frac{1}{7}+\frac{1}{9}\right)\frac{x^{10}}{5}+\dots.$$

[It is easy to see that

$$\begin{aligned}(1-x^4)\frac{dy}{dx}&=(1-x^2)(x+\tfrac{1}{3}x^3+\tfrac{1}{5}x^5+\dots)\\&\quad+(1+x^2)(x-\tfrac{1}{3}x^3+\tfrac{1}{5}x^5-\dots)\\&=2\{x-(\tfrac{1}{3}-\tfrac{1}{5})x^5-(\tfrac{1}{7}-\tfrac{1}{9})x^9-\dots\}.]\end{aligned}$$

16. If $y=(\sec x+\tan x)^m=\Sigma a_n x^n$, prove that

$$(n+1)a_{n+1}-\frac{1}{2!}(n-1)a_{n-1}+\frac{1}{4!}(n-3)a_{n-3}-\dots=ma_n.$$

$\left[\text{Here } \dfrac{dy}{dx}=my\sec x,\ \text{ or } \left(1-\dfrac{x^2}{2!}+\dfrac{x^4}{4!}-\dots\right)\dfrac{dy}{dx}=my.\right]$ [*Math. Trip.* 1896.]

17. Verify that

$$\tfrac{1}{2}(\arctan x)\,.\log(1+x^2)=S_2\frac{x^3}{3}-S_4\frac{x^5}{5}+S_6\frac{x^7}{7}-\dots,$$

where

$$S_{2n}=1+\frac{1}{2}+\frac{1}{3}+\dots+\frac{1}{2n};$$

and deduce that

$$\frac{\pi}{8}\log 2=\frac{1}{3}S_2-\frac{1}{5}S_4+\frac{1}{7}S_6-\dots.$$ [*Math. Trip.* 1897.]

[Here

$$\begin{aligned}(1+x^2)\frac{dy}{dx}&=x(\arctan x)+\tfrac{1}{2}\log(1+x^2)\\&=S_2x^2-(S_4-S_2)x^4+(S_6-S_4)x^6-\dots.\end{aligned}$$

For the second part, use Abel's theorem.]

Derangement of Expansions.

18. Prove that the coefficient of x^{2n} in the expansion of $(1+2px+qx^2)^{-m}$ is

$$(-1)^n\frac{m(m+1)\dots(m+n-1)}{n!}q^n\left[1-\frac{n(m+n)}{2!}P+\frac{n(n-1)(m+n)(m+n+1)}{4!}P^2-\dots\right],$$

where $P=4p^2/q$.

Shew that it is a multiple of the coefficient of t^n in the expansion of

$$\begin{aligned}&\frac{1}{\sqrt{(1+qt)}}\left(1-\frac{p^2t}{1+qt}\right)^{-(n+m)}\\&=\frac{1}{(1+qt)^{\frac{1}{2}}}+(n+m)\frac{p^2t}{(1+qt)^{\frac{3}{2}}}+\frac{(n+m)(n+m+1)}{2!}\frac{p^4t^2}{(1+qt)^{\frac{5}{2}}}+\dots.\end{aligned}$$

[*Math. Trip.* 1898.]

19. Expand $\exp(\arctan x)$ up to the term which contains x^5.

[*Math. Trip.* 1899.]

[This function satisfies $(1+x^2)\dfrac{dy}{dx}=y$, and so we find that if

$$y=1+\Sigma a_n x^n/n!, \text{ then } a_1=1,\ a_{n+1}=a_n-n(n-1)a_{n-1}.$$

This gives $\qquad a_2=1,\ a_3=-1,\ a_4=-7, \ldots.$

The possibility of the expansion follows from Art. 36.]

20. Shew that, if $|x|<1$,

$$[1+\sqrt{(1+x)}]^a=2^a\left[1+a\left(\frac{x}{4}\right)+\frac{a(a-3)}{2!}\left(\frac{x}{4}\right)^2+\frac{a(a-4)(a-5)}{3!}\left(\frac{x}{4}\right)^3+\ldots\right].$$

[*Math. Trip.* 1902.]

[This is the expansion of $(2+y)^a$, where $x=2y+y^2$; use Lagrange's series.]

21. If $y=a_0+a_1x+a_2x^2+\ldots$, obtain the first and second terms in the power-series for

$$\frac{1}{y-a_0}-\frac{1}{a_1x} \text{ and } \frac{1}{y-a_0}-\frac{1}{x}\frac{dx}{dy}.$$

22. Apply the last example to prove that if

$$f(x)=(\sin x-\sin a)^{-1}-[(x-a)\cos a]^{-1}, \quad f(a)=\lim_{x\to a} f(x),$$

then
$$\frac{d}{da}[f(a)]-\lim_{x\to a} f'(x)=\tfrac{3}{4}\sec^3 a-\tfrac{5}{12}\sec a. \quad \text{[\textit{Math. Trip.} 1896.]}$$

23. Prove that the coefficient of y^n in the expansion of $(1-2xy+y^2)^{-\frac{1}{2}}$ is

$$\frac{1.3.5\ldots(2n-1)}{n!}\left[x^n-\frac{n(n-1)}{2(2n-1)}x^{n-2}+\frac{n(n-1)(n-2)(n-3)}{2.4(2n-1)(2n-3)}x^{n-4}-\ldots\right],$$

the number of terms being $\frac{1}{2}(n+1)$ or $\frac{1}{2}n+1$.

24. Determine the first three terms in the expansion of

$$x^2/[x-\log(1+x)].$$

25. If $[(1-xy)(1-xy^2)(1-x/y)(1-x/y^2)]^{-1}$ is expanded in powers of x, the part of the expansion which is independent of y is equal to

$$(1+x^3)/(1-x^3)(1-x^2)^2. \qquad \text{[\textit{Math. Trip.} 1903.]}$$

[If we expand $\{(1-xy)(1-x/y)\}$, we obtain

$$1+x(y+1/y)+x^2(y^2+1+1/y^2)+x^3(y^3+y+1/y+1/y^3)+\ldots$$
$$=(1-x^2)^{-1}\{1+x(y+1/y)+x^2(y^2+1/y^2)+\ldots\}.$$

It is then easy to pick out the specified terms in the form

$$(1-x^2)^{-2}(1+2x^3+2x^6+\ldots).]$$

26. Expand $\qquad (1+x)^{\frac{1}{x}}=\exp(1-\frac{1}{2}x+\frac{1}{3}x^2-\frac{1}{4}x^3+\ldots)$
up to and including the term in x^4.

[The first three terms are $e(1-\frac{1}{2}x+\frac{11}{24}x^2)$; the possibility of the expansion follows from Art. 36.]

27. Expand the series

$$\frac{1}{1-x}-\frac{x}{(1-x)^3}+\frac{1.3}{2!}\frac{x^2}{(1-x)^5}-\frac{1.3.5}{3!}\frac{x^3}{(1-x)^7}+\dots$$

in powers of x, shewing that the coefficient of x^n is

$$S_n=1+\sum_1^n(-1)^r\frac{(n+r)!}{(r!)^2(n-r)!}\frac{1}{2^r}.$$

Prove that the first series is equal to $1/\sqrt{(1+x^2)}$, and hence find an expression for S_n.

28. Prove that $\dfrac{1}{(1-x)(1-y)-\lambda xy}=\Sigma\Sigma F(-m,\ -n,\ 1,\ \lambda)x^m y^n$,

where $F(-m,\ -n,\ 1,\ \lambda)$ is a (terminated) hypergeometric series. [HARDY.]

29. Prove that $\displaystyle\sum_0^\infty\sum_0^\infty\frac{(m+n)!}{m!\,n!}\left(\frac{x}{2}\right)^{m+n}=\frac{1}{1-x},\quad \text{if } -2<x<1,$

and $\displaystyle\sum_0^\infty\sum_0^\infty\sum_0^\infty\frac{(m+n+p)!}{m!\,n!\,p!}\left(\frac{x}{3}\right)^{m+n+p}=\frac{1}{1-x},\quad \text{if } -3<x<1,$

and extend to any number of indices of summation. [*Math. Trip.* 1903.]

30. Shew that the following series are absolutely convergent, and by summing with respect to r first, shew that their values are as given:

$$\sum_{r=2}^\infty\sum_{s=2}^\infty\frac{1}{(p+s)^r}=\frac{1}{p+1}.\qquad \sum_{r=2}^\infty\sum_{s=1}^\infty\frac{1}{(2s)^r}=\log 2,$$

$$\sum_{r=1}^\infty\sum_{s=1}^\infty\frac{1}{(4s-1)^{2r+1}}=\frac{\pi}{8}-\frac{1}{2}\log 2,\qquad \sum_{r=1}^\infty\sum_{s=1}^\infty\frac{1}{(4s-1)^{2r}}=\frac{1}{4}\log 2,$$

$$\sum_{r=1}^\infty\sum_{s=1}^\infty\frac{1}{(4s-2)^{2r}}=\frac{\pi}{8}.$$

[STERN.*]

Special Series.

31. If $\displaystyle f(x,\ y)=\frac{1}{x}-y\frac{1}{x+1}+\frac{y(y-1)}{2!}\frac{1}{x+2}-\frac{y(y-1)(y-2)}{3!}\frac{1}{x+3}+\dots$

shew that the series $f(x,\ y)$ converges if $y+1$ is positive; and if x is also positive, prove that the series is equal to $f(y+1,\ x-1)$.

[If $x>0$ and $y>0$, we can prove by Art. 45 that $f(x,\ y)$ is equal to $\int_0^1 t^{x-1}(1-t)^y\,dt$; and this can be extended to cover the case $0>y>-1$, by Art. 175. Change the variable from t to $1-t$ to get the final result.]

32. If $(1+bx+ax^2)^{-1}=1+p_1x+p_2x^2+p_3x^3+\dots,$

then $\displaystyle 1+p_1^2x+p_2^2x^2+p_3^2x^3+\dots=\frac{1+ax}{(1-ax)[(1+ax)^2-b^2x]}.$

[*Math. Trip.* 1900.]

* See Dirichlet's *Vorlesungen über bestimmte Integrale* (ed. Meyer), § 117.

33. Shew that the sum of the squares of the coefficients in the binomial power-series is

$$\frac{\Gamma(1+2\nu)}{[\Gamma(1+\nu)]^2}, \quad \text{if } \nu > -\tfrac{1}{2}. \qquad [\textit{Math. Trip.}\ 1890.]$$

[Put $\alpha=\beta=-\nu$, $\gamma=1$ in Ex. 13, Ch. VI., or in Ex. 5 above.]

34. Prove that

$$\log(1+x)=x(1-x)+\tfrac{1}{2}x^2(1-x^2)+\tfrac{1}{3}x^3(1-x^3)+\dots.$$

Discuss the validity of this equation for $x=1$.

35. Shew that at r per cent., compound interest, a capital will increase to A times its original value in n years, approximately, where

$$n=\frac{\log A}{x}\left(1+\frac{x}{2}-\frac{x^2}{12}+\dots\right), \quad x=\frac{r}{100}.$$

In particular, the capital will be doubled in the time given by the approximation

$$n=\frac{69{\cdot}3}{r}+{\cdot}35.$$

36. If

$$\sum_0^\infty \frac{(n+a)^s}{n!}x^n=f_s(x)e^x,$$

prove that $f_s(x)$ is a polynomial of degree s in x which satisfies the equation

$$f_{s+1}=(x+a)f_s+x\frac{df_s}{dx}.$$

Shew that $f_1=x+a$, $f_2=(x+a)^2+x$, $f_3=(x+a)^3+3x(x+a)+x$, and that if a is positive all the roots of $f_s(x)=0$ are real and negative, and that they are separated by the roots of $f_{s-1}(x)=0$. [HARDY; and *Math. Trip.* 1902.]

37. If $\Sigma a_n x^n=(1-x)^{-\frac{1}{2}}$, prove that

$$a_0x+\tfrac{1}{3}a_1x^3+\tfrac{1}{5}a_2x^5+\dots=\text{arc}\sin x,$$

and deduce that

$$a_0a_n+\frac{1}{3}a_1a_{n-1}+\frac{1}{5}a_2a_{n-2}+\dots+\frac{1}{2n+1}a_na_0=\frac{1}{(2n+1)a_n}.$$

[For the latter part, use Ex. B, 2.] [*Math. Trip.* 1890.]

Trigonometrical Series.

38. By writing $r=-x\sin\theta$ in the power-series for

$$\text{arc}\tan\{r\sin\theta/(1-r\cos\theta)\},$$

shew that

$$\text{arc}\tan(a+x)-\text{arc}\tan a=(x\sin\theta)\sin\theta-\tfrac{1}{2}(x\sin\theta)^2\sin 2\theta+\tfrac{1}{3}(x\sin\theta)^3\sin 3\theta - \dots,$$

where

$$a=\cot\theta. \qquad [\text{EULER.}]$$

39. Prove that the series

$$\cos\theta\,.\,\sin\theta+\tfrac{1}{2}\cos^2\theta\,.\,\sin 2\theta+\tfrac{1}{3}\cos^3\theta\,.\,\sin 3\theta+\dots$$

is convergent and is equal to $\frac{1}{2}\pi-\theta$, when θ lies between 0 and π.

[Put $r=\cos\theta$ in Art. 65; or $x=\cot\theta$ in Ex. 38.]

40. Shew that $2\sin(\frac{1}{7}\pi)$ is approximately equal to $\frac{1}{2}\sqrt{3}$, the error being about $\frac{1}{5}$th per cent.

Deduce that the side of a regular heptagon inscribed in a circle is nearly equal to the height of an equilateral triangle whose side is equal to the radius.

41. If
$$\frac{\tan y}{\tan x} = \frac{1+\lambda}{1-\lambda},$$
deduce from Art. 65 that
$$y - x = \lambda \sin 2x + \tfrac{1}{2}\lambda^2 \sin 4x + \tfrac{1}{3}\lambda^3 \sin 6x + \dots.$$

42. Obtain the following definite integrals:
$$\int_0^\pi \log(1 - 2r\cos\theta + r^2)\cos n\theta\, d\theta = -\pi r^n/n,$$
$$\int_0^\pi \operatorname{arc\,tan}\left(\frac{r\sin\theta}{1-r\cos\theta}\right)\sin n\theta\, d\theta = \frac{\pi}{2}\frac{r^n}{n},$$
$$\int_0^\pi \frac{\cos n\theta}{1-2r\cos\theta+r^2}\,d\theta = \frac{\pi r^n}{1-r^2},$$
$$\int_0^\pi \frac{\sin n\theta \sin\theta}{1-2r\cos\theta+r^2}\,d\theta = \frac{\pi}{2} r^{n-1},$$
by expansion in series (Art. 65) and term-by-term integration (Art. 45). Here n is an integer and r lies between 0 and 1.

43. By integrating the equation (Art. 65)
$$\sin\theta + \tfrac{1}{2}\sin 2\theta + \tfrac{1}{3}\sin 3\theta + \dots = (\tfrac{1}{2} - x)\pi,$$
where $x = \theta/2\pi$ and x lies between 0 and 1, we can infer the results
$$\cos\theta + \frac{1}{2^2}\cos 2\theta + \frac{1}{3^2}\cos 3\theta + \dots = \left(x^2 - x + \frac{1}{6}\right)\pi^2,$$
$$\sin\theta + \frac{1}{2^3}\sin 2\theta + \frac{1}{3^3}\sin 3\theta + \dots = \left(\frac{2}{3}x^3 - x^2 + \frac{1}{3}x\right)\pi^3,$$
$$\cos\theta + \frac{1}{2^4}\cos 2\theta + \frac{1}{3^4}\cos 3\theta + \dots = \left(-\frac{1}{3}x^4 + \frac{2}{3}x^3 - \frac{1}{3}x^2 + \frac{1}{90}\right)\pi^4,$$
and so on. [D. BERNOULLI.]

[Note that the constants of integration can be determined thus:—The series $\Sigma(\cos n\theta)/n^2$, for example, is uniformly convergent, and therefore continuous, for all values of θ. Thus, if $1 + \frac{1}{2^2} + \frac{1}{3^2} + \dots = \lambda\pi^2$, we find the sum $(x^2 - x + \lambda)\pi^2$. Now write $\theta = \pi$, and we get
$$(\lambda - \tfrac{1}{4})\pi^2 = -\lambda\pi^2 + \frac{2}{2^2}\lambda\pi^2 = -\tfrac{1}{2}\lambda\pi^2,$$
which gives $\lambda = \frac{1}{6}$. See also Arts. 71, 93 for $\Sigma 1/n^2$, $\Sigma 1/n^4$.]

44. Shew that, if $|r|<1$,

$$r\cos\theta+\tfrac{1}{3}r^3\cos 3\theta+\tfrac{1}{5}r^5\cos 5\theta+\dots=\tfrac{1}{4}\log\left(\frac{1+2r\cos\theta+r^2}{1-2r\cos\theta+r^2}\right),$$

$$r\sin\theta+\tfrac{1}{3}r^3\sin 3\theta+\tfrac{1}{5}r^5\sin 5\theta+\dots=\tfrac{1}{2}\arctan\left(\frac{2r\sin\theta}{1-r^2}\right),$$

$$r\sin\theta-\tfrac{1}{3}r^3\sin 3\theta+\tfrac{1}{5}r^5\sin 5\theta-\dots=\tfrac{1}{4}\log\left(\frac{1+2r\sin\theta+r^2}{1-2r\sin\theta+r^2}\right),$$

$$r\cos\theta-\tfrac{1}{3}r^3\cos 3\theta+\tfrac{1}{5}r^5\cos 5\theta-\dots=\tfrac{1}{2}\arctan\left(\frac{2r\cos\theta}{1-r^2}\right).$$

45. Prove that

$$\cos\theta+\tfrac{1}{3}\cos 3\theta+\tfrac{1}{5}\cos 5\theta+\dots=\tfrac{1}{4}\log\cot^2\tfrac{1}{2}\theta,$$
$$\sin\theta+\tfrac{1}{3}\sin 3\theta+\tfrac{1}{5}\sin 5\theta+\dots=+\tfrac{1}{4}\pi,\quad (0<\theta<\pi)$$
$$\sin\theta-\tfrac{1}{3}\sin 3\theta+\tfrac{1}{5}\sin 5\theta-\dots=\tfrac{1}{4}\log(\sec\theta+\tan\theta)^2,$$
$$\cos\theta-\tfrac{1}{3}\cos 3\theta+\tfrac{1}{5}\cos 5\theta-\dots=+\tfrac{1}{4}\pi.\quad (-\tfrac{1}{2}\pi<\theta<\tfrac{1}{2}\pi)$$

The second series changes sign at $\theta=\pi, 2\pi, \dots$, and is discontinuous for these values of θ; while the fourth changes sign and is discontinuous at $\theta=\frac{1}{2}\pi, \frac{3}{2}\pi, \dots$.

46. Prove that

$$\cos\theta\cos\alpha+\tfrac{1}{2}\cos 2\theta\cos 2\alpha+\tfrac{1}{3}\cos 3\theta\cos 3\alpha+\dots=-\tfrac{1}{4}\log\left[4(\cos\theta-\cos\alpha)^2\right],$$
$$\cos\theta\cos\alpha-\tfrac{1}{2}\cos 2\theta\cos 2\alpha+\tfrac{1}{3}\cos 3\theta\cos 3\alpha-\dots=\tfrac{1}{4}\log\left[4(\cos\theta+\cos\alpha)^2\right],$$
$$\sin\theta\cos\alpha+\tfrac{1}{2}\sin 2\theta\cos 2\alpha+\tfrac{1}{3}\sin 3\theta\cos 3\alpha+\dots=f(\theta),$$
$$\cos\theta\sin\alpha+\tfrac{1}{2}\cos 2\theta\sin 2\alpha+\tfrac{1}{3}\cos 3\theta\sin 3\alpha+\dots=g(\theta),$$

where, assuming $0<\alpha<\pi$,

$$f(\theta)=-\tfrac{1}{2}\theta,\quad g(\theta)=\tfrac{1}{2}(\pi-\alpha),\quad \text{if } 0<\theta<\alpha,$$
$$f(\alpha)=\tfrac{1}{4}\pi-\tfrac{1}{2}\alpha=g(\alpha),$$
$$f(\theta)=\tfrac{1}{2}(\pi-\theta),\quad g(\theta)=-\tfrac{1}{2}\alpha,\quad \text{if } \alpha<\theta<\pi.$$

47. Prove that, if $0\leqq\theta\leqq\pi$,

$$\frac{\cos 4\theta}{1.2}+\frac{\cos 6\theta}{2.3}+\frac{\cos 8\theta}{3.4}+\dots=\cos 2\theta-\left(\frac{\pi}{2}-\theta\right)\sin 2\theta+\sin^2\theta\log(4\sin^2\theta),$$

$$\frac{\sin 4\theta}{1.2}+\frac{\sin 6\theta}{2.3}+\frac{\sin 8\theta}{3.4}+\dots=\sin 2\theta-(\pi-2\theta)\sin^2\theta-\sin\theta\cos\theta\log(4\sin^2\theta);$$

and find the sums of

$$\sum_2^\infty\frac{\cos n\theta}{n^2-1},\quad \sum_2^\infty\frac{\sin n\theta}{n^2-1}.$$

48. Series for π.

Since $\dfrac{1}{4}\pi=\arctan\dfrac{1}{2}+\arctan\dfrac{1}{3}$,

we find $$\frac{\pi}{4}=\left(\frac{1}{2}-\frac{1}{3.2^3}+\frac{1}{5.2^5}-\dots\right)+\left(\frac{1}{3}-\frac{1}{3.3^3}+\frac{1}{5.3^5}-\dots\right)$$

and $$\frac{\pi}{4}=\frac{4}{10}\left[1+\frac{2}{3}\frac{2}{10}+\frac{2.4}{3.5}\left(\frac{2}{10}\right)^2+\dots\right]+\frac{3}{10}\left[1+\frac{2}{3}\frac{1}{10}+\frac{2.4}{3.5}\left(\frac{1}{10}\right)^2+\dots\right];$$

both of these results are due to Euler.

Clausen gave the identity $\frac{1}{4}\pi=2\operatorname{arc\,tan}\frac{1}{3}+\operatorname{arc\,tan}\frac{1}{7}$,

which leads to Hutton's series

$$\frac{\pi}{4}=\frac{6}{10}\left[1+\frac{2}{3}\frac{1}{10}+\frac{2.4}{3.5}\left(\frac{1}{10}\right)^2+\dots\right]+\frac{14}{100}\left[1+\frac{2}{3}\frac{1}{100}+\frac{2.4}{3.5}\left(\frac{1}{100}\right)^2+\dots\right].$$

Euler also gave the result

$$\frac{1}{4}\pi=5\operatorname{arc\,tan}\frac{1}{7}+8\operatorname{arc\,tan}\frac{3}{79},$$

which leads to the highly convergent series

$$\frac{\pi}{4}=\frac{7}{10}\left[1+\frac{2}{3}\left(\frac{2}{100}\right)+\frac{2.4}{3.5}\left(\frac{2}{100}\right)^2+\dots\right]+\frac{7584}{10^5}\left[1+\frac{2}{3}\left(\frac{144}{10^5}\right)+\frac{2.4}{3.5}\left(\frac{144}{10^5}\right)^2+\dots\right].$$

In writing these series down, we note that Ex. B, 2, below can be put in the form

$$\operatorname{arc\,tan} x=\frac{x}{1+x^2}\left[1+\frac{2}{3}\frac{x^2}{1+x^2}+\frac{2.4}{3.5}\left(\frac{x^2}{1+x^2}\right)^2+\dots\right].$$

EXAMPLES B.

Euler's Transformation.

1. Shew by the same method as in Art. 24, that

$$a_1x+a_2x^3+a_3x^5+a_4x^7+\dots=\sqrt{(1+y^2)}[a_1y-(Da_1)y^3+(D^2a_1)y^5-\dots],$$

if $$x=y/\sqrt{(1+y^2)},\quad\text{or}\quad y=x/\sqrt{(1-x^2)}.$$

Similarly, prove that

$$a_1x-a_2x^3+a_3x^5-a_4x^7+\dots=\sqrt{(1-y^2)}[a_1y+(Da_1)y^3+(D^2a_1)y^5+\dots],$$

if $$x=y/\sqrt{(1-y^2)},\quad\text{or}\quad y=x/\sqrt{(1+x^2)}.$$ [EULER.]

2. By taking $a_1=1$, $a_2=\frac{1}{3}$, $a_3=\frac{1}{5}$, ... in Ex. 1, shew that if $|y|<1$,

$$\frac{1}{\sqrt{(1+y^2)}}\log[y+\sqrt{(1+y^2)}]=y-\frac{2}{3}y^3+\frac{2.4}{3.5}y^5-\dots,$$

$$\frac{1}{\sqrt{(1-y^2)}}\operatorname{arc\,sin} y=y+\frac{2}{3}y^3+\frac{2.4}{3.5}y^5+\dots.$$

Deduce that

$$\frac{2}{\sqrt{3}}\log\left(\frac{1+\sqrt{3}}{\sqrt{2}}\right)=1-\frac{2}{3}\frac{1}{2}+\frac{2.4}{3.5}\frac{1}{2^2}-\frac{2.4.6}{3.5.7}\frac{1}{2^3}+\dots,$$

$$\frac{\pi}{2}=1+\frac{2}{3}\frac{1}{2}+\frac{2.4}{3.5}\frac{1}{2^2}+\frac{2.4.6}{3.5.7}\frac{1}{2^3}+\dots.$$

Can we put $y=1$ in the first series? [EULER.]

3. Shew also that if

$$f(x)=b_0+b_1x+b_2x^2+\dots,$$

then $$a_0b_0+a_1b_1x+a_2b_2x^2+\dots=a_0f(x)-(Da_0)x\frac{df}{dx}+(D^2a_0)\frac{x^2}{2!}\frac{d^2f}{dx^2}-\dots.$$

[EULER.]

4. In particular, if $a_n=n^3$, we find

$$a_0=0,\ Da_0=-1,\ D^2a_0=6,\ D^3a_0=-6,\ D^4a_0=0.$$

Thus from Ex. 3, $$\Sigma\frac{n^3}{n!}x^n=(x+3x^2+x^3)e^x.$$

Similarly, $$\Sigma\frac{n^4}{n!}x^n=(x+7x^2+6x^3+x^4)e^x.$$

5. If $$S_r=\sum_1^\infty\frac{n^r}{n!},$$

S_r is an integral multiple of e, and in particular

$$S_1=e,\ S_2=2e,\ S_3=5e,\ S_4=15e,\ S_5=52e,$$
$$S_6=203e,\ S_7=877e,\ S_8=4140e.$$ [WOLSTENHOLME.]

6. By integrating the formulae of Ex. 2 above, prove that if $|y|<1$,

$$\tfrac{1}{2}[\log\{y+\sqrt{(1+y^2)}\}]^2=\frac{y^2}{2}-\frac{2}{3}\frac{y^4}{4}+\frac{2.4}{3.5}\frac{y^6}{6}-\dots,$$

$$\tfrac{1}{2}(\arcsin y)^2=\frac{y^2}{2}+\frac{2}{3}\frac{y^4}{4}+\frac{2.4}{3.5}\frac{y^6}{6}+\dots.$$

Are these equations valid for $y=1$? [*Math. Trip.* 1897 and 1905.]

7. From Ex. 4 above, prove that if

$$S_n=1^3+2^3+3^3+\dots+n^3,$$

$$\sum_1^\infty S_n\frac{x^n}{n!}=\frac{1}{4}xe^x(x^3+8x^2+14x+4),$$

and that $$\sum_1^\infty(-1)^{n-1}S_n\frac{2^n}{n!}=0.$$ [*Math. Trip.* 1904.]

Shew similarly that

$$\frac{1}{3}x+\frac{1.4}{3.6}2^3.x^2+\frac{1.4.7}{3.6.9}.3^3.x^3+\dots=\frac{1}{27}\frac{x}{(1-x)^{\frac{10}{3}}}(x^2+18x+9).$$

Obtain these results also by differentiating the exponential series and the series for $(1-x)^{-\frac{1}{3}}$.

8. Apply Euler's method to prove that

$$\frac{1}{m}-\frac{x}{m+1}+\frac{x^2}{m+2}-\frac{x^3}{m+3}+\dots$$
$$=\frac{1}{m(1+x)}\left[1+\frac{1}{m+1}\left(\frac{x}{1+x}\right)+\frac{1.2}{(m+1)(m+2)}\left(\frac{x}{1+x}\right)^2+\dots\right].$$

9. Apply Euler's method to prove that

$$F(\alpha,\beta,\gamma,x)=\frac{1}{(1-x)^\alpha}F\left(\alpha,\gamma-\beta,\gamma,\frac{x}{x-1}\right),$$

where $$F(\alpha,\beta,\gamma,x)=1+\frac{\alpha\beta}{1.\gamma}x+\frac{\alpha(\alpha+1)\beta(\beta+1)}{1.2.\gamma(\gamma+1)}x^2+\dots.$$ [GAUSS.]

Miscellaneous.

10. If $y=2x/(1+x^2)$ and $|x|$, $|y|$ are both less than 1, shew that

$$x=\frac{y}{2}+\frac{1}{2}\frac{y^3}{4}+\frac{1.3}{2.4}\frac{y^5}{6}+\frac{1.3.5}{2.4.6}\frac{y^7}{8}+\dots.$$

[$x=\{1-\sqrt{(1-y^2)}\}/y$; or by Lagrange's series.]

11. If $$y=2x/(1-x^2) \text{ and } z=x+\frac{x^3}{3^2}+\frac{x^5}{5^2}+\dots,$$

shew that $$z=\frac{1}{2}\left(y-\frac{2}{3}\frac{y^3}{3}+\frac{2.4}{3.5}\frac{y^5}{5}-\dots\right).$$

[First, $$x\frac{dz}{dx}=\frac{1}{2}\log\frac{1+x}{1-x}=\frac{1}{2}\log\{y+\sqrt{(1+y^2)}\}.$$

Also $$\frac{dy}{dx}=\frac{y}{x}\frac{1+x^2}{1-x^2}=\frac{y}{x}\sqrt{(1+y^2)},$$

so that (Ex. 2) $$\frac{dz}{dy}=\frac{1}{2}\left(1-\frac{2}{3}y^2+\frac{2.4}{3.5}y^4-\dots\right).]$$

12. From the expansion of $(1-\frac{1}{50})^{-\frac{1}{2}}$ determine the value of $\sqrt{2}$ to 12 decimal places. [1·414213562373.]

Obtain in the same way the cube root of 2 from the expansion of $(1+\frac{3}{125})^{\frac{1}{3}}$. [1·25992105.] [EULER.]

Theorems of Abel and Frobenius.

13. With the notation of Ex. 9, shew that

$$\lim_{x\to 1}\frac{F(\alpha,\beta,\gamma,x)}{(1-x)^{\gamma-\alpha-\beta}}=\frac{\Gamma(\gamma)}{\Gamma(\alpha)\Gamma(\beta)}, \quad \text{if } \gamma<\alpha+\beta,$$

$$\lim_{x\to 1}\frac{F(\alpha,\beta,\gamma,x)}{\log[1/(1-x)]}=\frac{\Gamma(\alpha+\beta)}{\Gamma(\alpha)\Gamma(\beta)}, \quad \text{if } \gamma=\alpha+\beta.$$ [GAUSS.]

14. If Σna_n is convergent, so also is Σa_n (Ex. 2, Ch. III.), and if $f(x)=\Sigma a_n x^n$, then

$$\Sigma na_n=\lim_{x\to 1}\{f(1)-f(x)\}/(1-x).$$ [STOLZ.]

[Note that $(1-x^n)/n(1-x)$ gives a decreasing sequence of factors, and apply Abel's theorem.]

15. If $v_n(x)$ decreases as n increases and $\lim_{x\to 1} v_n(x)=1$, extend Art. 51 to prove that if Σa_n is convergent or divergent (to $+\infty$),

$$\lim_{x\to 1}\Sigma a_n v_n(x)=\Sigma a_n \text{ or } \infty.$$

Also shew that if B_n/A_n tends to a definite limit l,

$$\lim_{x\to 1}\{\Sigma b_n v_n(x)\}/\{\Sigma a_n v_n(x)\}=l,$$

provided that A_n is always positive and that A_n tends to ∞.

16. If the coefficients a_n, b_n satisfy the conditions of Ex. 15, and $f_n(x)$ decreases as n increases (but is always positive), prove that $\Sigma b_n f_n(x)$ will converge provided that $\Sigma a_n f_n(x)$ does, if $\lim A_n f_n = 0$. Deduce that when $f_n(1)=0$, and $\lim_{x\to 1} \Sigma a_n f_n(x) = \sigma > 0$, then $\lim_{x\to 1} \Sigma b_n f_n(x) = \sigma l$.

[Apply the lemma of Art. 153.]

17. If $\Delta v_n = v_n - v_{n+1}$, $\Delta^2 v_n = v_n - 2v_{n+1} + v_{n+2}$, and $\lim_{n\to\infty}(nv_n)=0$, shew that

$$v_0 = \Delta^2 v_0 + 2\Delta^2 v_1 + 3\Delta^2 v_2 + \dots.$$

Writing $f_n(x) = \Delta^2 v_n$ in Ex. 16, shew that if $\Delta^3 v_n$ is positive, and if

$$\lim_{n\to\infty} n^2 \Delta^2 v_n = 0,$$

then the series $\Delta^2 v_0 + 2\Delta^2 v_2 + 3\Delta^2 v_4 + \dots$, $v_0 - 2v_1 + 3v_2 - 4v_3 + \dots$ converge and are equal. If further v_n has the limit 1 as x tends to 1, then the last series has the limit $\frac{1}{4}$ as x tends to 1.

18. Use the method of Ex. 17 to shew that if $\Delta^2 v_n$ is positive, and if $\lim_{n\to\infty} n\Delta v_n = 0$ and $\lim_{x\to 1} v_n = 1$, then $\lim_{x\to 1}(v_0 - v_1 + v_2 - v_3 + \dots) = \frac{1}{2}$.

[For another method see Ex. 3, Art. 24.]

19. From Ex. 15, prove that

$$\lim_{x\to 1} \sum_1^\infty (-1)^{n-1} \frac{x^n}{n(1+x^n)} = \frac{1}{2}\log 2 = \lim_{x\to 1} \sum_1^\infty (-1)^{n-1} \frac{x^n(1-x)}{1-x^{2n}}.$$

20. Establish the asymptotic formulae (as $x \to 1$),

$$\sum_1^\infty \frac{x^n}{n(1+x^n)} \sim \frac{1}{2}\log\left(\frac{1}{1-x}\right) \quad \text{and} \quad \sum_1^\infty \frac{n(-x)^n}{1-x^{2n}} \sim -\frac{1}{4}\frac{1}{1-x}.$$

[The difference between the two sides of the first is less than $\frac{1}{2}$; in the second, multiply by $1-x$ and use Ex. 18.]

21. On the lines of Exs. 17–20 establish the following asymptotic formulae (as $x \to 1$):

$$\left(\frac{x}{1-x} + \frac{x^2}{1-x^2} + \frac{x^3}{1-x^3} + \dots\right) \sim \frac{1}{1-x}\log\left(\frac{1}{1-x}\right),$$

$$\left(\frac{x}{1-x} - \frac{x^3}{1-x^3} + \frac{x^5}{1-x^5} - \dots\right) \sim \frac{\pi}{4}\frac{1}{1-x},$$

where C is Euler's constant. [CESÀRO.]

Lagrange's Series.

22. Shew that, if n is a positive integer,

$$(t+a)^n = t^n + na(t+b)^{n-1} + \dots + n_r a(a-rb)^{r-1}(t+rb)^{n-r} + \dots + a(a-nb)^{n-1},$$

where n_r is the ordinary binomial coefficient

$$n(n-1)\dots(n-r+1)/r! \qquad \text{[ABEL.]}$$

[Take the result of Ex. 4, Art. 56, multiply by e^{tx}, and equate coefficients of $x^n/n!$.

Several authors have considered the validity of the equation, also due to Abel,

$$\phi(t+a)=\phi(t)+a\phi'(t+b)+\frac{a(a-2b)}{2!}\phi''(t+2b)+\dots,$$

but their results cannot be given here. We may remark, however, that the theorem fails if $\phi(t)$ is $\log t$ or a *negative* power of t. The most recent results are due to Pincherle (*Acta Math.*, Bd. 28, 1904, p. 225).]

23. Prove that the coefficient of x^{n-1} in the expansion of $[x/(e^x-1)]^n$ is $(-1)^{n-1}$. [WOLSTENHOLME; and *Math. Trip.* 1904.]

Prove that the coefficient of x^{n-1} in the expansion of

$$(1+x)^{2n-1}(2+x)^{-n}$$

is $\frac{1}{2}$. [*Math. Trip.* 1906.]

[Use Lagrange's series (1) for $y=e^x-1$; and (2) for $\log(1-y)$, where

$$y=x(2+x)/(1+x)^2=1-1/(1+x)^2.]$$

24. Expand t^m and $\log t$ in powers of x, where

$$t^{-\beta}-t^{-\alpha}=(\alpha-\beta)x,$$

and determine the interval of convergence.

[Write $t^{\alpha-\beta}=1+y$ and apply Lagrange's series; or otherwise.]

25. If $f(x)$ is a power-series in x, whose lowest term is x, shew that the coefficient of $1/x$ in the expansion of $[1/f(x)]^n$, in ascending powers of x, is n times the coefficient of x^n in the expansion of $g(x)$, the function inverse to $f(x)$.

Determine the coefficient for the following forms of $f(x)$:

(1) $\sin x$; (2) $\tan x$; (3) $\log(1+x)$; (4) $1+x-\sqrt{(1+x^2)}$;
(5) $\sinh x$; (6) $\tanh x$. [WOLSTENHOLME.]

[The results are:

(1) 0 or $\dfrac{1.3.5\dots(n-2)}{2.4.6\dots(n-1)}$; (2) 0 or $(-1)^{(n-1)/2}$; (3) $1/(n-1)!$;

(4) $\frac{1}{2}n$; (5) 0 or $(-1)^{(n-1)/2}\dfrac{1.3.5\dots(n-2)}{2.4.6\dots(n-1)}$; (6) 0 or 1.

The values 0 occur when n is even.]

26. If $x=y(a+y)$, prove that

$$y=\frac{x}{a}-\frac{x^2}{a^3}+\frac{4}{1.2}\frac{x^3}{a^5}-\frac{5.6}{1.2.3}\frac{x^4}{a^7}+\dots+(-1)^{n-1}\frac{(2n-2)!}{n!(n-1)!}\frac{x^n}{a^{2n-1}}+\dots.$$

[Use Lagrange's series; or expand $\frac{1}{2}\{-a+\sqrt{(a^2+4x)}\}$.]

27. Use Lagrange's series to establish the equation

$$(1-x)^\nu=1-\nu t+\frac{\nu(\nu-3)}{2!}t^2-\frac{\nu(\nu-4)(\nu-5)}{3!}t^3+\dots,$$

where $t=x(1-x)$ and $|t|<\frac{1}{4}$.

28. Prove that

$$\arc\tan x = t - \frac{4}{1}\frac{t^3}{3} + \frac{6.7}{1.2}\frac{t^5}{5} - \frac{8.9.10}{1.2.3}\frac{t^7}{7} + \dots,$$

where $t = x(1+x^2)$ and $|t|^2 < \frac{4}{27}$.

29. Shew that

$$2^{n-1} + n.2^{n-2} + \frac{n(n+1)}{2!}2^{n-3} + \dots + \frac{n(n+1)\dots(2n-2)}{(n-1)!} = 4^{n-1}.$$

[*Math. Trip.* 1903.]

[This is the coefficient of $1/x$ in the expansion of $\{x(1-x)\}^{-n}(1-2x)^{-1}$, and is therefore equal to na_n if

$$\Sigma a_n y^n = -\tfrac{1}{2}\log(1-2x), \text{ where } y = x(1-x).$$

Hence $\Sigma a_n y^n = -\tfrac{1}{4}\log(1-4y)$, or $a_n = 4^{n-1}/n$.]

An alternative way of stating the result is to say that *the sum of the first n terms in the binomial series for* $(1-\frac{1}{2})^{-n}$ *is equal to the remainder.*

30. Extend the method of Art. 55 to prove that if

$$y = a_2x^2 + a_3x^3 + a_4x^4 + \dots,$$

there are two expansions for x of the form

$$x_1 = b_1y^{\frac{1}{2}} + b_2y + b_3y^{\frac{3}{2}} + \dots, \quad x_2 = -b_1y^{\frac{1}{2}} + b_2y - b_3y^{\frac{3}{2}} + \dots.$$

Shew also that if $g(x) = c_0 + c_1x + c_2x^2 + \dots,$

$$g(x_1) + g(x_2) = 2c_0 + d_1y + d_2y^2 + d_3y^3 + \dots,$$

where nd_n is the coefficient of $1/x$ in the expansion of $g'(x)/y^n$.

31. As a particular case of the last example, shew that if

$$y(1 + ax + bx^2 + cx^3 + \dots) = x^2,$$

then

$$x_1 + x_2 = ay + (ab + c)y^2 + \dots.$$

32. If $y = x^2(1+x)^{-m}$, we find $x_1 + x_2 = \Sigma a_n y^n$,

where

$$a_1 = m, \quad a_2 = m(2m-1)(2m-2)/3!,$$

$$a_3 = m(3m-1)(3m-2)(3m-3)(3m-4)/5!, \text{ etc.}$$

33. It is easy to write down the general forms for the expansion of $x_1 + x_2$ in the following cases:

$$y = x^2 + ax^m; \quad y = x^2e^{ax}.$$

Differential Equations.

34. If $P = p_0 + p_1x + p_2x^2 + \dots, \quad Q = q_0 + q_1x + q_2x^2 + \dots$

are two power-series which converge for $|x| < R$, prove that the differential equation

$$\frac{d^2y}{dx^2} = P\frac{dy}{dx} + Qy$$

has solutions of the type $y = A_0 + A_1x + A_2x^2 + \dots$, which also converge for $|x| < R$. Here A_0 and A_1 are arbitrary, while $A_2, A_3, A_4, \dots$ are linear combinations of A_0 and A_1.

[Assuming the convergence, we obtain at once the conditions

$$n(n-1)A_n=(n-1)A_{n-1}p_0+(n-2)A_{n-2}p_1+\ldots+A_1p_{n-2}$$
$$+A_{n-2}q_0 \qquad +\ldots+A_1q_{n-3}+A_0q_{n-2}.$$

Thus, as in Art. 55, we see that $B_n \geqq |A_n|$, if $B_1=|A_1|$, $B_0=|A_0|$ and

$$n(n-1)B_n=nB_{n-1}M+(n-1)B_{n-2}\frac{M}{r}+\ldots+2B_1\frac{M}{r^{n-2}}+B_0\frac{M}{r^{n-1}},$$

where $r<R$ and M is such that $|p_n|<M/r^n$, $|q_n|<M/r^{n+1}$.

We then find $\quad n(n-1)B_n-\frac{1}{r}(n-1)(n-2)B_{n-1}=nB_{n-1}M,$

so that $$\lim(B_{n-1}/B_n)=r.$$

Thus $\Sigma B_n x^n$ converges if $|x|<r$; and so we find that $\Sigma A_n x^n$ converges if $|x|<R$ by taking $2r=|x|+R$.]

35. With the same notation for P, Q as in the last example, prove that the differential equation

$$x^2\frac{d^2y}{dx^2}=xP\frac{dy}{dx}+Qy$$

has solutions of the type

$$x^t(A_0+A_1x+A_2x^2+\ldots),$$

where $t(t-1)=tp_0+q_0$ and A_0 is arbitrary, the other coefficients being multiples of A_0.

[If t' is the other root of the quadratic, we find

$$n(n+t-t')A_n=\{(n-1+t)p_1+q_1\}A_{n-1}+\ldots+(tp_n+q_n)A_0,$$

which may be compared with

$$n(n-\delta)B_n=M\left\{(n+\tau)\frac{B_{n-1}}{r}+(n-1+\tau)\frac{B_{n-2}}{r^2}+\ldots+(1+\tau)\frac{B_0}{r^n}\right\},$$

where $\delta=|t-t'|$, $\tau=|t|$, $|p_n|<M/r^n$, $|q_n|<M/r^n$, and $n>\delta$.

If we take $B_0=|A_0|$ and $B_p=|A_p|$ so long as $p\leqq\delta$, we shall have $|A_n|\leqq B_n$, when $n>\delta$. Further,

$$r\{n(n-\delta)B_n\}-(n-1)(n-1-\delta)B_{n-1}=M(n+\tau)B_{n-1},$$

so that $$\lim(B_{n-1}/B_n)=r.$$

Thus we prove that $\Sigma A_n x^n$ converges if $|x|<R$, as in the last example.

The general method requires modification if δ is an integer, when dealing with the *smaller* root of the quadratic, since then $(n+t-t')$ vanishes when $n=\delta$; a description of the changes will be found in Forsyth's *Differential Equations* (3rd ed.), pp. 235-243.]

36. Suppose that $f(x, y)$ is defined for values of x, y such that

$$|x-x_0|\leqq a, \quad |y-y_0|\leqq b,$$

and that $$y_n=y_0+\int_{x_0}^{x}f(x, y_{n-1})dx, \quad (n=1, 2, 3, \ldots).$$

Then y_n approaches as its limit a continuous function of x, say η, which satisfies

$$\frac{d\eta}{dx}=f(x, \eta)$$

for sufficiently small values of $|x-x_0|$. [PICARD.]

[It is assumed that for these values of x, y

$$|f|<M, \quad |f(x, y')-f(x, y)|<A\,|y'-y|,$$

where M and A are constants; and so in particular $f(x, y)$ is a continuous function of y.

Then $|y_n-y_0|<M|x-x_0|$, and so, *if c is the smaller of a and b/M*, the values of y_n fall within the prescribed limits, provided that $|x-x_0|<c$.

Suppose now, for brevity of statement, that $x>x_0$: then

$$|y_{n+1}-y_n|\leqq\int_{x_0}^{x}|f(x, y_n)-f(x, y_{n-1})|\,dx<A\int_{x_0}^{x}|y_n-y_{n-1}|\,dx$$

and

$$|y_1-y_0|<M(x-x_0)<Mc.$$

Thus, by integration we find

$$|y_2-y_1|<\tfrac{1}{2}MA(x-x_0)^2<\tfrac{1}{2}MAc^2,$$

and generally

$$|y_n-y_{n-1}|<\frac{1}{n!}MA^{n-1}(x-x_0)^n<\frac{1}{n!}MA^{n-1}c^n.$$

Thus, by Weierstrass's rule, y_n and (similarly) $\frac{dy_n}{dx}$ converge *uniformly* to their limits η, $\frac{d\eta}{dx}$.

Also,

$$\frac{d\eta}{dx}=\lim\frac{dy_n}{dx}=\lim f(x, y_{n-1})=f(x, \eta),$$

because $f(x, y)$ is continuous as regards y.]

CHAPTER IX.

TRIGONOMETRICAL INVESTIGATIONS.

66. Expressions for $\cos n\theta$ **and** $(\sin n\theta/\sin\theta)$ **as polynomials in** $\cos\theta$.

We have seen (Art. 65) that

$$\log(1-2r\cos\theta+r^2)=-2(r\cos\theta+\tfrac{1}{2}r^2\cos 2\theta+\tfrac{1}{3}r^3\cos 3\theta+\ldots).$$

But

$$\log(1-2r\cos\theta+r^2)=-[(r\gamma-r^2)+\tfrac{1}{2}(r\gamma-r^2)^2+\tfrac{1}{3}(r\gamma-r^2)^3+\ldots]$$

where $\gamma=2\cos\theta$; and, further, the latter series may be re-arranged in powers of r, without alteration of value, provided that $0<r\leqq\frac{2}{5}$. It is therefore evident that $\dfrac{2}{n}\cos n\theta$ is the coefficient of r^n in the expression

$$\frac{1}{n}(r\gamma-r^2)^n+\frac{1}{n-1}(r\gamma-r^2)^{n-1}+\ldots+(r\gamma-r^2),$$

because $(r\gamma-r^2)^{n+1}$, $(r\gamma-r^2)^{n+2}$, ... contain no terms in r^n.

Thus

$$\begin{aligned}\frac{2}{n}\cos n\theta=\frac{\gamma^n}{n}-\frac{(n-1)\gamma^{n-2}}{n-1}+\ldots\\ +(-1)^s\frac{(n-s)(n-s-1)\ldots(n-2s+1)}{(n-s)\,.\,s!}\gamma^{n-2s}\\ +\ldots,\end{aligned}$$

the number of terms being either $\frac{1}{2}(n+1)$ or $\frac{1}{2}(n+2)$.

Hence

$$\begin{aligned}2\cos n\theta=\gamma^n-n\gamma^{n-2}+\frac{n(n-3)}{2!}\gamma^{n-4}-\ldots\\ +(-1)^s\frac{n(n-s-1)(n-s-2)\ldots(n-2s+1)}{s!}\gamma^{n-2s}\\ +\ldots.\end{aligned}$$

Similarly, we have seen that

$$\frac{r\sin\theta}{1-2r\cos\theta+r^2}=r\sin\theta+r^2\sin 2\theta+r^3\sin 3\theta+\dots.$$

Hence we deduce that $\dfrac{\sin n\theta}{\sin\theta}$ is the coefficient of r^{n-1} in the series

$$1+(r\gamma-r^2)+(r\gamma-r^2)^2+(r\gamma-r^2)^3+\dots.$$

Thus

$$\frac{\sin n\theta}{\sin\theta}=\gamma^{n-1}-(n-2)\gamma^{n-3}+\frac{(n-3)(n-4)}{2!}\gamma^{n-5}$$

$$-\dots+(-1)^s\frac{(n-s-1)(n-s-2)\dots(n-2s)}{s!}\gamma^{n-2s-1}$$

$$+\dots,$$

where the number of terms is either $\frac{1}{2}n$ or $\frac{1}{2}(n+1)$. We note that this formula can be deduced from the last by differentiation.

It is therefore evident that both $\cos n\theta$ and $\sin n\theta/\sin\theta$ are polynomials in $\cos\theta$, of degrees n and $(n-1)$ respectively. But for some purposes it is more useful to express the functions of $n\theta$ in terms of $\sin\theta$. This we shall do in the following article.

Before leaving the formulae above, it is worth while to notice that if we write $\gamma=t+1/t$, instead of $2\cos\theta$, then $1-r\gamma+r^2=(1-rt)(1-r/t)$.

Hence $\log(1-r\gamma+r^2)=\log(1-rt)+\log(1-r/t)=-\Sigma\frac{1}{n}r^n(t^n+t^{-n})$,

and so, from the foregoing argument, we get the algebraic identity

$$t^n+t^{-n}=\gamma^n-n\gamma^{n-2}+\dots \text{ as above.}$$

Similarly, $$\frac{r(t-1/t)}{1-r\gamma+r^2}=\frac{1}{1-rt}-\frac{1}{1-r/t}=\Sigma r^n(t^n-t^{-n}),$$

and so we find $$\frac{t^n-t^{-n}}{t-t^{-1}}=\gamma^{n-1}-(n-2)\gamma^{n-3}+\dots \text{ as above.}$$

67. Further forms for $\cos n\theta$ and $\sin n\theta$.

In the formulae of the last article change θ to $(\frac{1}{2}\pi-\theta)$; then $\gamma=2\sin\theta$, and we find

$$(-1)^m 2\cos 2m\theta=\gamma^{2m}-2m\gamma^{2m-2}+\dots \text{ to } (m+1) \text{ terms,}$$

$$(-1)^{m-1}\frac{\sin 2m\theta}{2\sin\theta\cos\theta}=\gamma^{2m-2}-(2m-2)\gamma^{2m-4}+\dots \text{ to } m \text{ terms,}$$

in case n is even and equal to $2m$.

But if n is odd and equal to $(2m+1)$, we have

$$(-1)^m 2\sin(2m+1)\theta = \gamma^{2m+1} - (2m+1)\gamma^{2m-1} + \dots \text{ to } (m+1) \text{ terms},$$

$$(-1)^m \frac{\cos(2m+1)\theta}{\cos\theta} = \gamma^{2m} - (2m-1)\gamma^{2m-2} + \dots \text{ to } (m+1) \text{ terms}.$$

However, these formulae take a more elegant shape when arranged according to *ascending* powers of $\sin\theta$; of course it is not specially difficult to rearrange the expressions algebraically, but it is instructive to obtain the results in another way.

If $y = \sin n\theta$ or $\cos n\theta$, we have

$$\frac{d^2y}{d\theta^2} + n^2 y = 0.$$

Now write $x = \sin\theta$, and we have

$$(1-x^2)\frac{d^2y}{dx^2} - x\frac{dy}{dx} + n^2 y = 0.$$

Now, if we consider the expression given above for $\cos 2m\theta$, we see that *when n is even*, $\cos n\theta$ can be expressed as a polynomial of degree n in x, containing only even powers; thus we can write

$$\cos n\theta = 1 + A_2x^2 + A_4x^4 + \dots + A_nx^n,$$

the constant term being 1, because $\theta = 0$ gives $\cos n\theta = 1$, $x = 0$.

If we substitute this expression in the differential equation, we find

$$\begin{aligned}(1-x^2)&(1.2.A_2 + 3.4A_4x^2 + \dots + (n-1)nA_nx^{n-2})\\ &-x^2(2A_2 + 4A_4x^2 + \dots + nA_nx^{n-2})\\ &+n^2(1 + A_2x^2 + A_4x^4 + \dots + A_nx^n) = 0.\end{aligned}$$

Thus $\quad 1.2.A_2 + n^2 = 0,\ 3.4.A_4 + (n^2-2^2)A_2 = 0,$

$\quad 5.6.A_6 + (n^2-4^2)A_4 = 0, \dots,$

and so $A_2 = -\dfrac{n^2}{2!},\ A_4 = \dfrac{n^2(n^2-2^2)}{4!},\ A_6 = -\dfrac{n^2(n^2-2^2)(n^2-4^2)}{6!},$

etc.

Hence $\cos n\theta = 1 - \dfrac{n^2}{2!}x^2 + \dfrac{n^2(n^2-2^2)}{4!}x^4 - \dots$ to $\tfrac{1}{2}(n+2)$ terms

when n is *even*.

Similarly, *when n is odd*, we find that $\sin n\theta$ is a polynomial of degree n, which contains only odd powers of x; thus we write

$$\sin n\theta = nx + A_3x^3 + A_5x^5 + \dots + A_nx^n,$$

the first coefficient being determined by considering that for $\theta=0$,

$$x=0,\ \frac{dx}{d\theta}=1,\ \frac{d}{d\theta}(\sin n\theta)=n.$$

Hence, on substitution, we find

$$(1-x^2)(3.2.A_3x+5.4.A_5x^3+\ldots+n(n-1)A_nx^{n-2})$$
$$-x(n+3A_3x^2+5A_5x^4+\ldots+nA_nx^{n-1})$$
$$+n^2(nx+A_3x^3+A_5x^5+\ldots+A_nx^n)=0.$$

Thus $3.2.A_3+(n^2-1)n=0,\ 5.4.A_5+(n^2-3^2)A_3=0,\ldots,$ giving

$$\sin n\theta=nx-\frac{n(n^2-1^2)}{3!}x^3+\frac{n(n^2-1^2)(n^2-3^2)}{5!}x^5-\ldots$$

to $\frac{1}{2}(n+1)$ terms,

n being *odd*.

To verify the algebraic identity between these results and those of Art. 66, consider the case $n=6$. Then Art. 66 gives

$$2\cos 6\theta=\gamma^6-6\gamma^4+9\gamma^2-2$$

or

$$\cos 6\theta=32\cos^6\theta-48\cos^4\theta+18\cos^2\theta-1.$$

Change from θ to $(\frac{1}{2}\pi-\theta)$, and we get

$$\cos 6\theta=1-18\sin^2\theta+48\sin^4\theta-32\sin^6\theta$$

$$=1-\frac{6^2}{2!}\sin^2\theta+\frac{6^2(6^2-2^2)}{4!}\sin^4\theta-\frac{6^2(6^2-2^2)(6^2-4^2)}{6!}\sin^6\theta,$$

in agreement with the formula for $\cos n\theta$.

Again, consider the case $n=7$; we have, from Art. 66,

$$2\cos 7\theta=\gamma^7-7\gamma^5+14\gamma^3-7\gamma$$

or

$$\cos 7\theta=64\cos^7\theta-112\cos^5\theta+56\cos^3\theta-7\cos\theta.$$

Hence, changing θ to $\frac{1}{2}\pi-\theta$, we have

$$\sin 7\theta=7\sin\theta-56\sin^3\theta+112\sin^5\theta-64\sin^7\theta$$

$$=7\sin\theta-\frac{7(7^2-1^2)}{3!}\sin^3\theta+\frac{7(7^2-1^2)(7^2-3^2)}{5!}\sin^5\theta$$
$$-\frac{7(7^2-1^2)(7^2-3^2)(7^2-5^2)}{7!}\sin^7\theta.$$

By differentiating the formulae just obtained for $\sin n\theta$ and $\cos n\theta$, we find

$$\frac{\cos n\theta}{\cos\theta}=1-\frac{n^2-1^2}{2!}\sin^2\theta+\frac{(n^2-1^2)(n^2-3^2)}{4!}\sin^4\theta-\ldots$$

to $\frac{1}{2}(n+1)$ terms,

when n is *odd*; and

$$\frac{\sin n\theta}{\cos\theta} = n\sin\theta - \frac{n(n^2-2^2)}{3!}\sin^3\theta + \frac{n(n^2-2^2)(n^2-4^2)}{5!}\sin^5\theta - \dots$$

to $\frac{1}{2}n$ terms,

when n is *even*.

The reader may shew that

$$\cos 7\theta/\cos\theta = 1 - 24\sin^2\theta + 80\sin^4\theta - 64\sin^6\theta$$

and that $\sin 6\theta/\cos\theta = 6\sin\theta - 32\sin^3\theta + 32\sin^5\theta$,

and compare these formulae with those of Art. 66.

68. The expressions obtained in the last article are restricted by certain conditions on the form of n. Let us now see if these conditions can be removed in any way.

Take for example the series

$$y = 1 - \frac{n^2}{2!}x^2 + \frac{n^2(n^2-2^2)}{4!}x^4 - \dots, \quad (x = \sin\theta)$$

which was proved to terminate and to represent $\cos n\theta$, when n is *even*.

If n is an odd integer, or is not an integer, the series does not terminate. It is natural to consider whether it is convergent, and if so, to investigate its sum.

The test (5) of Art. 12 shews at once that the series converges absolutely when $|x|=1$; and so, as we have explained in Art. 50, the series converges *absolutely* and *uniformly* for $|x| \leqq 1$.

It follows that y is continuous for all real values of θ, and as on p. 179, it satisfies the equation $\frac{d^2y}{d\theta^2} + n^2 y = 0$.

From this equation it follows* that y is of the form

$$A\cos n\theta + B\sin n\theta.$$

Now for $\theta = 0$, we have $y = 1$, $\frac{dy}{d\theta} = 0$.

Thus $A = 1$, $B = 0$, and so

$$\cos n\theta = 1 - \frac{n^2}{2!}x^2 + \frac{n^2(n^2-2^2)}{4!}x^4 - \dots \text{ to } \infty$$

for any value of n.

* This obviously satisfies the equation; and since the solution contains *two* arbitrary constants, no other solution can be found (Forsyth, *Differential Equations*, Art. 8).

Similarly, we find that all the other formulae of Art. 67 are valid for any value of n, provided that $|x| < 1$ and that the series are continued to infinity.

It should however be noticed that the third and fourth series are not convergent when $|x| = 1$; but they converge absolutely for $|x| < 1$, and uniformly for $|x| \leqq k < 1$.

69. Various deductions from Art. 67.

We know that we can write.

$$\frac{\sin n\theta}{\sin\theta} = n + A_2\sin^2\theta + \dots + A_{n-1}\sin^{n-1}\theta, \qquad (n \text{ odd})$$

or
$$\frac{\sin n\theta}{\sin\theta\cos\theta} = n + B_2\sin^2\theta + \dots + B_{n-2}\sin^{n-2}\theta, \qquad (n \text{ even})$$

where the coefficients are the same as those worked out in Art. 67, but are not needed in an explicit form at present.

Now the left-hand side vanishes for

$$\theta = \pm\pi/n,\ \pm 2\pi/n,\ \pm 3\pi/n, \dots,$$

so that the right-hand side (regarded as a polynomial in $\sin\theta$) must have roots

$$\sin\theta = \pm\sin(\pi/n),\ \pm\sin(2\pi/n),\ \pm\sin(3\pi/n), \dots.$$

When n is odd, there must be $(n-1)$ of these roots whic. are all different; and these can be taken as

$$\sin\theta = \pm\sin\frac{\pi}{n},\ \pm\sin\frac{2\pi}{n}, \dots,\ \pm\sin\frac{(n-1)\pi}{2n}.$$

But if n is even, there are only $(n-2)$ different roots, namely

$$\sin\theta = \pm\sin\frac{\pi}{n},\ \pm\sin\frac{2\pi}{n}, \dots,\ \pm\sin\frac{(n-2)\pi}{2n}.$$

Thus we have the formulae

(n *odd*)

$$\frac{\sin n\theta}{\sin\theta} = n\left(1 - \frac{\sin^2\theta}{\sin^2\alpha}\right)\left(1 - \frac{\sin^2\theta}{\sin^2 2\alpha}\right)\dots\left[1 - \frac{\sin^2\theta}{\sin^2\frac{1}{2}(n-1)\alpha}\right],$$

(n *even*)

$$\frac{\sin n\theta}{\sin\theta\cos\theta} = n\left(1 - \frac{\sin^2\theta}{\sin^2\alpha}\right)\left(1 - \frac{\sin^2\theta}{\sin^2 2\alpha}\right)\dots\left[1 - \frac{\sin^2\theta}{\sin^2\frac{1}{2}(n-2)\alpha}\right],$$

where, in each case, $\alpha = \pi/n$.

If we compare these with the explicit forms given in Art. 67, we can deduce other identities, such as

$$\frac{n^2-1}{6}=\frac{1}{\sin^2 a}+\frac{1}{\sin^2 2a}+\dots+\frac{1}{\sin^2\frac{1}{2}(n-1)a}, \quad (n \textit{ odd})$$

$$\frac{n^2-4}{6}=\frac{1}{\sin^2 a}+\frac{1}{\sin^2 2a}+\dots+\frac{1}{\sin^2\frac{1}{2}(n-2)a}, \quad (n \textit{ even}).$$

In a similar way we prove the identities

$$(n \textit{ odd}) \quad \frac{\cos n\theta}{\cos\theta}=\left(1-\frac{\sin^2\theta}{\sin^2\beta}\right)\left(1-\frac{\sin^2\theta}{\sin^2 3\beta}\right)\dots\left[1-\frac{\sin^2\theta}{\sin^2(n-2)\beta}\right],$$

$$(n \textit{ even}) \quad \cos n\theta=\left(1-\frac{\sin^2\theta}{\sin^2\beta}\right)\left(1-\frac{\sin^2\theta}{\sin^2 3\beta}\right)\dots\left[1-\frac{\sin^2\theta}{\sin^2(n-1)\beta}\right],$$

where $\beta=\pi/2n$ and only the *odd* multiples of β appear.

On comparing these with the forms of Art. 67, we see that

$$\frac{n^2-1}{2}=\frac{1}{\sin^2\beta}+\frac{1}{\sin^2 3\beta}+\dots+\frac{1}{\sin^2(n-2)\beta}, \quad (n \textit{ odd})$$

$$\frac{n^2}{2}=\frac{1}{\sin^2\beta}+\frac{1}{\sin^2 3\beta}+\dots+\frac{1}{\sin^2(n-1)\beta}, \quad (n \textit{ even}).$$

Again, if we consider the formulae of Art. 66, it is evident that $(\cos n\theta-\cos n\lambda)$ may be expressed as a polynomial of degree n in $\cos\theta$, the term of highest degree being $2^{n-1}\cos^n\theta$. But the expression $(\cos n\theta-\cos n\lambda)$ is zero if

$$n\theta=\pm n\lambda,\ 2\pi\pm n\lambda,\ 4\pi\pm n\lambda,\dots.$$

Thus the factors of the polynomial in question will be n different expressions of the form

$$\cos\theta-\cos\lambda,\ \cos\theta-\cos(\lambda\pm 2a),\ \cos\theta-\cos(\lambda\pm 4a),\dots,$$

where a, as before, is used for π/n.

It is easily seen that the n different factors can be taken as

$$\cos\theta-\cos\lambda,\ \cos\theta-\cos(\lambda+2a),\ \cos\theta-\cos(\lambda+4a),\dots$$
$$\cos\theta-\cos(\lambda+2(n-1)a),$$

because $\cos(\lambda-2ra)=\cos(\lambda+2(n-r)a)$.

Hence we have the identity

$$\cos n\theta-\cos n\lambda=2^{n-1}\prod_{r=0}^{n-1}[\cos\theta-\cos(\lambda+2ra)].$$

If we write $\theta=0$ in this expression we have

$$\sin^2\tfrac{1}{2}n\lambda=2^{2n-2}\prod_{r=0}^{n-1}\sin^2(\tfrac{1}{2}\lambda+ra),$$

or, with a change of notation,

$$\sin n\theta = \pm 2^{n-1} \prod_{r=0}^{n-1} \sin(\theta + ra).$$

But the sign must be +, because, if $0 < \theta < a$, all the factors are positive, and it is easily seen that both sides change sign together (as θ passes through any multiple of a).

70. Expression of $\sin\theta$ as an infinite product.

We have seen in the last article that, if n is an odd integer,

$$\frac{\sin n\phi}{n \sin\phi} = \prod_{r=1}^{\frac{1}{2}(n-1)} \left(1 - \frac{\sin^2\phi}{\sin^2 ra}\right),$$

where $a = \pi/n$. Thus if we write $n\phi = \theta$ we have

$$\frac{\sin\theta}{n\sin(\theta/n)} = \prod_{r=1}^{\frac{1}{2}(n-1)} \left[1 - \frac{\sin^2(\theta/n)}{\sin^2(r\pi/n)}\right].$$

To this equation we apply the general theorem of Art. 49; we have, in fact,*

$$\left|\frac{\sin^2(\theta/n)}{\sin^2(r\pi/n)}\right| < \frac{\theta^2}{4r^2},$$

because $r\pi/n$ is less than $\frac{1}{2}\pi$. Now this expression is independent of n, and the series $\sum_{r=1}^{\infty} \theta^2/4r^2$ is convergent; consequently, the theorem applies. But we have

$$\lim_{n\to\infty} n\sin(\theta/n) = \theta,$$

$$\lim_{n\to\infty} \frac{\sin^2(\theta/n)}{\sin^2(r\pi/n)} = \lim_{n\to\infty} \frac{n^2\sin^2(\theta/n)}{n^2\sin^2(r\pi/n)} = \frac{\theta^2}{r^2\pi^2}.$$

Consequently,
$$\frac{\sin\theta}{\theta} = \prod_{r=1}^{\infty}\left(1 - \frac{\theta^2}{r^2\pi^2}\right).$$

The special value $\theta = \frac{1}{2}\pi$ leads at once to *Wallis's Theorem*:

$$\frac{2}{\pi} = \prod_{1}^{\infty}\left(1 - \frac{1}{4r^2}\right) = \prod_{1}^{\infty} \frac{(2r-1)(2r+1)}{2r \cdot 2r}$$

or
$$\frac{\pi}{2} = \frac{2}{1}\cdot\frac{2}{3}\cdot\frac{4}{3}\cdot\frac{4}{5}\cdot\frac{6}{5}\cdot\frac{6}{7}\dots \text{ to } \infty.$$

* We see, by differentiation or otherwise, that $\sin x/x$ decreases as x increases from 0 to $\frac{1}{2}\pi$: thus

$$1 > (\sin x)/x > 2/\pi, \quad \text{if } 0 < x < \tfrac{1}{2}\pi.$$

Consequently, $\sin(r\pi/n) > 2r/n$, if $r < \frac{1}{2}n$.

Also $|\sin\theta/n| < |\theta/n|$, for any value of θ; and so the inequality follows.

By combining the formula for $\sin\theta$ with the results of Art. 42, we see that

$$\Pi(x-1).\Pi(-x)=\frac{\pi}{\sin(\pi x)}=\Gamma(x).\Gamma(1-x).$$

It is perhaps worth while to refer briefly to an incomplete "proof" given in some of the older books. Since $\sin\theta$ vanishes for $\theta=0$ and for $\theta=\pm r\pi$, and since $\lim_{\theta\to 0}(\sin\theta/\theta)=1$, it is urged that $\sin\theta/\theta$ must be of the form given above; but exactly the same argument would apply equally to the function $a^\theta\sin\theta$, where a is any real number, so that this "proof" only suggests that $\prod_{r=1}^{\infty}\left(1-\frac{\theta^2}{r^2\pi^2}\right)$ may possibly be of the form $a^\theta\sin\theta/\theta$; it cannot prove that a is 1. In this connexion, it may be noted that if we separate $1-\frac{\theta^2}{r^2\pi^2}$ into factors $\left(1-\frac{\theta}{r\pi}\right)$, $\left(1+\frac{\theta}{r\pi}\right)$ and then take more positive than negative factors (say p positive to every q negative factors), the value of the product is $\left(\frac{p}{q}\right)^{\frac{\theta}{\pi}}\left(\frac{\sin\theta}{\theta}\right)$. [See Art. 41, and write $a_{2r-1}=a_{2r}=\frac{\theta}{r\pi}$.]

A satisfactory proof on these lines involves some knowledge of function-theory; a good proof is given by Harkness and Morley (*Introduction to the Theory of Functions*, Art. 112).

We have already pointed out the danger of applying the theorem of Art. 49 to cases when the M-test does not hold good. An additional illustration of this risk may be given here.

Since $\sin(\pi-\phi)=\sin\phi$, it follows that the values of $\sin(r\pi/n)$ when r ranges from $\frac{1}{2}(n+1)$ to $(n-1)$ are the same as those when r ranges from 1 to $\frac{1}{2}(n-1)$, but in the reverse order.

Hence
$$\frac{\sin^2\theta}{[n\sin(\theta/n)]^2}=\prod_{r=1}^{n-1}\left[1-\frac{\sin^2(\theta/n)}{\sin^2(r\pi/n)}\right],$$
and if we apply the theorem here, we appear to get
$$\frac{\sin^2\theta}{\theta^2}=\prod_{r=1}^{\infty}\left(1-\frac{\theta^2}{r^2\pi^2}\right),$$
which contradicts the result obtained before. Of course the explanation is that the inequality $\frac{\sin^2(\theta/n)}{\sin^2(r\pi/n)}<\frac{\theta^2}{4r^2}$ is no longer true, since $r\pi/n$ may be greater than $\frac{1}{2}\pi$; and it is, in fact, impossible to construct a convergent comparison-series ΣM_r.

71. Expressions for $\cos\theta$, $\cot\theta$.

The reader should find no difficulty in expressing $\cos\theta$ as an infinite product, following the lines of the preceding article.

We have in fact (Art. 69)
$$\cos\theta=\prod_{r=1}^{n-1}\left[1-\frac{\sin^2(\theta/n)}{\sin^2(r\pi/2n)}\right],$$
where n is *even* and r is *odd*.

Here the comparison-series is $\Sigma\theta^2/r^2$, and the result is

$$\cos\theta = \left(1-\frac{4\theta^2}{\pi^2}\right)\left(1-\frac{4\theta^2}{3^2\pi^2}\right)\left(1-\frac{4\theta^2}{5^2\pi^2}\right)\ldots.$$

Alternative methods are to write $\cos\theta = \frac{\sin 2\theta}{2\theta}\Big/\frac{\sin\theta}{\theta}$, and appeal directly to the product of Art. 70; or to write $\frac{1}{2}\pi-\theta$ for θ in that product and rearrange the factors.

To express $\cot\theta$, we know that

$$\sin\theta = n\sin\frac{\theta}{n}\prod_{r=1}^{\frac{1}{2}(n-1)}\left[1-\frac{\sin^2(\theta/n)}{\sin^2(r\pi/n)}\right], \qquad (n \text{ being } odd)$$

and so if we take logarithms and then differentiate, we have the identity

$$\cot\theta = \frac{1}{n}\cot\frac{\theta}{n} - \sum_{r=1}^{\frac{1}{2}(n-1)}\frac{2}{n}\,\frac{\sin(\theta/n)\cos(\theta/n)}{\sin^2(r\pi/n)-\sin^2(\theta/n)},$$

assuming that θ is not a multiple of π.

To this identity we apply the theorem of Art. 49, and our first step is to obtain a comparison-series ΣM_r.

Now (see footnote, Art. 70, p. 184)

$$\sin^2(r\pi/n) > 4r^2/n^2, \qquad \text{provided that } r < \tfrac{1}{2}n;$$

also

$$\sin^2(\theta/n) < \theta^2/n^2.$$

Thus,

$$\sin^2(r\pi/n) - \sin^2(\theta/n) > (4r^2-\theta^2)/n^2$$

for all values of r such that $n > 2r > \theta$; and consequently

$$\left|\frac{2}{n}\,\frac{\sin(\theta/n)\cos(\theta/n)}{\sin^2(r\pi/n)-\sin^2(\theta/n)}\right| < \frac{2|\theta|}{4r^2-\theta^2}, \qquad \text{if } \tfrac{1}{2}\theta < r < \tfrac{1}{2}n.$$

Since θ is *fixed*, there may be a fixed number of terms (*i.e.* a number independent of n) at the beginning of the expression for $\cot\theta$ for which the last inequality is not valid. But for the rest of the terms we can use the comparison-series

$$\sum_{r>\frac{1}{2}\theta}^{\infty} 2|\theta|/(4r^2-\theta^2),$$

which is obviously convergent and independent of n.

To these terms we can apply the theorem of Art. 49; for the *fixed* number of terms preceding them no special test is necessary.*

---

* No test of this character is requisite for any *fixed* number of terms; the object of Art. 49 is to enable us to handle a sum the number of whose terms tends to ∞ with n.

Now, $$\lim_{n\to\infty} \frac{1}{n}\cot\frac{\theta}{n} = \frac{1}{\theta}\lim_{n\to\infty}\left(\frac{\theta}{n}\Big/\tan\frac{\theta}{n}\right) = \frac{1}{\theta}$$

and $$\lim_{n\to\infty} \frac{1}{n}\frac{\sin(\theta/n)\cos(\theta/n)}{\sin^2(r\pi/n)-\sin^2(\theta/n)} = \frac{\theta}{r^2\pi^2-\theta^2}.$$

Thus, $$\cot\theta = \frac{1}{\theta} - \sum_1^\infty \frac{2\theta}{r^2\pi^2-\theta^2} = \frac{1}{\theta} + \sum_1^\infty \frac{2\theta}{\theta^2-r^2\pi^2}.$$

This result can also be derived at once from the product for $\sin\theta$, by differentiating logarithmically; we should then use the theorem at the end of Art. 48 to justify the operation.

In exactly the same way the identities (p. 183),

$$\frac{n^2-1}{6n^2} = \sum_1^{\frac{1}{2}(n-1)} \frac{1}{n^2\sin^2(r\pi/n)} \qquad (n \textit{ odd})$$

and $$\frac{1}{2} = \sum_1^{n-1} \frac{1}{n^2\sin^2(r\pi/2n)} \qquad (r \textit{ odd}, n \textit{ even})$$

give, on applying Art. 49, the two series

$$\frac{1}{6} = \sum_1^\infty \frac{1}{r^2\pi^2}, \qquad \frac{1}{2} = \sum_1^\infty \frac{4}{(2s-1)^2\pi^2},$$

or $$\frac{1}{1^2}+\frac{1}{2^2}+\frac{1}{3^2}+\frac{1}{4^2}+\dots = \frac{\pi^2}{6}$$

and $$\frac{1}{1^2}+\frac{1}{3^2}+\frac{1}{5^2}+\frac{1}{7^2}+\dots = \frac{\pi^2}{8}.$$

Of course there is no difficulty in deducing the second of these from the first, because

$$\frac{1}{1^2}+\frac{1}{3^2}+\frac{1}{5^2}+\frac{1}{7^2}+\dots = \left(\frac{1}{1^2}+\frac{1}{2^2}+\frac{1}{3^2}+\frac{1}{4^2}+\dots\right) - \left(\frac{1}{2^2}+\frac{1}{4^2}+\frac{1}{6^2}+\dots\right)$$
$$= \left(1-\frac{1}{4}\right)\left(\frac{1}{1^2}+\frac{1}{2^2}+\frac{1}{3^2}+\frac{1}{4^2}+\dots\right),$$

the transformations being justified by Art. 26. [Compare Exs. 2, 3, Ch. IV.]

To elaborate further details of analytical trigonometry (although of great interest) would lead us too far afield. We shall therefore content ourselves with making a reference to Chrystal's *Algebra*, vol. 2, ch. XXX., in which will be found a large number of useful and interesting identities.

EXAMPLES.

1. From the formula (of Art. 68)

$$\sin(m\theta) = m\sin\theta - \frac{m(m^2-1)}{3!}\sin^3\theta + \dots \text{ to } \infty,$$

obtain a power-series for $(\arcsin x)^3$: namely

$$\frac{1}{6}(\arcsin x)^3 = \frac{1}{2}\frac{x^3}{3} + \left(\frac{1}{1^2}+\frac{1}{3^2}\right)\frac{1.3}{2.4}\frac{x^5}{5} + \left(\frac{1}{1^2}+\frac{1}{3^2}+\frac{1}{5^2}\right)\frac{1.3.5}{2.4.6}\frac{x^7}{7} + \dots \text{ to } \infty.$$

2. From the formula for $\sin n\theta$ given at the end of Art. 69, prove that, if $a = \pi/n$,

$$\cot\theta + \cot(\theta+a) + \dots + \cot\{\theta+(n-1)a\} = n\cot n\theta,$$

$$\operatorname{cosec}^2\theta + \operatorname{cosec}^2(\theta+a) + \dots + \operatorname{cosec}^2\{\theta+(n-1)a\} = n^2\operatorname{cosec}^2 n\theta.$$

[*Math. Trip.* 1900.]

3. If n is odd and equal to $2m+1$, shew that

$$\sum_{r=1}^{m}\tan^4(r\pi/n) = \tfrac{1}{6}n(n-1)(n^2+n-3). \quad \text{[\textit{Math. Trip.} 1903.]}$$

4. If n is odd, shew that

$$\sum_{r=1}^{n-1}\operatorname{cosec}^2(r\pi/n) = \tfrac{1}{3}(n^2-1).$$

If $n = abc \dots k$, where $a, b, c, \dots, k$ are primes, shew that the above sum, if extended *only to values of r which are prime to n*, is equal to

$$\tfrac{1}{3}(a^2-1)(b^2-1)(c^2-1)\dots(k^2-1). \quad \text{[\textit{Math. Trip.} 1902.]}$$

5. Shew that the roots of the equation $x^3 - x^2 + \frac{1}{7} = 0$ are given by

$$x\sqrt{7} = 2\sin(\tfrac{2}{7}\pi),\ 2\sin(\tfrac{4}{7}\pi),\ 2\sin(\tfrac{8}{7}\pi).$$

[Write $y = 2\sin\theta$ in the formula of Art. 67 for $\sin 7\theta$. Then we get

$$y^6 = 7(y^2-1)^2 \text{ or } y^3 = \pm\sqrt{7}(y^2-1).$$

To distinguish between the two sets of three roots note that $2\sin(\frac{2}{7}\pi)$, $2\sin(\frac{4}{7}\pi)$ are both greater than 1, while $2\sin\frac{8}{7}\pi$ lies between -1 and 0.]

6. Shew that the roots of the equation

$$x^3 + x^2 - 2x - 1 = 0$$

are

$$2\cos(\tfrac{2}{7}\pi),\ 2\cos(\tfrac{4}{7}\pi),\ 2\cos(\tfrac{8}{7}\pi).$$

[The formula of Art. 66 for $\sin 7\theta$ shews that

$$4\cos^2\frac{\pi}{7},\ 4\cos^2\frac{2\pi}{7},\ 4\cos^2\frac{4\pi}{7}$$

are the roots of

$$y^3 - 5y^2 + 6y - 1 = 0.$$

The substitution $y = x + 2$ leads at once to the required result.]

7. Shew that, if $|x| < 1$,

$$\frac{(1-x)\cos\theta}{1-2x\cos 2\theta + x^2} = \cos\theta + x\cos 3\theta + x^2\cos 5\theta + \dots + x^n\cos(2n+1)\theta + \dots,$$

and deduce that, if $\gamma = 2\cos 2\theta$,

$$\frac{\cos 13\theta}{\cos\theta} = \gamma^6 - \gamma^5 - 5\gamma^4 + 4\gamma^3 + 6\gamma^2 - 3\gamma - 1.$$

8. Sum the series $S(x)$ which is obtained by omitting all terms for which n is a multiple of k from the series

$$\sin x + \tfrac{1}{2}\sin 2x + \tfrac{1}{3}\sin 3x + \ldots.$$

Shew that if $0 < t < k$, the greatest integer in t is equal to

$$\frac{1}{2}(k-1) - \frac{k}{\pi} S\left(\frac{2\pi t}{k}\right). \qquad \text{[Eisenstein.]}$$

9. Sum the series

$$\frac{4}{\pi^2}\left(\sin x - \frac{1}{3^2}\sin 3x + \frac{1}{5^2}\sin 5x - \ldots\right) + \frac{2}{\pi}\left(\sin x - \frac{1}{2}\sin 2x + \frac{1}{3}\sin 3x - \ldots\right).$$

$\left[\text{The sum is the smaller of } \frac{2x}{\pi} \text{ and 1 if } 0 \leqq x < \pi.\right]$

10. If

$$u_n = \frac{(n+a_1)(n+a_2)\ldots(n+a_k)}{(n+b_1)(n+b_2)\ldots(n+b_k)},$$

where $\Sigma a = \Sigma b$ and none of $b_1, b_2, \ldots, b_k$ are zero, then

$$\prod_{-\infty}^{\infty} u_n = \frac{\sin(a_1\pi)\sin(a_2\pi)\ldots\sin(a_k\pi)}{\sin(b_1\pi)\sin(b_2\pi)\ldots\sin(b_k\pi)} = P. \qquad \text{[Euler.]}$$

If $\Sigma b - \Sigma a = \delta$, shew that, if the accent implies that $n=0$ is excluded,

$$u_0 \prod_{-\infty}^{\infty}{}' (u_n e^{\delta/n}) = \lim_{\nu\to\infty} \prod_{-\nu}^{+\nu} u_n = P.$$

Further, if $\lim(\mu/\nu) = k$, we have

$$\lim \prod_{-\nu}^{\mu} u_n = k^{-\delta} P.$$

Determine the value of $\prod_{-\infty}^{\infty}{}'(u_n e^{\delta/n})$, where $n=0$ is excluded and some of $b_1, b_2, \ldots, b_k$ are zero.

11. Prove that

$$\frac{x+c}{c} \prod_{-\infty}^{\infty}{}' \left(1 - \frac{x}{n-c}\right) e^{x/n} = \frac{\sin\pi(x+c)}{\sin\pi c}$$

and

$$\prod_{-\infty}^{\infty}\left[1 - \frac{x^2}{(n-c)^2}\right] = 1 - \frac{\sin^2\pi x}{\sin^2\pi c}. \qquad \text{[Euler.]}$$

Why cannot the first product be written without an exponential factor?

12. Shew that

$$(1-x)(1+\tfrac{1}{3}x)(1-\tfrac{1}{5}x)(1+\tfrac{1}{7}x)\ldots = \cos(\tfrac{1}{4}\pi x) - \sin(\tfrac{1}{4}\pi x).$$

[Group the terms in pairs and apply Ex. 21, Ch. VI., obtaining

$$\frac{\Gamma\left(\frac{1}{4}\right)\Gamma\left(\frac{3}{4}\right)}{\Gamma\left(\frac{1-x}{4}\right)\Gamma\left(\frac{3+x}{4}\right)};$$

or write out the product form for $\sin\{\tfrac{1}{4}\pi(1-x)\}/\sin(\tfrac{1}{4}\pi)$.]

13. Prove that

$$\prod_{-\infty}^{\infty}\left[1 - \frac{4x^2}{(n\pi+x)^2}\right] = -\frac{\sin 3x}{\sin x}. \qquad \text{[Euler.]}$$

14. Determine the limit of the product

$$\prod_{-\nu}^{\mu}\left(1+\frac{x}{a+n}\right),$$

where μ, ν tend to ∞ in such a way that $\lim(\mu/\nu)=k$.

15. Shew directly from the products for $\sin x$ and $\cos x$ that

$$\sin(x+\tfrac{1}{2}\pi)=\cos x, \quad \cos(x+\tfrac{1}{2}\pi)=-\sin x,$$

and deduce the periodic properties of the sine and cosine.

16. Deduce the infinite product for $\sin x$ from the equation

$$\sin(\pi x)=2x\int_0^{\frac{1}{2}\pi}\cos(2xt)\,dt$$

by means of the series for $\cos(2xt)$ in powers of $\sin t$ (Art. 68).

17. Shew that

$$\sin(\pi x)=\pi x\prod_{-\infty}^{\infty}{}'\left[\left(1-\frac{x}{n}\right)e^{x/n}\right],$$

$$\pi\cot(\pi x)=\frac{1}{x}+\sum_{-\infty}^{\infty}{}'\left(\frac{1}{x-n}+\frac{1}{n}\right)=\frac{1}{x}+\sum_{-\infty}^{\infty}{}'\frac{x}{n(x-n)},$$

$$\pi[\cot(\pi x)-\cot(\pi a)]=\sum_{-\infty}^{\infty}\left(\frac{1}{x-n}-\frac{1}{a-n}\right)=\sum_{-\infty}^{\infty}\frac{a-x}{(x-n)(a-n)},$$

$$\pi^2\operatorname{cosec}^2(\pi x)=\sum_{-\infty}^{\infty}\frac{1}{(x-n)^2},$$

$$\pi^3\cot(\pi x)\operatorname{cosec}^2(\pi x)=\sum_{-\infty}^{\infty}\frac{1}{(x-n)^3},$$

$$\pi^4\left[\operatorname{cosec}^4(\pi x)-\frac{2}{3}\operatorname{cosec}^2(\pi x)\right]=\sum_{-\infty}^{\infty}\frac{1}{(x-n)^4}.$$

[EULER.]

In the first and second equations the accent implies that $n=0$ is excluded.

18. Deduce from the last example that

$$\lim_{\nu\to\infty}\sum_{n=-\nu}^{+\nu}\frac{1}{x-n}=\pi\cot(\pi x)$$

and

$$\lim_{\nu\to\infty}\left[\sum_{m=-\nu}^{\nu}\sum_{n=-\nu}^{\nu}\frac{1}{(x-m)(x-n)}\right]=-\pi^2,$$

where in the double summation all values $m=n$ are excluded.

19. Shew that

$$\operatorname{cosec} x=\frac{1}{x}-\frac{2x}{x^2-\pi^2}+\frac{2x}{x^2-2^2\pi^2}-\frac{2x}{x^2-3^2\pi^2}+\dots$$

and

$$\sec x=\frac{\pi}{\frac{1}{4}\pi^2-x^2}-\frac{3\pi}{\frac{9}{4}\pi^2-x^2}+\frac{5\pi}{\frac{25}{4}\pi^2-x^2}-\dots.$$

[EULER.]

[We can get the expansion for $\operatorname{cosec} x$ from the identity

$$\operatorname{cosec} x=\cot(\tfrac{1}{2}x)-\cot x.$$

The second series can be derived from the series (equivalent to the first)

$$\operatorname{cosec} x=\frac{1}{x}-\left(\frac{1}{x-\pi}+\frac{1}{x+\pi}\right)+\left(\frac{1}{x-2\pi}+\frac{1}{x+2\pi}\right)-\dots$$

by writing $x+\frac{1}{2}\pi$ for x and grouping the terms differently.]

20. Shew that

$$\frac{\pi}{2}\tan\left(\frac{\pi x}{2}\right)=\frac{2x}{1^2-x^2}+\frac{2x}{3^2-x^2}+\frac{2x}{5^2-x^2}+\dots. \qquad \text{[EULER.]}$$

Deduce the value of

$$\sum_{-\infty}^{\infty}{}'\left[\frac{1}{n}-\frac{1}{n+x+(-1)^{n-1}y}\right],$$

$n=0$ being excluded from the summation. [*Math. Trip.* 1896.]

21. Prove that

$$\frac{\pi}{4}\left(\sec\frac{\pi}{2n}-1\right)=\frac{1}{n^2-1}-\frac{1}{3^3n^2-3}+\frac{1}{5^3n^2-5}-\frac{1}{7^3n^2-7}+\dots. \qquad \text{[EULER.]}$$

22. If the general term of a series Σu_n can be divided into partial fractions in the form

$$u_n=\sum_a\frac{A}{n+a}, \qquad \text{where } \Sigma A=0,$$

then

$$\sum_{-\infty}^{\infty}u_n=\sum_a A\pi\cot(a\pi),$$

where all the numbers a are supposed different from zero.

23. Find the sum of the series

$$\sum_{-\infty}^{\infty}{}'\frac{5n-2}{6n^3-5n^2+n},$$

the value $n=0$ being excluded.

$$\left[\text{Here } \frac{5n-2}{6n^3-5n^2+n}=\left(\frac{2}{2n-1}-\frac{1}{n}\right)+\left(\frac{3}{3n-1}-\frac{1}{n}\right).\right]$$

24. Find the value of $\sum_1^{\infty}u_n$, where

$$u_n=\frac{A}{(2n-1)^3}-\frac{7A}{(2n)^3}+\frac{B}{(2n-1)^2}+\frac{C}{(2n)^2};$$

and determine B, C so that $(2n-1)^3(2n)^3u_n$ may be quadratic in n.

25. We have seen (Ch. VI., Ex. 11) that if

$$S_t=\left[\frac{\Gamma(1)\Gamma(t)}{\Gamma(1+t)}\right]^2+\left[\frac{\Gamma(2)\Gamma(t)}{\Gamma(2+t)}\right]^2+\left[\frac{\Gamma(3)\Gamma(t)}{\Gamma(3+t)}\right]^2+\dots,$$

then

$$S_{t+1}=\frac{2}{t}(2t-1)S_t-\frac{3}{t^2}, \qquad \text{if } t>\tfrac{1}{2}.$$

Now $S_1=\frac{1}{6}\pi^2$ by Art. 71, so we find

$$S_2=\frac{1}{3}(\pi^2-9), \quad S_3=\pi^2-\frac{39}{4}.$$

Thus

$$\left(\frac{1}{1.2.3}\right)^2+\left(\frac{1}{2.3.4}\right)^2+\left(\frac{1}{3.4.5}\right)^2+\dots=\frac{1}{16}(4\pi^2-39).$$

CHAPTER X.

COMPLEX SERIES AND PRODUCTS.

72. The algebra of complex numbers.

We assume that the reader has already become acquainted with the leading features of the algebra of complex numbers. The fundamental laws of operation are:

If $$x=\xi+i\eta,\quad x'=\xi'+i\eta',$$

then
$$x+x'=\xi+\xi'+i(\eta+\eta'),\qquad (\textit{addition})$$
$$x-x'=\xi-\xi'+i(\eta-\eta'),\qquad (\textit{subtraction})$$
$$xx'=\xi\xi'-\eta\eta'+i(\xi\eta'+\xi'\eta),\qquad (\textit{multiplication})$$
$$\frac{x'}{x}=\frac{\xi\xi'+\eta\eta'}{\xi^2+\eta^2}+i\,\frac{\xi\eta'-\xi'\eta}{\xi^2+\eta^2},\qquad (\textit{division}).$$

It is easily seen that these laws include those of real numbers as a special case; and that these four operations can be carried out without exception (excluding division by zero). Further, these laws are consistent with the relations with which we are familiar for real numbers, such as

$$x(y+z)=xy+xz,$$
$$xy=yx,\quad x(yz)=(xy)z.$$

Thus any of the ordinary algebraic identities, which are established in the first instance for real numbers, are still true if the letters are supposed to represent complex numbers.

It is natural to ask whether other assumptions might not be made which would be equally satisfactory. Thus the analogy for addition might suggest for multiplication $$xx'=\xi\xi'+i\eta\eta'.$$

But this is inconsistent with the relation $x+x=2x$, since $x'=2$ would then give
$$2x=2\xi+i(0)=2\xi,$$
whereas
$$x+x=\xi+\xi+i(\eta+\eta)=2\xi+2i\eta.$$

Since the assumption $i^2=-1$ together with the ordinary associative, commutative and distributive laws are sufficient to fix the law of multiplication, we might try to find some other law of multiplication, by assuming that $i^2=a+i\beta$, where a, β are some fixed real numbers. It can then be shewn (see Stolz, *Allgemeine Arithmetik*, Bd. II., pp. 8–12) that we are either led back to the assumptions made above, or else we are forced to admit the existence of numbers x_1, x_2 such that the product x_1x_2 is zero without either x_1 or x_2 being zero.

73. Argand's diagram.

The reader is doubtless also familiar with the usual representation of the complex number

$$x=\xi+i\eta$$

by a point with rectangular coordinates (ξ, η).

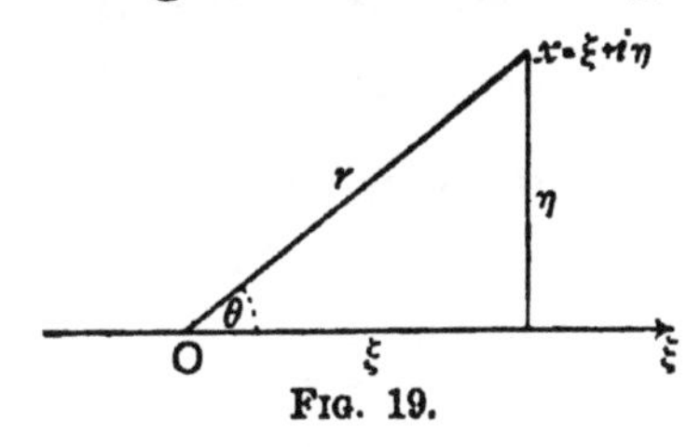

Fig. 19.

But it may be convenient to give a brief summary of the method.

If we introduce polar coordinates r, θ, we can write

$$x=r(\cos\theta+i\sin\theta).$$

We shall call $r=(\xi^2+\eta^2)^{\frac{1}{2}}$ the *absolute value* of x (it is sometimes also called the *modulus* or the *norm* of x); and we shall denote it by the symbol $|x|$. This, of course, is quite consistent with the notation used previously; for if x is *real*, $|x|$ will be either $+x$ or $-x$, according as x is positive or negative.

We call θ the *argument* of x: it is sometimes called the *phase* or *amplitude* of x.

From the diagram the meaning of $x+x'$, $x-x'$ is now evident.

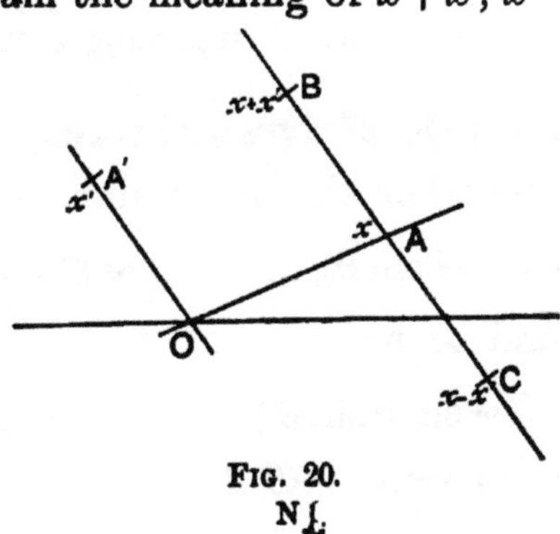

Fig. 20.

If we draw through A, AB, CA equal and parallel to OA', then B, C are respectively $x+x'$ and $x-x'$. The fact that $x+x'=x'+x$ is represented by the geometrical theorem that $A'B$ is equal and parallel to OA (Euclid I. 33).

Since $\quad OB < OA + AB \quad$ or $\quad OB < OA + OA'$,

we have the relation $\quad |x+x'| < |x| + |x'|$,

and similarly, $\quad |x-x'| < |x| + |x'|$.

Again, supposing $OA' < OA$, we have

$$OB + AB > OA \quad \text{or} \quad OB > OA - OA'.$$

Thus,

and so also

$$\left.\begin{array}{l} |x+x'| > |x| - |x'| \\ |x-x'| > |x| - |x'| \end{array}\right\}, \quad \text{if } |x| > |x'|.$$

It is easy to prove similarly that

$$|x+y+z+w| < |x| + |y| + |z| + |w|,$$

and generally, that $\quad |\Sigma x| < \Sigma |x|$.

These facts can be proved algebraically, thus, consider the first inequality and write $\quad R = |x+x'|$, so that $R^2 = (\xi+\xi')^2 + (\eta+\eta')^2$.

Then we have $\quad R^2 = r^2 + r'^2 + 2(\xi\xi' + \eta\eta')$.

Hence $\quad (r+r')^2 - R^2 = 2(rr' - \xi\xi' - \eta\eta')$,

and this is certainly positive if $\xi\xi' + \eta\eta'$ is zero or negative. But if $\xi\xi' + \eta\eta'$ is positive, we have

$$(rr')^2 - (\xi\xi' + \eta\eta')^2 = (\xi\eta' - \xi'\eta)^2,$$

so that $\quad rr' \geqq \xi\xi' + \eta\eta'$,

the sign of equality only occurring if $\xi\eta' - \xi'\eta = 0$.

Thus in all cases $\quad (r+r')^2 - R^2 \geqq 0$

or $\quad r + r' - R \geqq 0$,

and the sign of equality can only appear if $\xi\eta' - \xi'\eta = 0$ and $\xi\xi' + \eta\eta' > 0$; which is represented geometrically by supposing that OA' falls along OA.

74. Multiplication; de Moivre's theorem.

If we multiply together the two numbers

$$x = r(\cos\theta + i\sin\theta), \quad x' = r'(\cos\theta' + i\sin\theta'),$$

the product is found to be

$$\begin{aligned} xx' &= rr'(\cos\theta\cos\theta' - \sin\theta\sin\theta') + irr'(\cos\theta\sin\theta' + \sin\theta\cos\theta') \\ &= rr'[\cos(\theta+\theta') + i\sin(\theta+\theta')]. \end{aligned}$$

Thus the absolute value of xx' is rr'; or

$$|xx'|=|x|.|x'|,$$

and the argument of xx' is $\theta+\theta'$, the sum of the arguments of x and x'.

In particular, if $r'=1$, the product $x(\cos\theta'+i\sin\theta')$ is equal to $r[\cos(\theta+\theta')+i\sin(\theta+\theta')]$.

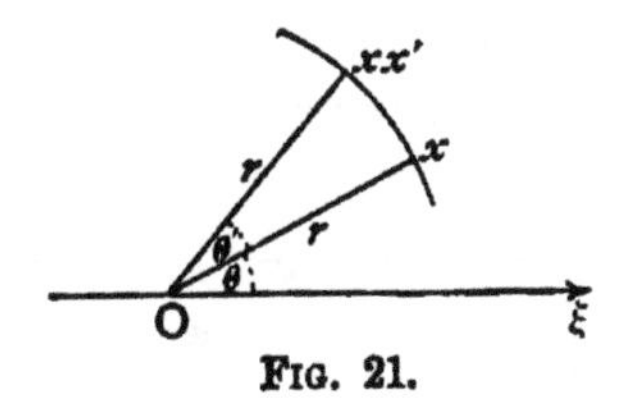

FIG. 21.

It is therefore clear that if $t=\cos\theta+i\sin\theta$, t^2 is equal to $\cos 2\theta+i\sin 2\theta$; t^3 to $\cos 3\theta+i\sin 3\theta$; t^{-1} to $\cos(-\theta)+i\sin(-\theta)$; and so on, as indicated in the diagram.

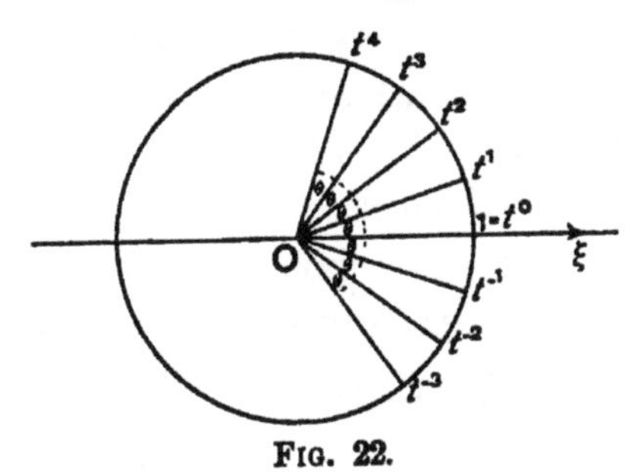

FIG. 22.

Hence if n is any whole number, positive or negative, we have

$$(\cos\theta+i\sin\theta)^n=t^n=\cos n\theta+i\sin n\theta.$$

To interpret $t^{m/n}$, where m and n are whole numbers, we agree that the law of indices $(t^{m/n})^n=t^m$ is still true.

Hence if $$t^{m/n}=\rho(\cos\phi+i\sin\phi),$$

we have $$\rho^n(\cos n\phi+i\sin n\phi)=\cos m\theta+i\sin m\theta;$$

thus, $$\rho^n=1, \text{ and } n\phi=m\theta+2k\pi,$$

where k is any whole number.

That is,

$$t^{m/n}=\cos\left\{\frac{1}{n}(m\theta+2k\pi)\right\}+i\sin\left\{\frac{1}{n}(m\theta+2k\pi)\right\};$$

thus $t^{m/n}$ has n different values, given by taking $k=0, 1, \ldots, n-1$. The case $k=0$ constitutes de Moivre's theorem.

A large number of elegant geometrical applications have been given by Morley, some of which will be found in Harkness and Morley's *Introduction to the Theory of Analytic Functions*, chapter II. We note a few samples:

1. The triangles x, y, z and x', y', z' are directly similar if

$$\begin{vmatrix} x, & y, & z \\ x', & y', & z' \\ 1, & 1, & 1 \end{vmatrix} = 0.$$

2. If the triangles x, y, z and x', y', z' are directly similar, any three points dividing xx', yy', zz' in the same ratio form a third similar triangle.

3. The conditions

$$\begin{vmatrix} x, & y, & z \\ \alpha, & \beta, & \gamma \\ 1, & 1, & 1 \end{vmatrix} = 0 \text{ and } \begin{vmatrix} yz+xw, & zx+yw, & xy+zw \\ \alpha, & \beta, & \gamma \\ 1, & 1, & 1 \end{vmatrix} = 0$$

(where α, β, γ are *real* numbers) are respectively the conditions that x, y, z should be collinear, and that x, y, z, w should be concyclic.

75. General principle of convergence for complex sequences.

If $S_n = X_n + iY_n$, we say that the sequence (S_n) converges, when *both* (X_n) and (Y_n) are convergent; and if $X_n \to X$, and $Y_n \to Y$, we write $S_n \to X + iY$.

The necessary and sufficient test for the convergence of the sequence (S_n) is that, corresponding to any real positive number ϵ, however small, we can find an index m such that

$$|S_n - S_m| < \epsilon, \quad \text{if } n > m.$$

To prove this statement, we observe that

$$|S_n - S_m|^2 = (X_n - X_m)^2 + (Y_n - Y_m)^2,$$

so that
$$|S_n - S_m| \leqq |X_n - X_m| + |Y_n - Y_m|$$

and
$$|S_n - S_m| \geqq |X_n - X_m|, \quad |S_n - S_m| \geqq |Y_n - Y_m|.$$

Now, if (S_n) converges, (X_n) and (Y_n) are also convergent, so that we can find m to satisfy

$$|X_n - X_m| < \tfrac{1}{2}\epsilon, \quad |Y_n - Y_m| < \tfrac{1}{2}\epsilon, \quad \text{if } n > m,$$

and consequently
$$|S_n - S_m| < \epsilon, \quad \text{if } n > m;$$

so that the condition is *necessary*.

On the other hand, when this inequality is satisfied, we have also

$$|X_n - X_m| < \epsilon, \quad |Y_n - Y_m| < \epsilon, \quad \text{if } n > m,$$

and therefore the sequences (X_n), (Y_n) are convergent.

Thus (S_n) converges; and therefore the condition is *sufficient*.

As an application of this principle we consider the interpretation of t^a, where $t = \cos\theta + i\sin\theta$ and a is *irrational*. We note as a preliminary that

$$|(\cos\phi + i\sin\phi) - (\cos\psi + i\sin\psi)| = 2|\sin\tfrac{1}{2}(\phi - \psi)| < |\phi - \psi|.$$

Now, if (a_n) is any sequence of *rational* numbers which has a as its limit, we can find m so that

$$|a_n - a_m| < \epsilon / |\theta|, \quad \text{if } n > m.$$

Thus

$$|\{\cos(a_n\theta) + i\sin(a_n\theta)\} - \{\cos(a_m\theta) + i\sin(a_m\theta)\}| < |a_n - a_m| . |\theta| < \epsilon,$$

if $n > m$; and so the sequence $t^{a_n} = \cos(a_n\theta) + i\sin(a_n\theta)$ is convergent. It is therefore natural to define t^a as $\lim t^{a_n}$; but it is of course to be remembered that *all* the limits*

$$\lim\,[\cos a_n(\theta + 2k\pi) + i\sin a_n(\theta + 2k\pi)] \quad (k = \pm 1, \pm 2, \pm 3, \text{ to } \infty)$$

may equally well be regarded as included in the symbol t^a. Thus special care must be taken to specify the meaning to be attached to t^a; for most purposes it is sufficient to retain the value which reduces to 1 when θ is zero (that is, the value given by $k = 0$).

Convergence and divergence of a series of complex terms.

If $a_n = \xi_n + i\eta_n$, we have

$$a_1 + a_2 + \ldots + a_n = (\xi_1 + \xi_2 + \ldots + \xi_n) + i(\eta_1 + \eta_2 + \ldots + \eta_n).$$

Then if $\xi_1 + \xi_2 + \ldots$ to ∞ and $\eta_1 + \eta_2 + \ldots$ to ∞ are separately convergent to the sums s, t respectively, we say that

$$a_1 + a_2 + \ldots \text{ to } \infty$$

converges to the sum $s + it$.

But if *either* $\Sigma\xi_n$ *or* $\Sigma\eta_n$ diverges (or oscillates), we say that Σa_n diverges (or oscillates).

It is easy to see from this definition and the foregoing discussion for sequences that *the necessary and sufficient condition for convergence* is simply that, corresponding to the real positive fraction ϵ, we can find m such that

$$|a_{m+1} + a_{m+2} + \ldots + a_{m+p}| < \epsilon,$$

no matter how great p may be.

In like manner we obtain the definition of convergence of an infinite product of complex factors; but in ordinary work we need only absolutely convergent products, which we shall discuss in Art. 77.

---

* All these values are unequal, because no integer k makes ka equal to an integer; of course k must not be allowed to vary with n.

76. Absolute convergence of a series of complex terms.

If $a_n = \xi_n + i\eta_n$ and if $\Sigma|a_n|$ is convergent, we shall say that Σa_n is *absolutely convergent.* It is evident in this case, from Art. 18, that the separate series $\Sigma\xi_n$, $\Sigma\eta_n$ are both absolutely convergent, because

$$|\xi_n| \leqq |a_n| \text{ and } |\eta_n| \leqq |a_n|.$$

It follows therefore that Σa_n is convergent in virtue of our definition (Art. 75); and by Art. 26 the sums $\Sigma\xi_n$, $\Sigma\eta_n$ are independent of the order of the terms. Hence also, *the sum of an absolutely convergent series is independent of the order of arrangement.*

It is perhaps worth while to give a graphical illustration of the method of inferring the convergence of Σa_n from $\Sigma|a_n|$.

Let
$$s_n = a_1 + a_2 + \ldots + a_n,$$
$$\sigma_n = |a_1| + |a_2| + \ldots + |a_n|,$$
and represent the numbers in Argand's diagram, as below:

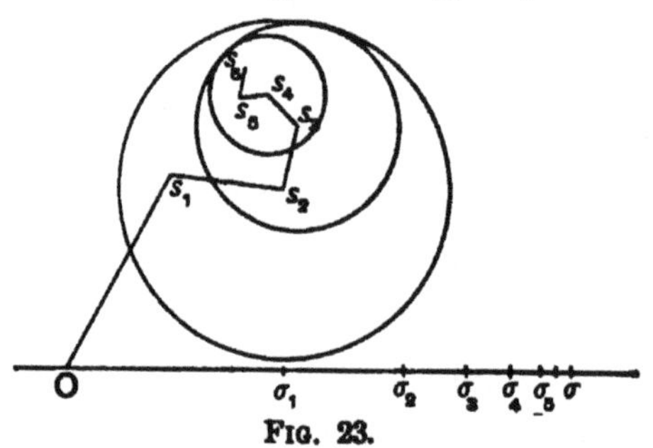

FIG. 23.

By definition we have

$$Os_1 = O\sigma_1, \quad s_1s_2 = \sigma_1\sigma_2, \quad s_2s_3 = \sigma_2\sigma_3, \quad \text{etc.}$$

Thus $\quad s_ns_p \leqq \sigma_n\sigma_p, \quad$ by elementary geometry.

But we have $\quad \sigma_n\sigma_p \leqq \sigma_n\sigma, \quad$ if $p > n$,

where σ represents the sum $\Sigma|a_n| = \lim \sigma_n$.

Now, by the definition of convergence, we can choose n to correspond to ϵ, so that
$$\sigma_n\sigma = \sigma - \sigma_n < \epsilon.$$

Consequently, n can be found, so that
$$s_ns_p < \epsilon, \quad \text{if } p > n;$$
or, geometrically, *all the points s_p must lie within a circle of radius ϵ, whose centre is s_n.* Since ϵ can be taken as small as we please, the last result shews (see Art. 75) that the points s_p must cluster about some limiting point s, which may be *on* the circle but cannot be outside it.

The diagram indicates the circles with centres s_2, s_3, s_4.

An alternative form of diagram has been proposed by G. H. Ling, *Annals of Mathematics* (2), vol. 5, 1904.

It is of course obvious from what has been said as to series of real terms, that absolute convergence is not *necessary* for convergence. An example of a complex series which converges, although not absolutely, is

$$i+\tfrac{1}{2}(i)^2+\tfrac{1}{3}(i)^3+\tfrac{1}{4}(i)^4+\dots$$
$$=-(\tfrac{1}{2}-\tfrac{1}{4}+\tfrac{1}{6}-\tfrac{1}{8}+\dots)+i(1-\tfrac{1}{3}+\tfrac{1}{5}-\tfrac{1}{7}+\dots).$$

For both real and imaginary parts converge, by Art. 21; but the series of absolute values is

$$1+\tfrac{1}{2}+\tfrac{1}{3}+\tfrac{1}{4}+\dots,$$

which diverges by Art. 7 (Ex. 2) or Art. 11.

It is evident from chap. IV. that the sum of a non-absolutely convergent series may be altered by derangement.

77. Absolute convergence of an infinite product of complex factors.

The infinite product $\Pi(1+a_n)$ is said to be *absolutely convergent* if the product $\Pi(1+|a_n|)$ is convergent. It follows at once that *if* Σa_n *is absolutely convergent, so also is* $\Pi(1+a_n)$.

For we know that $\Sigma|a_n|$ converges: and so by Art. 39 the two products $\Pi(1+|a_n|)$ and $\Pi(1-|a_n|)$ are convergent.

But we have still to shew that (in the case of complex products) *simple convergence can be deduced from absolute convergence.* Now, if $|a_n|<1$, we have

$$1-|a_n|\leqq|1+a_n|\leqq1+|a_n|,$$

and so $$\prod_{m+1}^{m+p}(1-|a_n|)\leqq\left|\prod_{m+1}^{m+p}(1+a_n)\right|\leqq\prod_{m+1}^{m+p}(1+|a_n|).$$

Also, since $\Pi(1+|a_n|)$ converges, it follows from Art. 39 that $\Sigma|a_n|$ converges, and therefore also $\Pi(1-|a_n|)$ is convergent; thus we can find m, so that

$$\prod_{m+1}^{m+p}(1-|a_n|)>1-\epsilon \text{ and } \prod_{m+1}^{m+p}(1+|a_n|)<1+\epsilon,$$

however small ϵ may be.

Hence we can find m, so that

$$1-\epsilon<\left|\prod_{m+1}^{m+p}(1+a_n)\right|<1+\epsilon,$$

however small ϵ may be, and however large p may be.

That is, $\Pi(1+a_n)$ satisfies the general condition for convergence; and so the product is convergent.

It is easy to modify the proof of Art. 41 to shew that *if* Σa_n *is absolutely convergent the value of* $\Pi(1+a_n)$ *is unaltered by changing the order of the factors.*

78. Pringsheim's tests for absolute convergence of a complex series.*

Of course the conditions of chap. II. can be applied at once to the series of *positive* terms $\Sigma|a_n|$; but since

$$|a_n|=\sqrt{(\xi_n^2+\eta_n^2)}, \quad \text{if } a_n=\xi_n+i\eta_n,$$

it is evident that the square-root complicates some of the tests.

Of course the tests of Art. 9 can be at once changed to

$$\overline{\lim}\, C_n^2 . |a_n|^2 < \infty, \quad (\textit{convergence})$$
$$\underline{\lim}\, D_n^2 . |a_n|^2 > 0, \quad (\textit{divergence});$$

but the same transformation cannot, as a rule, be applied to the ratio-tests of Art. 12.

Thus, the condition

$$\underline{\lim}\left[D_n^2\left|\frac{a_n}{a_{n+1}}\right|^2-D_{n+1}^2\right]>0$$

is by no means sufficient to ensure the convergence of $\Sigma|a_n|$; because whenever $\lim D_n=\infty$ (which is usually the case), the above condition does not exclude the possibility

$$\underline{\lim}\left[D_n\left|\frac{a_n}{a_{n+1}}\right|-D_{n+1}\right]=0,$$

which may occur with a divergent series.

For instance, with $1/a_n=n\log n$ and $D_n=n \quad (n>2)$, we have

$$D_n^2\left|\frac{a_n}{a_{n+1}}\right|^2-D_{n+1}^2=(n+1)^2\left[\left\{\frac{\log(n+1)}{\log n}\right\}^2-1\right]$$
$$=\frac{(n+1)^2}{(\log n)^2}\log\left(1+\frac{1}{n}\right)[\log(n+1)+\log n]$$
$$>2(n+1)/\log n,$$

because $\log(1+1/n)>1/(n+1)$.

Thus, $$\lim\left[D_n^2\left|\frac{a_n}{a_{n+1}}\right|^2-D_{n+1}^2\right]=\infty,$$

but yet $\Sigma|a_n|$ diverges. [See Art. 11 (2).]

* *Archiv für Math. u. Phys.* (3), Bd. 4, 1902, pp. 1-19.

It can be seen that here

$$\lim\left[D_n\left|\frac{a_n}{a_{n+1}}\right|-D_{n+1}\right]=0.$$

Pringsheim therefore introduces the conditions

$$\underline{\lim}\left[D_n\left|\frac{a_n}{a_{n+1}}\right|^2-\frac{D_{n+1}^2}{D_n}\right]>0, \quad (convergence)$$

$$\overline{\lim}\left[D_n\left|\frac{a_n}{a_{n+1}}\right|^2-\frac{D_{n+1}^2}{D_n}\right]<0, \quad (divergence)$$

which are substantially equivalent to the conditions of Art. 12.

For if the first condition is satisfied, we can find ρ and m so that

$$D_n\left|\frac{a_n}{a_{n+1}}\right|^2-\frac{D_{n+1}^2}{D_n}\geqq\rho>0, \quad \text{if } n>m.$$

Thus
$$D_n\left|\frac{a_n}{a_{n+1}}\right|-D_{n+1}\geqq\rho\Big/\left[\left|\frac{a_n}{a_{n+1}}\right|+\frac{D_{n+1}}{D_n}\right].$$

Now in all cases of practical interest,* it is possible to assign an upper limit to $\left|\frac{a_n}{a_{n+1}}\right|$ and $\frac{D_{n+1}}{D_n}$, say l; then $D_n\left|\frac{a_n}{a_{n+1}}\right|-D_{n+1}\geqq\rho/2l$, and so the condition of Art. 12 is satisfied.

But if the second condition holds, we can find m so that

$$\left[D_n\left|\frac{a_n}{a_{n+1}}\right|-D_{n+1}\right]\left[\left|\frac{a_n}{a_{n+1}}\right|+\frac{D_{n+1}}{D_n}\right]<0, \quad \text{if } n>m$$

so that the first factor must be negative.

79. Applications.

It is easy to transform Pringsheim's conditions by writing $D_n=f(n)$ as in Art. 12 (3), and then we find

$$\underline{\lim}\,\kappa_n>0, \ \textit{convergence}; \quad \overline{\lim}\,\kappa_n<0, \ \textit{divergence},$$

where
$$\left|\frac{a_n}{a_{n+1}}\right|^2=1+2\frac{f'(n)}{f(n)}+\frac{\kappa_n}{f(n)}.$$

The only fresh point to notice is that $[f'(n)]^2/f(n)\to 0$.

For
$$\lim_{x\to\infty}\frac{[f'(x)]^2}{f(x)}=\lim_{x\to\infty}\frac{2f'(x)f''(x)}{f'(x)}=\lim_{x\to\infty}2f''(x),$$

in virtue of L'Hospital's rule (Art. 152); now we assumed that $f''(x)\to 0$, and so $[f'(x)]^2/f(x)$ also tends to zero.

In particular, let us take

$$(1)\ f(n)=1, \quad (2)\ f(n)=n, \quad (3)\ f(n)=n\log n.$$

* Pringsheim admits slightly more general conditions.

We obtain, after a few transformations, the following conditions:*

$$(1)\quad \overline{\lim}\left|\frac{a_n}{a_{n+1}}\right|^2 \begin{matrix}>1, & (convergence)\\ <1, & (divergence).\end{matrix}$$

$$(2)\quad \overline{\lim}\, n\left\{\left|\frac{a_n}{a_{n+1}}\right|^2-1\right\} \begin{matrix}>2, & (convergence)\\ <2, & (divergence).\end{matrix}$$

$$(3)\quad \overline{\lim}\,(\log n)\left[n\left\{\left|\frac{a_n}{a_{n+1}}\right|^2-1\right\}-2\right] \begin{matrix}>2, & (convergence)\\ <2, & (divergence).\end{matrix}$$

The most interesting case in practice is one corresponding to (5) of Art. 12, where we can write

$$(4)\quad \frac{a_n}{a_{n+1}}=1+\frac{\mu}{n}+\frac{\omega}{n^\lambda}, \quad \begin{matrix}(\lambda>1)\\ (|\omega|<A).\end{matrix}$$

It is easy to see that if $\mu=a+i\beta$ in (4),

$$(5)\quad \left|\frac{a_n}{a_{n+1}}\right|^2=1+\frac{2a}{n}+\frac{\rho'}{n^\kappa}, \quad (\kappa>1, |\rho'|<B)$$

and so from (2) we see that the series $\Sigma|a_n|$ converges if $a>1$ and diverges if $a<1$; in case $a=1$, we apply (3), and find divergence. Thus we have the rule:†

When $$\frac{a_n}{a_{n+1}}=1+\frac{\mu}{n}+\frac{\omega}{n^\lambda}, \quad \begin{matrix}(\lambda>1)\\ (|\omega|<A)\end{matrix}$$

the series $\Sigma|a_n|$ converges if the real part of μ is greater than 1, and otherwise diverges.

On account of the importance of the series $\Sigma a_n x^n$ when (4) holds, we proceed to consider some further results.

An application of (1) above shews that the series converges absolutely if $|x|<1$, diverges if $|x|>1$; and we have just examined the case when $|x|=1$ and $a>1$, proving that then $\Sigma a_n x^n$ is absolutely convergent.

But we do not know yet whether the series may converge for $|x|=1$ (although not absolutely) even when $a\leqq 1$.

If a is negative, the sequence $|a_n|$ will increase with n, at least after a certain stage, in virtue of (5) above; and

---

* The inequalities $$\overline{\lim}\, P \begin{matrix}>1\\ <1\end{matrix}$$ are here used as equivalent to the two $\underline{\lim}\, P>1$, $\overline{\lim}\, P<1$.

† Weierstrass, *Ges. Werke*, Bd. 1, p. 185.

consequently, convergence of $\Sigma a_n x^n$ is impossible, since the terms do not tend to zero as n increases.

Further, if $a=0$, it is easy to see (by an argument similar to Ex. 3, Art. 39) that $|a_n|$ tends to a finite limit,* so that again $\Sigma a_n x^n$ cannot converge.

On the other hand, if $a>0$, it follows from Ex. 3, Art. 39, that $|a_n|$ tends to zero as a limit; so that convergence *may* occur.

We prove next that *when condition* (4) *holds and* $0<a\leqq 1$, *the series* $\Sigma|a_n-a_{n+1}|$ *is convergent.*

In the first place it is evident that (at least after a certain value of n) the sequence $|a_n|$ decreases as n increases; hence the series $\Sigma\{|a_n|-|a_{n+1}|\}$ consists finally of positive terms only, and it is obviously convergent. Further, we have, from (5),

$$\frac{|a_n-a_{n+1}|}{|a_n|-|a_{n+1}|}=\left|\frac{a_n}{a_{n+1}}-1\right|\Big/\left\{\left|\frac{a_n}{a_{n+1}}\right|-1\right\}=\frac{|\mu+\omega n^{1-\lambda}|}{2a+\rho' n^{1-\kappa}}\left\{\left|\frac{a_n}{a_{n+1}}\right|+1\right\}.$$

Thus
$$\lim_{n\to\infty}\frac{|a_n-a_{n+1}|}{|a_n|-|a_{n+1}|}=\frac{|\mu|}{a},$$

and it follows, as in Art. 18, that the series $\Sigma|a_n-a_{n+1}|$ is convergent.

Now, we have identically,

$$\sum_m^p a_n x^n=\sum_m^{p-1}(a_n-a_{n+1})\frac{x^m-x^{n+1}}{1-x}+a_p\frac{x^m-x^{p+1}}{1-x},$$

and
$$\left|\frac{x^m-x^{n+1}}{1-x}\right|\leqq\frac{2}{|1-x|},\quad\text{since } |x|=1.$$

Thus
$$\left|\sum_m^p a_n x^n\right|\leqq\frac{2}{|1-x|}\left\{\sum_m^{p-1}|a_n-a_{n+1}|+|a_p|\right\},$$

and consequently, if $|x|=1$ and x is different from 1, the series $\sum_0^\infty a_n x^n$ converges, because $\Sigma|a_n-a_{n+1}|$ converges and $\lim a_n=0$.

This result may also be derived from Dirichlet's test (Art. 80).

We have obtained no information as to the behaviour of Σa_n itself, when $0<a\leqq 1$; as a matter of fact the series is not convergent, but the proof is a little tedious, and we simply

* It is to be noted that a_n itself does not approach a limit unless $\beta=0$; if β is not zero, a_n *oscillates*. The reader should not find much difficulty in proving this; it is not needed for the argument in the text.

refer to Pringsheim for the details.* Of course we have already proved that the series diverges in the corresponding case when a_n is *real* (see Art. 12).

The final conclusion is thus:

If
$$\frac{a_n}{a_{n+1}} = 1 + \frac{\mu}{n} + \frac{\omega}{n^\lambda}, \qquad \begin{matrix}(\lambda > 1)\\ (|\omega| < A)\end{matrix}$$
then the series $\Sigma a_n x^n$ converges absolutely for $|x|<1$, and cannot converge for $|x|>1$.

If $|x|=1$ and $\mu = a + i\beta$, there are three cases:

(i) *when $a > 1$, the series converges absolutely.*

(ii) *when $0 < a \leqq 1$, the series converges (but not absolutely), except for $x=1$.*

(iii) *when $0 \geqq a$, the series does not converge.* (Weierstrass, *l.c.*)

Ex. As a special example we may take the hypergeometric series (Art. 12), in which
$$\frac{a_n}{a_{n+1}} = \frac{(n+1)(n+\gamma)}{(n+a)(n+\beta)},$$
so that
$$\mu = 1 + \gamma - (a+\beta).$$
Then we see that, for $|x|=1$:

(i) if the real part of $(\gamma - a - \beta)$ is positive, the hypergeometric series converges absolutely;

(ii) if this real part is between -1 and 0, the series converges except for $x=1$;

(iii) if the real part is -1 or less than -1, the series does not converge.

Special cases of this have been found already in Arts. 12 and 21.

80. Further tests for convergence.

The reader will find no difficulty in modifying the proof of Art. 18 to establish the theorem:

(1) *If Σa_n is absolutely convergent, so also is $\Sigma a_n v_n$, provided that $|v_n|$ never exceeds a fixed number k.*

---

* *L.c.* pp. 13-17; the proof shews that (except for $\mu=1$, when the series is easily seen to diverge) the behaviour of Σa_n is the same as that of
$$1 + \frac{1-\mu}{1} + \frac{(1-\mu)(2-\mu)}{1.2} + \frac{(1-\mu)(2-\mu)(3-\mu)}{1.2.3} + \ldots.$$
And the sum of n terms of this series is
$$\frac{(2-\mu)(3-\mu)\ldots(n-\mu)}{1.2.3\ldots(n-1)} \quad\text{or}\quad \left(1+\frac{1-\mu}{1}\right)\left(1+\frac{1-\mu}{2}\right)\ldots\left(1+\frac{1-\mu}{n-1}\right).$$
It can easily be seen that the absolute value of this expression tends to ∞ with n, and that its argument has no definite limit.

It is, however, obvious that Arts. 19, 20 will need a little alteration to include cases in which v_n is complex. We shall prove that:

(2) *If* Σa_n *converges, and* $\Sigma|v_n - v_{n+1}|$ *is convergent, then* $\Sigma a_n v_n$ *is convergent.*

First consider the modified form of Abel's Lemma (Art. 23).

Write for brevity U_n to denote the sum of the convergent series
$$|v_n - v_{n+1}| + |v_{n+1} - v_{n+2}| + \dots \text{ to } \infty,$$
so that
$$|v_n - v_{n+1}| = U_n - U_{n+1}.$$

Thus, since
$$v_m - v_n = (v_m - v_{m+1}) + (v_{m+1} - v_{m+2}) + \dots + (v_{n-1} - v_n),$$
we have
$$|v_m - v_n| \leqq (U_m - U_{m+1}) + (U_{m+1} - U_{m+2}) + \dots + (U_{n-1} - U_n),$$
so that
$$|v_m - v_n| \leqq U_m - U_n, \text{ if } n > m.$$

Now (see Art. 76) the series $\Sigma(v_n - v_{n+1})$ is convergent because $\Sigma|v_n - v_{n+1}|$ is convergent. And
$$(v_1 - v_2) + (v_2 - v_3) + \dots + (v_n - v_{n+1}) = v_1 - v_{n+1},$$
so that v_n has a definite limit g as $n \to \infty$; and let us write
$$G = |g| = \lim |v_n|.$$

Again, we have
$$|v_m| \leqq |v_n| + |v_m - v_n| \leqq |v_n| + U_m - U_n,$$
so that if we take the limit of this inequality as $n \to \infty$, we find
$$|v_m| \leqq G + U_m.$$

Thus, if we write $V_n = G + U_n$, V_n is always greater than $|v_n|$. It is also obvious that the sequence (V_n) is a *decreasing* sequence, and that $V_n - V_{n+1}$ is equal to $|v_n - v_{n+1}|$.

Next, if $\Sigma_n = a_1 v_1 + a_2 v_2 + \dots + a_n v_n$, we find, as in Art. 23, that
$$\Sigma_n = s_1(v_1 - v_2) + s_2(v_2 - v_3) + \dots + s_{n-1}(v_{n-1} - v_n) + s_n v_n.$$

Thus, if H is not less than the upper limit of $|s_1|, |s_2|, \dots |s_n|$, we find
$$|\Sigma_n| \leqq H[(V_1 - V_2) + (V_2 - V_3) + \dots + (V_{n-1} - V_n) + V_n] \leqq HV_1,$$
which is the form of Abel's Lemma for complex terms.

Similarly, if η is not less than the upper limit of $|s_r - \sigma|$, as r ranges from 1 to $m-1$, and η' is not less than the upper limit as r ranges from m to n, we find
$$|\Sigma_n - \sigma v_1| \leqq \eta(V_1 - V_m) + \eta' V_m,$$
where σ is any conveniently chosen number.

If the sequence (v_n) is real, positive and *decreasing*, we find that $G=\lim v_n$ and $V_n=v_n$, so that these inequalities are almost the same as those of Art. 23.

Now apply the Lemma to the sum

$$a_{m+1}v_{m+1}+a_{m+2}v_{m+2}+\ldots+a_{m+p}v_{m+p}.$$

It is then evident that this is less, in absolute value, than $H'V_{m+1}$, where H' is the upper limit of

$$|a_{m+1}|,\ |a_{m+1}+a_{m+2}|,\ |a_{m+1}+a_{m+2}+a_{m+3}|,\ \ldots.$$

Now since Σa_n converges, m can be chosen so large as to make $H'<\epsilon$, however small ϵ may be; and then we have

$$|a_{m+1}v_{m+1}+\ldots+a_{m+p}v_{m+p}|<\epsilon V_{m+1}<\epsilon V_1,$$

which proves that the series $\Sigma a_n v_n$ is convergent.

This may be called **Abel's test for complex series.**

If v_n is a function of x, Abel's test also enables us to ensure the *uniform convergence* of $\Sigma a_n v_n$ at all points x within an area for which $|v_1|$ and U_1 remain less than a fixed value. For

$$V_1=G+U_1, \text{ and } G\leqq|v_1|+U_1.$$

Similarly, we can establish **Dirichlet's test for complex series.**

(3) *If* $\Sigma|v_n-v_{n+1}|$ *converges and* $\lim v_n=0$, *then* $\Sigma a_n v_n$ *will converge if* Σa_n *oscillates between finite limits.*

In fact G is here zero, and (just as in Art. 20) we can find a constant l (*independent of* m) so that $H'<2l$. Thus

$$\left|\sum_{m+1}^{m+p} a_n v_n\right|<2lV_{m+1},$$

which can be made less than ϵ by proper choice of m because $V_n\to 0$.

A special case of this test has been already used in Art. 79.

81. Uniform convergence.

After what has been explained in Art. 43, there will be no difficulty in appreciating the idea of uniform convergence for a series $\Sigma f_n(x)$, when x is *complex*; the only essential point of novelty being that the region of uniform convergence now usually consists of an area in the (ξ, η) plane (if $x=\xi+i\eta$) instead of an interval (or segment) of the real axis. It is also sometimes convenient to use the idea of *uniform convergence along a line*, which should present no fresh difficulty to the reader.

Weierstrass's M-test for uniform convergence can be retained almost unaltered, thus:

If for all points ($x=\xi+i\eta$) *within a certain area A, the function* $f_n(x)$ *has the property that*

$$|f_n(x)|\leqq M_n,$$

where M_n *is a positive* **constant,** *and if the series* ΣM_n *converges, then* $\Sigma f_n(x)$ *converges absolutely and uniformly at all points within A.*

Abel's and Dirichlet's tests for uniform convergence (for a complex series) are obtained at once from the last article.

The proof of Art. 45 (1) can be easily modified to shew that $\Sigma f_n(x)$ *is a continuous function of* x *within the region of uniform convergence, provided that the separate functions* $f_n(x)$ *are continuous in the region.*

The discussion of differentiation and integration with respect to the *complex* variable x falls beyond the scope of this book; but it is not out of place to mention that (when the fundamental notions have been made clear) the results of Arts. 45-47 remain practically unchanged.

It is evident also that Art. 48 is still correct, when x is a complex variable.

There is no difficulty in seeing that the two theorems of Art. 49 remain valid, when the functions $v_r(n)$ are complex.

It is often important to integrate a complex function with respect to a *real* variable; in particular it is useful to consider **the mean value of a continuous function** $f(x)$ **along a circle** $|x|=\rho$, which is defined by the equation

$$\mathfrak{M}f(x)=\frac{1}{2\pi}\int_0^{2\pi}f(x)\,d\theta,\quad \text{where } x=\rho(\cos\theta+i\sin\theta).$$

The existence of a definite mean value is inferred at once from the continuity of $f(x)$, just as in Art. 163 of the Appendix; and the following conclusions are immediate consequences of the definition:

(i) $\mathfrak{M}a=a$, if a is a constant,

(ii) $|\mathfrak{M}f(x)|<A$, if $|f(x)|<A$ on the circle,

(iii) $\mathfrak{M}x^k=0$, $\mathfrak{M}x^{-k}=0$, if k is an integer (not zero),

because $x^k=\rho^k(\cos k\theta+i\sin k\theta)$ and

$$\int_0^{2\pi}\cos(k\theta)\,d\theta=0,\quad \int_0^{2\pi}\sin(k\theta)d\theta=0.$$

Further, from Art. (45 2), we deduce that *if* $F(x)=\Sigma f_n(x)$ *converges uniformly on the circle* $|x|=\rho$, *then*

$$\mathfrak{M}F(x)=\Sigma\mathfrak{M}f_n(x).$$

We can define the mean value without using the Integral Calculus, by supposing the circumference divided into ν equal parts at $x_1, x_2, \dots x_\nu$, and writing

$$\mathfrak{M}f(x)=\lim_{\nu\to\infty}\frac{1}{\nu}[f(x_1)+f(x_2)+\dots+f(x_\nu)].$$

This method leads to the same results as those just proved; and thus Cauchy's inequalities (p. 209) can be established without the Calculus.

82. Circle of convergence of a power-series $\Sigma a_n x^n$.

From Art. 10 it is evident that the series is absolutely convergent if

$$\overline{\lim}\,|a_n x^n|^{\frac{1}{n}}<1,$$

and the series certainly cannot converge if

$$\overline{\lim}\,|a_n x^n|^{\frac{1}{n}}>1,$$

because then $a_n x^n$ cannot tend to zero as a limit.

Hence, if we write (as in Art. 50)

$$\overline{\lim}\,|a_n|^{\frac{1}{n}}=l$$

(where, of course, l is real and positive), the power-series converges absolutely if $|x|<1/l$; and cannot converge if $|x|>1/l$.

To interpret this geometrically, let a circle of radius $1/l$ be drawn in Argand's diagram; then the series is absolutely convergent at any point within the circle, and cannot converge at

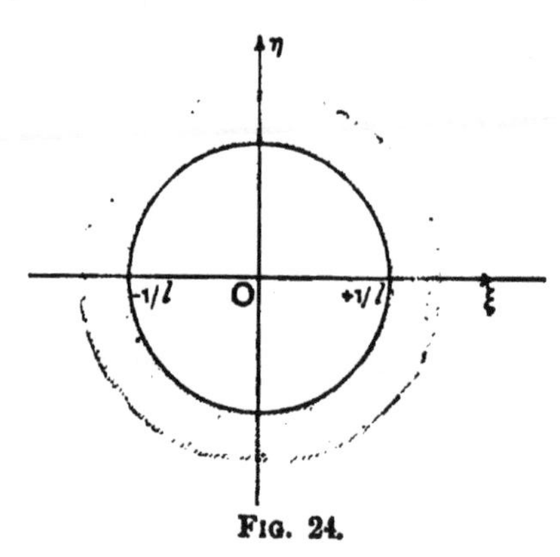

FIG. 24.

any point outside the circle. The circle is called *the circle of convergence*; and it will be seen that, if a_n is real, the interval of convergence obtained in Art. 50 is a diameter of the circle.

Of course we know nothing at present with respect to points *on* the circle of convergence; but when $l=1$ (a case to which every other can be reduced, excepting $l=0$ and ∞) we can usually obtain information by using Weierstrass's rule, discussed in Art. 79. It must not, however, be supposed that the three types there discussed exhaust all possibilities as to the points of convergence on the circle; indeed, Pringsheim* has constructed series which converge at all points on the circle, but not absolutely.

The region of uniform convergence, when $\Sigma|a_n|$ is convergent and $l=1$, is the *whole* of the circle $|x|=1$, *including the circumference*; this is evident from the M-test.

In the other cases, we can at present only say that the series converges uniformly within *and on* any circle $|x|=k$, where k lies between 0 and 1. We shall, however, consider this point more fully in Art. 83.

The reader will find little difficulty in seeing that the theorems of Arts. 52-55 hold for complex power-series, certain small verbal alterations being made.

Since a power-series converges uniformly on every circle $|x|=\rho$, for which† $\rho<1/l$, we can readily obtain its mean-value along the circle by integrating term-by-term.

Thus, if $f(x)=\sum_0^\infty a_n x^n$, we have

$$\mathfrak{M}f(x)=\sum_0^\infty a_n \mathfrak{M}x^n=a_0=f(0),$$

so that *the mean-value of a power-series along a circle*

$$|x|=\rho\ (<1/l)$$

is equal to its value at the centre.

Similarly, we find

$$\mathfrak{M}[f(x)/x^n]=a_n.$$

Thus, if M is the maximum value of $|f(x)|$ on the circle $|x|=\rho$, we have *Cauchy's inequalities*

$$|a_0|<M, \quad |a_n|<M/\rho^n.$$

---

* *Math. Annalen*, Bd. 25, 1885, p. 419. One type is given in Ex. B. 24 at the end of the chapter.

† If the circle $|x|=1/l$ belongs to the region of uniform convergence, we may of course take $\rho=1/l$.

Again, since the series

$$\frac{x}{x-c}=1+\frac{c}{x}+\frac{c^2}{x^2}+\ldots,\quad |c|<\rho,$$

converges *uniformly* on the circle $|x|=\rho$, we find

$$\mathfrak{M}\frac{xf(x)}{x-c}=\sum_0^\infty c^n\mathfrak{M}\frac{f(x)}{x^n}=\sum_0^\infty a_nc^n=f(c),\quad |c|<\rho.$$

Similarly, we find

$$\frac{x}{x-c}=-\left(\frac{x}{c}+\frac{x^2}{c^2}+\frac{x^3}{c^3}+\ldots\right),\quad \mathfrak{M}\frac{xf(x)}{x-c}=0,\qquad \text{if } |c|>\rho.$$

We shall now consider the question : *Can a power-series be determined so as to have given values along a definite circle, say* $|x|=1$?

Let us write the coefficients a_n in the form $\alpha_n+i\beta_n$ where α_n, β_n are real; then write $\Sigma\alpha_nx^n=f_1(x)$, $\Sigma i\beta_nx^n=f_2(x)$, so that $f(x)=\Sigma a_nx^n=f_1(x)+f_2(x)$. Now suppose that when $x=\cos\theta+i\sin\theta$, we have $f_1(x)=u_1+iv_1$, $f_2(x)=u_2+iv_2$ and $f(x)=u+iv$, where u_1, v_1, etc., are all functions of θ such that $u=u_1+u_2$, $v=v_1+v_2$. Then, if $|c|<1$, we find as above (assuming the uniform convergence of Σa_nx^n on the circle $|x|=1$)

$$f_1(c)=\frac{1}{2\pi}\int_0^{2\pi}(u_1+iv_1)\frac{x}{x-c}\,d\theta,\quad 0=\frac{1}{2\pi}\int_0^{2\pi}(u_1+iv_1)\frac{x\,d\theta}{x-(1/c)}.$$

In the second integral put $1/x$ for x: this will change u_1+iv_1 to u_1-iv_1, and so we have

$$0=\frac{1}{2\pi}\int_0^{2\pi}(u_1-iv_1)\frac{c}{c-x}\,d\theta.$$

If we subtract the last result from the formula for $f_1(c)$, we obtain

$$f_1(c)=\frac{1}{2\pi}\int_0^{2\pi}\frac{x+c}{x-c}u_1\,d\theta+\frac{i}{2\pi}\int_0^{2\pi}v_1\,d\theta.$$

Similarly, by addition we get

$$f_1(c)=\frac{1}{2\pi}\int_0^{2\pi}u_1\,d\theta+\frac{i}{2\pi}\int_0^{2\pi}\frac{x+c}{x-c}v_1\,d\theta.$$

In the same way we can find integrals for $f_2(c)$ in terms of u_2, v_2: the only essential change in the argument being that when x is changed to $1/x$, u_2+iv_2 becomes $-u_2+iv_2$. This, however, does not alter the final formulae; and so by addition we see that *these formulae remain true when the suffixes are omitted throughout.* Thus $f(c)$ is completely determined (save for a constant) by a knowledge of *either* u *or* v. But, given an arbitrary continuous function for u (or v), we do not yet know that it is actually possible to determine $f(c)$ so that its real (or imaginary) part does assume the given values on the circle; this problem will be discussed in Art. 83.

83. Abel's theorem and allied theorems.

Suppose that $|x|=1$ is the circle of convergence for Σa_nx^n and that Σa_n is known to converge, although not absolutely.

If we take $v_n = x^n$ in Art. 80, we find that $G = \lim |v_n| = 0$, and

$$|v_n - v_{n+1}| = |1-x| . |x|^n,$$

so that $\quad V_0 = \Sigma |v_n - v_{n+1}| = |1-x| / \{1 - |x|\} \geqq 1.$

Thus, since Σa_n is convergent, the series $\Sigma a_n x^n$ will converge uniformly in any area for which we have

$$|1-x| \leqq \lambda \{1 - |x|\},$$

where λ is any assigned number greater than 1, and of course $|x| < 1$. (See Abel's test on p. 206.)

To interpret this inequality we observe that it may be written

$$\rho \leqq \lambda(1-r) \quad \text{or} \quad (\lambda - \rho)^2 \geqq \lambda^2 r^2,$$

where $\quad 1 - x = \rho(\cos\phi + i \sin\phi), \quad$ (see fig. 25).

Thus $\quad \lambda^2 - 2\lambda\rho + \rho^2 \geqq \lambda^2(1 - 2\rho\cos\phi + \rho^2)$

or $\quad (\lambda^2 - 1)\rho \leqq 2(\lambda^2 \cos\phi - \lambda).$

In this condition, ϕ lies between $\pm\frac{1}{2}\pi$, and the equation

$$(\lambda^2 - 1)\rho = 2(\lambda^2 \cos\phi - \lambda), \quad (-\tfrac{1}{2}\pi < \phi < \tfrac{1}{2}\pi),$$

gives the inner arc of a limaçon (with a node at $\rho = 0$), indicated roughly in figure 25 for the case $\lambda = 3$.

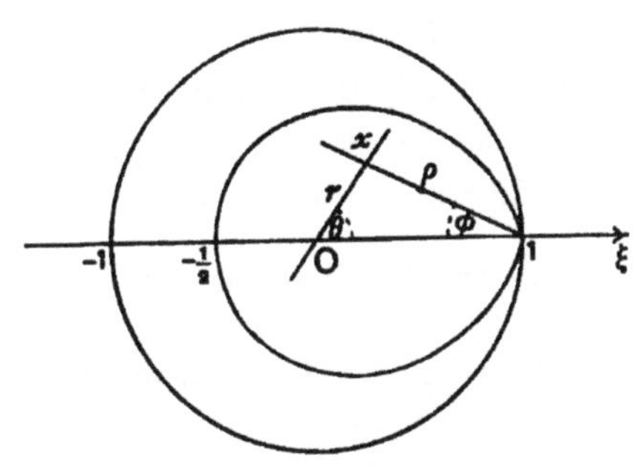

FIG. 25.

It is easy to see that the arc of the limaçon approaches the more nearly to the circle $|x| = 1$, the larger λ is taken.

Thus *the region of uniform convergence of $\Sigma a_n x^n$ may be taken as the inner arc and the contained area of any one of these limaçons.**

If any regular curve is drawn from a point inside the circle to the point $x = 1$, then, provided the curve cuts the circle at a

* Stolz and Gmeiner (*Einleitung in die Funktionentheorie*, 1905, p. 287) use a limaçon which in our notation would be represented by $\rho = 2\cos\phi - 2/\lambda$. This limaçon lies within the one used above.

finite angle, we can draw a limaçon to enclose the whole of the curve: that is, the series will converge uniformly along the curve. *Hence* $\lim \Sigma a_n x^n = \Sigma a_n$, *where* x *approaches* 1 *along any regular curve which cuts the circle at a finite angle.** This is the extension of Abel's theorem to complex variables.

The theorems in Art. 51 relating to the divergence of Σa_n cannot be extended so as to hold for complex variables quite so easily, because the lemma of Art. 80 gives less precise information than the lemma of Art. 23, and it is necessary to assume that the series $\Sigma a_n x^n$ possesses some further property in addition to the divergence of Σa_n. For as a matter of fact, even if a_n is real and positive, Pringsheim has shewn that the divergence of Σa_n does *not* ensure†

$$\lim |\Sigma a_n x^n| = \infty$$

for *all* paths defined as above.

The condition introduced by Pringsheim is that of *uniform divergence*, which implies

$$|\Sigma a_n x^n| / \Sigma a_n |x|^n \geqq a > 0,$$

where $a_n > 0$ and the point x lies within the limaçon.

It is then obvious from Art. 51 that $\lim |\Sigma a_n x^n| = \infty$.

The reader will find no great difficulty in modifying the proofs given in Art. 51 so as to apply for complex variables when Pringsheim's condition is satisfied.

We proceed now to find the limiting values of the integrals, given at the end of Art. 82.

For example, suppose that v is an arbitrary real continuous function; then the second formula gives a value for $f(c)$ which can be expanded as a power-series in c, convergent if $|c|<1$. We shall now prove that if this function is denoted by $U+iV$, then V tends to v as c moves up to any point on the circle; so that we have determined a power-series whose imaginary part has an assigned continuous value v along the circle $|c|=1$.

Clearly it is sufficient to establish the result for *any* point on the circle; so we shall calculate the limit of V as c moves up to 1.

---

* Picard, *Traité d'Analyse*, t. 2, p. 73.

† For consider $E\left[\frac{1}{(1-x)^2}\right] = E\left[\frac{1}{\rho^2}(\cos 2\phi - i \sin 2\phi)\right]$, where $E(x) = e^x$.

There is no difficulty in seeing that, when this series is put in the form $\Sigma a_n x^n$, the coefficients a_n are positive and that Σa_n diverges. However, if $\frac{1}{4}\pi < \phi < \frac{1}{2}\pi$, $\cos 2\phi$ is *negative*, and so $\lim_{\rho \to 0} E\left[\frac{1}{(1-x)^2}\right] = 0$.

It will be seen that

$$\frac{x+c}{x-c}=\frac{1-r^2+2ir\sin(\theta-\omega)}{1-2r\cos(\theta-\omega)+r^2},$$

where now $\quad x=\cos\omega+i\sin\omega \quad$ and $\quad c=r(\cos\theta+i\sin\theta)$,

so that
$$V=\frac{1}{2\pi}\int_0^{2\pi}\frac{v(1-r^2)d\omega}{1-2r\cos(\theta-\omega)+r^2}.$$

Also, from Art. 65,

$$\frac{1}{2\pi}\int_0^{2\pi}\frac{(1-r^2)d\omega}{1-2r\cos(\theta-\omega)+r^2}=\frac{1}{2\pi}\int_0^{2\pi}[1+2r\cos(\theta-\omega)+2r^2\cos 2(\theta-\omega)+\ldots]d\omega=1,$$

and we note further that, since the subject of integration is positive, the value of the integral taken over any *smaller* range must be less than 1.

Thus, if v_0 is the value of v for $\omega=0$, we find

$$V-v_0=\frac{1}{2\pi}\int_0^{2\pi}\frac{(v-v_0)(1-r^2)d\omega}{1-2r\cos(\theta-\omega)+r^2};$$

and since v is a continuous function of ω, we can determine a so that

$$|v-v_0|<\tfrac{1}{2}\epsilon, \quad \text{if } |\omega|<2a.$$

Thus
$$\frac{1}{2\pi}\left(\int_0^{2a}+\int_{2\pi-2a}^{2\pi}\right)\frac{|v-v_0|(1-r^2)d\omega}{1-2r\cos(\theta-\omega)+r^2}<\tfrac{1}{2}\epsilon.$$

We have next to consider the integral from $\omega=2a$ to $\omega=2\pi-2a$; here, provided that $|\theta|<a$, $\cos(\theta-\omega)$ is not greater than $\cos a$, and so

$$1-2r\cos(\theta-\omega)+r^2 \geqq 1-2r\cos a+r^2 \geqq \sin^2 a$$

while
$$1-r^2<2(1-r).$$

Thus, if H is the upper limit to the values of $|v|$ on the circle, we have

$$\left|\frac{(v-v_0)(1-r^2)}{1-2r\cos(\theta-\omega)+r^2}\right|<\frac{4H(1-r)}{\sin^2 a}, \quad \text{if } |\theta|<a.$$

Consequently

$$\frac{1}{2\pi}\int_{2a}^{2\pi-2a}\frac{|v-v_0|(1-r^2)d\omega}{1-2r\cos(\theta-\omega)+r^2}<\frac{4H(1-r)}{\sin^2 a}, \quad \text{if } |\theta|<a.$$

It is therefore possible to find a, δ, so that

$$|V-v_0|<\epsilon, \quad \text{if } |\theta|<a, \text{ and } 1-r<\delta,$$

that is to say, *V approaches the limit v_0 as the point (r, θ) moves up towards the point 1 by any path.*

If v is continuous except at $\omega=0$ and is there discontinuous, the integral still gives a power-series for $f(c)$ and the preceding work is valid as c approaches any point on the circle except 1. To deal with $c=1$, suppose that v has the limit l, when $\omega\to 0$ through positive values; and the limit m when $\omega\to 0$ through negative values. Then, if we write

$$v'=v-\frac{l-m}{\pi}\Sigma\frac{1}{n}\sin n\omega,$$

it is evident from Art. 65 that v' becomes continuous at $\omega=0$, if we assign to it the value $\tfrac{1}{2}(l+m)$ for $\omega=0$.

Further,
$$V'=V-\frac{l-m}{\pi}\Sigma\frac{r^n}{n}\sin n\theta=V-\frac{l-m}{\pi}\phi,$$

where ϕ represents the same angle as is indicated in the diagram on p. 211.

Now from the first case considered we have

$$\lim_{(r,\theta)} V' = \lim_{(\omega)} v' = \tfrac{1}{2}(l+m),$$

so that

$$\lim_{(r,\theta)} V = \tfrac{1}{2}(l+m) + (l-m)\frac{\phi_0}{\pi},$$

where ϕ_0 is the limiting value of ϕ as (r, θ) approaches 1.

In the particular case when v is given in the form $\Sigma a_n \sin n\omega$, we shall have $m = -l$, and then the result is

$$\lim_{(r,\theta)} \Sigma a_n r^n \sin n\theta = 2l\phi_0/\pi.$$

It will be noted that in this case the series Σa_n cannot be convergent; for if it were convergent we should have

$$\lim_{(r,\theta)} \Sigma a_n r^n (\cos n\theta + i \sin n\theta) = \Sigma a_n$$

in virtue of the extension of Abel's theorem (at the beginning of this article); that is,

$$\lim_{(r,\theta)} \Sigma a_n r^n \sin n\theta = 0,$$

which is not the case.

84. Taylor's theorem for a power-series.

We have seen that a power-series $\Sigma a_n x^n$ represents a continuous function of x, say $f(x)$, within its circle of convergence $|x| = R$; let us now attempt to express $f(x+h)$ as a power-series in h. Draw the circle of convergence, and mark a point x

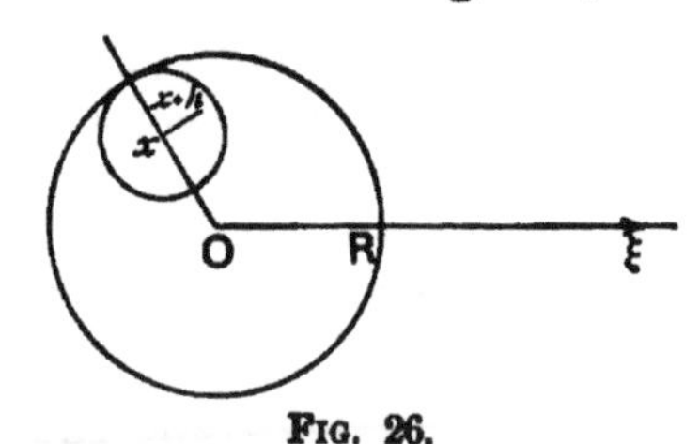

FIG. 26.

inside it, such that $|x| = r$; draw a second circle (of radius $R - r$), with centre x, to touch the first, and mark a point $x+h$ within the second circle.

We shall now see that $f(x+h)$ can be expressed as a power-series in h.

In fact $f(x+h)$ is the sum, by columns, of the double series

$$\begin{aligned}
a_0 + a_1 x + \ & a_2 x^2 + \ a_3 x^3 + \dots \\
+ a_1 h + \ & 2a_2 xh + 3a_3 x^2 h + \dots \\
+ \ & a_2 h^2 + 3a_3 x h^2 + \dots \\
& \qquad + \ a_3 h^3 + \dots \\
& \qquad\qquad + \dots .
\end{aligned}$$

But this series is absolutely convergent, because, if we replace each term by its absolute value, we get the series

$$|a_0|+|a_1|(r+\rho)+|a_2|(r+\rho)^2+|a_3|(r+\rho)^3+\dots,$$

where $\rho=|h|$. Now this series is convergent, because $r+\rho<R$ by the construction; and therefore the double series converges absolutely. That is, we can sum the double series by rows, without altering its value (Art. 33).

Hence $f(x+h)=f(x)+hf_1(x)+\frac{h^2}{2!}f_2(x)+\frac{h^3}{3!}f_3(x)+\dots,$

where

$$f_1(x)=a_1+2a_2x+3a_3x^2+\dots,$$
$$f_2(x)=1.2a_2+2.3a_3x+3.4a_4x^2+\dots,$$
$$f_3(x)=1.2.3a_3+2.3.4a_4x+3.4.5a_5x^2+\dots,$$

so that these series may be obtained from $f(x)$ by simply applying the formal rules for successive differentiation, without paying any attention to the meaning of the process.

The series in h may be called *Taylor's series*.

It may be useful to remark that the circle of convergence for the new series often reaches *beyond* the circle $|x|=R$; we know that it *must* reach as far as this circle, but there is no evidence that it may not extend further.

For instance, it is easy to see that if we write

$$f(x)=1+x+x^2+x^3+\dots,$$

then

$$f(\tfrac{1}{2}i+h)=\frac{1}{1-\frac{1}{2}i}+\frac{h}{(1-\frac{1}{2}i)^2}+\frac{h^2}{(1-\frac{1}{2}i)^3}+\dots,$$

which converges if $|h|<|1-\frac{1}{2}i|$

or if $|h|<\frac{1}{2}\sqrt{5}$.

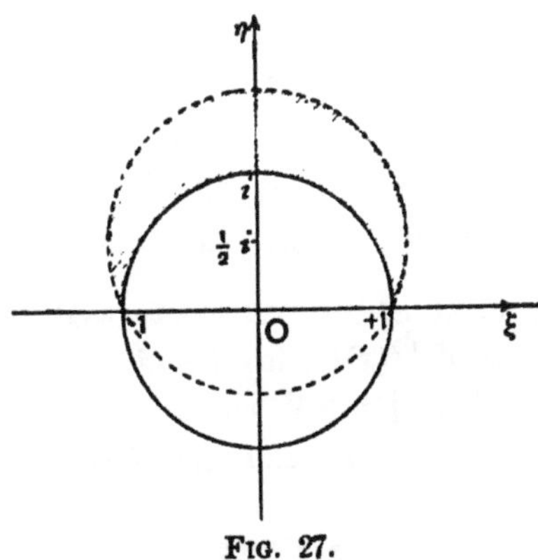

FIG. 27.

Thus the Taylor's series converges in the shaded area outside the original circle of convergence. We have thus a new power-series which *continues* the function $f(x)$ beyond the area of its original definition: this idea of *continuation* is fundamental in Weierstrass's theory of Functions, but further details lie outside our province. The reader may consult Harkness and Morley's *Introduction to the Theory of Analytic Functions* for a good account of this theory.

We can now obtain an extended form of Cauchy's inequalities (Art. 82). Suppose that the exact radius of convergence of $f(x)$ is not known, but that $|x|=r$ is known to be *within* the circle of convergence; and further that for all points on the circle $|x|=r$, the h-series $f(x+h)$ converges uniformly on $|h|=s$. Then, if M is the maximum of $|f(x+h)|$ for all points such that $|x|=r$, $|h|=s$, we have by applying Cauchy's inequality to the h-series

$$\left|\frac{f_n(x)}{n!}\right| < \frac{M}{s^n} = M', \text{ say.}$$

Applying the same inequality to the x-series for $f_n(x)$ we see that

$$\frac{(m+n)(m+n-1)\ldots(m+1)}{n!}|a_{m+n}| < \frac{M'}{r^m} = \frac{M}{r^m s^n}.$$

Thus in the expansion of $(r+s)^{m+n}|a_{m+n}|$ *every term is less than* M*; and therefore*

$$(r+s)^{m+n}|a_{m+n}| < (m+n+1)M.$$

It follows that *the radius of convergence of* $\Sigma a_n x^n$ *is at least equal to* $(r+s)$.

This leads at once to the theorem* that *there is at least one singular point on the circle of convergence of a power-series*; that is, a point in the neighbourhood of which Taylor's theorem cannot be applied.

Baker † has also used this result to shew that *the circle of convergence of the reciprocal of a power-series is either the same as that of the original series, or else reaches up to the zero of the given series which is nearest to the origin.*

In fact the argument of Art. 54 shews that if L, l are the maximum and minimum values of $|f(x)|$ within and on any circle $|x|=R'<R$, then the power-series for $[f(x)]^{-1}$ will converge if

$$|x|<|a_0|R'/\{|a_0|+L\} \text{ and therefore if } |x| \leqq lR'/(l+L).$$

* Harkness and Morley, *l.c.*, Art. 102.

† *Proc. Lond. Math. Soc.* (1), vol. 34, 1902, p. 296; the discussion given there is for series in two variables, and of course can be extended to any number.

By transferring now to a point x_1 such that $|x_1|=r_1=lR'/(l+L)$, we can infer that $[f(x)]^{-1}$ expressed as a series in $(x-x_1)$ will certainly converge if

$$\frac{|x-x_1|}{R'-r_1} \leqq \frac{l}{l+L}.$$

Thus by means of the result established above we see that the radius of convergence of $[f(x)]^{-1}$ is at least equal to r_2, where

$$r_2=r_1+(R'-r_1)l/(l+L),$$

so that
$$R'-r_2=(R'-r_1)L/(l+L)=R'[L/(l+L)]^2.$$

Continuing the process, the radius is seen to be not less than r_n, where

$$R'-r_n=R'[L/(l+L)]^n,$$

and consequently, so long as l is not zero, the radius of convergence cannot be less than R'.

85. The exponential power-series.

There is no difficulty in modifying the proof of Art. 57 to shew that

$$\lim_{\nu \to \infty}(1+\xi)^{\nu}=1+x+\frac{x^2}{2!}+\frac{x^3}{3!}+\ldots=E(x),$$

where
$$\lim(\nu\xi)=x,$$

and ν is real, although ξ is complex.

By multiplication of series, or by an argument similar to that of Art. 58, we find

$$E(x)\times E(y)=E(x+y),$$

which is the fundamental equation of the exponential power-series.

As a kind of converse theorem, we shall now obtain the most general power-series,

$$f(x)=a_0+a_1x+a_2\frac{x^2}{2!}+a_3\frac{x^3}{3!}+\cdots,$$

which converges within a circle $|x|=R$, say, and satisfies the equation

$$f(x+y)=f(x)f(y),$$

provided that $|x|$, $|y|$, $|x+y|$ are all less than R (which certainly holds good if $|x|$ and $|y|$ are less than $\frac{1}{2}R$). Since this condition requires the equation to hold for *real* values of x, y in the interval $(-\frac{1}{2}R, +\frac{1}{2}R)$, we shall consider these values first.* Now put $y=0$;

* We restrict x, y to be *real* so as to avoid the difficulty of differentiating with respect to a *complex* independent variable. The fact that the *coefficients* in $f(x)$ may be complex does not affect the application of Art. 52 (3), because we can differentiate the real and imaginary parts separately.

then
$$f(x)\times f(0)=f(x)$$
or
$$f(0)=a_0=1.$$

Hence $a_0=1$, and so
$$f(x)=1+a_1x+a_2\frac{x^2}{2!}+\dots.$$

Again,
$$\frac{f(x+y)-f(x)}{y}=f(x)\frac{f(y)-1}{y}=f(x)\left(a_1+\frac{a_2y}{2!}+\frac{a_3y^2}{3!}+\dots\right),$$
and if we take the limit of both sides as y tends to zero, we get at once $\qquad f'(x)=a_1f(x).$

Or, applying Art. 52 (3), we have
$$a_1+a_2x+a_3\frac{x^2}{2!}+\dots=a_1\left(1+a_1x+a_2\frac{x^2}{2!}+\dots\right),$$
and since this equation must hold for *all* values of x in the interval $(-\frac{1}{2}R,\ +\frac{1}{2}R)$, we must have
$$a_2=a_1^2,\quad a_3=a_1a_2,\quad a_4=a_1a_3,\ \dots. \quad \text{See Art. 52 (5).}$$

That is, $a_2=a_1^2,\ a_3=a_1^3,\ a_4=a_1^4,\ \dots,\ a_n=a_1^n,\ \dots,$
and so
$$f(x)=1+a_1x+a_1^2\frac{x^2}{2!}+a_1^3\frac{x^3}{3!}+\dots=E(a_1x).$$

We do not know from this argument that $f(x)$ satisfies *all* the conditions of the problem; but we see that *if* there is such a power-series, it can be no other than $E(a_1x)$. Now $E(a_1x)$ does satisfy the relation
$$E(a_1x)\times E(a_1y)=E[a_1(x+y)]$$
for any real or complex values of x, y.

Consequently our problem has been solved;* and
$$f(x)=E(a_1x),$$
where a_1 is the coefficient of x in the power-series for $f(x)$.

It is usual, and in many respects convenient, to write e^x for $E(x)$ even when x is complex. But it must be remembered that this is merely a convention; and that in such an equation as $e^{\frac{1}{2}\pi i}=i$ (see below, Art. 86) the index does not denote an ordinary power.

---

* It does not follow from the foregoing that no other *function* can satisfy the relation $f(x)\times f(y)=f(x+y)$, because we have assumed $f(x)$ to be a power-series. But, if we assume that $f'(x)$ is continuous, there is no difficulty in shewing that $f(x)$ has the exponential form.

86. Connexion between the exponential and circular functions.

If the complex variable x in the exponential series $E(x)$ depends on a *real* variable t, we may differentiate term-by-term with respect to t, and obtain the same formula as if x were real:

$$\frac{d}{dt}[E(x)] = E(x)\frac{dx}{dt}.$$

For, suppose that corresponding to a change δt in t, x changes to $x+\delta x$; then by Art. 85

$$\frac{1}{\delta t}[E(x+\delta x)-E(x)] = E(x)[E(\delta x)-1]/\delta t.$$

And
$$E(\delta x)-1 = \delta x\left[1+\frac{1}{2!}(\delta x)+\frac{1}{3!}(\delta x)^2+\dots\right],$$

so that
$$\left|\frac{1}{\delta t}\{E(x+\delta x)-E(x)\}-E(x)\frac{\delta x}{\delta t}\right| < \frac{1}{2}\left|\frac{\delta x}{\delta t}\right|\frac{|\delta x|}{1-\frac{1}{3}|\delta x|}.$$

Now, as $\delta t\to 0$, $\delta x/\delta t$ approaches the limit dx/dt, so that $|\delta x|\to 0$; and so it follows from the last inequality that

$$\lim \frac{1}{\delta t}[E(x+\delta x)-E(x)] = E(x)\frac{dx}{dt}.$$

In particular, suppose that x is a pure imaginary and equal to $i\eta$, where η is real; then we have

$$\frac{d}{d\eta}[E(i\eta)] = iE(i\eta),$$

or, if
$$E(i\eta) = r(\cos\theta + i\sin\theta),$$

we have
$$\frac{dr}{d\eta} + ir\frac{d\theta}{d\eta} = ir$$

or
$$\frac{dr}{d\eta} = 0, \quad \frac{d\theta}{d\eta} = 1.$$

Thus r and $\theta-\eta$ are independent of η; but for $\eta=0$, $E(i\eta)=1$, and so $r=1$, $\theta=0$.

Hence generally $r=1$, $\theta=\eta$,

and so
$$E(i\eta) = \cos\eta + i\sin\eta,$$

which is confirmed by the remark that

$$|E(i\eta)|^2 = E(i\eta)\times E(-i\eta) = E(i\eta - i\eta) = E(0) = 1.$$

Another method of establishing the last result is given by observing that

$$\cos\eta + i\sin\eta = (\cos\phi + i\sin\phi)^n, \quad \text{if } \phi = \eta/n.$$

Now write
$$\cos\phi + i\sin\phi = 1 + \kappa_n,$$

and we see that
$$\lim_{n\to\infty} n\kappa_n = i\eta,$$

because $$\lim (n \sin \phi) = \eta \lim (\sin \phi/\phi) = \eta,$$

and $$\lim n(1-\cos \phi) = \eta \lim [(1-\cos\phi)/\phi] = 0.$$

Hence $$\cos \eta + i \sin \eta = \lim_{n\to\infty} (1+\kappa_n)^n = E(i\eta)$$

by Art. 85.

Still another method, analogous to that of Art. 59, can be used. In fact, write

$$\cos\eta + i\sin\eta - \left[1 + i\eta + \frac{1}{2!}(i\eta)^2 + \ldots + \frac{1}{n!}(i\eta)^n\right] = y_n.$$

Then $$\frac{dy_n}{d\eta} = iy_{n-1} \text{ and } \frac{dy_0}{d\eta} = i(\cos\eta + i\sin\eta).$$

But, if $y = r(\cos\theta + i\sin\theta)$,

we have $$\frac{dy}{d\eta} = \left(\frac{dr}{d\eta} + ir\frac{d\theta}{d\eta}\right)(\cos\theta + i\sin\theta)$$

or $$\left|\frac{dy}{d\eta}\right| = \left[\left(\frac{dr}{d\eta}\right)^2 + r^2\left(\frac{d\theta}{d\eta}\right)^2\right]^{\frac{1}{2}} \geqq \left|\frac{dr}{d\eta}\right|.$$

Hence $$\left|\frac{dr_n}{d\eta}\right| \leqq \left|\frac{dy_n}{d\eta}\right| = |y_{n-1}| = r_{n-1}, \quad \left|\frac{dr_0}{d\eta}\right| \leqq 1,$$

and $y_0, y_1, \ldots$ are all zero for $\eta = 0$. Hence we find, if η is positive, the sequence of equations

$$r_0 = |y_0| \leqq \eta, \quad r_1 = |y_1| \leqq \frac{1}{2!}\eta^2, \quad r_2 = |y_2| \leqq \frac{1}{3!}\eta^3, \ldots, \quad r_{n-1} = |y_{n-1}| \leqq \frac{1}{n!}\eta^n.$$

Thus $$\lim_{n\to\infty} y_n = 0.$$

If we substitute $i\eta$ in the exponential series, we find

$$1 + i\eta - \frac{\eta^2}{2!} - i\frac{\eta^3}{3!} + \frac{\eta^4}{4!} + i\frac{\eta^5}{5!} - \ldots$$

$$= \left(1 - \frac{\eta^2}{2!} + \frac{\eta^4}{4!} - \ldots\right) + i\left(\eta - \frac{\eta^3}{3!} + \frac{\eta^5}{5!} - \ldots\right),$$

and so we have now a new method of finding the sine and cosine power-series (Art. 59).

If we write $\eta = \frac{1}{2}\pi$ and π, we get the equations

$$E(\tfrac{1}{2}\pi i) = i, \quad E(\pi i) = -1.$$

Using the notation explained in Art. 85, we may write

$$\cos\eta + i\sin\eta = e^{i\eta},$$

and changing the sign of η, we find

$$\cos\eta - i\sin\eta = e^{-i\eta};$$

thus $$\cos\eta = \frac{1}{2}(e^{i\eta} + e^{-i\eta}), \quad \sin\eta = \frac{1}{2i}(e^{i\eta} - e^{-i\eta}).$$

We have at present no definitions of $\cos x$ and $\sin x$ when x is complex; but it is usual and convenient to define them by the power-series already established when x is real. Then the equations

$$\cos x = \frac{1}{2}(e^{ix} + e^{-ix}), \quad \sin x = \frac{1}{2i}(e^{ix} - e^{-ix})$$

are true for complex values of x as well as real ones.

It follows also that any trigonometrical formulae which depend only on the addition-theorems remain unaltered for complex variables; thus in particular the formulae of Arts. 66, 67, 69 remain true.

If we write $x = \xi + i\eta$, it will be seen that

$$\cos x = \cos\xi \cosh\eta - i \sin\xi \sinh\eta,$$
$$\sin x = \sin\xi \cosh\eta + i \cos\xi \sinh\eta,$$

where $\cosh\eta = \frac{1}{2}(e^{\eta} + e^{-\eta}), \quad \sinh\eta = \frac{1}{2}(e^{\eta} - e^{-\eta}).$

We shall not elaborate the details of the analysis of the sinh and cosh functions; the results can be found in many text-books (for instance, Chrystal's *Algebra*, ch. XXIX.).

It is to be noticed that *when x is complex, the inequalities*

$$|\sin x| < |x|, \quad |\cos x| < 1$$

are no longer valid. We can, however, replace them by others, thus:

$$|\sin x| \leqq \sinh|x| = |x| + \frac{|x|^3}{3!} + \frac{|x|^5}{5!} + \dots,$$

and so, if $|x| < 1$, we have

$$|\sin x| < |x| \left\{1 + \frac{1}{6} + \frac{1}{6^2} + \dots\right\} < \frac{6}{5}|x|.$$

Similarly, we have

$$|\cos x| \leqq \cosh|x|;$$

and, if $|x| < 1$, we find

$$|\cos x| < \left(1 + \frac{1}{2!} + \frac{1}{4!} + \dots\right) < 2.$$

87. The logarithm.

We have already seen that if η is a real angle

$$E(i\eta) = \cos\eta + i \sin\eta.$$

Hence if n is any integer (positive or negative),

$$E(2n\pi i) = 1,$$

and since $E(\xi+i\eta)=e^{\xi}(\cos\eta+i\sin\eta)$ there are no solutions of the equation
$$E(\xi+i\eta)=1,$$
other than $\xi=0,\ \eta=2n\pi$.

It follows that if we wish to solve the equation $E(y)=x$, so as to obtain the function inverse to the exponential function, the value obtained is not single-valued, but is of the form
$$y=y_0+2n\pi i, \quad (n=0,\ \pm 1,\ \pm 2, \ldots),$$
where y_0 is one solution of the equation.

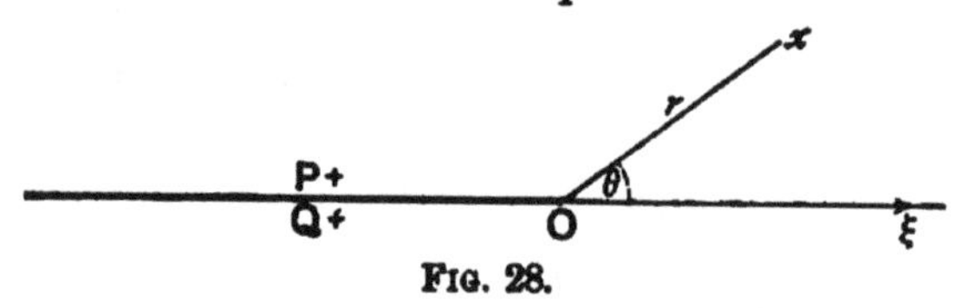

FIG. 28.

If we represent x geometrically in Argand's diagram, we have
$$x=r(\cos\theta+i\sin\theta)=rE(i\theta).$$

But if $\log r$ is the logarithm of the real number r, defined as in Art. 157 of the Appendix, we have
$$r=E(\log r),$$
and consequently
$$x=E(\log r+i\theta).$$

Thus we can take $y_0=\log r+i\theta$, and then the general solution is
$$y=\log x=\log r+i(\theta+2n\pi), \quad (n=0,\ \pm 1,\ \pm 2, \ldots).$$

We define the logarithmic function as consisting of all the inverses of the exponential function; and we can specify a one-valued branch of the logarithm by supposing a cut made along the negative part of the real axis, and regarding x as prevented from crossing the cut. Then we shall have
$$\log x=\log r+i\theta, \text{ where } -\pi<\theta\leqq\pi.$$

With this determination, $\log x$ is real when x is real, which is generally the most convenient assumption. But it should be observed that then such formulae as
$$\log(xx')=\log x+\log x'$$
can only be employed with caution, since it may easily happen that $(\theta+\theta')$ is greater than π, in which case we ought to write
$$\log(xx')=\log x+\log x'-2\pi i.$$

The reader will note that for two points such as P, Q in the diagram (Q being the reflexion of P in the negative half of the real axis),
$$\lim_{P\to Q}(\log x_P-\log x_Q)=2\pi i.$$

But, except at the cut, the branch selected for $\log x$ is obviously continuous over the whole plane of x.

88. The logarithmic power-series.

We know from Arts. 58 and 62, that if x is real and $|x| < 1$ the series

(1) $$y = x - \tfrac{1}{2}x^2 + \tfrac{1}{3}x^3 - \ldots$$

represents the function inverse to the exponential function

(2) $$1 + x = E(y) = 1 + y + \frac{y^2}{2!} + \frac{y^3}{3!} + \ldots.$$

In other words, if we substitute the series (1) in the series (2), and then arrange according to powers of x, the result* must be $1 + x$. But this transformation is merely algebraical, and, as such, is equally true whether x is real or complex.

Since the series (2) converges absolutely for all values of y, the derangement implied in this transformation is legitimate (see Art. 36), provided that the series (1) is absolutely convergent. Hence, if $|x| < 1$, equation (1) gives **one** value of y satisfying equation (2); and further, from (1), y is real when x is real. Thus, using the branch of the logarithm defined in the last article, we have

$$\log(1 + x) = x - \tfrac{1}{2}x^2 + \tfrac{1}{3}x^3 - \ldots \quad (\text{if } |x| < 1).$$

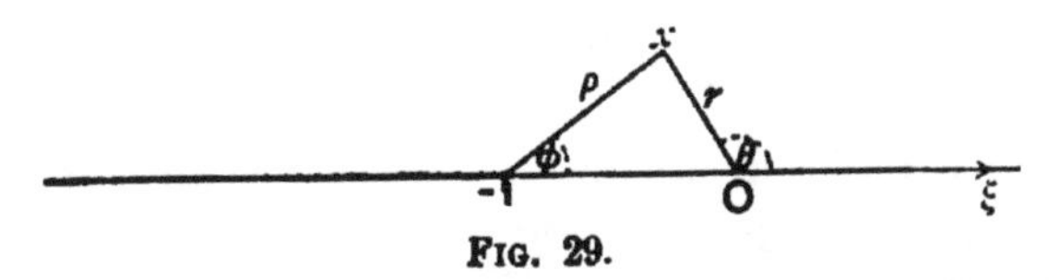

FIG. 29.

From the figure, it is evident that this equation gives

$$\log \rho + i\phi = x - \tfrac{1}{2}x^2 + \tfrac{1}{3}x^3 - \ldots,$$

where $$-\pi < \phi \leqq +\pi \quad (\text{see Art. 87}).$$

This result can be at once confirmed by reference to Art. 65, where we proved that (if $0 < r < 1$)

$$\tfrac{1}{2}\log(1 + 2r\cos\theta + r^2) = r\cos\theta - \tfrac{1}{2}r^2\cos 2\theta + \tfrac{1}{3}r^3\cos 3\theta - \ldots,$$

$$\arctan\frac{r\sin\theta}{1 + r\cos\theta} = r\sin\theta - \tfrac{1}{2}r^2\sin 2\theta + \tfrac{1}{3}r^3\sin 3\theta - \ldots.$$

If we write $x = r(\cos\theta + i\sin\theta)$ in the power-series (1), we get

$$x - \tfrac{1}{2}x^2 + \tfrac{1}{3}x^3 - \ldots = r\cos\theta - \tfrac{1}{2}r^2\cos 2\theta + \tfrac{1}{3}r^3\cos 3\theta - \ldots$$
$$+ i(r\sin\theta - \tfrac{1}{2}r^2\sin 2\theta + \tfrac{1}{3}r^3\sin 3\theta - \ldots),$$

* It is a good exercise to verify this by numerical computation up to, say, x^6.

and obviously
$$\rho^2 = 1 + 2r\cos\theta + r^2,$$
$$\tan\phi = r\sin\theta/(1 + r\cos\theta).$$

Thus our results are in agreement with those of Art. 65, except that we now see that ϕ actually lies between $-\frac{1}{2}\pi$ and $+\frac{1}{2}\pi$ (because $r < 1$) instead of $-\pi$ and π.

We shall obtain an independent proof of the equation
$$\log(1+x) = x - \tfrac{1}{2}x^2 + \tfrac{1}{3}x^3 - \dots \quad (\text{if } |x| < 1)$$
in the following article.

The series for arc sin x **and** arc tan x.

Again, by Art. 64, the series

$$(3) \qquad y = x + \frac{1}{2}\frac{x^3}{3} + \frac{1.3}{2.4}\frac{x^5}{5} + \dots$$

represents the function inverse to the sine-function (Art. 59).

$$(4) \qquad x = y - \frac{y^3}{3!} + \frac{y^5}{5!} - \dots$$

for *real* values of x, y, such that $|x| < 1$. Since the series (4) is absolutely convergent for all values of y, and the series (3) for $x| < 1$, the algebraic relation between these series is now seen to persist for complex values of x, and we can accordingly write

$$\text{arc}\sin x = x + \frac{1}{2}\frac{x^3}{3} + \frac{1.3}{2.4}\frac{x^5}{5} + \dots \quad (\text{if } |x| < 1),$$

since the series (4) is taken as defining the sine for complex values of the variable (Art. 86).

Similarly the pair of functions

$$(5) \qquad y = x - \tfrac{1}{3}x^3 + \tfrac{1}{5}x^5 - \tfrac{1}{7}x^7 + \dots,$$

$$(6) \qquad x = \left(y - \frac{y^3}{3!} + \frac{y^5}{5!} - \dots\right) \Big/ \left(1 - \frac{y^2}{2!} + \frac{y^4}{4!} - \dots\right)$$

are inverse to one another for real values of x, such that $|x| < 1$, and we may therefore write for complex values of x

$$\text{arc}\tan x = x - \tfrac{1}{3}x^3 + \tfrac{1}{5}x^5 - \dots \quad (\text{if } |x| < 1).$$

In these equations the values of the inverse functions are determined by the conditions

$$|\text{arc}\sin x| \leqq |x| + \frac{1}{2}\frac{|x|^3}{3} + \frac{1.3}{4}\frac{|x|^5}{5} + \dots = \text{arc}\sin|x| \leqq \tfrac{1}{2}\pi,$$

$$|\text{arc}\tan x| \leqq |x| + \tfrac{1}{3}|x|^3 + \tfrac{1}{5}|x|^5 + \dots = \tfrac{1}{2}\log\{1+|x|\}/\{1-|x|\}.$$

89. The binomial power-series.

Consider the series

$$f(\nu, x) = 1 + \nu x + \nu(\nu-1)\frac{x^2}{2!} + \nu(\nu-1)(\nu-2)\frac{x^3}{3!} + \dots,$$

where both ν and x may be complex. The series is absolutely convergent when $|x| < 1$, and thus we have the identity

$$f(\nu, x) \times f(\nu', x) = f(\nu+\nu', x) \quad (|x| < 1).$$

For, if we pick out the coefficient of x^r in the product

$$f(\nu, x) \times f(\nu', x),$$

it is seen by the ordinary rule (Art. 54) to be a polynomial of degree r in both ν and ν'; thus the coefficient of x^r in

$$f(\nu, x) \times f(\nu', x) - f(\nu+\nu', x)$$

is also a polynomial N_r of the same degree.

But, when ν, ν' are *any* two integers, N_r is zero, because then $f(\nu, x) = (1+x)^\nu$; and consequently N_r must be identically zero, because, when ν' is any assigned integer, N_r is zero for an infinity of different values of ν (namely, 1, 2, 3, ... to ∞).

Thus, identically,

$$f(\nu, x) \times f(\nu', x) = f(\nu+\nu', x), \quad (|x| < 1).$$

From this relation we can apply the method indicated in Art. 61 (2) to prove that

$$f(\nu, x) = (1+x)^\nu$$

when ν is a rational number.

But to deal with the case of complex values of ν, we proceed somewhat differently. In the first place $f(\nu, x)$ can be expressed as a power-series in ν; for $f(\nu, x)$ can be regarded as the sum by columns of the double series

$$\begin{array}{l} 1 + \nu x - \frac{\nu}{2}x^2 + \frac{\nu}{3}x^3 - \frac{\nu}{4}x^4 + \dots \\ \qquad + \frac{\nu^2}{2}x^2 - \frac{\nu^2}{2}x^3 + \frac{11\nu^2}{24}x^4 - \dots \\ \qquad\qquad + \frac{\nu^3}{6}x^3 - \frac{\nu^3}{4}x^4 + \dots \\ \qquad\qquad\qquad + \frac{\nu^4}{24}x^4 - \dots \\ \qquad\qquad\qquad\qquad + \dots. \end{array}$$

Now this double series is absolutely convergent because, if $|\nu|=\nu_0$ and $|x|=x_0$, the sum of the absolute values of the terms in the $(p+1)$th column is

$$\frac{\nu_0(\nu_0+1)\ldots(\nu_0+p-1)}{1.2\ldots p}x_0^p,$$

which is the $(p+1)$th term of the series $f(\nu_0, x_0)$; and $f(\nu_0, x_0)$ converges if $x_0<1$ by Art. 12.

The double series being absolutely convergent its sum is not altered (see Art. 33) by changing the mode of summation to rows, which gives

$$f(\nu, x)=1+\nu X_1+\nu^2 X_2+\ldots,$$

where $$X_1=x-\tfrac{1}{2}x^2+\tfrac{1}{3}x^3-\ldots.$$

Thus, since $f(\nu, x)\times f(\nu', x)=f(\nu+\nu', x)$, we can apply Art. 85 above, and deduce that*

$$f(\nu, x)=E(\nu X_1).$$

In order to determine X_1, let us write $\nu=1$, which gives

$$1+x=E(X_1).$$

Thus X_1 is a value of $\log(1+x)$; and since X_1 is real when x is real, it is the value defined in Art. 87. We have thus a new investigation of the logarithmic series.

Hence $$X_1=\log\rho+i\phi$$
(see fig. 29, Art. 88), so that

$$f(\nu, x)=E[\nu\log(1+x)]$$
$$=\rho^{\alpha}e^{-\beta\phi}[\cos(\alpha\phi+\beta\log\rho)+i\sin(\alpha\phi+\beta\log\rho)],$$

where $$\nu=\alpha+i\beta,$$

a result which is due to Abel. The investigation above is based on the proof given by Goursat (*Cours d'Analyse Mathématique*, § 275).

The method given in the example of Art. 36 (p. 89) applies to complex indices; and the following method was suggested in 1903 by Prof. A. C. Dixon:

The relation $$f(\nu, x)\times f(\nu', x)=f(\nu+\nu', x)$$

gives at once $$f(\nu, x)=\left[f\left(\frac{\nu}{n}, x\right)\right]^n=(1+\xi)^n, \text{ say},$$

where n is a positive integer.

* Of course ν corresponds here to x of that article; and X_1 corresponds to a_1.

Now
$$n\xi = \nu\left[x - \left(1-\frac{\nu}{n}\right)\frac{x^2}{2} + \left(1-\frac{\nu}{n}\right)\left(1-\frac{\nu}{2n}\right)\frac{x^3}{3} - \left(1-\frac{\nu}{n}\right)\left(1-\frac{\nu}{2n}\right)\left(1-\frac{\nu}{3n}\right)\frac{x^4}{4} + \dots\right].$$

But the series in square brackets has each of its terms less, in absolute value, than the corresponding term of

$$x_0 + (1+\nu_0)\frac{x_0^2}{2} + (1+\nu_0)\left(1+\frac{\nu_0}{2}\right)\frac{x_0^3}{3} + \dots,$$

which is a convergent series, *independent of* n; and consequently the limit of $n\xi$ as n tends to ∞ can be found by taking the limit of each term (Art. 49).

Hence
$$\lim_{n\to\infty}(n\xi) = \nu(x - \tfrac{1}{2}x^2 + \tfrac{1}{3}x^3 - \tfrac{1}{4}x^4 + \dots) = \nu \log(1+x).$$

But (Art. 85) $\lim_{n\to\infty}(1+\xi)^n = E(x')$, if $x' = \lim_{n\to\infty}(n\xi)$.

Thus,
$$f(\nu, x) = \lim(1+\xi)^n = E[\nu\log(1+x)].$$

The discussion given above applies only to points within the circle $|x| = 1$. To examine the convergence at points *on* the circle, we can refer back to Weierstrass's rule (Art. 79); but on account of the importance of the binomial series, we shall give an independent treatment of case (ii).

We have at once

$$\left|\frac{a_n x^n}{a_{n+1}x^{n+1}}\right| = \frac{n+1}{|n-\nu|} = \frac{n+1}{[(n-a)^2+\beta^2]^{\frac{1}{2}}} = 1 + \frac{a+1}{n} + \frac{\omega}{n^2},$$

where $\nu = a + i\beta$ and $|\omega| < A$.

Thus the series converges absolutely *on* the circumference if a is positive; and so $\Sigma a_n x^n$ is continuous up to and including the circumference. Thus, since $E[\nu\log(1+x)]$ is also continuous,* it is evident that

$$E[\nu\log(1+x)] = f(\nu, x) \quad (\text{if } a > 0)$$

at all points *on* the circumference $|x| = 1$.

On the other hand, if $a + 1 \leqq 0$, after a certain stage the absolute values of the terms of the series never decrease, and so $f(\nu, x)$ *cannot* converge when $|x| = 1$.

---

* Except at $x = -1$; but there is no difficulty in seeing that (if $a > 0$)
$$\lim_{x\to -1} E[\nu\log(1+x)] = 0.$$
For the absolute value of this exponential is $\rho^a e^{-\beta\phi}$, where $|\phi| < \pi$.

Again, if $-1<\alpha\leqq 0$, we see from Ex. 3, Art. 39, that $\lim_{n\to\infty}|a_n|=0$; also we have $f(1, x)=1+x$, so that

$$(1+x)f(\nu, x)=f(\nu+1, x).$$

If we take only the terms up to x^n in this identity, we see that

$$(1+x)S_n=S_n'+\frac{\nu(\nu-1)\dots(\nu-n+1)}{n!}x^{n+1},$$

where S_n, S_n' are the sums of the first $(n+1)$ terms in $f(\nu, x)$ and $f(\nu+1, x)$ respectively.

Now the real part of $\nu+1$ is $\alpha+1$, and is accordingly positive; and so it follows from the previous argument that S_n' tends to a definite limit as n increases to ∞.

And since $\lim_{n\to\infty}\left|\frac{\nu(\nu-1)\dots(\nu-n+1)}{n!}\right|=\lim_{n\to\infty}|a_n|=0$, it follows that $\lim_{n\to\infty}(1+x)S_n$ is definite and equal to $E[(\nu+1)\log(1+x)]$; thus, *unless* $x=-1$, we have $f(\nu, x)=\lim_{n\to\infty}S_n=E[\nu\log(1+x)]$, and the series converges on the circle, except at $x=-1$.

For points on the circumference, it is evident that

$$\rho=2\cos\tfrac{1}{2}\theta,\quad \phi=\tfrac{1}{2}\theta,\quad (-\pi<\theta<\pi)$$

and so we have

$$f(\nu, x)=(2\cos\tfrac{1}{2}\theta)^{\alpha}e^{-\frac{1}{2}\beta\theta}[\cos\{\tfrac{1}{2}\alpha\theta+\beta\log(2\cos\tfrac{1}{2}\theta)\}$$
$$+i\sin\{\tfrac{1}{2}\alpha\theta+\beta\log(2\cos\tfrac{1}{2}\theta)\}],$$

provided that $\alpha>-1$.

For the special value $x=-1$, we have the identity (see p. 152)

$$S_n=(1-\nu)\left(1-\frac{\nu}{2}\right)\dots\left(1-\frac{\nu}{n}\right).$$

It follows, as in Arts. 42 and 61, that

$$S_n\to 0,\quad \text{if } \alpha>0,\qquad |S_n|\to\infty,\quad \text{if } \alpha<0,$$

but that S_n oscillates if $\alpha=0$.

90. Differentiation of Trigonometrical Series.

In some cases of interest, it is found that although the series

$$f(x)=\Sigma a_n e^{inx}=\Sigma a_n(\cos nx+i\sin nx)$$

is uniformly convergent, yet the series of differential coefficients ceases to converge. If this occurs, it is often possible to obtain the value of $f'(x)$ by differentiating the series

$$f(x)(1-e^{ix})=a_0-(a_0-a_1)e^{ix}-(a_1-a_2)e^{2ix}-\dots.$$

This gives

$$f'(x)(1-e^{ix})-if(x)e^{ix}=-i[(a_0-a_1)e^{ix}+2(a_1-a_2)e^{2ix}+3(a_2-a_3)e^{3ix}+\ldots],$$

provided that the series $\Sigma n(a_{n-1}-a_n)e^{inx}$ is uniformly convergent, which is certainly the case (in any interval from which $x=0, 2\pi$ are excluded) if $n(a_{n-1}-a_n)$ is a real positive decreasing sequence, or in any interval if $\Sigma n|a_{n-1}-a_n|$ is convergent (see Arts. 44, 46).

If we substitute the value of $f(x)$ as a series, we find the equation

$$f'(x)(1-e^{ix})=i[a_1e^{ix}-(a_1-2a_2)e^{2ix}-(2a_2-3a_3)e^{3ix}-\ldots].$$

In practice it will usually be found best to differentiate the series for $f(x)$ first, and obtain the *formal* equation

$$f'(x)=i(a_1e^{ix}+2a_2e^{2ix}+3a_3e^{3ix}+\ldots),$$

which is to be interpreted by multiplication by $(1-e^{ix})$ and rearrangement according to powers of e^{ix}, as if the differentiated series were convergent. This rule will be seen in Chapter XI. to be more than an accidental coincidence [Arts. 103 (4) and 110 (2)].

It is easy to see that exactly the same method can be used, if necessary, to establish a similar rule to interpret the series for $f'(x)$ by using the factor $(1+e^{ix})$ or $(1\pm e^{irx})$, where r is an integer.*

Another process, which in practice is almost the same as the foregoing, is due to Stokes (*Math. and Phys. Papers*, vol. 1, pp. 256–260); the first case of Stokes's rule may be reduced to the form

$$\frac{d}{dx}(\Sigma a_ne^{inx})=i\Sigma(na_n-A)e^{inx}-Aie^{ix}/(e^{ix}-1),$$

where A is a constant determined by the condition that $\Sigma(na_n-A)e^{inx}$ is convergent; and another case can be written

$$\frac{d}{dx}(\Sigma a_ne^{inx})=i\Sigma[na_n+(-1)^nB]e^{inx}+Bie^{ix}/(e^{ix}+1),$$

* Methods substantially equivalent to this have been given by Lerch (*Ann. de l'École Normale Sup.* (3), t. 12, 1895, p. 351) and Brenke (*Annals of Mathematics* (2), vol. 8, 1907, p. 87). But the foregoing process seems simpler, both in practice and in principle.

if B is a constant which makes the series on the right convergent. These results of course depend on the fact that

$$\frac{d}{dx}[\Sigma(1/n)e^{inx}] = -\tfrac{1}{2}(i+\cot\tfrac{1}{2}x) = -ie^{ix}/(e^{ix}-1)$$

in virtue of Art. 65 (see also Ex. 1, below).

Ex. 1. Consider $\qquad f(x)=e^{ix}+\frac{1}{2}e^{2ix}+\frac{1}{3}e^{3ix}+\dots.$

This gives the *formal* equation

$$f'(x)=i(e^{ix}+e^{2ix}+e^{3ix}+\dots),$$

which leads to the *real* equation

$$(1-e^{ix})f'(x)=ie^{ix}$$

or
$$f'(x)=-\tfrac{1}{2}(i+\cot\tfrac{1}{2}x), \quad 0<x<2\pi.$$

Thus
$$f(x)=-\log(\sin\tfrac{1}{2}x)-\tfrac{1}{2}ix+\text{const.}$$

But for $x=\pi$, we find $\quad f(\pi)=-\log 2,$

so that
$$f(x)=-\log(2\sin\tfrac{1}{2}x)+\tfrac{1}{2}i(\pi-x), \quad 0<x<2\pi,$$

in agreement with Art. 65.

Ex. 2. Similarly we can prove that

$$e^{ix}+\tfrac{1}{3}e^{3ix}+\tfrac{1}{5}e^{5ix}+\dots=-\tfrac{1}{2}\log(\tan\tfrac{1}{2}x)+\tfrac{1}{4}\pi i, \qquad 0<x<\pi,$$

and
$$e^{ix}-\tfrac{1}{3}e^{3ix}+\tfrac{1}{5}e^{5ix}-\dots=\tfrac{1}{4}\pi+\tfrac{1}{2}i\log(\sec x+\tan x), \qquad -\tfrac{1}{2}\pi<x<\tfrac{1}{2}\pi.$$

These give, if $0<x<\frac{1}{2}\pi$,

$$\cos x+\tfrac{1}{5}\cos 5x+\tfrac{1}{9}\cos 9x+\dots=\tfrac{1}{8}\pi-\tfrac{1}{4}\log(\tan\tfrac{1}{2}x),$$
$$\sin x+\tfrac{1}{5}\sin 5x+\tfrac{1}{9}\sin 9x+\dots=\tfrac{1}{8}\pi+\tfrac{1}{4}\log(\sec x+\tan x).$$

Ex. 3. Again consider $\qquad f(x)=\sum_{-\infty}^{\infty}\frac{e^{inx}}{t-n}.$

Here we get the *formal* equation

$$f'(x)-itf(x)=-i\sum_{-\infty}^{\infty}e^{inx},$$

which gives the *real* equation

$$[f'(x)-itf(x)](1-e^{ix})=0.$$

Thus $f(x)=Ae^{itx}$, where A does not involve x, and x ranges from 0 to 2π. Putting $x=\pi$ we obtain the result

$$Ae^{i\pi t}=\sum_{-\infty}^{\infty}\frac{(-1)^n}{t-n}=\frac{1}{t}-\frac{1}{t-1}+\frac{1}{t-2}-\frac{1}{t-3}+\dots$$
$$-\frac{1}{t+1}+\frac{1}{t+2}-\frac{1}{t+3}+\dots$$

or
$$Ae^{i\pi t}=\frac{1}{t}-\frac{2t}{t^2-1}+\frac{2t}{t^2-2^2}-\dots=\pi\operatorname{cosec}(\pi t) \text{ (see Art. 92).}$$

That is,
$$A=\frac{2\pi i}{e^{2\pi it}-1},$$

so that
$$\frac{2\pi ie^{itx}}{e^{2\pi it}-1}=\sum_{-\infty}^{\infty}\frac{e^{inx}}{t-n}, \quad (0<x<2\pi).$$

Thus, by dividing by e^{itx}, we get

$$\pi\cot(\pi t)=\sum_{-\infty}^{\infty}\frac{\cos(t-n)x}{t-n},\quad \pi=\sum_{-\infty}^{\infty}\frac{\sin(t-n)x}{t-n}.$$

91. The infinite products for $\sin x$ **and** $\cos x$.

The identities of Art. 69 remain true for complex values of x, and we deduce, as in Art. 70,

$$\frac{\sin x}{n\sin(x/n)}=\prod_{r=1}^{\frac{1}{2}(n-1)}\left[1-\frac{\sin^2(x/n)}{\sin^2(r\pi/n)}\right].$$

Now, since n is to tend to ∞, we can always ensure that n is greater than $|x|$, and so Art. 86 gives

$$|\sin(x/n)|<\tfrac{6}{5}|x/n|;$$

and, since $r<\frac{1}{2}n$, $\quad\sin(r\pi/n)>2r/n$

(see footnote, p. 184).

Hence $$\left|\frac{\sin^2(x/n)}{\sin^2(r\pi/n)}\right|<\frac{9}{25}\frac{|x|^2}{r^2},\quad \text{if } n>|x|;$$

and consequently we can take

$$M_r=\frac{9}{25}\frac{|x|^2}{r^2}$$

in the theorem of Art. 49. Hence, as in Art. 70, we find

$$\frac{\sin x}{x}=\prod_{r=1}^{\infty}\left(1-\frac{x^2}{r^2\pi^2}\right)=\prod_{-\infty}^{\infty}{}'\left(1-\frac{x}{r\pi}\right)e^{\frac{x}{r\pi}}.$$

In the same way we find

$$\cos x=\prod_{r=1}^{\infty}\left[1-\frac{4x^2}{(2r-1)^2\pi^2}\right].$$

92. The series of fractions for $\cot x$, $\tan x$, $\operatorname{cosec} x$.

The investigation given in Art. 71 for real angles, can be extended without difficulty to a complex argument, by making the following modifications:

We have, of course, the identity

$$\cot x=\frac{1}{n}\cot\frac{x}{n}-\sum_{r=1}^{\frac{1}{2}(n-1)}\frac{2}{n}\frac{\sin(x/n)\cos(x/n)}{\sin^2(r\pi/n)-\sin^2(x/n)},$$

since this identity is merely an algebraical deduction from the trigonometrical addition-theorems, which are true, whether the arguments are real or complex.

Now here, as in Art. 91, we have

$$\sin(r\pi/n) > 2r/n, \quad |\sin(x/n)| < \tfrac{6}{5}|x|/n,$$

assuming that $n > |x|$.

Also we have (Art. 86)

$$|\cos(x/n)| < 2.$$

Hence

$$|2n\sin(x/n)\cos(x/n)| < \tfrac{24}{5}|x|$$

and

$$n^2|\sin^2(r\pi/n) - \sin^2(x/n)| > 4r^2 - \tfrac{36}{25}|x|^2.$$

Thus, provided that $5r > 3|x|$, we have

$$\left|\frac{2\sin(x/n)\cos(x/n)}{n\{\sin^2(r\pi/n) - \sin^2(x/n)\}}\right| < \frac{30|x|}{25r^2 - 9|x|^2}.$$

Consequently we can use the comparison-series

$$\sum^{\infty} \frac{30|x|}{25r^2 - 9|x|^2}, \quad (r > \tfrac{3}{5}|x|)$$

and so the theorem of Art. 49 can be applied just as in Art. 71; further,

$$\lim\left(n\sin\frac{x}{n}\right) = x, \quad \lim\left(\cos\frac{x}{n}\right) = 1,$$

exactly as if x were *real* instead of complex.*

Hence

$$\lim\left(\frac{1}{n}\cot\frac{x}{n}\right) = \frac{1}{x}$$

and

$$\lim\frac{2}{n}\,\frac{\sin(x/n)\cos(x/n)}{\sin^2(r\pi/n) - \sin^2(x/n)} = \frac{2x}{r^2\pi^2 - x^2}.$$

Thus we have, for all values of x, real or complex (except multiples of π),

$$\cot x = \frac{1}{x} + \sum_1^{\infty} \frac{2x}{x^2 - n^2\pi^2} = \frac{1}{x} + {\sum_{-\infty}^{\infty}}' \left(\frac{1}{x - n\pi} + \frac{1}{n\pi}\right),$$

where n is now the variable of summation, instead of r.

Now the following identities hold:†

$$\tan x = \cot x - 2\cot 2x,$$

$$\operatorname{cosec} x = \cot\tfrac{1}{2}x - \cot x.$$

---

* These statements follow at once from the power-series for the sine and cosine.

† The identities are familiar results when x is real; for other values, they follow from the formulae obtained in Art. 86.

Hence we find, on subtraction,

$$\tan x = \sum_1^\infty \frac{8x}{(2n-1)^2\pi^2 - 4x^2} = \sum_1^\infty \frac{2x}{(n-\frac{1}{2})^2\pi^2 - x^2},$$

$$\operatorname{cosec} x = \frac{1}{x} + \sum_1^\infty (-1)^n \frac{2x}{x^2 - n^2\pi^2}.$$

Since $$\cot x = \frac{\cos x}{\sin x} = i\frac{e^{ix}+e^{-ix}}{e^{ix}-e^{-ix}} = i\frac{e^{2ix}+1}{e^{2ix}-1},$$

by writing $y = 2ix$, we see that

$$\frac{e^y+1}{e^y-1} = \frac{2}{y} + \sum_1^\infty \frac{4y}{y^2+4n^2\pi^2}$$

or $$\frac{1}{e^y-1} = \frac{1}{2}\left(\frac{e^y+1}{e^y-1} - 1\right) = \frac{1}{y} - \frac{1}{2} + \sum_1^\infty \frac{2y}{y^2+4n^2\pi^2}.$$

93. The power-series for $x/(e^x-1.)$

The exponential series gives at once

$$(e^x-1)/x = 1 + \frac{x}{2!} + \frac{x^2}{3!} + \dots,$$

and consequently (as in Art. 54) the reciprocal function $x/(e^x-1)$ can be expanded in powers of x, provided that $|x| < \rho$, where

$$\frac{\rho}{2!} + \frac{\rho^2}{3!} + \dots \leqq 1.$$

This last condition is certainly satisfied by taking

$$\frac{\rho}{2}\bigg/\left(1-\frac{\rho}{3}\right) = 1$$

or by taking $$\rho = \tfrac{6}{5} = 1{\cdot}2.$$

Thus we can write

$$\frac{x}{e^x-1} = 1 - \frac{x}{2} + A_2x^2 + A_3x^3 + A_4x^4 + \dots, \quad \text{if } |x| < 1{\cdot}2.$$

By changing the sign of x, we find

$$\frac{x}{1-e^{-x}} = 1 + \frac{x}{2} + A_2x^2 - A_3x^3 + A_4x^4 - \dots.$$

If we subtract the first of these expansions from the second, we obtain the identity

$$x = x - 2A_3x^3 - 2A_5x^5 - \dots,$$

so that $$A_3 = 0,\ A_5 = 0,\ A_7 = 0, \dots.$$

Consequently we can write

$$\frac{x}{e^x-1}=1-\frac{x}{2}+B_1\frac{x^2}{2!}-B_2\frac{x^4}{4!}+B_3\frac{x^6}{6!}-\dots$$

where $B_1, B_2, B_3, \dots$ are *Bernoulli's numbers.*

It is easy to verify by direct division that

$$B_1=\tfrac{1}{6},\ B_2=\tfrac{1}{30},\ B_3=\tfrac{1}{42},\ B_4=\tfrac{1}{30},\ B_5=\tfrac{5}{66},$$

but the higher numbers become very complicated.*

Again, from the last Article we see that

$$\frac{1}{e^x-1}=\frac{1}{x}-\frac{1}{2}+\sum_1^\infty\frac{2x}{x^2+4n^2\pi^2}.$$

Now if $|x|<2\pi$, each fraction can be expanded in powers of x, giving

$$\frac{2x}{x^2+4n^2\pi^2}=\frac{x}{2n^2\pi^2}\left[1-\frac{x^2}{4n^2\pi^2}+\frac{x^4}{16n^4\pi^4}-\dots\right].$$

And the resulting double series is absolutely convergent, since the series of absolute values is obtained by expanding the convergent series

$$\sum_1^\infty\frac{2|x|}{4n^2\pi^2-|x|^2}.$$

It is therefore permissible to arrange the double series in powers of x, and then we obtain

$$\frac{1}{e^x-1}=\frac{1}{x}-\frac{1}{2}+\frac{x}{2\pi^2}\left(\sum_1^\infty\frac{1}{n^2}\right)-\frac{x^3}{2^3\pi^4}\left(\sum_1^\infty\frac{1}{n^4}\right)+\frac{x^5}{2^5\pi^6}\left(\sum_1^\infty\frac{1}{n^6}\right)-\dots,$$

which is now seen to be valid for $|x|<2\pi$.†

By comparison with the former expression, we see that

$$B_1=\frac{1}{\pi^2}\sum\frac{1}{n^2},\quad B_2=\frac{3}{\pi^4}\sum\frac{1}{n^4},\quad B_3=\frac{45}{2\pi^6}\sum\frac{1}{n^6},$$

and generally

$$B_r=\frac{(2r)!}{2^{2r-1}\pi^{2r}}\sum\frac{1}{n^{2r}}.$$

---

* The numbers (as decimals) and their logarithms have been tabulated by Glaisher (*Trans. Camb. Phil. Soc.*, vol. 12, p. 384); and B_1 to B_{62} are given by Adams (*Scientific Papers*, vol. 1, pp. 453 and 455). (For more details, see Chrystal's *Algebra*, Ch. XXVIII. § 6.)

† That 2π is the radius of convergence may be seen from Baker's theorem (Art. 84); for the zeros of e^x-1 are $x=2n\pi i$, and the least distance of any of these from the origin is 2π.

We obtain thus the results

$$\sum \frac{1}{n^2} = \frac{\pi^2}{6} \text{ (Art. 71)}, \quad \sum \frac{1}{n^4} = \frac{\pi^4}{90}, \quad \sum \frac{1}{n^6} = \frac{\pi^6}{945}, \quad \sum \frac{1}{n^8} = \frac{\pi^8}{9450}.$$

It is instructive to notice that, when x is *real*, we have (for *any* value of x)

$$\frac{2x}{x^2+4n^2\pi^2} = \frac{x}{2n^2\pi^2}\left[1 - \frac{x^2}{4n^2\pi^2} + \dots + (-1)^r \left(\frac{x^2}{4n^2\pi^2}\right)^r \right.$$
$$\left. + (-1)^{r+1} \left(\frac{x^2}{4n^2\pi^2}\right)^r \frac{x^2}{x^2+4n^2\pi^2}\right].$$

Thus, by addition, we see that $\frac{1}{e^x-1}$ is represented by the first $(r+3)$ terms of the series with an error which is less than the following term of the series.

For instance, for any real positive value of x, we have

$$0 < \frac{1}{e^x-1} - \left(\frac{1}{x} - \frac{1}{2}\right) < \frac{x}{12},$$

$$0 > \frac{1}{e^x-1} - \left(\frac{1}{x} - \frac{1}{2} + \frac{x}{12}\right) > -\frac{x^3}{720},$$

and so on.

Ex. By means of the identity

$$\frac{1}{e^x-1} - \frac{2}{e^{2x}-1} = \frac{1}{e^x+1},$$

we can shew that

$$\frac{1}{e^x+1} = \frac{1}{2} - B_1(2^2-1)\frac{x}{2!} + B_2(2^4-1)\frac{x^3}{4!} - B_3(2^6-1)\frac{x^5}{6!} + \dots$$

and
$$\frac{e^x}{e^x+1} = \frac{1}{2} + B_1(2^2-1)\frac{x}{2!} - B_2(2^4-1)\frac{x^3}{4!} + B_3(2^6-1)\frac{x^5}{6!} - \dots.$$

94. Bernoullian functions.

The Bernoullian function of degree n, denoted by $\phi_n(x)$, is the coefficient of $t^n/n!$ in the expansion of

$$t\frac{e^{xt}-1}{e^t-1},$$

which, by the foregoing, can be expanded in powers of t if $|t| < 2\pi$.

Thus we have

$$\Sigma\phi_n(x)\frac{t^n}{n!}=\left(xt+\frac{x^2t^2}{2!}+\frac{x^3t^3}{3!}+\dots\right)\left(1-\frac{1}{2}t+B_1\frac{t^2}{2!}-B_2\frac{t^4}{4!}+\dots\right),$$

so that $\phi_n(x)$

$$=x^n-\frac{n}{2}x^{n-1}+\frac{n(n-1)}{2!}B_1x^{n-2}-\frac{n(n-1)(n-2)(n-3)}{4!}B_2x^{n-4}+\dots,$$

where the polynomial terminates with either x or x^2.

From this formula, or by direct multiplication, we find that the first six polynomials are:

$$\begin{aligned}
\phi_1(x)&=x,\\
\phi_2(x)&=x^2-x=y,\\
\phi_3(x)&=x^3-\tfrac{3}{2}x^2+\tfrac{1}{2}x=yz,\\
\phi_4(x)&=x^4-2x^3+x^2=y^2,\\
\phi_5(x)&=x^5-\tfrac{5}{2}x^4+\tfrac{5}{3}x^3-\tfrac{1}{6}x=yz(y-\tfrac{1}{3}),\\
\phi_6(x)&=x^6-3x^5+\tfrac{5}{2}x^4-\tfrac{1}{2}x^2=y^2(y-\tfrac{1}{2}),
\end{aligned}$$

where
$$y=x(x-1),\quad z=x-\tfrac{1}{2}=\tfrac{1}{2}\frac{dy}{dx}.$$

Again, $\phi_n(x+1)-\phi_n(x)$ is the coefficient of $t^n/n!$ in the expansion of

$$\frac{t}{e^t-1}[e^{(1+x)t}-e^{xt}]=te^{xt},$$

so that
$$\phi_n(x+1)-\phi_n(x)=nx^{n-1}.$$

If we write $x=1, 2, 3, \dots$ in the last equation and add the results, we see that, *if x is any positive integer,*

$$1+2^{n-1}+3^{n-1}+\dots+x^{n-1}=\frac{1}{n}\phi_n(x+1),$$

which gives one application of the polynomials.

Further, by differentiation we see that $\phi_n'(x)$ is the coefficient of $t^n/n!$ in the expansion of

$$\frac{t^2e^{xt}}{e^t-1}=t\left[\frac{t(e^{xt}-1)}{e^t-1}+\frac{t}{e^t-1}\right].$$

Hence we find

$$\phi_2'(x)=2[\phi_1(x)-\tfrac{1}{2}],\qquad \phi_3'(x)=3[\phi_2(x)+B_1],$$
$$\phi_4'(x)=4\phi_3(x),\qquad \phi_5'(x)=5[\phi_4(x)-B_2],$$

and generally $\phi'_{2m}(x)=2m\phi_{2m-1}(x)$, $(m>1)$

$$\phi'_{2m+1}(x)=(2m+1)[\phi_{2m}(x)+(-1)^{m-1}B_m], \quad (m\geqq 1).$$

If we change x to $1-x$ and t to $-t$, we see that

$$\sum_1^\infty \phi_n(1-x)\frac{(-t)^n}{n!}=t\frac{e^{xt}-e^t}{e^t-1}=\sum_1^\infty \phi_n(x)\frac{t^n}{n!}-t.$$

Thus $\qquad \phi_n(1-x)=(-1)^n\phi_n(x), \quad (n>1).$

Since $\phi_n(0)=0$, it follows from the last equation that $x=1$ is a root of $\phi_n(x)$ and that $x=\frac{1}{2}$ is a root of $\phi_{2m+1}(x)$; and a glance at the five functions $\phi_2, \ldots, \phi_6$ leads to the conjecture that $\phi_{2m}(x)$ *has no root between* 0 *and* 1, *while* $\phi_{2m+1}(x)$ *has only the root* $\frac{1}{2}$.

Suppose that this conjecture has been established for all values of m up to, say, μ: then since

$$\phi'_{2\mu+2}(x)=(2\mu+2)\phi_{2\mu+1}(x),$$

$\phi_{2\mu+2}(x)$ is numerically greatest at $x=\frac{1}{2}$, and cannot vanish between 0 and 1.

Consequently $\phi'_{2\mu+3}(x)=(2\mu+3)\phi_{2\mu+2}(x)+\text{const.}$ can change sign once, *at most*, between $x=0$ and $\frac{1}{2}$. Hence $\phi_{2\mu+3}(x)$ can have no zero between $x=0$ and $\frac{1}{2}$, and therefore none between $\frac{1}{2}$ and 1. Thus the theorem has been extended to the value $m=\mu+1$; and it is accordingly always true. The diagram indicates the relations between $\phi_2(x)$, $\phi_3(x)$, $\phi_4(x)$, $\phi_5(x)$, and so illustrates the general argument.

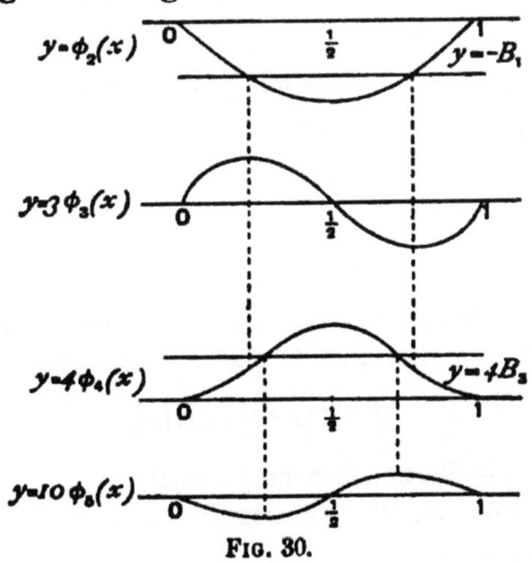

FIG. 30.

It will be seen that $\phi_{2m}(x)$ *contains the factors* $x^2(x-1)^2$ $(m>1)$, *because* $\phi'_{2m}(x)=0$, *for* $x=0, 1$; *and that the sign of* $\phi_{2m}(x)$ *is the same as that of* $(-1)^{m-1}$, *in the interval* $(0, 1)$.

95. Euler's summation formula.

We have seen in Art. 94 that if x and n are positive integers,

$$1+2^{n-1}+3^{n-1}+\ldots+(x-1)^{n-1}=\frac{1}{n}\phi_n(x)$$

$$=\frac{1}{n}x^n-\tfrac{1}{2}x^{n-1}+\frac{n-1}{2!}B_1x^{n-2}-\ldots,$$

the polynomial containing $\frac{1}{2}(n+2)$ or $\frac{1}{2}(n+3)$ terms.

It is obvious that if $f(x)$ is a polynomial in x, we can obtain the value of the sum

$$f(1)+f(2)+\ldots+f(x-1)$$

by the addition of suitable multiples of the Bernoullian functions of proper degrees. But to obtain a single formula, we must utilise the Calculus; and so we observe that we can write the foregoing expression in the form

$$\frac{1}{n}\phi_n(x)=\int x^{n-1}dx-\tfrac{1}{2}x^{n-1}+\frac{1}{2!}B_1\frac{d}{dx}(x^{n-1})-\frac{1}{4!}B_2\frac{d^3}{dx^3}(x^{n-1})+\ldots.$$

Hence *when $f(x)$ is a polynomial, we have Euler's formula*,

$$f(1)+f(2)+\ldots+f(x-1)$$

$$=\int f(x)dx-\tfrac{1}{2}f(x)+\frac{1}{2!}B_1f'(x)-\frac{1}{4!}B_2f'''(x)+\ldots,$$

where there is no term on the right-hand side (in its final form) which is not divisible by x.

However, the most interesting applications of this formula arise when $f(x)$ is a rational, algebraic, or transcendental function, and then of course the foregoing method of proof cannot be used; and the right-hand side becomes an infinite series which may not converge.

To obtain the formula in these cases, we shall apply the method of integration by parts.* Consider in fact the integral

$$\int_0^1 f^{2n}(x+t)\phi_{2n}(t)dt,$$

where ϕ denotes the Bernoullian polynomial.

Remembering that $\phi_{2n}(t)$ vanishes for $t=0$ and $t=1$, we find that this integral is equal to

$$-\int_0^1 f^{2n-1}(x+t)\phi'_{2n}(t)dt=-2n\int_0^1 f^{2n-1}(x+t)\phi_{2n-1}(t)dt.$$

* Seliwanoff, *Differenzenrechnung*, §§ 38, 39.

Integrating by parts again, we obtain similarly

$$2n(2n-1)\int_0^1 f^{2n-2}(x+t)[\phi_{2n-2}(t)+(-1)^n B_{n-1}]dt.$$

Hence if we write

$$X_n = \frac{1}{(2n)!}\int_0^1 f^{2n}(x+t)\phi_{2n}(t)dt,$$

we have the result

$$X_n - X_{n-1} = (-1)^n \frac{B_{n-1}}{(2n-2)!}[f^{2n-3}(x+1) - f^{2n-3}(x)], \quad (n>1).$$

Let us now examine the case $n=1$; we have then

$$X_1 = \frac{1}{2!}\int_0^1 f''(x+t)\phi_2(t)dt.$$

Thus, since $\phi_2(t) = t^2 - t$, we have

$$\begin{aligned} X_1 &= -\int_0^1 f'(x+t)(t-\tfrac{1}{2})dt \\ &= -\tfrac{1}{2}[f(x+1)+f(x)] + \int_0^1 f(x+t)dt. \end{aligned}$$

That is,

$$\tfrac{1}{2}[f(x)+f(x+1)] = \int_x^{x+1} f(t)dt - X_1$$

and

$$X_2 - X_1 = +\frac{B_1}{2!}[f'(x+1) - f'(x)],$$

$$X_3 - X_2 = -\frac{B_2}{4!}[f'''(x+1) - f'''(x)],$$

etc.

$$\begin{aligned} \text{Thus } \tfrac{1}{2}[f(x)+f(x+1)] &= \int_x^{x+1} f(t)dt - X_1 \\ &= \int_x^{x+1} f(t)dt + \frac{B_1}{2!}[f'(x+1)-f'(x)] - X_2 \\ &= \int_x^{x+1} f(t)dt + \frac{B_1}{2!}[f'(x+1)-f'(x)] \\ &\quad - \frac{B_2}{4!}[f'''(x+1)-f'''(x)] - X_3, \end{aligned}$$

and so on.

Let x take a succession of values $a, a+1, a+2, \ldots, b-1$, and add the corresponding equations. Then we have, if we stop at X_3 for instance,

$$\begin{aligned} f(a)+f(a+1)+\ldots+f(b-1)+f(b) &= \int_a^b f(t)dt + \tfrac{1}{2}[f(b)+f(a)] \\ &\quad + \frac{B_1}{2!}[f'(b)-f'(a)] - \frac{B_2}{4!}[f'''(b)-f'''(a)] \\ &\quad - \frac{1}{6!}\int_0^1 \phi_6(t)[f^6(a+t)+\ldots+f^6(b-1+t)]dt. \end{aligned}$$

And obviously we can introduce as many terms as we please on the right-hand side.

For a further discussion of Euler's formula, see Chapter XI. Art. 131.

Ex. 1. If $\psi_n(x)$ is the coefficient of $t^n/n!$ in the expansion of $e^{xt}/(e^t+1)$, prove that

$$\psi_n(x)+\psi_n(x+1)=x^n,$$

and if x is a positive integer,

$$-\psi_n(0)\pm\psi_n(x)=1^n-2^n+3^n-\ldots\mp(x-1)^n.$$

Deduce that if $f(x)$ is a polynomial in x,

$$f(1)-f(2)+f(3)-f(4)+\ldots+(-1)^{x-1}f(x)$$

$$=(-1)^{x-1}\left[\tfrac{1}{2}f(x)+\frac{2^2-1}{2!}B_1f'(x)-\frac{2^4-1}{4!}B_2f'''(x)+\frac{2^6-1}{6!}B_3f^5(x)-\ldots\right]+\text{const.}$$

Ex. 2. As particular cases of Ex. 1, we find

$$\psi_1(x)=\tfrac{1}{2}x-\tfrac{1}{4},\quad \psi_2(x)=\tfrac{1}{2}x(x-1),$$

$$\psi_3(x)=\tfrac{1}{2}x^3-\tfrac{3}{4}x^2+\tfrac{1}{8},\quad \psi_4(x)=\tfrac{1}{2}x^4-x^3+\tfrac{1}{2}x.$$

These give, when x is a positive integer,

$$1-2+3-4+\ldots+(-1)^{x-1}x=-\tfrac{1}{2}x \text{ or } +\tfrac{1}{2}(x+1)$$

$$1-2^2+3^2-4^2+\ldots+(-1)^{x-1}x^2=(-1)^{x-1}\tfrac{1}{2}(x^2+x),$$

$$1-2^3+3^3-4^3+\ldots+(-1)^{x-1}x^3=-(\tfrac{1}{2}x^3+\tfrac{3}{4}x^2) \text{ or } +(\tfrac{1}{2}x^3+\tfrac{3}{4}x^2-\tfrac{1}{4}),$$

$$1-2^4+3^4-4^4+\ldots+(-1)^{x-1}x^4=(-1)^{x-1}(\tfrac{1}{2}x^4+x^3-\tfrac{1}{2}x),$$

and so on. In the first and third cases the alternatives are to be chosen according as x is even or odd.

Ex. 3. It is easy to see that

$$\psi_n'(x)=n\psi_{n-1}(x),\quad \psi_n(1-x)=(-1)^n\psi_n(x),$$

$$\psi_{2n-1}(0)=-\psi_{2n-1}(1)=(-1)^nB_n\frac{2^{2n}-1}{2n},\quad \psi_{2n}(0)=\psi_{2n}(1)=0.$$

From the foregoing equations and from those of Ex. 3 prove that

$$\psi_5(x)=\tfrac{1}{2}x^5-\tfrac{5}{4}x^4+\tfrac{5}{4}x^2-\tfrac{1}{4},\quad \psi_6(x)=\tfrac{1}{2}x^6-\tfrac{3}{2}x^5+\tfrac{5}{2}x^3-\tfrac{3}{2}x.$$

Shew also that $(x-\tfrac{1}{2})$ is a factor of the odd polynomials, and $x(x-1)$ of the even polynomials.

Ex. 4. Prove that if n is odd and k is an integer,

$$\sum_{r=0}^{k-1}\phi_n(x+r/k)=\phi_n(kx)/k^{n-1},$$

and obtain the corresponding result when n is even.

EXAMPLES A.

Complex Numbers.

1. If the numbers a, x are both complex, shew that when the points a^x are marked in Argand's diagram, they lie on an equiangular spiral whose angle depends only on x and not on a. [*Math. Trip.* 1899.]

Examine the special cases when x is (1) real, (2) pure imaginary; and in particular, if $x=i$ and a is real, prove that if $b=a^i$,

$$|b+1/b|=\sqrt{[2\{\cosh(4k\pi)+\cos 2(\log a)\}]},$$

where k is an arbitrary integer.

2. If x and y are complex, prove that

$$|x+y|^2+|x-y|^2=2\{|x|^2+|y|^2\},$$

and interpret this equation in Argand's diagram. Deduce that

$$|x+y|+|x-y|=|x+\sqrt{(x^2-y^2)}|+|x-\sqrt{(x^2-y^2)}|.$$

[HARKNESS and MORLEY.]

3. If A, B are the points in Argand's diagram which represent the roots of $ax^2+2bx+c=0$, and A', B' represent the roots of $a'x^2+2b'x+c'=0$, shew that the condition $ac'+a'c-2bb'=0$ is equivalent to the conditions

$$OA^2=OA'.OB',\quad A'\hat{O}A=A\hat{O}B',$$

where O is the mid point of AB. [*Math. Trip.* 1901.]

[Transfer to O as origin, which gives $b=0$.]

4. If $|\cos x|=1$, where $x=\xi+i\eta$, shew that $\sinh\eta=\pm\sin\xi$; and that if we write

$$\cos x=\cos\theta+i\sin\theta,$$

where θ is real, then $\sin\theta=\pm\sin^2\xi$.

5. Shew in a diagram the roots of the equation $32x^5=(x+1)^5$, and prove that they are concyclic.

6. If the equation

$$a_0x^4+4a_1x^3+6a_2x^2+4a_3x+a_4=0$$

has real coefficients, and if its roots in Argand's diagram are concyclic (two being real and two complex), then

$$a_0a_2a_4+2a_1a_2a_3-a_0a_3^2-a_1^2a_4-a_2^3=0.$$

7. If a, b, c are complex, α, β, γ real constants, the point

$$x=\frac{at^2+2bt+c}{\alpha t^2+2\beta t+\gamma}$$

traces a conic or straight line, when t takes all real values. Examine the conditions for the various cases. [STOLZ und GMEINER.]

8. If t represents a complex number such that $|t|=1$, shew that as t varies, the point

$$x=\frac{at+b}{t-c}$$

describes a circle, unless $|c|=1$, when it moves along a straight line.

[MORLEY.]

9. If t varies so that $|t|=1$, shew that the point
$$2x=at+b/t$$
in general describes an ellipse whose axes are $|a|+|b|$ and $|a|-|b|$, and whose foci are given by $x^2=ab$.

If $|a|=|b|$, prove that the point x traces out the portion of a straight line which is terminated by the two points $x^2=ab$.

10. If t varies so that $|t|=1$, prove that the point
$$x=at^2+2bt+c$$
in general describes a limaçon, whose focus is $c-b^2/a$. Find the node; and if $|a|=|b|$, shew that the limaçon reduces to a cardioid. [MORLEY.]

11. If t varies so that $|t|=1$, shew that the point
$$x=\frac{a}{t-1}+\frac{b}{t+1}+c$$
in general describes a hyperbola, and find its asymptotes. Under what conditions is the origin (1) the centre, (2) a focus of the curve?

Prove also that the point $x=\frac{a}{(t-1)^2}+\frac{b}{t-1}+c$ describes a parabola.

12. Constructions for trisecting an angle.

If $a=\cos\alpha+i\sin\alpha$, the determination of $\frac{1}{3}\alpha$ is equivalent to the solution of the equation in t, $\qquad t^3=a.$

To effect this geometrically we use the intersections of a conic with the circle $|t|=1$; the form of the conic is largely arbitrary, but we shall give three typical constructions, the first and second of which, at any rate, were known to the later Greek geometers (*e.g.* Pappus).

(1) *A rectangular hyperbola.*

If we write our equation in the form
$$t^2=a/t,$$
and then put $t=\xi+i\eta$, $1/t=\xi-i\eta$, we find that the points trisecting the angle are given by three of the intersections with the circle $\xi^2+\eta^2=1$ of the two rectangular hyperbolas
$$\xi^2-\eta^2-(\xi\cos\alpha+\eta\sin\alpha)=0,\quad 2\xi\eta-\xi\sin\alpha+\eta\cos\alpha=0.$$
Of course the fourth intersection of the hyperbolas is the origin and so is not on the circle.

Either of these hyperbolas solves the problem, but the second is the easier to construct; its asymptotes are parallel to the axes (the one axis being an arm of the angle to be trisected), its centre is the point $(-\frac{1}{2}\cos\alpha,\ \frac{1}{2}\sin\alpha)$, and it passes through the centre of the circle (that is, the vertex of the angle to be trisected). Since a hyperbola is determined by its asymptotes and a point on the curve, we can now construct the hyperbola.

(2) *A hyperbola of eccentricity* 2.

The first hyperbola in (1) cuts the circle $\xi^2+\eta^2=1$ in the same points as the hyperbola $\xi^2-3\eta^2-2(\xi\cos\alpha+\eta\sin\alpha)+1=0$.

This hyperbola has eccentricity 2, and one focus at $(\cos a, \sin a)$, and $\eta=0$ is the corresponding directrix; $(-\cos a, \sin a)$ is the vertex on the other branch of the curve. From the present point of view, this hyperbola presents itself less naturally than those given in (1); but the reverse is the case if we use geometrical properties of conics, and this was of course the method used by the Greeks.

(3) *A parabola.*

Again, we find that the first hyperbola of (1) cuts the circle $\xi^2+\eta^2=1$ in the same points as the parabola

$$2\eta^2+\xi \cos a+\eta \sin a-1=0.$$

This parabola has its axis parallel to $\eta=0$, passes through the points

$$(\cos a, \tfrac{1}{2}\sin a), \quad (\cos a, -\tfrac{1}{2}\sin a),$$

and touches the line $\xi \cos a+\eta \sin a-1=0$ at the point $(\sec a, 0)$.

13. If $x=\exp(2\pi i/a)$ and $X=\sum_{n=0}^{a-1} x^{n^2}$, shew that

$$\text{(i) } a=7 \text{ gives } X=i\sqrt{7},$$
$$\text{(ii) } a=11 \text{ ...... } X=i\sqrt{11},$$
$$\text{(iii) } a=13 \text{ ...... } X=\sqrt{13}.$$

[Taking case (i), we find at once that $X=1+2S$, where

$$S=x+x^4+x^2, \quad S'=x^6+x^3+x^5.$$

It is easily proved that $S+S'=-1$, since $(x^7-1)/(x-1)=0$, and

$$SS'=3+S+S'=2.$$

Thus S is a root of $S^2+S+2=0$, which gives

$$X^2=-7, \text{ or } X=\pm i\sqrt{7}.$$

It is easily proved by considering

$$\sin(2\pi/7)+\sin(8\pi/7)+\sin(4\pi/7)$$

that the sign must be $+$; compare Ex. 5, p. 188.

In like manner we deal with case (ii).

In case (iii) we write again $X=1+2S$, where now

$$S=x+x^4+x^9+x^3+x^{12}+x^{10},$$
$$S'=x^2+x^8+x^5+x^6+x^{11}+x^7.$$

Here again $S+S'=-1$, but $SS'=3(S+S')=-3$.

Thus $S^2+S=3$ and $X^2=13$. That S (and therefore X) must be positive is obvious by considering that

$$\tfrac{1}{2}S=\cos(2\pi/13)+\cos(8\pi/13)+\cos(6\pi/13),$$

in which the only negative term is the second, and that term is less than the first (in numerical value).]

14. With the same notation as in the last example, shew that

$$\text{(i) } a=4 \text{ gives } X=(1+i)2,$$
$$\text{(ii) } a=8 \text{ ...... } X=(1+i)2\sqrt{2},$$
$$\text{(iii) } a=12 \text{ ...... } X=(1+i)2\sqrt{3}.$$

[In the first case we have $x=i$.

In the second case, we have $x^2=i$, $x=(1+i)/\sqrt{2}$.

In the third case, $x^3=i$, $x=\tfrac{1}{2}(\sqrt{3}+i)$.]

15. The value of the more general sum $y=\sum_{n=0}^{a-1} x^{bn^2}$, where b is an integer prime to a, can now be inferred in these special cases. We find, in fact,

$$\text{(i) } a=7, \quad \begin{cases} y = i\sqrt{7}, & \text{if } b=1,\ 2,\ 4, \\ \text{or } \;= -i\sqrt{7}, & \text{if } b=3,\ 5,\ 6. \end{cases}$$

$$\text{(ii) } a=11, \quad \begin{cases} y = i\sqrt{11}, & \text{if } b=1,\ 3,\ 4,\ 5,\ 9, \\ \text{or } \;= -i\sqrt{11}, & \text{if } b=2,\ 6,\ 7,\ 8,\ 10, \end{cases}$$

$$\text{(iii) } a=13, \quad \begin{cases} y = \sqrt{13}, & \text{if } b=1,\ 3,\ 4,\ 9,\ 10,\ 12, \\ \text{or } \;= -\sqrt{13}, & \text{if } b=2,\ 5,\ 6,\ 7,\ 8,\ 11. \end{cases}$$

$$\text{(iv) } a=8, \quad \begin{cases} y=(1+i)2\sqrt{2}, & \text{if } b=1, \\ \text{or } \;(-1+i)2\sqrt{2}, & \text{if } b=3, \\ \text{or } \;(-1-i)2\sqrt{2}, & \text{if } b=5, \\ \text{or } \;(1-i)2\sqrt{2}, & \text{if } b=7. \end{cases}$$

16. It will be seen from a consideration of the special cases discussed in Exs. 13, 14, that the set of values x^{-n^2} may, or may not, be equivalent to the set x^{n^2}. In the former case, a is of the form $4k+1$, where k is an integer; and the sum S consists of k pairs of terms, whose indices are complementary (that is, of the form ν, $a-\nu$). On multiplying SS' out, it is easily seen to be the same as $k(S+S')$.

Thus we find $\quad S^2+S-k=0 \;$ or $\; X^2=4k+1=a$.

Similarly, if a is of the form $4k+3$, we find that the terms x^{-n^2} belong to S', and then we find $\quad SS'=(2k+1)+k(S+S')=k+1$.

Thus $\quad S^2+S+k+1=0 \;$ or $\; X^2=-(4k+3)=-a$.

[*Math. Trip.* 1895.]

A general determination of the sign of X (and indeed a complete discussion of the distribution of indices between S and S') belongs to the problem of quadratic residues in the Theory of Numbers.*

17. When a is an even integer $a=2k$, where k is *odd*, we note that $x^{(n+k)^2}=-x^{n^2}$, so that X is identically zero.

When $a=4k$, the results of Ex. 14 suggest that $X=(1+i)\sqrt{a}$, but a complete proof of this requires some further discussion.†

18. Deduce from Ex. 13 that

$$\tan\frac{3\pi}{11}+4\sin\frac{2\pi}{11}=\sqrt{11}. \qquad \text{[\textit{Math. Trip.} 1895.]}$$

[In fact,

$$i\tan\frac{3\pi}{11}=\frac{x^3-1}{x^3+1}=\frac{x^3-x^{33}}{x^3+1}=x^3-x^6+x^9-x+x^4-x^7+x^{10}-x^2+x^5-x^8,$$

since $x^{11}=1$. Thus, in the notation of Ex. 13,

$$i\tan\frac{3\pi}{11}=S-S'-2(x-x^{10}).]$$

---

* For example, see Gauss, *Disq. Arithm.*, Art. 356; *Werke*, Bd. 1, p. 441; *Werke*, Bd. 2, p. 11; G. B. Mathews, *Theory of Numbers*, pt. 1, pp. 200-212; H. Weber, *Algebra*, Bd. 1, § 179; Dirichlet, *Zahlentheorie*, §§ 111-117.

† Gauss, *Werke*, Bd. 2, pp. 34-45.

19. The results of Ex. 13 lead to an easy geometrical construction for the regular heptagon inscribed in a circle. In fact, we see at once that x, x^2, x^4 are the roots of the cubic $t^3-St^2+S't-1=0$, or of

$$t^3-\tfrac{1}{2}t^2(i\sqrt{7}-1)-\tfrac{1}{2}t(i\sqrt{7}+1)-1=0,$$

so that $1, x, x^2, x^4$ are roots of

$$t^2+\frac{1}{t^2}-\frac{1}{2}\left(t+\frac{1}{t}\right)-1-\frac{1}{2}i\sqrt{7}\left(t-\frac{1}{t}\right)=0.$$

If we write $t=\xi+i\eta$, $1/t=\xi-i\eta$, we find from the last equation

$$2(\xi^2-\eta^2)-\xi+\eta\sqrt{7}-1=0,$$

which represents a rectangular hyperbola passing through the vertices $1, x, x^2, x^4$ of a regular heptagon inscribed in the circle $\xi^2+\eta^2=1$.

Another construction is given by either of the parabolas

$$4\xi^2-\xi+\eta\sqrt{7}-3=0, \quad 4\eta^2+\xi-\eta\sqrt{7}-1=0.$$

[*Oxford Sen. Schol.*, 1904.]

20. If $\omega=\pi/n$, shew that

$$\sum\cot(\theta+r\omega)=n\cot n\theta, \qquad (r=0, 1, 2, \dots n-1),$$

and deduce by differentiation that

$$\sum\operatorname{cosec}^2(\theta+r\omega)=n^2\operatorname{cosec}^2 n\theta,$$

$$\sum\cot(\theta+r\omega)\operatorname{cosec}^2(\theta+r\omega)=n^3\cot n\theta\operatorname{cosec}^2 n\theta.$$

Deduce that, if $\alpha=\frac{1}{4}\omega=\pi/4n$,

$$\cot\alpha\operatorname{cosec}^2\alpha-\cot 3\alpha\operatorname{cosec}^2 3\alpha+\cot 5\alpha\operatorname{cosec}^2 5\alpha-\dots \text{ to } n \text{ terms}$$

is equal to $2n^3$. [*Math. Trip.* 1901.]

21. If p is an odd integer and q is any integer prime to p, shew that

$$\sum_{1}^{\frac{1}{2}(p-1)}\sin(2\lambda n\theta)\cot(n\theta)=\tfrac{1}{2}p-\lambda,$$

where $\theta=\pi q/p$, and λ is any integer from 1 to $p-1$ (both included). Determine the value of the sum when λ is greater than p.

[Eisenstein and *Math. Trip.* 1897.]

[Write $t=e^{2i\theta}$, then from the theory of partial fractions

$$\frac{px^{\lambda-1}}{x^p-1}-\frac{1}{x-1}=\sum_{1}^{p-1}\frac{t^{n\lambda}}{x-t^n}, \quad \text{if } 1\leqq\lambda\leqq p.$$

Take the limit of both sides as $x\to 1$, and we get

$$\tfrac{1}{2}(p+1)-\lambda=\sum t^{n\lambda}/(t^n-1) \quad (n=1, 2, \dots p-1).$$

Also $\qquad -1=\sum t^{n\lambda}$, if we now suppose $\lambda<p$,

so that $\qquad \tfrac{1}{2}p-\lambda=\tfrac{1}{2}\sum_{1}^{p-1}t^{n\lambda}\dfrac{t^n+1}{t^n-1}.$]

22. Prove similarly that, with the same notation as in the last example,

$$\sum_{0}^{p-1}\frac{\cos k(\alpha+n\theta)}{\sin(\alpha+n\theta)}=p\cot p\alpha, \quad \sum_{0}^{p-1}\frac{\sin k(\alpha+n\theta)}{\sin(\alpha+n\theta)}=p,$$

where k is odd and not greater than $2p-1$, but p need not be odd.

[*Royal Univ. of Ireland*, 1900.]

[Write $\lambda=\frac{1}{2}(k+1)$, $x=e^{-2i\alpha}$ in the partial fractions used in Ex. 21.]

Convergence of complex sequences.

23. If (a_n) is a sequence of complex numbers, which converges to a as a limit, and if b is another complex number, shew that values of b^{a_n} can be selected so as to form a convergent sequence, whose limit is one of the values of b^a.

24. If x is real, prove that any value of x^i oscillates finitely both as x tends to 0 and to ∞.

25. A straight line can be drawn in the plane of the complex variable x, so that the series

$$1-x+\frac{x(x-1)}{2!}-\frac{x(x-1)(x-2)}{3!}+\dots$$

converges to 0 on one side of the line; and its *modulus* tends to infinity on the other side of the line. [*Math. Trip.* 1905.]

26. If an infinite set of points is taken within a square, the set has at least one limiting point (that is, a point in whose neighbourhood there is an infinity of points of the set).

[For if the square is subdivided into four by bisecting the sides, at least one of the four contains an infinity of points of the set; repeating this argument, there is an infinity within at least one square whose side is $a/2^n$, where a is the side of the original square, and n is any integer. It is then not difficult to see that we can select a sequence of squares, each within the preceding, and each containing an infinity of points of the set; the centres of these squares then define a sequence of points which can be proved to have a limiting point. Finally, we can shew that within any square whose centre is at this limiting point, there is an infinity of points of the set.]

27. Suppose that $S_n(x)=f_0(x)+f_1(x)+f_2(x)+\dots+f_n(x)$, and let the roots of $S_n(x)=0$ be marked in Argand's diagram for all values of n: if these roots have $x=a$ as a limiting point, the series

$$f_0(x)+f_1(x)+f_2(x)+\dots$$

has $x=a$ as a zero, provided that the series converges uniformly within an area including $x=a$. [HURWITZ.]

EXAMPLES B.

Power-series.

1. If R, R' are the radii of convergence of $\Sigma a_n x^n$ and $\Sigma b_n x^n$ respectively, then:

(1) RR' is the radius of convergence of $\Sigma a_n b_n x^n$.

(2) If R is less than R', R is the radius of convergence of $\Sigma(a_n+b_n)x^n$; but if $R=R'$ the radius is at least equal to R and may be greater.

[Apply the method of Art. 82.]

2. If a power-series is zero at all points of a set which has the origin as a limiting point, then the series is identically zero. [Compare Art. 52.

3. A power-series cannot be purely real (or purely imaginary) at all points within a circle whose centre is the origin.

[Use the last example.]

4. We have seen that in many cases the radius of convergence of $\Sigma a_n x^n$ can be determined from

$$\lim (a_n/a_{n+1});$$

and Fabry has proved (see Hadamard, *La Série de Taylor*, pp. 19–25) that if this limit is equal to λ, then $x=\lambda$ is a singular point of the power-series. Sometimes when the limit λ does not exist, we can determine a relation such as

$$a_{n+2}=p_n a_{n+1}+q_n a_n,$$

where p_n, q_n tend to definite limits p, q as $n\to\infty$.

Then the radius of convergence is equal to the modulus of the least root of $1-px-qx^2=0$.

[For details, see Van Vleck, *Trans. Amer. Math. Soc.*, vol. 1, 1900, p. 293.]

5. An illustration of the last example is afforded by the series for

$$\log[(1+x)/(1-x)]=2(x+\tfrac{1}{3}x^3+\tfrac{1}{5}x^5+\ldots),$$

in which

$$(n+2)a_{n+2}=na_n,$$

and the associated quadratic is $1-x^2$.

Similarly for the series

$$2x+\tfrac{1}{2}x^2+\tfrac{1}{4}x^4+\tfrac{2}{5}x^5+\tfrac{1}{6}x^6+\tfrac{1}{8}x^8+\tfrac{2}{9}x^9+\ldots$$

obtained from $-\log(1-x)+\arctan x$, we get

$$(n+3)a_{n+3}-(n+2)a_{n+2}+(n+1)a_{n+1}-na_n=0,$$

and the associated cubic is $1-x+x^2-x^3=0$.

Generally, if $\Sigma a_n x^n$, $\Sigma b_n x^n$, $\Sigma c_n x^n$ have the same radius of convergence but different singular points, the series $\Sigma(a_n+b_n+c_n)x^n$ may be expected to come under Van Vleck's rule.

6. If $\Sigma a_n x^n$ converges within the circle $|x|=R(>0)$, shew that $\Sigma a_n \dfrac{x^n}{n!}$ converges for all values of x; and examine the relation between the regions of convergence of $\Sigma a_n x^n/n^s$ and $\Sigma a_n x^n$.

7. If M is the maximum value of $|\Sigma a_n x^n|$ on a circle $|x|=r$, M is also greater than the value of $|\Sigma a_n x^n|$ at any point within the circle; here of course r is less than the radius of convergence.

Shew further that $\Sigma a_n x^n$ has no zero within the circle $|x|=Ar/(A+M)$, where $A=|a_0|$.

[Use Cauchy's inequalities of Art. 82.]

8. If $f(x)=\Sigma a_n x^n$ converges for $|x|<R$, then (see Art. 82)

$$\mathfrak{M}|f(x)|^2=\Sigma|a_n|^2 r^{2n}, \quad \text{where } |x|=r<R.$$

Deduce Cauchy's inequalities. [GUTZMER.]

[For we have
$$\mathfrak{M}x^{-n}f(x)=a_n,$$
and if a_n' is the conjugate to a_n,
$$\Sigma a_n' r^{2n}/x^n = f_1(r^2/x) \text{ is the conjugate to } f(x).$$
Thus
$$|f(x)|^2 = f(x)f_1(r^2/x)$$
and
$$\mathfrak{M}|f(x)|^2 = \Sigma a_n' r^{2n} \mathfrak{M} x^{-n} f(x) \quad \text{(Art. 82)}$$
$$= \Sigma a_n a_n' r^{2n} = \Sigma |a_n|^2 r^{2n}.]$$

9. If $f(x)=\Sigma a_n x^n$ converges for $|x|<R$, then
$$|a_1| = |f'(0)| \leqq \tfrac{1}{2}D/r,$$
where D is the maximum of $|f(x)-f(-x)|$ on the circle $|x|=r<R$.

[LANDAU and TOEPLITZ.]

[In fact,
$$a_1 = \mathfrak{M}x^{-1}f(x), \quad -a_1 = \mathfrak{M}x^{-1}f(-x),$$
so that
$$2a_1 = \mathfrak{M}[x^{-1}\{f(x)-f(-x)\}],$$
which gives the desired result.]

10. Shew that if ρ is the radius of convergence of $\Sigma a_n x^n$, the series $\Sigma a_n x^{ni}$ will converge absolutely, provided that the argument of x is greater than $\log(1/\rho)$.

11. Obtain from the binomial series, or otherwise, the equation
$$(2\cos\theta)^\nu = \cos\nu\theta + \nu\cos(\nu-2)\theta + \frac{\nu(\nu-1)}{2!}\cos(\nu-4)\theta + \dots,$$
where ν is real and greater than -1. What restrictions are required as to the value of θ?

Shew that the equation ceases to be true for $\nu=\frac{1}{3}$, $\theta=\pi$, and explain why.

12. Find the sum of
$$1+\frac{1}{2}\cos\theta + \frac{1.3}{2.4}\cos 2\theta + \frac{1.3.5}{2.4.6}\cos 3\theta + \dots$$
and of
$$\frac{1}{2}\sin\theta + \frac{1.3}{2.4}\sin 2\theta + \frac{1.3.5}{2.4.6}\sin 3\theta + \dots.$$

[Apply Art. 89, putting $\nu=-\frac{1}{2}$, $x=-e^{i\theta}$.]

13. If m is positive, shew that
$$2^{\frac{m}{2}}\cos\left(\frac{1}{4}m\pi\right) = 1 - \frac{m(m-1)}{2!} + \frac{m(m-1)(m-2)(m-3)}{4!} - \dots,$$
and examine the special form of the result when $m=1/10$.

[Take $x=i$ in the expansion of $(1+x)^m$.]

14. Examine the convergence of the power-series
$$\Sigma\frac{1.3\dots(2n-1)}{2.4\dots 2n}\left(1+\frac{1}{2}+\dots+\frac{1}{n}\right)x^{2n}, \quad \text{when } |x|=1.$$

[Apply Weierstrass's rule, Art. 79.]

15. Discuss the convergence of the power-series
$$\Sigma\frac{x^n}{(n+a)^s}, \quad \Sigma\frac{x^n}{n\log n}, \quad \Sigma\left(1+\frac{1}{2}+\dots+\frac{1}{n}\right)\frac{x^n}{n+1}, \quad \text{for } |x|=1.$$

[In the third, the coefficient of x^n steadily decreases; see Ex. 2, Art. 34.]

16. Discuss the convergence of

$$\left(\frac{1}{3}\right)^2 x+\left(\frac{1\,.\,2}{3\,.\,5}\right)^2 x^2+\left(\frac{1\,.\,2\,.\,3}{3\,.\,5\,.\,7}\right)^2 x^3+\dots$$

and of

$$\sum_1^\infty \frac{x^n}{(n^n+n)^{\frac{1}{n}}}.$$

17. Determine the expansion of $e^{-x\cos\theta}\cos(x\sin\theta)$ in powers of x, and deduce that

$$\int_0^x \frac{\sin t}{t}\,dt=\frac{\pi}{2}-\int_0^{\frac{\pi}{2}} e^{-x\cos\theta}\cos(x\sin\theta)\,d\theta.$$

[Put $xe^{i\theta}$ for x in the exponential series.]

18. Shew that (compare Ex. 17)

$$e^{ax}\cos bx=\sum_0^\infty \frac{x^n}{n!}\left\{a^n-\frac{n(n-1)}{2!}a^{n-2}b^2+\frac{n(n-1)(n-2)(n-3)}{4!}a^{n-4}b^4-\dots\right\},$$

where there are $\frac{1}{2}(n+1)$ or $\frac{1}{2}(n+2)$ terms in the brackets.

Determine a similar series for $e^{ax}\sin bx$.

19. Shew that if $|\theta|<1/e$,

$$\cos\theta=1-\theta\sin\theta+\frac{1}{2!}\theta^2\cos2\theta+\frac{2^2}{3!}\theta^3\sin3\theta-\frac{3^3}{4!}\theta^4\cos4\theta-\dots,$$

$$\sin\theta=\qquad\theta\cos\theta+\frac{1}{2!}\theta^2\sin2\theta-\frac{2^2}{3!}\theta^3\cos3\theta-\frac{3^3}{4!}\theta^4\sin4\theta+\dots.$$

[*Math. Trip.* 1891.]

[Write $a=b=i$ in the formula of Ex. 4, Art. 56. The introduction of complex numbers in the place of real ones may be justified by an argument of the same type as that used in Art. 88.]

20. If $x=y\surd(1+ax)$, shew that Lagrange's series for one root is

$$x=y+\tfrac{1}{2}ay^2+\sum_1^\infty(-1)^{n-1}\frac{1\,.\,3\dots(2n-3)}{2\,.\,4\dots2n}\left(\frac{a}{2}\right)^{2n}y^{2n+1},$$

and that the series converges if $|ay|<2$. [*Math. Trip.* 1902.]

21. If $t=\cos\theta+i\sin\theta$ and $0<r<1$, shew that

$$\frac{1}{2\pi}\int_0^{2\pi}(1+rt)^n(1+r/t)^n d\theta=1+r^2n_1^2+r^4n_2^2+r^6n_3^2+\dots$$

and

$$\frac{1}{2\pi}\int_0^{2\pi}(1+rt)^n(1-r/t)^n d\theta=1-r^2n_1^2+r^4n_2^2-r^6n_3^2+\dots,$$

where $n_1, n_2, n_3, \dots$ are the coefficients in the binomial series.

Deduce from Abel's theorem that, if $n>0$,

$$1+n_1^2+n_2^2+\dots=\frac{2^n}{\pi}\int_0^\pi(1+\cos\theta)^n d\theta=\frac{\Gamma(2n+1)}{[\Gamma(n+1)]^2},$$

$$1-n_1^2+n_2^2-\dots=\frac{2^n}{\pi}\cos(\tfrac{1}{2}n\pi)\int_0^\pi(\sin\theta)^n d\theta=\frac{\Gamma(n+1)}{[\Gamma(\frac{1}{2}n+1)]^2}\cos(\tfrac{1}{2}n\pi).$$

[Note that the argument of $(1-r^2+2ir\sin\theta)^n$ approaches the limit $+\frac{1}{2}n\pi$ (as $r\to1$) when $\sin\theta$ is positive, and $-\frac{1}{2}n\pi$ when $\sin\theta$ is negative. The first summation is valid if $n>-\frac{1}{2}$, and the second if $n>-1$; but the proofs become rather more difficult when n is negative.]

22. If $x=\cos\theta+i\sin\theta$, $a_n\to 0$ and $\Sigma|a_n-a_{n+1}|$ is convergent, prove by Dirichlet's test (Art. 80) that $\Sigma a_n x^n$ is convergent if $0<\theta<2\pi$.

If $\Sigma b_n x^n$ is convergent, under similar conditions, then shew that

$$(\Sigma a_n x^n)\times(\Sigma b_n x^n)=\Sigma c_n x^n,$$

where
$$c_n=a_0 b_n+a_1 b_{n-1}+\ldots+a_n b_0,$$

provided that $\Sigma A_n B_n$ converges, where

$$A_n=|a_n-a_{n+1}|+|a_{n+1}-a_{n+2}|+\ldots,$$
$$B_n=|b_n-b_{n+1}|+|b_{n+1}-b_{n+2}|+\ldots.$$

[Note that the sequences (A_n), (B_n) are decreasing sequences and that $|a_n|<A_n$, $|b_n|<B_n$ (as in Art. 80). Then apply the method given in Art. 35 for establishing Pringsheim's theorem on the multiplication of series.]

23. If $f_n=1/n\log n$ when $n>1$, and $f_1=1$, the series

$$\Sigma v_n=f_1-f_2-f_3+f_4+f_5+f_6+f_7-\ldots,\quad \text{where } v_n=\pm f_n,$$

is convergent, but not absolutely, while $\Sigma|v_n-v_{n+1}|$ is convergent. Note that the signs of v_n are the same in groups of 1, 2, 4, 8, ... terms.

[For we have
$$f_5<f_4<\tfrac{1}{2}f_2 \quad\text{or}\quad f_4+f_5<f_2$$
and
$$f_7<f_6<\tfrac{1}{2}f_3 \quad\text{or}\quad f_6+f_7<f_3.$$

Thus $f_2+f_3<f_4+f_5+f_6+f_7$: and so on for each group of 2^k terms with the same sign. The convergence then follows from Art. 21.

Again,
$$\Sigma|v_n-v_{n+1}|=(f_1+f_2)+(f_2+f_4)+(f_4+f_8)+\ldots$$
$$<2(f_1+f_2+f_4+f_8+\ldots)$$
$$<2\left(1+\frac{1}{2}+\frac{1}{2^2}+\frac{1}{2^3}+\ldots\right),$$

so that $\Sigma|v_n-v_{n+1}|$ converges.] [PRINGSHEIM.]

24. It follows from the previous example that the series

$$\Sigma v_n x^n=f_1x-f_2x^2-f_3x^3+f_4x^4+f_5x^5+f_6x^6+f_7x^7-\ldots$$

converges, but not absolutely, at *every* point of the circle $|x|=1$.

[The convergence of $\Sigma|v_n-v_{n+1}|$, combined with $\lim v_n=0$, establishes the convergence of $\Sigma v_n x^n$ at all points except $x=1$; and the last example enables us to include $x=1$. As regards the absolute convergence, we note that $\Sigma|v_n x^n|=\Sigma|v_n|=\Sigma f_n$ is divergent.] [PRINGSHEIM.]

25. If $B_n=b_0+b_1+\ldots+b_n$, and if K, K_m are the upper limits of $|B_n|$ as n ranges from 0 to $m-1$, and from m to ∞, respectively, it follows from Arts. 81, 83 that

$$|\Sigma b_n x^n|\leqq\lambda[K\{1-|x|^m\}+K_m|x|^m].$$

Thus, if Σa_n is a series of positive terms satisfying the condition of uniform divergence (Art. 83), we have

$$\lim_{x\to 1}(\Sigma b_n x^n)/(\Sigma a_n x^n)=0,\quad \text{if } \lim(B_n/A_n)=0,$$

the path of approach lying within the limaçon of Art. 83.

Hence generally, for such paths,

$$\lim_{x\to 1}(\Sigma b_n x^n)/(\Sigma a_n x^n)=\lim(B_n/A_n),$$

the right-hand limit being supposed to exist. [PRINGSHEIM.]

26. The last result enables us to immediately extend Exs. 2, 3, 4, Art. 51, to the complex variable; and in particular to extend Frobenius's theorem so as to apply to any path of approach lying within the limaçon of Art. 83. We note also the following result:

If $a_0+a_1+\ldots+a_n \backsim \log n$, then, as $x \to 1$,

$$\Sigma a_n x^n \backsim \log\left(\frac{1}{1-|x|}\right).$$

A further extension is quoted on p. 133, above.

27. Converse of Abel's theorem.

If $\lim\limits_{x\to 1}(\Sigma a_n x^n)$ exists and is equal to a finite number A, it is not possible to infer the convergence of Σa_n without further restriction on the coefficients. In two simple cases we can make this inference:

(1) When the coefficients a_n are all positive after a certain stage. [PRINGSHEIM.]

(2) When $\lim na_n=0$, and x approaches 1 by any path within the limaçon of Art. 83. [TAUBER.]

[Since in case (1) x can approach 1 by real values, we can infer from the existence of $\lim \Sigma a_n x^n$ that Σa_n cannot diverge; further, Σa_n cannot oscillate. Hence, in case (1) Σa_n converges, and is therefore equal to A, by Abel's theorem.

In case (2), write $n|a_n|=c_n$, then we find

$$\left|\sum_0^{\nu-1} a_n(1-x^n)\right| < |1-x| \sum_0^{\nu-1} c_n,$$

because $$|(1-x^n)/(1-x)|=|1+x+x^2+\ldots+x^{n-1}| \leqq n;$$

also, if H_ν is the upper limit to $c_\nu, c_{\nu+1}, \ldots$ to ∞, we have

$$\left|\sum_\nu^\infty a_n x^n\right| < H_\nu \frac{|x|^\nu}{\nu}(1+|x|+|x|^2+\ldots) < H_\nu \frac{|x|^\nu}{\nu\{1-|x|\}}.$$

Take then x as a point on the given path such that $|x|=1-1/\nu$; we have then, as in Art. 83, $$|1-x|<\lambda/\nu,$$

and so $$\left|\sum_0^{\nu-1} a_n - \sum_0^\infty a_n x^n\right| < (\lambda/\nu)\sum_0^{\nu-1} c_n + H_\nu.$$

As $\nu\to\infty$, each of the terms on the right tends to 0 (the first in virtue of Art. 154); and so $$\lim_{\nu\to\infty} \sum_0^{\nu-1} a_n = \lim_{x\to 1} \sum_0^\infty a_n x^n = A.]$$

28. In case na_n has no definite limit, we can infer the convergence of Σa_n from the existence of $\lim\limits_{x\to 1} \Sigma a_n x^n$ (for some path within the limaçon), and from the condition

$$\lim \frac{1}{n}(a_1+2a_2+3a_3+\ldots+na_n)=0.$$

These conditions are both necessary for the convergence of Σa_n, and, taken together, they are sufficient.

[Tauber, *Monatshefte f. Math. u. Phys.*, Bd. 8, 1897, p. 273; Pringsheim, *Münchener Sitzungsberichte*, Bd. 30, 1900, p. 37, and Bd. 31, 1901, p. 507.]

29. Applications of Art. 84.

If
$$f(x)=1+\nu x+\nu(\nu-1)\frac{x^2}{2!}+\nu(\nu-1)(\nu-2)\frac{x^3}{3!}+\dots,$$
it is easily verified that, with the notation of Art. 84,
$$f_1(x)=\nu\left[1+(\nu-1)x+(\nu-1)(\nu-2)\frac{x^2}{2!}+\dots\right]=\nu f(x)/(1+x),$$
$$f_2(x)=\nu(\nu-1)\left[1+(\nu-2)x+(\nu-2)(\nu-3)\frac{x^2}{2!}+\dots\right]=\nu(\nu-1)f(x)/(1+x)^2,$$
and so on. Thus, we obtain the transformation
$$f(x_1)=f(x)+\frac{\nu f(x)}{1+x}(x_1-x)+\frac{\nu(\nu-1)f(x)}{(1+x)^2}\frac{(x_1-x)^2}{2!}+\dots$$
or
$$f(x_1)=f(x)f\left(\frac{x_1-x}{1+x}\right).$$

The two series on the right-hand are both convergent if $|x|<1$ and $|x_1-x|<|1+x|$, and the latter condition is satisfied for some points x_1 which are outside the circle $|x_1|=1$; we have thus obtained a *continuation* of the binomial series. Repeating the process, we obtain
$$f(x_n)=f(x)f\left(\frac{x_1-x}{1+x}\right)f\left(\frac{x_2-x_1}{1+x_1}\right)\dots f\left(\frac{x_n-x_{n-1}}{1+x_{n-1}}\right),$$
where we assume that the broken line from x to x_n is drawn so that
$$|x_{r+1}-x_r|<|1+x_r| \quad (r=0,\ 1,\dots,\ n-1).$$

For example, by taking
$$1+x=1,\ 1+x_1=\frac{1+i}{\sqrt{2}},\ 1+x_2=i,\ 1+x_3=\frac{-1+i}{\sqrt{2}},\ 1+x_4=-1,$$
we find
$$\frac{x_{r+1}-x_r}{1+x_r}=\frac{1+i}{\sqrt{2}}-1,\quad \left|\frac{x_{r+1}-x_r}{1+x_r}\right|^2=2-\sqrt{2}<1,$$
so that
$$f\left(\frac{x_{r+1}-x_r}{1+x_r}\right)=e^{-\frac{1}{4}\beta\pi}[\cos(\tfrac{1}{4}\alpha\pi)+i\sin(\tfrac{1}{4}\alpha\pi)],$$
where
$$\nu=\alpha+i\beta.$$

Thus we are led to $f(-2)=e^{-\beta\pi}[\cos(\alpha\pi)+i\sin(\alpha\pi)]$.

But it should be noticed that if we take a broken line passing *below* the real axis, we find $f(-2)=e^{\beta\pi}[\cos(\alpha\pi)-i\sin(\alpha\pi)]$; we thus obtain two different values for $f(-2)$ by approaching -2 along different paths. This indicates (what we know to be the case) that $f(x)$ is many-valued unles $\beta=0$ and α is an integer.

30. A method similar to the last example can be applied to
$$\phi(x)=x-\tfrac{1}{2}x^2+\tfrac{1}{3}x^3-\tfrac{1}{4}x^4+\dots,$$
for which we find
$$\phi_1(x)=(1+x)^{-1},\ \phi_2(x)=-(1+x)^{-2},\ \phi_3(x)=(2!)(1+x)^{-3},\dots,$$
and so
$$\phi(x_1)=\phi(x)+\phi\left(\frac{x_1-x}{1+x}\right).$$

Using the same points x, x_1, x_2, x_3, x_4, as in the last example, we get $\phi(-2)=i\pi$. And with a broken line passing below the real axis, we get $\phi(-2)=-i\pi$.

31. If the coefficients of a power-series $\Sigma a_n x^n$ are all positive (at any rate after a certain stage), the series has a singular point at the point $x=R$, if R is the radius of convergence. [VIVANTI and PRINGSHEIM.]

[The following method of proof is due to Landau. Suppose, if possible, that (for $0<r<R$) the series $\Sigma f_n(r)\frac{(x-r)^n}{n!}$ has a larger radius of convergence than $R-r$; we can then choose a real number ρ $(>R)$, such that the last series converges for $x=\rho$. Now this series (as in Art. 84) can be arranged as a double series, *which contains here only positive terms*; it will therefore remain convergent when summed as $\Sigma a_n\{r+(\rho-r)\}^n$. That is, $\Sigma a_n x^n$ will converge for $x=\rho$, contrary to the original hypothesis; and so $x=R$ must be a singular point.]

32. Weierstrass's double-series theorem.*

Suppose that the series

$$f_m(x)=\sum_{n=0}^{\infty} a_{m,n}x^n \qquad (m=0, 1, 2, \dots \infty)$$

are all convergent for $|x|<R$, and further that the series

$$F(x)=\sum_{m=0}^{\infty} f_m(x)$$

converges *uniformly* along every circle whose radius is less than R. Then

(1) the series $\sum_{m=0}^{\infty} a_{m,n}$ converges for every value of n.

(2) if $A_n=\sum_{m=0}^{\infty} a_{m,n}$, then $F(x)=\sum_{n=0}^{\infty} A_n x^n, \quad |x|<R.$

[For

$$\sum_{m=0}^{\infty} a_{m,n}=\sum_{m=0}^{\infty} \mathfrak{M}\{f_m(x)/x^n\},$$

the mean being taken along any circle $|x|=r_1<R$. Now on this circle $F(x)=\Sigma f_m(x)$ is uniformly convergent, and so the series $\Sigma a_{m,n}$ must be convergent and equal to $\mathfrak{M}\{F(x)/x^n\}$.

Again, if μ is any integer and

$$G(x)=\sum_{m=\mu}^{\infty} f_m(x), \quad B_n=\sum_{m=\mu}^{\infty} a_{m,n},$$

we have similarly

$$B_n=\mathfrak{M}\{G(x)/x^n\}, \text{ and so } |B_n|<M_1/r_1^n$$

if M_1 is the maximum of $G(x)$ on the circle $|x|=r_1$.

Hence, if $|x|=r(<r_1)$, we have

$$\left|\sum_{n=0}^{\infty} B_n x^n\right|<\sum_{n=0}^{\infty} M_1(r/r_1)^n=M_1r_1/(r_1-r),$$

and by Ex. 7, p. 247, $|G(x)|<M_1$, so that

$$\left|G(x)-\sum_{n=0}^{\infty} B_n x^n\right|<M_1(2r_1-r)/(r_1-r), \quad \text{if } |x|=r.$$

---

* Weierstrass, *Ges. Werke*, Bd. II., p. 205.

Now, we have identically

$$F(x)-G(x)=\sum_{n=0}^{\infty}(A_n-B_n)x^n,$$

because this equation contains only a *finite* number (μ) of series: and so we find

$$|F(x)-\sum_{n=0}^{\infty}A_n x^n|<M_1(2r_1-r)(r_1-r), \quad \text{if } |x|=r.$$

But, since $F(x)$ converges uniformly on the circle $|x|=r$, we can make M_1 as small as we please by proper choice of μ. Thus, since $F(x)$ and $\Sigma A_n x^n$ are *independent* of μ, we must have $F(x)=\sum_{n=0}^{\infty}A_n x^n$.]

EXAMPLES C.

Miscellaneous Series.

1. If

$$F(x, n)=\frac{n!}{x(x+1)\dots(x+n)},$$

the series $\Sigma a_n F(x, n)$ converges absolutely, provided that $\Sigma|a_n|n^{-\xi}$ does so, where ξ is the real part of x. Thus, in particular, if $\Sigma a_n x^n$ has a radius of convergence greater than 1, $\Sigma a_n F(x, n)$ is absolutely convergent for all values of x (other than real negative integers). But if the radius of convergence is less than 1, $\Sigma a_n F(x, n)$ cannot converge.

Finally, if the radius of convergence is equal to 1, suppose that

$$\left|\frac{a_n}{a_{n+1}}\right|=1+\frac{a}{n}+\frac{\omega_n}{n^\lambda},$$

where $\lambda>1$ and $|\omega_n|<A$: then $\Sigma a_n F(x, n)$ is absolutely convergent if $\xi>1-a$. [KLUYVER, see also NIELSEN, *Gammafunktion*, §§ 93, 94.]

[Note that $F(x, n)\sim\Gamma(x)n^{-x}$.]

2. Shew that, with the notation of the last example, the two series $\Sigma(a_n/n^x)$, $\Sigma a_n F(x, n)$ converge for the same values of x. [LANDAU.]

[Apply Art. 80 (2), taking $v_n=n^x F(x, n)$; the series $\Sigma|v_n-v_{n+1}|$ and $\Sigma|1/v_n-1/v_{n+1}|$ are then easily proved to be convergent.]

3. Shew that in the notation of Ex. 1,

$$F(x-1, n)-F(x-1, n+1)=F(x, n),$$

and deduce that $\Sigma F(x, n)$ converges only when $\xi>1$; that is, $\Sigma F(x, n)$ can only converge absolutely. Shew also that $\Sigma(-1)^n F(x, n)$ converges if $\xi>0$; and apply Ex. 2 to deduce the corresponding results for Σn^{-x}, $\Sigma(-1)^n n^{-x}$.

4. The series (see Ex. 19, Ch. I.)

$$\frac{x}{1-x^2}+\frac{x^2}{1-x^4}+\frac{x^4}{1-x^8}+\frac{x^8}{1-x^{16}}+\dots$$

represents the function $x/(1-x)$, if $|x|<1$, and $1/(1-x)$, if $|x|>1$.

[J. TANNERY.]

5. Shew that the series

$$\sum_0^\infty\left[\left(x+n+\frac{1}{2}\right)\log\left(1+\frac{1}{x+n}\right)-1\right], \quad \sum_0^\infty\left[\frac{1}{x+n}-\log\left(1+\frac{1}{x+n}\right)\right]$$

are both convergent for all values of x, except $0, -1, -2, -3, \dots$.

[For applications, see NIELSEN, *Gammafunktion*, §§ 33, 34.]

6. If (c_n) is a sequence of complex numbers such that $|c_n|$ tends steadily to ∞, shew that the series

$$\Sigma \frac{x^n}{c_n{}^n(x-c_n)}$$

converges absolutely for all values of x, except for $c_1, c_2, c_3, \ldots$. The series converges uniformly within the area bounded externally by the circle $|x|=R$, and internally by those circles $|x-c_n|=r$, which are contained within the circle $|x|=R$, the number r being taken small enough to prevent any overlapping of the circles.

7. The series of the last example can be simplified in case the points c_n lie along a straight line, and are such that $|c_{n+1}-c_n| \geqq k > 0$, where k is a constant; under these circumstances we can make similar statements with respect to

$$\Sigma \frac{x}{c_n(x-c_n)}, \quad \Sigma \frac{1}{(x-c_n)^2}, \quad \Pi \left(1-\frac{x}{c_n}\right) e^{x/c_n}.$$

8. Again, if the points c_n, although not distributed along a straight line, are such that no two of them are at a less distance apart than a constant k, similar statements can be made with respect to

$$\Sigma \frac{x^2}{c_n{}^2(x-c_n)}, \quad \Sigma \left[\frac{1}{(x-c_n)^2} - \frac{1}{c_n{}^2}\right], \quad \Sigma \frac{1}{(x-c_n)^3}, \quad \Pi \left(1-\frac{x}{c_n}\right) e^{x/c_n + \frac{1}{2}(x/c_n)^2}.$$

A simple example of a set of this type occurs in the theory of elliptic functions, the points c_n being the vertices of a network of parallelograms.

[Here we note that not more than one point c_n can fall within a square of side $\frac{1}{2}k$; thus, if we draw squares, with centre at the origin, of sides $\frac{1}{2}k$, $\frac{3}{2}k$, $\frac{5}{2}k$, ..., not more than $8m$ points can lie between the two squares $(m-\frac{1}{2})k$, $(m+\frac{1}{2})k$. Hence $\Sigma |c_n|^{-3} < \Sigma 8m/(m-\frac{1}{2})^3 k^3$, and so $\Sigma |c_n|^{-3}$ converges.]

9. If (M_n) is a sequence of real numbers which tends steadily to ∞, and if x is a complex number whose real part is positive, the series

$$\Sigma | M_n^{-x} - M_{n+1}^{-x} |$$

is convergent.

[For, if $x=\xi+i\eta$ and $(M_n/M_{n+1})^{\xi}=\lambda_n$, the ratio of the general term of the given series to that of the series

$$\Sigma (M_n^{-\xi} - M_{n+1}^{-\xi})$$

is

$$R_n = |(1-\lambda_n^{1+i\eta/\xi})/(1-\lambda_n)\,.$$

Now, $\quad \lambda_n^{1+i\eta/\xi} = \lambda_n\{\cos(\kappa \log \lambda_n) + i \sin(\kappa \log \lambda_n)\}, \quad$ if $\kappa = \eta/\xi$,

and so if $\lambda_n \leqq \frac{1}{2}$, we see that $R_n \leqq (1+\frac{1}{2})/(1-\frac{1}{2}) \leqq 3$.

On the other hand, if $\lambda_n > \frac{1}{2}$, we can write (Art. 157 (3))

$$(\log \lambda_n)/(1-\lambda_n) < 1/\lambda_n < 2,$$

which leads to the result $R_n < \sqrt{(1+4\kappa^2)}$, because

$$R_n{}^2 = 1 + \frac{2\lambda_n}{(1-\lambda_n)^2}\{1-\cos(\kappa \log \lambda_n)\} < 1 + \kappa^2 \left(\frac{\log \lambda_n}{1-\lambda_n}\right)^2.$$

In either case there is a finite upper limit H to R_n, and so the given series converges because the comparison-series is convergent.]

10. If Σa_n is convergent, the series $\Sigma a_n M_n^{-x}$ is convergent if the real part of x is positive. Thus, *in general the region of convergence of $\Sigma a_n M_n^{-x}$ is bounded by a line parallel to the imaginary axis.*

Further, in case Σa_n is convergent, the series converges *uniformly* in a sector of the plane bounded by the lines $\eta = \pm\kappa\xi$, where κ is any assigned number.

[CAHEN.]

[For then we can use Abel's theorem (Arts. 80, 81), merely noting that we may write $V_n = HM_n^{-\xi}$ in virtue of the last example.]

11. It is not hard to modify the result of Ex. 27 (p. 251) so as to apply to series of the type $\Sigma a_n e^{-xf(n)}$, where $f(n)$ is a positive function which steadily increases to infinity (this is the same type as in the last example):

If as x tends to 0 by any path between the two lines $\eta = \pm\kappa\xi$ (where κ is an assigned number), the series has a limit A, and if $\lim\{a_n f(n)/f'(n)\} = 0$, then Σa_n converges to the sum A.

[Landau, *Monatshefte f. Math. u. Phys.*, Bd. 18, 1907, p. 19.]

12. Prove that if $0 < x < \frac{1}{2}\pi$,

$$\left(\frac{\pi}{2} - x\right)\left(\cos x - \frac{\cos 3x}{3} + \frac{\cos 5x}{5} - \dots\right) = \cos x + \frac{\cos 3x}{3^2} + \frac{\cos 5x}{5^2} + \dots.$$

13. From Ex. 2, Art. 90, deduce that, if $0 \leqq x \leqq \pi$,

$$\frac{\pi}{8}x(\pi - x) = \frac{\sin x}{1^3} + \frac{\sin 3x}{3^3} + \frac{\sin 5x}{5^3} + \dots,$$

and obtain the sum of the series for all values of x. [EULER.]

14. From Ex. 13, shew that if $-\frac{1}{2}\pi \leqq x \leqq \frac{1}{2}\pi$,

$$\frac{\pi x}{8}\left(\frac{\pi^2}{4} - \frac{x^2}{3}\right) = \sin x - \frac{\sin 3x}{3^4} + \frac{\sin 5x}{5^4} - \dots,$$

and (by taking $x = \frac{1}{4}\pi$) shew that

$$\frac{11\pi^4}{1536} = \frac{1}{\sqrt{2}}\left(1 - \frac{1}{3^4} - \frac{1}{5^4} + \frac{1}{7^4} + \frac{1}{9^4} - \dots\right).$$

15. Deduce from Ex. 2, Art. 90, that, if $-\frac{1}{2}\pi \leqq x \leqq \frac{1}{2}\pi$,

$$\frac{\cos 3x}{1.3.5} - \frac{\cos 5x}{3.5.7} + \frac{\cos 7x}{5.7.9} - \dots = \frac{\pi}{8}\cos^2 x - \frac{1}{3}\cos x,$$

and find the sum of the series for all values of x.

16. Shew that
$$\frac{\theta e^{\theta y}}{e^\theta - 1} = 1 + \sum_{-\infty}^{\infty}{}' \frac{\theta e^{2n\pi i y}}{\theta - 2n\pi i},$$

where $n = 0$ is excluded from the summation and $0 < y < 1$.

Deduce that

$$2\sum_1^\infty \frac{\sin(2n\pi y)}{2n\pi} = \frac{1}{2} - y, \quad 2\sum_1^\infty \frac{\cos(2n\pi y)}{(2n\pi)^2} = \frac{1}{2!}[\phi_2(y) + B_1],$$

$$2\sum_1^\infty \frac{\cos(2n\pi y)}{(2n\pi)^{2k}} = (-1)^{k-1}\frac{1}{(2k)!}\phi_{2k}(y) + B_k,$$

$$2\sum_1^\infty \frac{\sin(2n\pi y)}{(2n\pi)^{2k+1}} = -\frac{1}{(2k+1)!}\phi_{2k+1}(y), \text{ etc.}$$

[RAABE and SCHLÖMILCH.]

[Put $t = \theta/2\pi i$, $x = 2\pi y$ in Ex. 3, Art. 90, and apply Art. 94.]

17. From Ex. 3, Art. 90, deduce that if λ lies between the two even integers $2r$, $2(r+1)$,

$$\sum_{-\infty}^{\infty} \frac{e^{i\lambda(\xi+n\pi)}}{\xi+n\pi} = \frac{e^{(2r+1)i\xi}}{\sin\xi},$$

and examine the case $\lambda=2r$.

[Write $x=(\lambda-2r)\pi$, $t=-\xi/\pi$.]

18. Shew that

$$\sum_{0}^{\infty}(-1)^n \frac{e^{inx}}{(n+1)(n+2)} = (e^{-ix}+e^{-2ix})\log(1+e^{ix})-e^{-ix},$$

and divide this equation into real and imaginary parts.

19. Shew that $$\sum_{-\infty}^{\infty}(-1)^n \frac{e^{2inv\pi}}{t-n} = \frac{\pi e^{2i\theta t\pi}}{\sin t\pi},$$

where θ is the difference between v and the integer nearest to v.

[Write $x=\pi(1+2\theta)$ in the series of Ex. 3, Art. 90, and observe that $-\frac{1}{2}<\theta<+\frac{1}{2}$.]

20. If n is an integer and $0<x<1$, we have

$$\sum_{m=1}^{\infty}{}' \frac{\cos(2m\pi x)}{m^2-n^2} = \frac{1}{2n^2} + \frac{\cos(2n\pi x)}{4n^2} + \pi(x-\tfrac{1}{2})\frac{\sin(2n\pi x)}{n},$$

where $m=n$ is excluded from the summation. [Ex. 3, Art. 90.]

21. If both x, y are between 0 and $\frac{1}{2}$, we have the following results:

$$\begin{aligned} \sum_{1}^{\infty}\frac{1}{n}\sin(2n\pi x)\cos(2n\pi y) &= -\pi(x-\tfrac{1}{2}), \quad \text{if } x>y,\\ &\text{or } -\pi(x-\tfrac{1}{4}), \quad \text{if } x=y,\\ &\text{or } -\pi x, \quad \text{if } x<y;\end{aligned}$$

$$\begin{aligned} \sum_{1}^{\infty}\frac{1}{n^2}\cos(2n\pi x)\cos(2n\pi y) &= \pi^2[\phi_2(x)+\phi_2(y)+y-\tfrac{1}{6}], \quad \text{if } x\geqq y,\\ &\text{or } \pi^2[\phi_2(x)+\phi_2(v)+x-\tfrac{1}{6}], \quad \text{if } x\leqq y,\end{aligned}$$

where $\phi_2(x)=x^2-x+\frac{1}{6}$ is the Bernoullian function of order 2.

22. By writing $t=ia$ and $-ia$ in Ex. 3, Art. 90, or otherwise, shew that

$$\pi\frac{\cosh a(\pi-x)}{\sinh a\pi} = \frac{1}{a}+2a\sum_{1}^{\infty}\frac{\cos nx}{n^2+a^2}, \qquad 0\leqq x\leqq 2\pi,$$

$$\pi\frac{\sinh a(\pi-x)}{\sinh a\pi} = 2\sum_{1}^{\infty}\frac{n\sin nx}{n^2+a^2}, \qquad 0<x<2\pi.$$

Use Stokes's rule (Art. 90) to deduce each of these from the other by differentiating. [*Math. Trip.* 1902.]

23. Deduce from the last example, that

$$\pi\frac{\cosh ax}{\sinh a\pi} = \frac{1}{a}+2a\sum_{1}^{\infty}\frac{(-1)^n\cos nx}{n^2+a^2}, \qquad -\pi\leqq x\leqq\pi,$$

$$\pi\frac{\sinh ax}{\sinh a\pi} = 2\sum_{1}^{\infty}\frac{(-1)^{n-1}n\sin nx}{n^2+a^2} \qquad -\pi<x<\pi,$$

$$\frac{\pi}{2}e^{ax} = \sum_{1}^{\infty}\frac{n}{n^2+a^2}[1+(-1)^{n-1}e^{a\pi}]\sin nx, \qquad 0<x<\pi,$$

and obtain a cosine-series for e^{ax}.

24. Shew that the product

$$\prod_1^\infty [(1-e^{x/n})/\log(1-x/n)]$$

converges absolutely except when x is a positive integer.

[If the general term is $1-a_n$, $\lim(n^2 a_n)=\frac{1}{8}x^2$.]

25. Evaluate $\quad x^2\prod_1^\infty(1+x^4/n^4)$ and $x^3\prod_1^\infty(1+x^6/n^6)$.

[Glaisher, *Proc. Lond. Math. Soc.* (1), vol. 7, p. 23.]

26. From Ex. 10, Ch. IX., find the values of

$$\prod_{-\infty}^\infty\left[1-\frac{x^4}{(n-c)^4}\right] \text{ and } \prod_{-\infty}^\infty\left[1+\frac{x^4}{(n-c)^4}\right].$$ [*Math. Trip.* 1902.]

27. Prove that

$$\frac{e^x-c}{1-c}=e^{\frac{1}{2}x}\lim_{\nu\to\infty}\prod_{-\nu}^{\nu}\left(1-\frac{x}{\log_n c}\right),$$

where $\qquad \log_n c=\log c+2n\pi i,$

any determination of $\log c$ being taken. [HARDY.]

28. Shew that $\qquad \prod_2^\infty \frac{n^3-1}{n^3+1}=\frac{2}{3}.$ [GRAM.]

[If $t=\frac{1}{2}(-1+i\sqrt{3})$, so that $t^3=1$, we have, as in Ex. 21, Ch. VI.,

$$\prod_1^\infty\frac{n^3-x^3}{n^3+x^3}=\frac{\Gamma(1+x)\Gamma(1+tx)\Gamma(1+t^2x)}{\Gamma(1-x)\Gamma(1-tx)\Gamma(1-t^2x)}.$$

Thus $\qquad \frac{1+x+x^2}{1+x^3}\prod_2^\infty\frac{n^3-x^3}{n^3+x^3}=\frac{\Gamma(1+x)\Gamma(1+tx)\Gamma(1+t^2x)}{\Gamma(2-x)\Gamma(1-tx)\Gamma(1-t^2x)}.$

Now write $x=1$, and observe that

$$\Gamma(1-t)\Gamma(1-t^2)=\Gamma(2+t^2)\Gamma(2+t)=(1+t)(1+t^2)\Gamma(1+t)\Gamma(1+t^2)$$
$$=\Gamma(1+t)\Gamma(1+t^2).]$$

29. Shew that

$$\sum_{-\infty}^\infty\frac{1}{n^4+x^4}=\frac{\pi}{x^3\sqrt{2}}\frac{\sinh(\pi x\sqrt{2})+\sin(\pi x\sqrt{2})}{\cosh(\pi x\sqrt{2})-\cos(\pi x\sqrt{2})}.$$

[*Math. Trip.* 1888.]

[We have $\qquad \frac{1}{n^4+x^4}=\sum_{(t)}\frac{t}{4x^3}\left(\frac{1}{n+tx}-\frac{1}{n}\right),$

where $t^4=-1$.

Thus, as in Ex. 22, Ch. IX., the given sum is

$$\sum_{(t)}\frac{\pi t}{4x^3}\cot(\pi tx),$$

and this gives the required result.]

30. Apply a method similar to Ex. 29 to find

$$\sum_{-\infty}^\infty\frac{1}{n^3+x^3}.$$

31. Prove that

$$\sum_{-\infty}^{\infty}\frac{1}{(n+x)^2+y^2}=\frac{\pi}{y}\frac{\sinh(2\pi y)}{\cosh(2\pi y)-\cos(2\pi x)}.$$

Deduce that the least value of θ, when y is fixed and

$$\tan\theta=\sum_{-\infty}^{\infty}\frac{n+x}{[(n+x)^2+y^2]^2}\Big/\sum_{-\infty}^{\infty}\frac{y}{[(n+x)^2+y^2]^2},$$

is given by $\cos\theta=2\pi y/\sinh(2\pi y)$. [*Math. Trip.* 1892.]

[Here $$\frac{1}{(n+x)^2+y^2}=\frac{1}{2iy}\left\{\frac{1}{n+(x+iy)}-\frac{1}{n+(x-iy)}\right\},$$

and so the sum is $\dfrac{\pi}{2iy}\{\cot\pi(x+iy)-\cot\pi(x-iy)\}$.]

32. Shew that

$$\sum_{-\infty}^{\infty}(-1)^{n-1}\frac{4n^2-1}{16n^4+4n^2+1}=\pi\sqrt{2}\cosh\frac{\pi\sqrt{3}}{4}\operatorname{sech}\frac{\pi\sqrt{3}}{2}.$$

[*Math. Trip.* 1898.]

[We have $$\frac{1-4n^2}{16n^4+4n^2+1}=\frac{1}{2}\left(\frac{1}{2n-\omega}-\frac{1}{2n+\omega}+\frac{1}{2n-\omega^2}-\frac{1}{2n+\omega^2}\right),$$

where $\omega=\frac{1}{2}(-1+i\sqrt{3})$. Thus the given series is

$$-\tfrac{1}{2}\pi\{\operatorname{cosec}(\tfrac{1}{2}\pi\omega)+\operatorname{cosec}(\tfrac{1}{2}\pi\omega^2)\},$$

which is easily reduced to the given form.]

33. Shew that

$$\arctan\frac{x}{1-x^2}-\arctan\frac{3x}{3^2-x^2}+\arctan\frac{5x}{5^2-x^2}-\ldots=\arctan\left[\frac{\sinh(\frac{1}{4}\pi x)}{\cos(\frac{1}{4}\pi x\sqrt{3})}\right].$$

[*Math. Trip.* 1891.]

[It is easy to prove that

$$\arctan\frac{ax}{a^2-x^2}=\arctan\left(\frac{x}{2a+x\sqrt{3}}\right)+\arctan\left(\frac{x}{2a-x\sqrt{3}}\right),$$

and it follows that the given series is equal to

$$\arctan\frac{x}{2+y}+\sum_{-\infty}^{\infty}{}'(-1)^n\left\{\arctan\left(\frac{x}{4n+2+y}\right)-\frac{x}{4n}\right\},$$

where $y=x\sqrt{3}$. Applying the result of Ex. 34, we find the given formula.]

34. Shew that

$$\arctan\frac{y}{x}+\sum_{-\infty}^{\infty}{}'\left[\arctan\left(\frac{y}{n+x}\right)-\frac{y}{n}\right]=\arctan\left[\frac{\tanh(\pi y)}{\tan(\pi x)}\right],$$

$$\arctan\frac{y}{x}+\sum_{-\infty}^{\infty}{}'(-1)^n\left[\arctan\left(\frac{y}{n+x}\right)-\frac{y}{n}\right]=\arctan\left[\frac{\sinh(\pi y)}{\sin(\pi x)}\right].$$

In particular we find with $y=x$,

$$\sum_{1}^{\infty}\arctan\left(\frac{2x^2}{n^2}\right)=\frac{\pi}{4}-\arctan\left[\frac{\tanh(\pi x)}{\tan(\pi x)}\right],$$

$$\sum_{1}^{\infty}(-1)^{n-1}\arctan\left(\frac{2x^2}{n^2}\right)=-\frac{\pi}{4}+\arctan\left[\frac{\sinh(\pi x)}{\sin(\pi x)}\right].$$

[We have

$$\log \sin(\pi x) = \log(\pi x) + \sum_{-\infty}^{\infty}{}' \left\{ \log\left(1+\frac{x}{n}\right) - \frac{x}{n} \right\}$$

and $$\log \tan(\tfrac{1}{2}\pi x) = \log(\tfrac{1}{2}\pi x) + \sum_{-\infty}^{\infty}{}' (-1)^n \left\{ \log\left(1+\frac{x}{n}\right) - \frac{x}{n} \right\}.$$

In each of these, change x to $x+iy$, and equate the imaginary parts on the two sides.]

35. The points P, Q have coordinates (p, q), $(-p, q)$ respectively; N is the point with coordinates $(na, 0)$. Shew that if

$$\theta = \sum_{-\infty}^{\infty} P\hat{N}Q,$$

then $$\tan \tfrac{1}{2}\theta = \tan(\pi p/a) \coth(\pi q/a).$$ [*Math. Trip.* 1894.]

[If we write $p+iq=x$, $-p+iq=y$, it will be found that

$$i\theta = \log \sin(\pi x/a) - \log \sin(\pi y/a).]$$

36. Verify that, if x is a positive integer,

$$(a+b)^{n-1} + (a+2b)^{n-1} + \dots + (a+xb)^{n-1}$$
$$= [(a+xb+Bb)^n - (a+Bb)^n]/nb,$$

where we are to put $B^{2s} = B_s$, $B^{2s+1} = 0$ after expansion. [*Math. Trip.* 1897.]

[Apply Art. 94.]

37. Shew that

$$\frac{x}{2} \cot \frac{x}{2} = 1 - B_1 \frac{x^2}{2!} - B_2 \frac{x^4}{4!} - B_3 \frac{x^6}{6!} - \dots,$$

and that

$$\log \frac{x}{2\sin(\frac{1}{2}x)} = \frac{B_1}{2}\frac{x^2}{2!} + \frac{B_2}{4}\frac{x^4}{4!} + \frac{B_3}{6}\frac{x^6}{6!} + \dots.$$

38. In virtue of Baker's theorem (Art. 84), we can expand $\sec x$ in powers of x if $|x| < \frac{1}{2}\pi$, and if we write

$$\sec x = 1 + E_1 \frac{x^2}{2!} + E_2 \frac{x^4}{4!} + E_3 \frac{x^6}{6!} + \dots,$$

the coefficients E_n are called *Euler's numbers*. Prove that

$$E_1 = 1, \ E_2 = 5, \ E_3 = 61, \ E_4 = 1385,$$

and that $$1 - \frac{1}{3^{2n+1}} + \frac{1}{5^{2n+1}} - \frac{1}{7^{2n+1}} + \dots = \frac{E_n \pi^{2n+1}}{2^{2n+2}(2n)!}.$$

[See also Chrystal's *Algebra*, Chap. XXX., § 3.]

39. Shew from Ex. 37 that

$$\log \frac{\cosh x - \cos x}{x^2} = -\sum 2^{n+1} \cos(\tfrac{1}{2}n\pi) \frac{B_n}{2n} \frac{x^{2n}}{(2n)!}.$$

[*Math. Trip.* 1890.]

40. Assuming Stirling's formula (Art. 179), shew that

$$B_n \sim 4\sqrt{(\pi n)}(n/\pi e)^{2n},$$

when n is large.

CHAPTER XI.

NON-CONVERGENT AND ASYMPTOTIC SERIES.*

96. Short bibliography.

The following are the principal sources from which this chapter has been derived:

É. Borel, *Leçons sur les Séries Divergentes*, Paris, 1901.

" *Liouville's Journal de Mathématiques* (5ᵉ série), t. 2, 1896, p. 103.

" *Annales de l'École Normale Supérieure* (3ᵉ série), t. 16, 1899, p. 1.

E. Cesàro, *Bulletin des Sciences Mathématiques* (2ᵉ série), t. 14, 1890, p. 114.

G. H. Hardy, *Transactions of the Cambridge Philosophical Society*, vol. 19, 1904, p. 297 (published 1902).

" " *Quarterly Journal of Mathematics*, vol. 35, 1903, p. 22.

E. Le Roy, *Annales de la Faculté des Sciences de Toulouse* (2ᵉ série), t. 2, 1902, p. 317.

H. Poincaré, *Acta Mathematica*, t. 8, 1886, p. 295.

E. B. Van Vleck, *The Boston Colloquium*, New York, 1905, p. 75.

G. Vivanti, *Theorie der analytischen Funktionen*, Leipzig, 1906.

97. Historical introduction.

Before the methods of Analysis had been put on a sure footing, and in particular before the theory of convergence had been developed by Abel and Cauchy, mathematicians had little hesitation in using non-convergent series in both theoretical and numerical investigations.

In numerical work, however, they naturally used only series which are now called *asymptotic* (Art. 130 below); in such series the terms begin to decrease, and reach a minimum, afterwards increasing. If we take the sum to a stage at which the terms

---

* The majority of writers on these series use the word *divergent* as including *oscillatory* series; we shall, however, except in quotations, adopt the same distinction as in the previous part of the book.

are sufficiently small, we may hope to obtain an approximation with a degree of accuracy represented by the last term retained; and it can be proved that this is the case with many series which are convenient for numerical calculations (see Art. 130 for examples).

An important class of such series consists of the series used by astronomers to calculate the planetary positions: it has been proved by Poincaré* that these series do not converge, but yet the results of the calculations are confirmed by observation. The explanation of this fact may be inferred from Poincaré's theory of asymptotic series (Art. 133).

But mathematicians have often been led to employ series of a different character, in which the terms never decrease, and may increase to infinity. Typical examples of such series are:

(1) $$1-1+1-1+1-1+\dots;$$

(2) $$1-2+3-4+5-6+\dots;$$

(3) $$1-2+2^2-2^3+2^4-2^5+\dots;$$

(4) $$1-2!+3!-4!+5!-6!+\dots.$$

Euler considered the "sum" of a non-convergent series as the finite numerical value of the arithmetical expression from the expansion of which the series was derived. Thus he defined the "sums" of the series (1)–(3) as follows:

$$(1)=\frac{1}{1+1}=\frac{1}{2}; \quad (2)=\frac{1}{(1+1)^2}=\frac{1}{4}; \quad (3)=\frac{1}{1+2}=\frac{1}{3};$$

and his discussion of the series (4) will be found at the end of Art. 98 (see p. 267).

In principle, Euler's definition depends on the inversion of two limits, which, taken in one order, give a definite value, and taken in the reverse order give a non-convergent series. Thus series (1) is

$$\lim 1 - \lim x + \lim x^2 - \lim x^3 + \dots$$

as x tends to 1; Euler's definition replaces this by

$$\lim(1-x+x^2-x^3+\dots).$$

So, generally, if $\Sigma f_n(c)$ is not convergent, Euler defines the "sum" as $\lim_{x\to c} \Sigma f_n(x)$, when this limit is definite; a definition which should be compared with the principle formulated at the beginning of Art. 99.

* *Acta Mathematica*, t. 13, 1890; in particular § 13.

To this definition Callet raised the objection that the series (1) can also be obtained by writing $x=1$ in the series

$$(5)\qquad \frac{1+x}{1+x+x^2}=\frac{1-x^2}{1-x^3}=1-x^2+x^3-x^5+x^6-x^8+\ldots,$$

whereas the left-hand side then becomes $\frac{2}{3}$ instead of $\frac{1}{2}$.

This objection of Callet's was met by a remark of Lagrange's, who suggested that the series (5) should be written as

$$1+0-x^2+x^3+0-x^5+x^6+0-x^8+\ldots,$$

and that then the derived series would be

$$1+0-1+1+0-1+1+0-1+\ldots.$$

The last series gives the sums to 1, 2, 3, 4, 5, 6, ... terms as 1, 1, 0, 1, 1, 0, ...; so that the *average sum** is $\frac{2}{3}$, which is the value given by the left-hand side of (5). In the original series (1), the sums are 1, 0, 1, 0, 1, 0, ..., of which the average is $\frac{1}{2}$, agreeing with Euler's sum.

Having regard to the fact that Euler and other mathematicians made numerous discoveries by using series which do not converge, we may agree with Borel in the statement that the older mathematicians had sufficiently good *experimental* evidence that the use of such series as if they were convergent led to correct results † in the majority of cases when they presented themselves *naturally*.

A simple example of the use of a non-convergent series to obtain a correct result is afforded by a passage in Fourier's *Théorie Analytique de la Chaleur* (*Oeuvres*, t. 1, p. 206). Fourier is obtaining what we should now call a Fourier sine-series for the function $f(x)=\frac{\pi}{2}\frac{\sinh x}{\sinh \pi}$, and he finds that the coefficient of $\sin nx$ is

$$(-1)^{n-1}\left(\frac{1}{n}-\frac{1}{n^3}+\frac{1}{n^5}-\ldots\right).$$

* This remark of Lagrange's has been put on a more satisfactory basis by the theorem of Frobenius (*Crelle's Journal für Math.*, Bd. 89, 1880, p. 262), which was given in Art. 51, Ex. 2, above; namely, that

$$\lim_{x\to 1}(\Sigma a_n x^n)=\lim_{n\to\infty}\frac{s_0+s_1+\ldots+s_n}{n+1}.$$

In applying the theorem to the special series above we note that the sum $s_0+s_1+\ldots+s_n=(n+1)-k$, where k is the integral part of $\frac{1}{3}(n+1)$; thus

$$\lim\,(s_0+s_1+\ldots+s_n)/(n+1)=1-\lim k/(n+1)=\tfrac{2}{3}.$$

† Borel, *Leçons*, p. 9; the sketch given above is taken, with a few additions and slight changes, from pp. 1-10 of his book.

Thus the coefficient of $\sin x$ appears as $1-1+1-1+\dots$, and may therefore be expected to be $\frac{1}{2}$, if we adopt Euler's principle.

As a matter of fact this is correct, since

$$\int \sinh x \sin x \, dx = \tfrac{1}{2}(\cosh x \sin x - \sinh x \cos x),$$

so that

$$\frac{2}{\pi}\int_0^\pi f(x) \sin x \, dx = \tfrac{1}{2}.$$

Abel and Cauchy, however, pointed out that the use of non-convergent series had sometimes led to gross errors; and, in their anxiety to place mathematical analysis on the firmest foundations, they felt obliged to banish non-convergent series from their work. But this was not done without hesitation; thus Abel writes to his former teacher Holmboë in 1826 (*Oeuvres d'Abel*, 2me. éd. t. 2, p. 256): "Les séries divergentes sont, en général, quelque chose de bien fatal, et c'est une honte qu'on ose y fonder aucune démonstration ... la partie la plus essentielle des Mathématiques est sans fondement. Pour la plus grande partie, les résultats sont justes, il est vrai, mais c'est là une chose bien étrange. Je m'occupe à en chercher la raison, problème très intéressant."

And Cauchy, in the preface of his *Analyse Algébrique* (1821), writes: "J'ai été forcé d'admettre diverses propositions qui paraîtront peut-être un peu dures: par exemple, qu'une série divergente n'a pas de somme." *

Cauchy established the asymptotic property of Stirling's series (see Art. 132 below), by means of a method which can be applied to a large class of power-series. But the possibility of obtaining other useful asymptotic series was generally overlooked by later analysts; and after the time of Cauchy, workers in the regions of analysis for the most part abandoned all attempts at utilising non-convergent series. In England, however, Stokes published three remarkable papers† (dated 1850, 1857, 1868), in which Cauchy's method for dealing with Stirling's series was applied to a number of other problems, such as the calculation of Bessel's functions for large values of the variable.

---

* Of course no one would now regard Cauchy's statement as unusual, *in the sense in which he made it.*

† See the references of Arts. 133, 135 below.

But no general theory of non-convergent series was forthcoming until 1886, when papers discussing the subject were written by Stieltjes* and Poincaré.† Since that time a number of researches have been published on the theory, some of which were quoted in Art. 96.

In the following articles we shall for the most part confine our exposition to the theory of non-convergent series set forth by Borel in 1896, although we do not always adopt his methods.

98. General considerations on non-convergent series.

In view of the results obtained in the past by the use of non-convergent series, it seems probable that we can attach a perfectly precise meaning to a non-convergent series, so that such series may be used for purposes of formal calculation, under proper restrictions. Thus we attempt to formulate rules which enable us, given a series

$$u_0 + u_1 + u_2 + \dots$$

(convergent or not), to associate with it a number, which is a perfectly definite function of u_0, u_1, u_2, ..., and which we call the "sum" of the series. It is of course obvious that the definition chosen is to a large extent arbitrary; but it should be such that the resulting laws of calculation agree, as far as possible, with those of convergent series.

Of course it is evident that the "sum" associated with a non-convergent series is not to be confounded with the sum of a convergent series (in the sense of Art. 6); but it will avoid confusion if the definition is such that the same operation, when applied to a convergent series, yields the sum in the ordinary sense.

It ought to be pointed out that Euler, at any rate, was perfectly aware of the distinction between his "sum" of a non-convergent series and the sum of a convergent series. Thus he says (§§ 108-111 of the *Instit. Calc. Diff.*, 1755) that the series

$$1 - 2 + 2^2 - 2^3 + 2^4 - \dots = \frac{1}{1+2} = \frac{1}{3}$$

obviously cannot have the sum $\frac{1}{3}$ in the ordinary sense, since the sum of n terms is $S_n = \frac{1}{3}[1 - (-2)^n]$, and the larger n is, the more does S_n differ from $\frac{1}{3}$.

---

* *Annales de l'École Normale Supérieure* (3), t. 3, p. 201; we do not propose to set forth the theory of Stieltjes here. The reader may consult Van Vleck's book for a full account of this theory (see reference, Art. 96).

† *Acta Mathematica*, t. 8, p. 295 (for Poincaré's theory see Art. 133 below).

And he adds, after referring to various difficulties, that contradictions can be avoided by attributing a somewhat different meaning to the word *sum*. "Let us say, therefore, that the *sum* of any infinite series is the finite expression, by the expansion of which the series is generated. In this sense the sum of the infinite series $1-x+x^2-x^3+\ldots$ will be $\frac{1}{1+x}$, because the series arises from the expansion of the fraction, whatever number is put in place of x. If this is agreed, the new definition of the word *sum* coincides with the ordinary meaning when a series converges; and since divergent series have no sum, in the proper sense of the word, no inconvenience can arise from this new terminology. Finally, by means of this definition, we can preserve the utility of divergent series and defend their use from all objections."

In writing to N. Bernoulli (*L. Euleri Opera Posthuma*, t. 1, p. 536), Euler adds that he had had grave doubts as to the use of divergent series, but that he had never been led into error by using his definition of "sum." To this Bernoulli replies that the same series might arise from the expansion of more than one expression, and that if so, the "sum" would not be definite; to which Euler rejoins that he does not believe that any example of this could be given. However, Pringsheim (*Encyklopädie*, Bd. I., A. 3, 39) has given a number of examples to shew that Euler fell into error here; but in practice Euler used his definition almost exclusively in the form

$$\Sigma u_n = \lim_{x \to 1} (\Sigma u_n x^n),$$

and if restricted to this case, Euler's statement is correct.

It will be seen from these passages that Euler had views which do not differ greatly, at bottom, from those held by modern workers on this subject; although of course his attempted definition leaves something to be desired, in the light of modern analysis.

It is to be carefully borne in mind that **the legitimate use of non-convergent series is always symbolic**; the operations being merely convenient abbreviations of more complicated transformations in the background. Naturally this "shorthand representation" does not enable us to avoid the labour of justifying the various steps employed; but when general rules have been laid down and firmly established we may apply them with confidence in any particular case.

It may very likely be urged that we might just as well write the work in full, and so avoid all risk of misinterpretation. But experience shews that the use of the series frequently suggests profitable transformations which otherwise might never be thought of.

An example of this may be taken from Euler's correspondence with Nicholas Bernoulli (*L. Euleri Opera Posthuma*, t. 1, p. 547); where the real object of Euler's work is to shew how to attach a definite meaning to the series (4) of Art. 97.

He proves first that the series

$$x-(1\,!)x^2+(2\,!)x^3-(3\,!)x^4+\ldots$$

satisfies *formally* the differential equation

$$x^2\frac{dy}{dx}+y=x,$$

from which he obtains the integral

$$y=\int_0^x e^{\frac{1}{x}-\frac{1}{\xi}}\frac{d\xi}{\xi}.$$

Or, if
$$\frac{1}{\xi}-\frac{1}{x}=t,$$

we find
$$y=\int_0^\infty \frac{xe^{-t}}{1+xt}\,dt,$$

as given by Borel's method (Art. 136 below).

On the other hand, by using the rules which he had obtained for the transformation of convergent series into continued fractions, Euler gets

$$\frac{x}{1+}\;\frac{x}{1+}\;\frac{x}{1+}\;\frac{2x}{1+}\;\frac{2x}{1+}\;\frac{3x}{1+}\;\frac{3x}{1+}\;\ldots,$$

and it has since been proved by Laguerre* and Stieltjes that we have actually

$$\int_0^\infty \frac{xe^{-t}}{1+xt}\,dt=\frac{x}{1+}\;\frac{x}{1+}\;\frac{x}{1+}\;\frac{2x}{1+}\;\frac{2x}{1+}\;\ldots.$$

Now this relation does not suggest itself at all naturally without the use of the series; and, as already remarked, it is evident that Euler's work was entirely guided by the aim of evaluating the series (4).

Writing $x=1$, Euler obtains from the continued fraction the convergents

$$1>\tfrac{2}{3}>\tfrac{8}{13}>\tfrac{44}{73}>\ldots>\tfrac{20}{34}>\tfrac{4}{7}>\tfrac{1}{2},$$

and by using the 13th and 14th of these, he infers the numerical value 0·5963....

He then subtracts this decimal from 1 and infers that the value of the series (4) is 0·4036... (compare Art. 132 below).

99. Borel's integral; summable series.

The general considerations of the last article have now to be specialised by adopting some conventional meaning to be taken as the "sum" of a non-convergent series.

Hardy, in the papers quoted above, has formulated the following principle: *If two limiting processes, performed in a definite order on a function of two variables, lead to a definite value X, but, when performed in the reverse order, lead to a*

---

* *Bulletin de la Société Math. de France*, t. 7, 1879, p. 72.

meaningless expression Y, we may agree to interpret Y as meaning X.

For example, the equation

$$\lim_{n\to\infty}\int_0^\infty f(x, n)\,dx = \int_0^\infty [\lim_{n\to\infty} f(x, n)]\,dx$$

is true under certain restrictions* on the function $f(x, n)$. But it may happen that the right-hand side is perfectly definite while the left-hand oscillates between certain values (which may be $-\infty$ and $+\infty$). In such cases, Hardy's principle is that we may take the right-hand integral as *defining* the left-hand limit.

In particular, suppose that $f(x, n)$ is the sum to $n+1$ terms of a series of functions of x; say that

$$f(x, n) = \phi_0(x) + \phi_1(x) + \phi_2(x) + \ldots + \phi_n(x)$$

and that

$$\Phi(x) = \lim_{n\to\infty} f(x, n) = \phi_0(x) + \phi_1(x) + \phi_2(x) + \ldots \text{ to } \infty.$$

The equation then becomes

$$\sum_0^\infty \left[\int_0^\infty \phi_n(x)\,dx\right] = \int_0^\infty \Phi(x)\,dx,$$

and it may easily happen that the integral $\int_0^\infty \Phi(x)\,dx$ is convergent, when the series on the left is not; if so, *we have an integral which may be taken as defining the "sum" of a non-convergent series.*

The special type of function $\phi_n(x)$ which has led to the most interesting results is

$$\phi_n(x) = e^{-x} u_n x^n / n!,$$

where u_n is independent of x.

Then $$\int_0^\infty \phi_n(x)\,dx = u_n \qquad \text{(see Art. 178)}$$

and $$\Phi(x) = e^{-x}\left(u_0 + u_1 x + u_2 \frac{x^2}{2!} + u_3 \frac{x^3}{3!} + \ldots\right) = e^{-x} u(x),$$

let us say. We must assume† that the coefficients u_n are such that the series $u(x)$ converges for all values of x.

* A few simple cases will be found in the Appendix, Art. 172.

† At least for the present; but see Art. 136 below.

Then, *provided that the integral* $\int_0^\infty e^{-x}u(x)\,dx$ *is convergent, we may agree to associate its value with the series* $\sum_0^\infty u_n$, *if this series is not convergent*; this integral may then be called the "*sum*" of the series; and the series may be called *summable*. The sum may be denoted by the symbol

$$\mathscr{S}_0^\infty u_n.$$

This definition is due to Borel, who deduced it, however, from the definition given below in Art. 114; but since the integral has proved more serviceable in subsequent investigations, it seemed best to regard this as the fundamental definition.

100. Condition of consistency.

It is obvious that the definition given in the last article will lead to difficulties unless the sum $\mathscr{S}u_n$ agrees with the sum Σu_n whenever the series converges; we shall now prove that this "condition of consistency" is satisfied.

For, *if* Σu_n *is convergent*, we can find a number μ such that

$$|u_n| < \epsilon, \qquad \text{if } n \geqq \mu.$$

Then
$$\left|\sum_n^\infty u_r \frac{x^r}{r!} e^{-x}\right| < \epsilon\left[e^{-x}\sum_n^\infty \frac{x^r}{r!}\right] < \epsilon, \quad \text{if } n \geqq \mu;$$

and consequently the series $\sum_0^\infty u_r \frac{x^r}{r!}e^{-x}$ converges *uniformly* for all values of x within an interval $(0, \lambda)$, where λ may be arbitrarily large.

Hence
$$\int_0^\lambda e^{-x}u(x)\,dx = \sum_0^\infty u_n \int_0^\lambda e^{-x}\frac{x^n}{n!}\,dx$$

by Art. 45; that is,

$$\int_0^\lambda e^{-x}u(x)\,dx = \sum_0^\infty u_n\lambda_n,$$

where $\lambda_n = \int_0^\lambda e^{-x}\frac{x^n}{n!}\,dx = \left(\frac{\lambda^{n+1}}{(n+1)!} + \frac{\lambda^{n+2}}{(n+2)!} + \dots \text{ to } \infty\right)e^{-\lambda}$.

Now, since λ is positive, the sequence (λ_n) is a *decreasing* sequence; and consequently we can apply the extended form of Abel's lemma (Art. 23) to the sum $\Sigma u_n\lambda_n$, and we see that

$$h(\lambda_0 - \lambda_m) + h_m\lambda_m < \sum_0^\infty u_n\lambda_n < H(\lambda_0 - \lambda_m) + H_m\lambda_m,$$

where H, h are the upper and lower limits of $u_0+u_1+\ldots+u_r$ as r varies from 0 to $m-1$ and H_m, h_m are the limits as r varies from m to ∞.

Now *if* Σu_n *converges to the sum* s, we can choose m so that $h_m \geqq s-\epsilon$, $H_m \leqq s+\epsilon$, however small ϵ may be; and as λ tends to ∞, λ_0 and λ_m tend to the limit 1, because m is fixed.* Thus

$$\lim_{\lambda\to\infty} [h(\lambda_0-\lambda_m)+h_m\lambda_m]=h_m \geqq s-\epsilon$$

and
$$\lim_{\lambda\to\infty} [H(\lambda_0-\lambda_m)+H_m\lambda_m]=H_m \leqq s+\epsilon.$$

Hence we see that

$$s-\epsilon \leqq \overline{\lim_{\lambda\to\infty}} \sum_0^\infty u_n\lambda_n \leqq s+\epsilon.$$

But ϵ is arbitrarily small, and so (see Note (6), p. 5) the maximum and minimum limits of $\Sigma u_n\lambda_n$ must both be equal to s, so that

$$\lim_{\lambda\to\infty} \sum_0^\infty u_n\lambda_n = s,$$

or, what is the same thing, $\mathscr{S}_0^\infty u_n = \sum_0^\infty u_n$.

But *if* Σu_n *diverges* (say to $+\infty$), we can find m so that

$$h_m > N,$$

however great N may be.

Thus, since

$$\lim_{\lambda\to\infty} [h(\lambda_0-\lambda_m)+h_m\lambda_m]=h_m > N,$$

we see that
$$\underline{\lim_{\lambda\to\infty}} \sum_0^\infty u_n\lambda_n \geqq N.$$

Since N is arbitrarily great, this inequality implies that $\sum_0^\infty u_n\lambda_n$ tends to infinity with λ. That is, the integral for $\mathscr{S}_0^\infty u_n$ is divergent, when $\sum_0^\infty u_n$ is divergent. Of course this is only as it should be; otherwise there might be a risk of inconsistencies arising.

---

* For $\lambda_0=1-e^{-\lambda}$, $\lambda_m=1-\left(1+\lambda+\frac{1}{2}\lambda^2+\ldots+\frac{1}{m!}\lambda^m\right)e^{-\lambda}$ and each of these expressions tends to the limit 1.

101. Relation between $\mathscr{S}_0^\infty u_n$ and $\mathscr{S}_1^\infty u_n$.

It does not follow immediately from their definitions that the two sums

$$s = \mathscr{S}_0^\infty u_n \text{ and } s' = \mathscr{S}_1^\infty u_n$$

are related in such a way that the existence of either implies the existence of the other. The definition gives, however, the relation

$$s = \int_0^\infty e^{-x} u(x)\, dx, \quad s' = \int_0^\infty e^{-x} u'(x)\, dx,$$

because

$$u'(x) = \frac{du}{dx} = u_1 + u_2 x + u_3 \frac{x^2}{2!} + \dots$$

since the coefficients $u_0, u_1, u_2, u_3, \dots$ are such that $u(x)$ converges for any value of x; and consequently Art. 50 can be used to justify term-by-term differentiation of $u(x)$.

On integrating the expression for s' by parts, we find the relation

$$s' = \left[e^{-x} u(x)\right]_0^\infty + s.$$

This shews that when s, s' both exist, the limit

$$\lim_{x \to \infty} [e^{-x} u(x)]$$

exists also; and the value of this limit must be zero, since otherwise the integral for s could not converge. (See Art 166, Appendix.)

Hence, *provided that both integrals s, s' are convergent, we have* $s - s' = u_0$; thus the two series, *when summable*, possess properties, in this respect, analogous to those of convergent series. Of course any failure of the analogy here would lead to obvious difficulties and would greatly reduce the utility of the definition which we have selected for the "sum" of a summable series.

But it must be remembered that the convergence of

$$\int_0^\infty e^{-x} u(x) dx$$

by no means ensures the existence of a definite value for

$$\lim_{x \to \infty} e^{-x} u(x),$$

as is clear from Art. 166 of the Appendix. It follows that *the existence of a definite value for s need not imply the existence of any definite value for s'.*

An example of this has been given by Hardy (p. 30 of his second paper), by taking

$$u(x)=e^x \sin(e^x).$$

Then
$$s=\int_0^\infty \sin(e^x)dx=\int_1^\infty \frac{\sin y}{y}dy,$$

but the integral for s' is

$$\int_0^\infty e^{-x}\frac{d}{dx}[e^x \sin(e^x)]dx=\int_1^\infty \frac{1}{y}\frac{d}{dy}(y\sin y)dy=\int_1^\infty \left(\frac{\sin y}{y}+\cos y\right)dy,$$

which oscillates.

Thus s' does not converge, although s does converge.

The reader will find little difficulty in seeing that here

$$u_n=2^n\sum_{r=0}^{\infty}(-1)^r\frac{(r+1)^n}{(2r+1)!};$$

the algebraical transformation will be found in Hardy's paper.

Although the convergence of $\int_0^\infty e^{-x}u(x)dx$ does not ensure that $e^{-x}u(x)$ tends to a definite limit as x increases to ∞, *yet we can infer that* $\lim_{x\to\infty} e^{-x}u(x)=0$, *when the integral* $\int_0^\infty e^{-x}u'(x)dx$ *is convergent; and consequently that the integral* $\int_0^\infty e^{-x}u(x)dx$ *converges also.*

For, write
$$y=\int_0^x e^{-x}u(x)dx.$$

Then
$$\frac{dy}{dx}=e^{-x}u(x),\quad \frac{d^2y}{dx^2}=e^{-x}u'(x)-e^{-x}u(x),$$

and consequently

$$\int_0^x e^{-x}u'(x)dx=\int_0^x e^{-x}u(x)dx+\frac{dy}{dx}-u_0=y+y'-u_0.$$

The last expression tends to a finite limit l, as x increases to ∞; so write

$$\eta=y-(u_0+l),\quad \eta'=\frac{d\eta}{dx},$$

and then $(\eta+\eta')$ tends to the limit 0. Thus we can choose X, so that

$$|\eta+\eta'|<\epsilon,\quad \text{if } x>X.$$

If η has an unlimited number of extreme values, η' vanishes at each of these; hence at any extreme value beyond X, we have $|\eta|<\epsilon$. Thus

$$|\eta|<\epsilon,\quad \text{if } x>X,$$

or
$$\lim_{x\to\infty}\eta=0,\ \text{and hence } \lim_{x\to\infty}\eta'=0.$$

But if η has not an unlimited number of extreme values, η' must be finally of constant sign, and so η must approach a definite limit (Art. 2), say a; thus η' approaches the limit $-a$. Now, from Art. 152, it follows that if η' approaches the limit $-a$, then η/x approaches the same limit. Thus we have

$$\lim_{x\to\infty} \eta = a, \quad \lim_{x\to\infty} (\eta/x) = -a,$$

but η and η/x have the *same* sign, so that a must be zero.

Thus we must always have

$$\lim_{x\to\infty} \eta = 0, \quad \lim_{x\to\infty} \eta' = 0.$$

or

$$\int_0^\infty e^{-x} u(x)\, dx = u_0 + \int_0^\infty e^{-x} u'(x)\, dx,$$

and

$$\lim_{x\to\infty} e^{-x} u(x) = 0,$$

provided that the integral $\int_0^\infty e^{-x} u'(x)\, dx$ is convergent.

Accordingly we have the result: *If the series*

$$u_1 + u_2 + u_3 + \dots$$

is summable, then so also is the series

$$u_0 + u_1 + u_2 + u_3 + \dots;$$

and the relation between their sums

$$\overset{\infty}{\underset{0}{\mathscr{S}}} u_n = u_0 + \overset{\infty}{\underset{1}{\mathscr{S}}} u_n$$

is correct.

Another mode of stating these results is the following:

Any finite number of terms may be prefixed to a summable series, and the series will remain summable; and further, the sums will be related as if the two series were convergent.

But the removal of even a single term from the beginning of the series may destroy the property of summability.

102. Some obvious theorems.

It follows immediately from the definitions that if $\overset{\infty}{\underset{0}{\mathscr{S}}} u_n$ and $\overset{\infty}{\underset{0}{\mathscr{S}}} v_n$ are summable, then $\overset{\infty}{\underset{0}{\mathscr{S}}} (u_n + v_n)$ and $\overset{\infty}{\underset{0}{\mathscr{S}}} (u_n - v_n)$ are also summable, and

$$\overset{\infty}{\underset{0}{\mathscr{S}}} (u_n \pm v_n) = \overset{\infty}{\underset{0}{\mathscr{S}}} u_n \pm \overset{\infty}{\underset{0}{\mathscr{S}}} v_n.$$

Further, if C is any factor independent of n,

$$\overset{\infty}{\underset{0}{\mathscr{S}}} C u_n = C \overset{\infty}{\underset{0}{\mathscr{S}}} u_n.$$

We should note further that *the addition of a constant to any term of a summable series is equivalent to adding the same constant to its sum*: this of course is a special case of the theorem on the addition of summable series. For the series

$$k+0+0+0+0+0+\dots,$$
$$0+k+0+0+0+0+\dots,$$
$$0+0+k+0+0+0+\dots,$$
$$\dots\dots\dots\dots\dots\dots$$

are all summable and equal to k.

Thus $$k+\mathscr{S}_0^\infty u_n=(k+u_0)+u_1+u_2+u_3+\dots,$$

although, as we have just proved (Art. 101), $u_1+u_2+u_3+\dots$ is not necessarily summable when $0+u_1+u_2+u_3+\dots$ is so.

103. Examples of summable series.

Ex. 1. $$1-1+1-1+1-\dots.$$

Here $$u(x)=1-x+\frac{x^2}{2!}-\frac{x^3}{3!}+\dots=e^{-x},$$

and so $\int_0^\infty e^{-x}u(x)\,dx=\int_0^\infty e^{-2x}\,dx=\tfrac{1}{2}$. [Compare Art. 97.]

Assuming this series to be summable, and to have a sum s, we can evaluate s by prefixing a term -1 at the beginning of the series (Art. 101): this process leads to

$$s-1=-1+1-1+\dots=-s, \quad \text{(by Art. 102)}$$

so that $$s=\tfrac{1}{2}.$$

Ex. 2. $$1-t+t^2-t^3+\dots. \quad (t>1)$$

Here, just as in Ex. 1, we find $u(x)=e^{-tx}$, and then we get

$$\int_0^\infty e^{-x}u(x)\,dx=1/(1+t).$$

Here again, assuming the series summable, we obtain its value s, by prefixing -1 at the beginning of ts, giving

$$ts-1=-1+t-t^2+t^3-\dots=-s$$

or $$s=1/(1+t).$$

Ex. 3. $$1-2+3-4+5-\dots.$$

Here $$u(x)=1-2x+3\frac{x^2}{2!}-4\frac{x^3}{3!}+\dots=e^{-x}(1-x)$$

and so $$\int_0^\infty e^{-x}u(x)\,dx=\int_0^\infty e^{-2x}(1-x)\,dx=\tfrac{1}{2}-\tfrac{1}{4}=\tfrac{1}{4}.$$

Assuming the summability, let us write

$$s=1-2+3-4+5-6+\ldots,$$

and so $$s+0=0+1-2+3-4+5-\ldots \quad \text{(Art. 102).}$$

Adding the corresponding terms, we get

$$2s=1-1+1-1+1-1+\ldots=\tfrac{1}{2} \quad \text{(by Ex. 1)}$$

or $$s=\tfrac{1}{4}.$$

Ex. 4.
$$C=1+\cos\theta+\cos 2\theta+\cos 3\theta+\ldots,$$
$$S=0+\sin\theta+\sin 2\theta+\sin 3\theta+\ldots.$$

Thus $$C+iS=1+e^{i\theta}+e^{2i\theta}+e^{3i\theta}+\ldots,$$

and the associated function is

$$u(x)=1+xe^{i\theta}+\frac{x^2}{2!}e^{2i\theta}+\ldots=\exp(xe^{i\theta})$$

or $$u(x)=e^{x(\cos\theta+i\sin\theta)}.$$

Hence, provided that θ is not zero or a multiple of 2π, we find the sum

$$\int_0^\infty e^{-x(1-\cos\theta-i\sin\theta)}\,dx=\frac{1}{1-\cos\theta-i\sin\theta}=\frac{1}{2}\left(1+i\cot\frac{\theta}{2}\right).$$

Of course, assuming the summable property, the sum can be obtained by means of the same device as in Ex. 2, which gives the equation

$$e^{i\theta}(C+iS)+1=C+iS,$$

and therefore $$C+iS=1/(1-e^{i\theta}),$$

in agreement with the result just found.

Thus we have

$$\left.\begin{aligned} &1+\cos\theta+\cos 2\theta+\cos 3\theta+\ldots=\tfrac{1}{2}\\ \text{and}\quad &0+\sin\theta+\sin 2\theta+\sin 3\theta+\ldots=\tfrac{1}{2}\cot(\tfrac{1}{2}\theta).\end{aligned}\right\}\quad (0<\theta<2\pi)$$

In like manner we find that

$$e^{i\theta}+e^{2i\theta}+e^{3i\theta}+\ldots=e^{i\theta}/(1-e^{i\theta})=\frac{1}{2}\left(-1+i\cot\frac{\theta}{2}\right),$$

so that
$$\left.\begin{aligned} &\cos\theta+\cos 2\theta+\cos 3\theta+\ldots=-\tfrac{1}{2},\\ &\sin\theta+\sin 2\theta+\sin 3\theta+\ldots=\tfrac{1}{2}\cot(\tfrac{1}{2}\theta).\end{aligned}\right\}\quad (0<\theta<2\pi)$$

By changing from θ to $\pi-\theta$, we obtain

$$\left.\begin{aligned} &\cos\theta-\cos 2\theta+\cos 3\theta+\ldots=+\tfrac{1}{2},\\ &\sin\theta-\sin 2\theta+\sin 3\theta-\ldots=\tfrac{1}{2}\tan(\tfrac{1}{2}\theta).\end{aligned}\right\}\quad (-\pi<\theta<\pi)$$

Ex. 5. If $$C = \cos\theta + \cos 3\theta + \cos 5\theta + \dots$$
and $$S = \sin\theta + \sin 3\theta + \sin 5\theta + \dots,$$
we have $$C + iS = e^{i\theta}(1 + e^{2i\theta} + e^{4i\theta} + \dots) = e^{i\theta}/(1 - e^{2i\theta}),$$
using Ex. 4.

Hence $$C = 0, \quad S = \tfrac{1}{2}\operatorname{cosec}\theta \quad (0 < \theta < \pi).$$

It will be noticed that these two values are the same as are given (Ex. 4) by summing

$$\cos\theta + 0 + \cos 3\theta + 0 + \cos 5\theta + 0 + \dots$$

and $$\sin\theta + 0 + \sin 3\theta + 0 + \sin 5\theta + 0 + \dots;$$

but *there is no general theorem which would justify our equating these new series to C and S.*

By changing from θ to $\frac{1}{2}\pi - \theta$, we obtain

$$\left.\begin{aligned} \cos\theta - \cos 3\theta + \cos 5\theta - \dots &= \tfrac{1}{2}\sec\theta, \\ \sin\theta - \sin 3\theta + \sin 5\theta - \dots &= 0. \end{aligned}\right\} \quad (-\tfrac{1}{2}\pi < \theta < \tfrac{1}{2}\pi)$$

Ex. 6. It is easy to deduce from Ex. 4 that if θ is not equal to $\pm\phi$, we have

$$\cos\theta\cos\phi + \cos 2\theta\cos 2\phi + \dots = \tfrac{1}{2},$$
$$\sin\theta\sin\phi + \sin 2\theta\sin 2\phi + \dots = 0,$$
$$\cos\theta\sin\phi + \cos 2\theta\sin 2\phi + \dots = \tfrac{1}{2}\sin\phi/(\cos\theta - \cos\phi).$$

104. Absolutely summable series.

Following Borel, we say that *a series*

$$u_0 + u_1 + u_2 + \dots$$

is absolutely summable if the integrals

$$\int_0^\infty e^{-x} u(x)\,dx \quad \textit{and} \quad \int_0^\infty e^{-x} u^\lambda(x)\,dx,$$

where λ represents any index of differentiation, are all absolutely convergent.

It is at once evident that if the series $u_0 + u_1 + u_2 + \dots$ is absolutely summable, so also is the series obtained by removing any number of terms from the beginning of the series: for the integral function associated with the series

$$u_1 + u_2 + u_3 + \dots$$

is $$u_1 + u_2 x + u_3 \frac{x^2}{2!} + \dots = u'(x),$$

and by hypothesis $\int_0^\infty e^{-x} u'(x)\,dx$ is absolutely convergent.

It follows, in virtue of what was proved in Art. 101, that we must have

$$\lim_{x\to\infty} e^{-x}u(x)=0, \qquad \lim_{x\to\infty} e^{-x}u^{\lambda}(x)=0.$$

Accordingly, *if a series*

$$u_0+u_1+u_2+u_3+\dots$$

is absolutely summable, so also are all the series

$$u_\lambda+u_{\lambda+1}+u_{\lambda+2}+\dots,$$

and their sums are related as if they were convergent.

And, conversely, *if the series* $u_1+u_2+u_3+\dots$ *is absolutely summable, so also is the series formed by prefixing* u_0.

That is, we can infer the convergence of the integral

$$\int_0^\infty e^{-x}|u(x)|\,dx$$

from that of

$$\int_0^\infty e^{-x}|u'(x)|\,dx.$$

To prove this, let us write

$$\phi(x)=\int_0^x |u'(t)|\,dt \geqq \left|\int_0^x u'(t)\,dt\right|.$$

Thus $$\phi(x)\geqq|u(x)-u_0|,$$

so that $$|u(x)|\leqq\phi(x)+|u_0|.$$

Hence $\int_0^\infty e^{-x}|u(x)|\,dx$ certainly converges if $\int_0^\infty e^{-x}\phi(x)\,dx$ is convergent. But the convergence of the last integral can be inferred from that of $\int_0^\infty e^{-x}\phi'(x)dx$, by the argument of Art. 101 above. The proof, however, can be simplified here, in virtue of the monotonic nature of $\phi(x)$.

For we have the identity

$$\int_0^X e^{-x}\phi(x)\,dx=-e^{-X}\phi(X)+\int_0^X e^{-x}\phi'(x)\,dx,$$

and consequently, because $\phi(X)$ and $\phi'(x)$ are positive,

$$\int_0^X e^{-x}\phi(x)\,dx<\int_0^X e^{-x}\phi'(x)\,dx<\int_0^\infty e^{-x}\phi'(x)\,dx.$$

Hence, since the integrand $e^{-x}\phi(x)$ is positive, the integral

$$\int_0^\infty e^{-x}\phi(x)\,dx$$

is convergent, by Art. 166 of the Appendix.

It will be noted that this argument *cannot* be reversed; and that the convergence of $\int_0^\infty e^{-x}\phi'(x)\,dx$ is not deducible from the convergence of $\int_0^\infty e^{-x}\phi(x)\,dx$.

An example of a positive integral function $\phi(x)$ for which $\int_0^\infty e^{-x}\phi(x)\,dx$ is convergent, but $\lim_{x\to\infty} e^{-x}\phi(x)$ does not exist, may be obtained by taking $\phi(x)=\exp(x-x^3\sin^2 x)$, so that $\overline{\lim}\, e^{-x}\phi(x)=1$.

It follows from a general result due to du Bois Reymond that

$$\int_0^\infty \exp(-x^3\sin^2 x)\,dx$$

is convergent (see Ex. 3, Art. 166).

The function $\phi(x)$ can obviously be expanded in a power-series for any value of x, although the law of the coefficients is not simple.

As examples of absolutely summable series, we may refer to those given already in Art. 103. Thus, for instance, take Ex. 2; here it will be seen that

$$u(x)=e^{-tx} \quad\text{and}\quad \frac{d^\lambda u}{dx^\lambda}=(-t)^\lambda e^{-tx},$$

so that $\int_0^\infty e^{-x}\left|\frac{d^\lambda u}{dx^\lambda}\right|dx$ is obviously convergent, since $1+t>0$.

Similarly the other cases are proved to be absolutely summable.

105. Absolutely convergent series are absolutely summable.

For we have

$$|u(x)|\leqq \sum |u_n|\frac{x^n}{n!},$$

so that $$|u(x)|\leqq v(x), \quad \text{if } v_n=|u_n|.$$

Now, since the series Σv_n is convergent, the integral

$$\int_0^\infty e^{-x}v(x)\,dx$$

converges (Art. 100), and therefore the integral

$$\int_0^\infty e^{-x}u(x)\,dx$$

is absolutely convergent.

Further, the series $u_\lambda+u_{\lambda+1}+u_{\lambda+2}+\dots$ is absolutely convergent, and the same argument applied to this series shews that the integral

$$\int_0^\infty e^{-x}u^\lambda(x)\,dx$$

is also absolutely convergent. That is, an absolutely convergent series is absolutely summable.

But Borel's statement, that any convergent series is absolutely summable, is not correct, as has been proved by an example constructed by Hardy (in his second paper, pp. 25–28).

This example is given by taking

$$0-1+0+0+\tfrac{1}{2}+0+0+0+0-\tfrac{1}{3}+\dots,$$

in which $u_n=(-1)^\nu/\nu$ when $n=\nu^2$, and u_n is 0 when n is not a square.

Hardy then shews that if ξ is the largest integer contained in $\sqrt{x}$, and if $\sqrt{x}-\xi<\frac{1}{4}$, $|u(x)|$ is of the same order of magnitude as the term in which $\nu=\xi$; and that $e^{-x}|u(x)|>\kappa/x$, where κ is a constant, throughout all intervals $\{m^2, (m+\frac{1}{4})^2\}$, m being any integer. It is then easy to prove that $\int_0^\infty e^{-x}|u(x)|\,dx$ is divergent.

It is easy to construct simple series which are summable, but not absolutely; an example is given by the series

$$u=\mathop{\mathscr{S}}_0^\infty \frac{(1+i)^{n+1}}{n+1}.$$

For here $$u(x)=\frac{1}{x}[e^x(\cos x+i\sin x)-1],$$

and so $$\int_0^\infty e^{-x}u(x)\,dx=\int_0^\infty \frac{\cos x-e^{-x}+i\sin x}{x}\,dx=\frac{\pi i}{2}$$

[Art. 173 (1)].

But $$|u(x)|\geqq\frac{e^x-1}{x},$$

and so $$\int_0^X e^{-x}|u(x)|\,dx\geqq\int_0^X\frac{1-e^{-x}}{x}\,dx,$$

and thus $$\int_0^\infty e^{-x}|u(x)|\,dx \text{ is divergent.}$$

In regard to this series, we note that

$$e^{-x}u(x)=(e^{ix}-e^{-x})/x,$$

which tends to zero, as x tends to ∞; thus, as in Art. 101, we see that $\int_0^\infty e^{-x}u'(x)dx$ converges.

But the convergence of this integral is not absolute, because

$$e^{-x}u'(x)=(1+i)e^{ix}/x-(e^{ix}-e^{-x})/x^2,$$

so that

$$|e^{-x}u'(x)|\geqq\sqrt{2}/x-2/x^2, \quad \text{if } x>\sqrt{2}.$$

Thus the integral $\int_0^\infty e^{-x}|u'(x)|\,dx$ is divergent.

Similarly, we establish the convergence of $\int_0^\infty e^{-x}u^\lambda(x)dx$, λ being any index of differentiation, and the divergence of $\int_0^\infty e^{-x}|u^\lambda(x)|\,dx$.

Thus *the series* $u_\lambda+u_{\lambda+1}+u_{\lambda+2}+\dots$ *is summable for all values of* λ, *although not absolutely.*

106. Multiplication of absolutely summable series.

If the two series

$$u=u_0+u_1+u_2+\dots, \quad v=v_0+v_1+v_2+\dots$$

are both absolutely summable, and if

$$w_n=u_0v_n+u_1v_{n-1}+\dots+u_nv_0,$$

then the series

$$w=w_0+w_1+w_2+\dots$$

is also absolutely summable, and

$$w=uv:$$

For then we have

$$uv=\int_0^\infty e^{-x}u(x)\,dx\int_0^\infty e^{-y}v(y)\,dy$$

$$=\lim_{\lambda\to\infty}\iint e^{-(x+y)}u(x)v(y)\,dx\,dy,$$

where the latter double integral is taken over the area of a square of side λ.

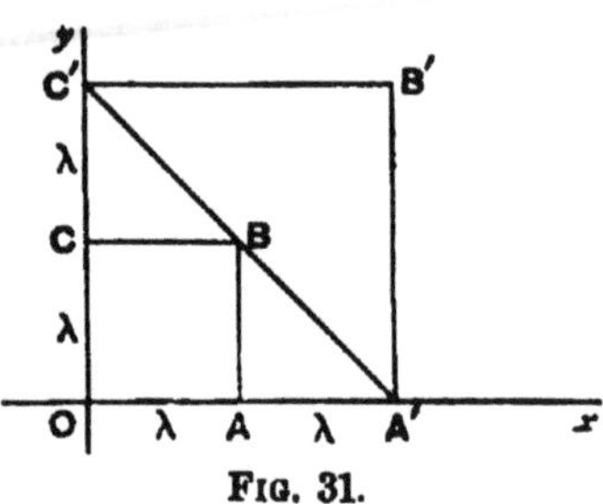

FIG. 31.

Now, in case $u(x)$ and $v(y)$ are both positive, the integral taken over a triangle such as $OA'C'$ lies between the integrals taken over the squares $OABC$ and $OA'B'C'$; so that, since the latter integrals both approach the limit uv when OA tends to ∞, the integral over $OA'C'$ will also approach the limit uv. But if $u(x)$ and $v(y)$ are not always positive, the difference between the integrals over $OABC$ and $OA'B'C'$ is less than the integral

$$\iint e^{-(x+y)} |u(x)| . |v(y)| \, dx \, dy$$

taken over the area $ABCC'B'A'$; and this is the difference of the integrals

$$\int_0^\lambda e^{-x} |u(x)| \, dx \int_0^\lambda e^{-y} |v(y)| \, dy$$

and
$$\int_0^{2\lambda} e^{-x} |u(x)| \, dx \int_0^{2\lambda} e^{-y} |v(y)| \, dy,$$

which tends to zero, since the integrals

$$\int_0^\infty e^{-x} |u(x)| \, dx \text{ and } \int_0^\infty e^{-y} |v(y)| \, dy$$

are both convergent.

Hence we can write

$$uv = \lim_{\lambda \to \infty} \iint e^{-(x+y)} u(x) v(y) \, dx \, dy,$$

where the integral is now to be taken over the area of a triangle such as $OA'C'$, whose side is 2λ.

Let us write $\qquad x + y = \xi, \quad y = \xi\eta,$

and then the integral becomes

$$\int_0^{2\lambda} e^{-\xi} \xi d\xi \int_0^1 u[\xi(1-\eta)] . v(\xi\eta) \, d\eta.$$

Owing to the fact that the series for $u(x)$, $v(y)$ are absolutely convergent for any values of x, y, we can obtain the product

$$u[\xi(1-\eta)] . v(\xi\eta)$$

by the rule for multiplication of series (Art. 34); and the product-series will obviously be uniformly convergent with respect to η, so that we may then integrate term-by-term.

The general term in the product is

$$\xi^n \sum_{r=0}^{n} \frac{u_r v_{n-r}}{r!(n-r)!} \eta^r (1-\eta)^{n-r},$$

and
$$\int_0^1 \eta^r (1-\eta)^{n-r} \, d\eta = \frac{r!(n-r)!}{(n+1)!},$$

so that
$$\int_0^1 u[\xi(1-\eta)] . v(\xi\eta) \, d\eta = \sum_0^\infty w_n \frac{\xi^n}{(n+1)!}.$$

Thus the integral over the triangle is

$$\int_0^{2\lambda} e^{-\xi} W(\xi)\, d\xi,$$

where

$$W(\xi)=\sum_0^\infty w_n \frac{\xi^{n+1}}{(n+1)!},$$

and accordingly we have the equation

$$uv=\int_0^\infty e^{-\xi} W(\xi)\, d\xi.$$

But we have not yet completed our investigation, because the last integral is the sum of the series

$$0+w_0+w_1+w_2+\dots,$$

and (Art. 101) it does not immediately follow that the series

$$w_0+w_1+w_2+\dots$$

is summable. It should be observed, however, that we *can* infer the absolute convergence of

$$\int_0^\infty e^{-\xi} | W(\xi) |\, d\xi;$$

because

$$| W(\xi) | \leqq \xi \int_0^1 | u\{\xi(1-\eta)\} | . | v(\xi\eta) |\, d\eta.$$

Hence, if we reverse the argument used above, we find

$$\int_0^\lambda e^{-\xi} | W(\xi) |\, d\xi < \int_0^\lambda e^{-x} | u(x) |\, dx . \int_0^\lambda e^{-y} | v(y) |\, dy$$

$$< \int_0^\infty e^{-x} | u(x) |\, dx . \int_0^\infty e^{-y} | v(y) |\, dy.$$

Thus, since the integrand is *positive*, the integral

$$\int_0^\infty e^{-\xi} | W(\xi) |\, d\xi$$

is convergent (see Art. 166).

To complete the demonstration of the theorem, let us multiply the two absolutely summable series

$$u_1+u_2+u_3+\dots, \quad v_0+v_1+v_2+\dots.$$

Their product is given (in virtue of the foregoing) by the absolutely convergent integral

$$\int_0^\infty e^{-\xi} \Omega(\xi)\, d\xi,$$

where

$$\Omega(\xi)=(u_1v_0)\xi+(u_1v_1+u_2v_0)\frac{\xi^2}{2!}+(u_1v_2+u_2v_1+u_3v_0)\frac{\xi^3}{3!}+\dots.$$

That is

$$\Omega(\xi)=w_0+w_1\xi+w_2\frac{\xi^2}{2!}+w_3\frac{\xi^3}{3!}+\dots-u_0\left(v_0+v_1\xi+v_2\frac{\xi^2}{2!}+\dots\right)$$

$$=W'(\xi)-u_0v(\xi).$$

Now

$$\int_0^\infty e^{-\xi}|\Omega(\xi)|\,d\xi$$

is convergent, and so also is

$$\int_0^\infty e^{-\xi}|v(\xi)|\,d\xi;$$

and therefore

$$\int_0^\infty e^{-\xi}W'(\xi)\,d\xi$$

is absolutely convergent.

Now

$$W'(\xi)=w(\xi)=w_0+w_1\xi+w_2\frac{\xi^2}{2!}+\dots$$

is the function associated with the series

$$w=w_0+w_1+w_2+\dots,$$

which is therefore summable; and its sum is equal to that of

$$0+w_0+w_1+w_2+\dots \qquad \text{(Art. 101).}$$

Hence

$$w=uv.$$

If we multiply similarly the absolutely summable series

$$u_2+u_3+u_4+\dots,\quad v_0+v_1+v_2+\dots,$$

we can establish the convergence of

$$\int_0^\infty e^{-\xi}|W''(\xi)|\,d\xi=\int_0^\infty e^{-\xi}|w'(\xi)|\,d\xi.$$

Continuing the process, we can prove that

$$\int_0^\infty e^{-\xi}|w^\lambda(\xi)|\,d\xi$$

converges for any integral value of λ; and accordingly the series w is absolutely summable.

By combining the foregoing results with the obvious theorem on the addition of absolutely summable series, we obtain the following:

If $P(u, v, w, \dots)$ is a polynomial in $u, v, w, \dots$, any finite number of absolutely summable series, we can obtain its value by combining the series as if they were convergent; the resulting series will be absolutely summable, and its sum will be equal to the value of P.

107. Examples of multiplication of absolutely summable series.

Ex. 1. $$(1-1+1-1+\dots)^2=1-2+3-4+\dots,$$
$$(1-1+1-1+\dots)^3=1-3+6-10+\dots.$$

Of course these results agree with the examples of Art. 103.

Ex. 2. Write $C=1+\cos\theta+\cos 2\theta+\cos 3\theta+\dots,$
$$S=0+\sin\theta+\sin 2\theta+\sin 3\theta+\dots;$$
then $$C+iS=1+e^{i\theta}+e^{2i\theta}+e^{3i\theta}+\dots.$$

Hence $(C+iS)^2=1+2e^{i\theta}+3e^{2i\theta}+4e^{3i\theta}+\dots$

or $$C^2-S^2=1+2\cos\theta+3\cos 2\theta+4\cos 3\theta+\dots,$$
$$2CS=0+2\sin\theta+3\sin 2\theta+4\sin 3\theta+\dots.$$

By using the values of C and S found in Art. 103, it is easy to deduce that
$$0+\cos\theta+2\cos 2\theta+3\cos 3\theta+\dots=-\tfrac{1}{4}\operatorname{cosec}^2(\tfrac{1}{2}\theta),$$
$$0+\sin\theta+2\sin 2\theta+3\sin 3\theta+\dots=0,$$
agreeing with the results found in Art. 110 below.

Ex. 3. Another exercise in multiplication is given by squaring C and S directly; this yields
$$C^2=\tfrac{1}{2}\operatorname{cosec}\theta(\sin\theta+\sin 2\theta+\sin 3\theta+\dots)$$
$$+\tfrac{1}{2}(1+2\cos\theta+3\cos 2\theta+\dots),$$
$$S^2=\tfrac{1}{2}\operatorname{cosec}\theta(\sin\theta+\sin 2\theta+\sin 3\theta+\dots)$$
$$-\tfrac{1}{2}(1+2\cos\theta+3\cos 2\theta+\dots).$$
These obviously agree with the value of C^2-S^2 found above. We can also find C^2+S^2 by multiplying $(C+iS)\times(C-iS)$; and we can find CS directly.

108. Multiplication of non-absolutely summable series.

Hardy (p. 43 of his second paper) has given two theorems, which are the extensions to summable series of the theorems of Mertens (Art. 35) and Abel (Art. 34). The second of these extensions will be found in Art. 111.

The first theorem is that *it is sufficient to suppose* one *of the series u, v to be absolutely summable*; then w will be summable and its sum will be equal to uv. We suppose that u is absolutely summable, while v is summable but not absolutely.

On reference to Art. 106 it is plain that the difference between the double integrals over the areas OAC, $OABC$ is
$$D=\int_0^\lambda e^{-x}u(x)\,dx\int_{\lambda-x}^\lambda e^{-y}v(y)\,dy.$$

Now write $\phi(x)=e^{-x}|u(x)|, \quad \psi(x)=\left|\int_{\lambda-x}^{\lambda} e^{-y}v(y)\,dy\right|,$

and then
$$|D| \leqq \int_0^\lambda \phi(x)\psi(x)\,dx. \quad \text{.................................(1)}$$

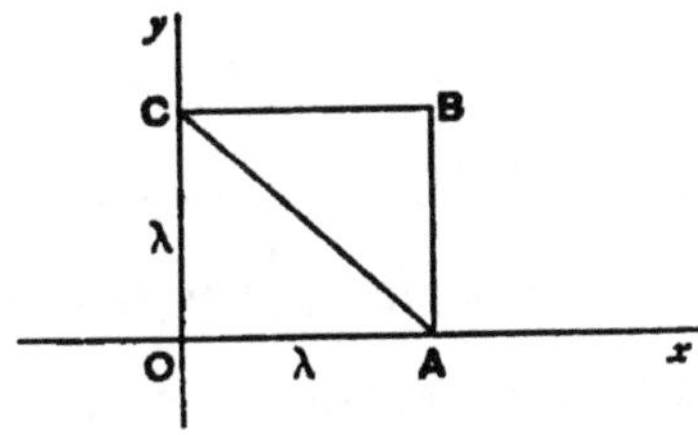

FIG. 32.

Since the integral $\int_0^\infty e^{-y}v(y)\,dy$ is supposed to be convergent, $\psi(x)$ has a finite upper limit H for all values of x and λ; further we can find μ so that
$$\psi(x)<\epsilon, \quad \text{if } \lambda-x \geqq \mu,$$
where ϵ may be arbitrarily small.

Now divide the integral (1) into two parts, from 0 to $\lambda-\mu$ and from $\lambda-\mu$ to λ; in the former $\psi(x)$ is less than ϵ, in the latter $\psi(x)$ does not exceed H. Thus, since $\phi(x)$ is positive, we find
$$|D| < \epsilon \int_0^{\lambda-\mu} \phi(x)\,dx + H \int_{\lambda-\mu}^{\lambda} \phi(x)\,dx.$$

Now $\int_0^\infty \phi(x)\,dx$ converges to some value U, and so if we take the limit of the last inequality as λ tends to infinity, we find
$$\overline{\lim_{\lambda\to\infty}}\,|D| \leqq \epsilon U.$$

But ϵ is arbitrarily small, and so (see Note (6), p. 5) the last inequality cannot be true unless $\overline{\lim}\,|D|$ is zero.

Thus $\lim\limits_{\lambda\to\infty} D=0$; and so the integral over the square $OABC$ may again be replaced by the integral over the triangle OAC.

Hence, as in Art. 106, we may write
$$uv=\lim_{\lambda\to\infty}\iint e^{-(x+y)}u(x)\,v(y)\,dx\,dy,$$
the integral being taken over the triangle OAC.

After this stage the proof follows exactly the same lines as before.

Thus *the series* $w_0+w_1+w_2+\ldots$ *is summable, and has the sum* uv, *provided that* **one** *of the series* u, v *is absolutely summable.*

It can be shewn that any series $w_a+w_{a+1}+w_{a+2}+\ldots$ is summable under the above hypothesis.

The series given at the end of Art. 105 is an example of the possibility that $w_a+w_{a+1}+w_{a+2}+\ldots$ may be summable for *any* value of a, without being *absolutely* summable.

109. Continuity, differentiation and integration of a summable series.

Suppose that the terms of the summable series are functions of a variable a, so that we may write

$$u(x, a)=\sum_0^\infty u_n(a)\frac{x^n}{n!};$$

then, in agreement with the previous use of the word *uniform*, we shall say that: *The series* $\mathscr{S}_0^\infty u_n(a)$ *is uniformly summable with respect to* a *in an interval* (β, γ), *provided that the integral* $\int_0^\infty e^{-x}u(x, a)\,dx$ *converges uniformly in that interval.*

Thus, in particular (Art. 171, Appendix), the series is certainly uniformly summable, if we can find a positive function $M(x)$, *independent of* a, such that

$$|u(x, a)| < M(x), \quad (\beta \leqq a \leqq \gamma)$$

while $\int_0^\infty e^{-x}M(x)\,dx$ is convergent.

Then the following theorems are true:

(i) *If all the terms* $u_n(a)$ *are continuous and the series* $\mathscr{S}_0^\infty u_n(a)$ *is uniformly summable, and* $\sum_0^\infty u_n(a)x^n/n!$ *is uniformly convergent for any finite value of* x, *when* a *lies in the interval* (β, γ), *the sum* $\mathscr{S}_0^\infty u_n(a)$ *is a continuous function of* a *in the same interval.*

For then, if a_0 is any particular value of a in the interval

$$\lim_{a\to a_0}\mathscr{S}_0^\infty u_n(a)=\lim_{a\to a_0}\int_0^\infty e^{-x}u(x, a)\,dx=\int_0^\infty e^{-x}\lim_{a\to a_0}[u(x, a)]\,dx,$$

this transformation being justified by the *uniform* convergence of the integral (Appendix, Art. 172).

Next $\lim_{a\to a_0}[u(x, a)]=\lim_{a\to a_0}\sum_0^\infty u_n(a)\frac{x^n}{n!}=\sum_0^\infty u_n(a_0)\frac{x^n}{n!}=u(x, a_0)$,

because $u_n(a)$ is a continuous function of a, and the series converges *uniformly* (Art. 45).

Thus $\lim_{a\to a_0}\mathscr{S}_0^\infty u_n(a)=\int_0^\infty e^{-x}u(x, a_0)\,dx=\mathscr{S}_0^\infty u_n(a_0)$,

which shews that the sum is a continuous function of a at any point of the interval (β, γ).

(ii) *Under the same conditions of uniformity as in* (i), *the equation*

$$\int_\beta^\gamma [\mathop{\mathscr{S}}_0^\infty u_n(a)]\, da = \mathop{\mathscr{S}}_0^\infty \int_\beta^\gamma u_n(a)\, da$$

will be correct.

This follows by an argument exactly similar to that of (i).

(iii) *If the same conditions of uniformity apply to the series* $\mathop{\mathscr{S}}_0^\infty u_n'(a)$, *then the equation*

$$\frac{d}{da}[\mathop{\mathscr{S}}_0^\infty u_n(a)] = \mathop{\mathscr{S}}_0^\infty u_n'(a)$$

will be correct.

This theorem is also an easy deduction from Arts. 46, 172.

110. Applications of Art. 109.

Ex. 1. If we consider the series $\mathop{\mathscr{S}}_0^\infty x^n$, where x is a complex number of absolute value not less than 1, we find that

$$\mathop{\mathscr{S}}_0^\infty x^n = \int_0^\infty e^{-t} e^{xt}\, dt = \frac{1}{1-x},$$

where we assume that the real part of x is less than 1.

It may be observed that $|e^{xt}| = e^{\rho t}$, where ρ is the real part of x, and consequently, if $\rho \leqq 1-a$, where a is arbitrarily small but positive, the integral for $\mathop{\mathscr{S}}_0^\infty x^n$ will converge uniformly with respect to x, because $|e^{-(1-x)t}| \leqq e^{-at}$; thus the sum $\mathop{\mathscr{S}}_0^\infty x^n$ is a continuous function of x within the region specified by $\rho \leqq 1-a$, as is obvious from the value of the sum.

Ex. 2. If in (1) we write $x = e^{i\theta}$, we find that

$$\frac{d}{d\theta}(x^n) = inx^n,$$

and so

$$\frac{d^p}{d\theta^p}(x^n) = (in)^p x^n.$$

Thus, the integral function associated with

$$\mathop{\mathscr{S}}_0^\infty \frac{d^p}{d\theta^p}(x^n)$$

is $u_p(xt) = i^p \sum_n n^p \frac{x^n t^n}{n!} = i^p e^{xt}[(xt)^p + A_1(xt)^{p-1} + \ldots + A_{p-1}(xt)],$

because we have the algebraical identity

$$n^p = n(n-1)\dots(n-p+1) + A_1 n(n-1)\dots(n-p+2) + \dots + A_{p-1} n,$$

where $A_1, A_2, \dots, A_{p-1}$ are certain coefficients which depend on p but not on n (compare Exs. 4, 5, p. 170).

Hence $|u_p(xt)| \leqq e^{t\cos\theta}(t^p + |A_1| t^{p-1} + \dots + |A_{p-1}| t)$,

and therefore the integral

$$\int_0^\infty e^{-t} u_p(xt)\, dt$$

converges uniformly if $\cos\theta \leqq 1 - a$.

Thus we may differentiate the series found in Art. 103 for $\mathscr{S}_1^\infty \cos n\theta$, $\mathscr{S}_1^\infty \sin n\theta$, as often as we please, provided that θ is not a multiple of 2π.

Hence we find, for instance,

$$\mathscr{S}_1^\infty n^{2s} \cos n\theta = 0, \qquad \mathscr{S}_1^\infty n^{2s+1} \sin n\theta = 0.$$

Taking $\theta = \pi$ in the first equation and $\frac{1}{2}\pi$ in the second, we find the results:

$$1^{2s} - 2^{2s} + 3^{2s} - \dots = 0, \quad 1^{2s+1} - 3^{2s+1} + 5^{2s+1} - \dots = 0.$$

Again,

$$\mathscr{S}_1^\infty (-1)^{n-1} n^{2s} \sin n\theta = (-1)^s \frac{d^{2s}}{d\theta^{2s}} (\tfrac{1}{2} \tan \tfrac{1}{2}\theta),$$

$$\mathscr{S}_1^\infty (-1)^{n-1} n^{2s+1} \cos n\theta = (-1)^s \frac{d^{2s+1}}{d\theta^{2s+1}} (\tfrac{1}{2} \tan \tfrac{1}{2}\theta),$$

provided that θ is not an odd multiple of π.

Thus, since

$$\tfrac{1}{2} \tan \tfrac{1}{2}\theta = \frac{2^2 - 1}{2!} B_1 \theta + \frac{2^4 - 1}{4!} B_4 \theta^3 + \dots,$$

we find

$$1^{2s+1} - 2^{2s+1} + 3^{2s+1} - \dots = (-1)^s \frac{2^{2s+2} - 1}{2s+2} B_{s+1},$$

[Art. 126, Ex. 5].

Thus in particular*

$$1 - 2 + 3 - \dots = \tfrac{1}{4} \quad \text{(Art. 103)},$$
$$1 - 2^3 + 3^3 - \dots = -\tfrac{1}{8},$$
$$1 - 2^5 + 3^5 - \dots = \tfrac{1}{4}.$$

---

* The second and third of these may also be obtained by means of devices similar to those used in Art. 103. See also Art. 126, Exs. 2–4.

In like manner the series

$$\cos\theta - \cos 3\theta + \cos 5\theta - \ldots = \tfrac{1}{2}\sec\theta$$

leads to the result

$$1^{2s} - 3^{2s} + 5^{2s} - \ldots = \tfrac{1}{2}(-1)^s E_s,$$

where E_s is Euler's number (Ch. X., Ex. C, 38). [CESÀRO.]

Hence in particular

$$1^2 - 3^2 + 5^2 - \ldots = -\tfrac{1}{2},$$
$$1^4 - 3^4 + 5^4 - \ldots = \tfrac{5}{2}.$$

Ex. 3. As examples of integration we may take the series

$$\sin\theta + \sin 3\theta + \sin 5\theta + \ldots = \tfrac{1}{2}\operatorname{cosec}\theta,$$
$$\sin 2\theta + \sin 4\theta + \sin 6\theta + \ldots = \tfrac{1}{2}\cot\theta,$$

which (as in 2) are uniformly summable in an interval $(\delta, \tfrac{1}{2}\pi)$ where $0 < \delta < \tfrac{1}{2}\pi$.

Thus we have

$$\int_\delta^{\frac{1}{2}\pi} \frac{\theta}{\sin\theta}\, d\theta = 2 \mathop{\mathscr{S}}_0^\infty \int_\delta^{\frac{1}{2}\pi} \theta \sin(2n+1)\theta \,.\, d\theta.$$

Now

$$\int_\delta^{\frac{1}{2}\pi} \theta\sin(2n+1)\theta\, d\theta = \frac{\delta\cos(2n+1)\delta}{2n+1} - \frac{\sin(2n+1)\delta}{(2n+1)^2} + \frac{(-1)^n}{(2n+1)^2},$$

and therefore the series on the right is convergent; also,

$$\Sigma\cos(2n+1)\delta/(2n+1) = \tfrac{1}{2}\log(\cot\tfrac{1}{2}\delta), \quad \text{(see Ex. 4, p. 290)}$$

and

$$\Sigma\sin(2n+1)\delta/(2n+1)^2$$

converges uniformly by Weierstrass's test.

Hence
$$\lim_{\delta\to 0} \sum_0^\infty \left[\frac{\delta\cos(2n+1)\delta}{2n+1} - \frac{\sin(2n+1)\delta}{(2n+1)^2}\right] = 0,$$

and therefore

$$\int_0^{\frac{1}{2}\pi} \frac{\theta}{\sin\theta}\, d\theta = 2\sum_0^\infty \frac{(-1)^n}{(2n+1)^2}.$$

In like manner we prove that

$$\int_\delta^{\frac{1}{2}\pi} \theta\cot\theta\, d\theta = 2 \mathop{\mathscr{S}}_1^\infty \int_\delta^{\frac{1}{2}\pi} \theta\sin 2n\theta \,.\, d\theta,$$

and hence

$$\int_0^{\frac{1}{2}\pi} \theta\cot\theta\, d\theta = 2\sum_1^\infty (-1)^{n-1}\frac{\pi}{4n} = \frac{\pi}{2}\log 2.$$

Ex. 4. As other examples of integration, we note that we may integrate the first series in Ex. 3 from θ to $\frac{1}{2}\pi$, and so obtain

$$\cos\theta+\tfrac{1}{3}\cos 3\theta+\tfrac{1}{5}\cos 5\theta+\ldots=\tfrac{1}{2}\log\cot\tfrac{1}{2}\theta, \quad (0<\theta<\pi)$$

where the resulting series converges. This series is summed independently in Art. 90, Ex. 2, and Ex. A, 43, Ch. VIII.

Similarly from the series

$$\cos\theta+\cos 2\theta+\cos 3\theta+\ldots=-\tfrac{1}{2},$$
$$\sin\theta+\sin 2\theta+\sin 3\theta+\ldots=\tfrac{1}{2}\cot\tfrac{1}{2}\theta,$$

we find, by integration from θ to π, the results

$$\left.\begin{aligned}\sin\theta+\tfrac{1}{2}\sin 2\theta+\tfrac{1}{3}\sin 3\theta+\ldots&=\tfrac{1}{2}(\pi-\theta)\\ \text{and } -(1+\cos\theta)+\tfrac{1}{2}(1-\cos 2\theta)-\tfrac{1}{3}(1+\cos 3\theta)+\ldots&\\ =\log\sin\tfrac{1}{2}\theta.&\end{aligned}\right\}\quad(0<\theta<2\pi)$$

These series converge, and the second one leads to the simpler form

$$\cos\theta+\tfrac{1}{2}\cos 2\theta+\tfrac{1}{3}\cos 3\theta+\ldots=-\tfrac{1}{2}\log(4\sin^2\tfrac{1}{2}\theta).$$

All these series agree with the results of Art. 65. Similarly we can establish the results

$$\sin\theta-\tfrac{1}{3}\sin 3\theta+\tfrac{1}{5}\sin 5\theta-\ldots=\tfrac{1}{2}\log(\sec\theta+\tan\theta)$$
$$(-\tfrac{1}{2}\pi<\theta<\tfrac{1}{2}\pi)$$

and $\sin\theta+\tfrac{1}{3}\sin 3\theta+\tfrac{1}{5}\sin 5\theta+\ldots=\tfrac{1}{4}\pi \qquad (0<\theta<\pi).$

Ex. 5. If
$$u=\sum_{-\infty}^{\infty}\frac{\cos x(a+n\pi)}{(a+n\pi)^2+b^2},$$

we get
$$-\frac{d^2u}{dx^2}+b^2u=\cos ax\,[1+2\mathop{\mathscr{S}}_{1}^{\infty}\cos(n\pi x)]=0.$$

Thus
$$u=Ae^{bx}+Be^{-bx}\quad(0\leqq x\leqq 2).$$

The values of u for $x=0$, $x=2$ are found in Ex. C, 31, Ch. X. (p. 259); from these it follows that

$$u=\frac{1}{b}\left(\frac{\cosh bx\sinh 2b}{\cosh 2b-\cos 2a}-\sinh bx\right).$$

For other applications, leading to the values of some interesting definite integrals, and for the investigation of conditions under which the equation

$$\int_{\beta}^{\infty}[\mathop{\mathscr{S}}_{0}^{\infty}u_n(a)]da=\mathop{\mathscr{S}}_{0}^{\infty}\int_{\beta}^{\infty}u_n(a)da$$

is true, we must refer to the latter part of Hardy's first paper (see the references given in Art. 96).

111. Analogue of Abel's theorem. (Art. 51.)

Suppose that $\overset{\infty}{\underset{0}{\mathscr{S}}} u_n$ is summable, and consider the possibility of summing $\overset{\infty}{\underset{0}{\mathscr{S}}} u_n t^n$, where t is real and lies between 0 and 1.

We have then, by definition,

$$\overset{\infty}{\underset{0}{\mathscr{S}}} u_n t^n = \int_0^\infty e^{-x} u(xt)\, dx,$$

and so, changing the independent variable from x to xt ($=y$, say), we find

$$\overset{\infty}{\underset{0}{\mathscr{S}}} u_n t^n = \frac{1}{t}\int_0^\infty e^{-y/t} u(y)\, dy.$$

Thus, if $1/t = 1+\theta$, so that θ is real and positive, we have

$$\overset{\infty}{\underset{0}{\mathscr{S}}} u_n t^n = (1+\theta)\int_0^\infty e^{-y(1+\theta)} u(y)\, dy.$$

Now, by hypothesis, the integral

$$\int_0^\infty e^{-y} u(y) dy$$

is convergent; and consequently (from Art. 171, Ex. 2, of the Appendix) the integral

$$\int_0^\infty e^{-y(1+\theta)} u(y)\, dy$$

converges uniformly with respect to θ, in any interval $(0, a)$, where a is any positive number.

Hence $\overset{\infty}{\underset{0}{\mathscr{S}}} u_n t^n$ can be summed uniformly with respect to t in the interval $(0, 1)$.

Thus in particular we have the result

$$\lim_{t\to 1} (\overset{\infty}{\underset{0}{\mathscr{S}}} u_n t^n) = \overset{\infty}{\underset{0}{\mathscr{S}}} u_n.$$

Further, provided that the series

$$u_\lambda + u_{\lambda+1} + u_{\lambda+2} + \dots$$

is summable for every integral value of λ, the series $\overset{\infty}{\underset{0}{\mathscr{S}}} u_n t^n$ is also absolutely summable for all positive values of t less than 1; for then we have

$$\lim_{x\to\infty} [e^{-x} u^\lambda(x)] = 0, \qquad (\lambda \geqq 0).$$

Thus we can find a constant A, such that

$$|u^\lambda(x)| < Ae^x \quad (0 \leqq x).$$

Now
$$\int_0^X e^{-x} \frac{d^\lambda}{dx^\lambda}[u(xt)]dx = t^{\lambda-1}\int_0^{Xt} e^{-y/t} u^\lambda(y)\,dy,$$

so that
$$\int_0^X e^{-x}\left|\frac{d^\lambda}{dx^\lambda}[u(xt)]\right|dx < At^{\lambda-1}\int_0^{Xt} e^{(1-1/t)y}\,dy$$

$$< At^\lambda/(1-t).$$

Thus, since the integrand is positive, the integral taken up to ∞ must converge (Art. 166); and so the conditions of absolute summability are satisfied by the series $\overset{\infty}{\underset{0}{\mathscr{S}}} u_n t^n$.

Thus we have the theorem:*

If $\overset{\infty}{\underset{0}{\mathscr{S}}} u_n$ *is summable, then* $\overset{\infty}{\underset{0}{\mathscr{S}}} u_n t^n$ *is uniformly summable for all values of* t *in the interval* (0, 1); *and if, further,* $\overset{\infty}{\underset{0}{\mathscr{S}}} u_{n+\lambda}$ *is summable for all positive integral values of* λ, *then* $\overset{\infty}{\underset{0}{\mathscr{S}}} u_n t^n$ *is absolutely summable for any value of* t *between* 0 *and* 1.

An immediate consequence from this result is Hardy's second theorem on multiplication of series:

If neither u *nor* v *is absolutely summable the equation* $uv = w$ *is still correct (using the notation of Art.* 106), *provided that* w *is summable.*

For then the three series $\overset{\infty}{\underset{0}{\mathscr{S}}} u_n t^n$, $\overset{\infty}{\underset{0}{\mathscr{S}}} v_n t^n$, $\overset{\infty}{\underset{0}{\mathscr{S}}} w_n t^n$ are absolutely summable, and therefore (Art. 106)

$$\left(\overset{\infty}{\underset{0}{\mathscr{S}}} u_n t^n\right).\left(\overset{\infty}{\underset{0}{\mathscr{S}}} v_n t^n\right) = \overset{\infty}{\underset{0}{\mathscr{S}}} w_n t^n \quad (0 < t < 1).$$

Hence, taking the limit as t approaches 1, we have

$$\left(\overset{\infty}{\underset{0}{\mathscr{S}}} u_n\right).\left(\overset{\infty}{\underset{0}{\mathscr{S}}} v_n\right) = \overset{\infty}{\underset{0}{\mathscr{S}}} w_n.$$

---

* Due in part to Phragmen (*Comptes Rendus*, t. 132, 1901, p. 1396) and to Hardy (p. 44 of his second paper).

112. An important class of summable power-series.

Suppose that
$$u_n = x^n \int_0^1 f(\xi)\xi^n \, d\xi,$$
where $f(\xi)$ is such that $\int_0^1 |f(\xi)| \, d\xi$ is convergent, so that the integral for u_n is also absolutely convergent.

Then, applying test A, Art. 175, we have
$$u(t) = \Sigma u_n \frac{t^n}{n!} = \int_0^1 e^{tx\xi} f(\xi) \, d\xi,$$
because the series for $e^{tx\xi}$ converges uniformly with respect to ξ, and $\int_0^1 |f(\xi)| \, d\xi$ is convergent.

Thus Borel's integral for the series is
$$\int_0^\infty e^{-t} u(t) \, dt = \int_0^\infty e^{-t} dt \int_0^1 e^{tx\xi} f(\xi) \, d\xi.$$

From Art. 177, it follows that the order of integration can be inverted in this integral, provided that the real part of x is not greater than a fixed number k (<1); for then the integral is seen to be absolutely convergent by comparing it with the integral
$$\int_0^\infty e^{-(1-k)t} dt \int_0^1 |f(\xi)| \, d\xi.$$

Thus Borel's integral reduces to
$$\int_0^1 \frac{f(\xi)}{1 - x\xi} d\xi \quad \text{(real part of } x \leqq k < 1).$$

It is evident that the last integral will be convergent for all values of x, except real values which are greater than 1; thus the last integral gives a larger region of summability than Borel's integral. We could, of course, have adopted this as a definition of the "sum" of Σu_n, by making another application of Hardy's principle (see Art. 99 above); but it would not have been evident to what extent the new definition could be deduced from Borel's.

Le Roy has given a number of applications of this integral in the paper quoted above (Art. 96), and the method has been extended by Hardy,* by the aid of contour-integrals.

* *Proc. Lond. Math. Soc.* (2), vol. 3, 1905, p. 381.

Ex. 1. We have $\dfrac{\Gamma(p)}{n^p}=\int_0^\infty e^{-nt}t^{p-1}\,dt=\int_0^1 \xi^{n-1}[\log(1/\xi)]^{p-1}\,d\xi, \quad (p>0),$

and accordingly the series

$$1+\frac{x}{2^p}+\frac{x^2}{3^p}+\frac{x^3}{4^p}+\dots$$

can be summed; and its sum is expressed by the integral

$$\frac{1}{\Gamma(p)}\int_0^1 \frac{[\log(1/\xi)]^{p-1}}{1-x\xi}\,d\xi. \qquad \text{[Hadamard.]}$$

Ex. 2. Again, $\dfrac{1}{n^2+1}=\int_0^\infty e^{-nt}\sin t\,dt=\int_0^1 \xi^{n-1}\sin[\log(1/\xi)]\,d\xi,$

so that the series

$$\frac{1}{1+1^2}+\frac{x}{1+2^2}+\frac{x^2}{1+3^2}+\dots$$

is summed by means of the integral

$$\int_0^1 \frac{\sin[\log(1/\xi)]}{1-x\xi}\,d\xi. \qquad \text{[Le Roy.]}$$

Ex. 3. If $b>a>0$, we have

$$\frac{a(a+1)\dots(a+n-1)}{b(b+1)\dots(b+n-1)}=\frac{\Gamma(b)}{\Gamma(a)\Gamma(b-a)}\int_0^1 \xi^{a+n-1}(1-\xi)^{b-a-1}\,d\xi,$$

so that the series

$$1+\frac{a}{b}x+\frac{a(a+1)}{b(b+1)}x^2+\frac{a(a+1)(a+2)}{b(b+1)(b+2)}x^3+\dots$$

can be summed by the integral

$$\frac{\Gamma(b)}{\Gamma(a)\Gamma(b-a)}\int_0^1 \frac{\xi^{a-1}(1-\xi)^{b-a-1}}{1-x\xi}\,d\xi. \qquad \text{[Hadamard.]}$$

In particular, if $b=1$, the integral can be evaluated by writing

$$v=\xi(1-x)/(1-\xi),$$

which gives $\qquad \dfrac{\sin(a\pi)}{\pi}\dfrac{1}{(1-x)^a}\int_0^\infty \dfrac{v^{a-1}\,dv}{1+v}=\dfrac{1}{(1-x)^a}.$

If x is complex the last transformation requires a little justification, which is beyond our range, as it depends on the theory of contour integration.

Ex. 4. Consider similarly the general binomial series

$$1-\nu x+\nu(\nu-1)\frac{x^2}{2!}-\nu(\nu-1)(\nu-2)\frac{x^3}{3!}+\dots,$$

where ν is contained between two integers m and $m+1$; then, if $m\geqq 0$, we can apply the method of Ex. 3 to the series obtained by omitting the first $(m+1)$ terms, and so prove that the original series is summable by applying Art. 101.

113. Application of Borel's process to power-series in general.

As a first example, consider the series

$$1+x+x^2+x^3+\dots.$$

We find by Borel's method

$$u(x,\,t)=e^{xt},$$

so that
$$\overset{\infty}{\underset{0}{\mathscr{S}}}\,x^n=\int_0^{\infty}e^{-t}e^{xt}\,dt=\frac{1}{1-x};$$

and this integral converges, provided that the real part of x is less than 1.

Similarly, by considering the series

$$\overset{\infty}{\underset{0}{\mathscr{S}}}\left(\frac{A}{a^n}+\frac{B}{b^n}+\frac{C}{c^n}\right)x^n,$$

we find the sum
$$\frac{Aa}{a-x}+\frac{Bb}{b-x}+\frac{Cc}{c-x},$$

provided that the real parts of x/a, x/b, x/c are all less than 1. The regions of summability in these two examples are as indicated below; the region to be excluded is shaded and the circles indicate the circles of convergence of the power-series.

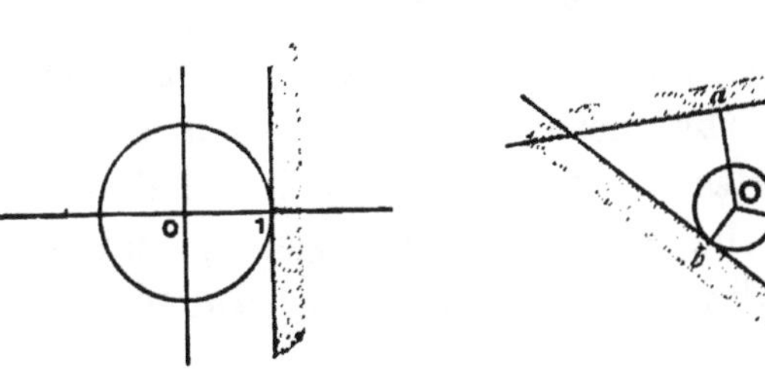

FIG. 33. FIG. 34.

Similarly, the special series examined in the last article are summable in the region indicated in figure 33.

These special cases lead to the conjecture that:

The region of summability is bounded by straight lines drawn at right angles to the radii from the origin to the singularities of the power-series.

For suppose any point P taken in this region; then the circle drawn on OP as diameter has no singularity of the power-series within or on the circumference. It is therefore possible to draw a concentric circle of radius r, say, which encloses O, P and still excludes the singularities.

Let $f(x)=\Sigma a_n x^n$ be the series; then it follows from Art. 82 that if k is the mid-point of OP, and x is any point within the second circle,

$$f(x)=\mathfrak{M}\frac{\xi-k}{\xi-x}f(\xi),$$

where the mean value is taken round the circle $|\xi-k|=r$.

Thus
$$a_n=\mathfrak{M}(\xi-k)f(\xi)/\xi^{n+1},$$
and so we get
$$u(x, t)=\Sigma a_n\frac{(xt)^n}{n!}=\mathfrak{M}\frac{\xi-k}{\xi}f(\xi)\exp\left(\frac{xt}{\xi}\right),$$

where the interchange of summation and integration is permissible because the series converges uniformly at all points on the circle since

$$|\xi|\geqq r-|k|>0.$$

Hence the integral $\int_0^\infty |u^\lambda(x, t)|e^{-t}dt$ converges, provided that the real part of $(\xi-x)/\xi=1-x/\xi$ is positive at all points ξ on the circle; now this is satisfied when x is at P, because OP subtends an acute angle θ at all points ξ of the circle, and so $(\xi-x)/\xi=\rho e^{i\theta}$, where ρ is real and positive.

Thus *the series is absolutely summable at any point within the polygon specified.*

By the aid of complex integrals and a slight modification of the method of Art. 111, it may be proved that if a power-series is *absolutely* summable at P, it can have no singularity within the circle described on OP as diameter.* This is the converse of the last theorem.

It will be evident that, whenever the number of singular points is *finite*, the method of summation enables the value of the power-series to be found at points belonging to certain regions outside the circle of convergence; and we have thus a process for finding an *analytical continuation* for the power-series.

It follows from Art. 111 that *the power-series is uniformly summable within any area inside the polygon of summability.* This property completes the analogy with the circle of convergence.

Borel has also proved in a later paper † that if $F(xt)$ is the integral function associated with the given power-series, then the straight lines (drawn as above) determine on each radius through O, the limit of the points x for which

$$\lim_{t\to\infty} e^{-t}F(xt)=0.$$

---

* Borel, *Leçons*, p. 108.

† Borel, *Math. Annalen*, Bd. 55, 1902, p. 74.

OTHER METHODS OF SUMMATION.

114. Borel's primary definition of the sum of an oscillatory series.

Write $$s_n = u_0 + u_1 + u_2 + \dots + u_n,$$
and then consider the expression

$$\frac{d}{dx}\left[e^{-x}\left(\sum_0^\infty s_n \frac{x^n}{n!}\right)\right] = e^{-x}\sum_0^\infty (s_{n+1} - s_n)\frac{x^n}{n!} = e^{-x}\sum_0^\infty u_{n+1}\frac{x^n}{n!} = e^{-x}u'(x).$$

Hence (1) $$e^{-x}\left(\sum_0^\infty s_n \frac{x^n}{n!}\right) - s_0 = \int_0^x e^{-x}u'(x)\,dx.$$

Thus, since the integral $\int_0^\infty e^{-x}u'(x)\,dx$ is associated with the series

$$u_1 + u_2 + u_3 + \dots = \mathop{\mathscr{S}}_1^\infty u_n,$$

we see that if the series $\mathop{\mathscr{S}}_1^\infty u_n$ is summable, then the limit

$$\lim_{x\to\infty}\left[e^{-x}\left(\sum_0^\infty s_n \frac{x^n}{n!}\right) - s_0\right]$$

exists and is equal to the sum $\mathop{\mathscr{S}}_1^\infty u_n$.

But we have proved that if $\mathop{\mathscr{S}}_1^\infty u_n$ can be summed, so also can $\mathop{\mathscr{S}}_0^\infty u_n$ (see Art. 101); and since $u_0 = s_0$, we see that *when* $\mathop{\mathscr{S}}_1^\infty u_n$ *is summable, the equation*

(2) $$\lim_{x\to\infty}\left[e^{-x}\left(\sum_0^\infty s_n \frac{x^n}{n!}\right)\right] = \mathop{\mathscr{S}}_0^\infty u_n$$

is true; and conversely, when the limit exists, $\mathop{\mathscr{S}}_1^\infty u_n$ *and* $\mathop{\mathscr{S}}_0^\infty u_n$ *are both summable, and the equation* (2) *is true.*

For the existence of the limit implies the convergence of $\int_0^\infty e^{-x}u'(x)\,dx$, and this again leads to equation (2).

The limit just obtained was the original definition suggested by Borel;* but, as pointed out by Hardy, the integral definition

* Of course this method can be applied to numerical calculation, whereas the other cannot (at least not *directly*). Taking the series for $(1+t)^{\frac{1}{2}}$, and writing $t=2$, Borel has calculated $e^{-8}\sum^{34} s_n \frac{8^n}{n!}$, and verified that this expression gives $\sqrt{3}$ to three decimals; the work shews that it is possible to use the limiting process to obtain numerical values for oscillatory series, even though the labour is considerable. The details of the calculation are given by Borel (*Liouville's Journal* (5), t. 2, 1896, p. 119). See also Ex. 2, Art. 121.

(deduced by Borel from this limit) can be applied to cases in which the limiting process gives no definite value (see for instance the example of Art. 101).

It is almost evident from Art. 100 that the limit is equal to the sum, if the original series is convergent; but a direct proof is easy. For if $\sum_0^\infty u_n$ converges to a sum s, we can choose m so that

$$|s_n - s| < \epsilon, \quad \text{if } n \geqq m.$$

Hence we find

$$\left| e^{-x} \sum_m^\infty s_n \frac{x^n}{n!} - e^{-x} \sum_m^\infty s \frac{x^n}{n!} \right| < \epsilon \left[e^{-x} \sum_m^\infty \frac{x^n}{n!} \right] < \epsilon,$$

and therefore

$$\left| e^{-x} \sum_0^\infty s_n \frac{x^n}{n!} - s \right| < \epsilon + He^{-x} \sum_0^{m-1} \frac{x^n}{n!},$$

where H is the upper limit of $|s_n - s|$ as n ranges from 0 to $m-1$.

Now
$$\lim_{x \to \infty} \left(e^{-x} \sum_0^{m-1} \frac{x^n}{n!} \right) = 0,$$

and so we see that

$$\overline{\lim_{x \to \infty}} \left| e^{-x} \sum_0^\infty s_n \frac{x^n}{n!} - s \right| \leqq \epsilon,$$

But ϵ is arbitrarily small, and so (see Note (6), p. 5) this maximum limit must be zero, and therefore

$$\lim_{x \to \infty} \left(e^{-x} \sum_0^\infty s_n \frac{x^n}{n!} \right) = s.$$

It is easy to modify the argument so as to prove that when Σu_n is divergent,

$$\lim_{x \to \infty} \left(e^{-x} \sum_0^\infty s_n \frac{x^n}{n!} \right) = \infty.$$

Ex. If
$$u_0 + u_1 + u_2 + \ldots = 1 - t + t^2 - t^3 + \ldots,$$

we have
$$s_n = [1 + (-1)^n t^{n+1}]/(1+t).$$

Thus
$$\sum_0^\infty s_n \frac{x^n}{n!} = \frac{1}{1+t}(e^x + te^{-xt})$$

and
$$e^{-x} \sum_0^\infty s_n \frac{x^n}{n!} = \frac{1}{1+t}(1 + te^{-x(1+t)}),$$

giving
$$\lim_{x \to \infty} \left(e^{-x} \sum_0^\infty s_n \frac{x^n}{n!} \right) = \frac{1}{1+t}, \quad \text{if } t > -1.$$

The reader will find it a good exercise to discuss similarly the other series given in Art. 103.

115. Le Roy's extension of Borel's definition.

Le Roy has given an extension of Borel's integral by modifying the series in the form

$$u_0+0+0+\ldots+0+u_1+0+0+\ldots+0+u_2+0+\ldots,$$

where $(p-1)$ zero terms are placed between u_n and u_{n+1}.

The analytical formula for the sum is then

$$\int_0^\infty e^{-x}\,U_p(x)\,dx,$$

where
$$U_p(x)=\sum_0^\infty \frac{u_n x^{np}}{(np)!}.$$

The object of using this definition is to extend Borel's integral to cases in which the series $u(x)$ could never converge: such cases are illustrated by the example

$$1-(2!)^2+(3!)^2-(4!)^2+\ldots.$$

Of course if Σu_n is convergent, its value is equal to that of the modified series, and so Le Roy's extension obviously satisfies the condition of consistency (in virtue of Art. 100).

116. Le Roy's independent definition.

The expression here taken for the sum of an oscillatory series $u_0+u_1+u_2+\ldots$ is the limit

$$\lim_{t\to 1}\sum_0^\infty \frac{\Gamma(nt+1)}{\Gamma(n+1)}u_n.$$

Now, assuming that the series used in this definition is absolutely convergent, we can write

$$\sum_0^\infty \frac{\Gamma(nt+1)}{\Gamma(n+1)}u_n=\int_0^\infty e^{-x}\left(\sum_0^\infty \frac{u_n x^{nt}}{n!}\right)dx.$$

For
$$\int_0^\infty \left|e^{-x}\frac{u_n x^{nt}}{n!}\right|dx=|u_n|\cdot\frac{\Gamma(nt+1)}{\Gamma(n+1)},$$

and therefore the series
$$\Sigma\int_0^\infty \left|e^{-x}\frac{u_n x^{nt}}{n!}\right|dx$$

is convergent, so that the test B of Art. 176, Appendix, can be applied to justify the inversion of the order of integration and summation.

Hence
$$\sum_0^\infty\left(\int_0^\infty e^{-x}\frac{u_n x^{nt}}{n!}\,dx\right)=\int_0^\infty\left(e^{-x}\sum_0^\infty\frac{u_n x^{nt}}{n!}\right)dx.$$

This integral is the same as

$$\int_0^\infty e^{-x} u(x^t)\, dx,$$

and so Le Roy's definition becomes

$$\lim_{t\to 1} \frac{1}{t}\int_0^\infty x^{1/t-1} \exp(-x^{1/t})\, u(x) dx$$

$$= \lim_{a\to 0} \int_0^\infty x^a \exp(-x^{1+a}) u(x)\, dx.$$

If Borel's limit exists, this definition gives the same value; for the function

$$x^a \exp(x - x^{1+a})$$

decreases* as x increases, if $x>1$ and $a>0$, and is equal to 1 for $x=1$; and we can therefore apply Abel's test for uniform convergence given in Art. 171 of the Appendix, which gives

$$\lim_{a\to 0} \int_1^\infty x^a \exp(x - x^{1+a}) \,.\, e^{-x} u(x)\, dx = \int_1^\infty e^{-x} u(x)\, dx,$$

because the latter integral converges.

Further,

$$\lim_{a\to 0} \int_0^1 x^a \exp(x - x^{1+a}) \,.\, e^{-x} u(x)\, dx = \int_0^1 e^{-x} u(x)\, dx,$$

because the range of integration is finite.

Hence Le Roy's definition coincides with Borel's, whenever the latter is convergent.

117. Borel's third method.

This method of summing differs from the first method in the fact that the terms are arranged in groups of k before applying the method. It is however of less importance than the other methods, at any rate from the arithmetical point of view.

Thus we obtain from $\sum_0^\infty u_n$ the integral

$$\int_0^\infty e^{-x} u_k(x)\, dx,$$

* Because the logarithmic derivate is

$$1-(1+a)x^a + a/x = -(x^a-1) - a(x^a - 1/x),$$

which is plainly negative when $x>1$ and $a>0$.

where

$$u_k(x) = (u_0 + u_1 + \ldots + u_{k-1}) + (u_k + u_{k+1} + \ldots + u_{2k-1})x$$
$$+ (u_{2k} + u_{2k+1} + \ldots + u_{3k-1})\frac{x^2}{2!} + \ldots.$$

If Σu_n is convergent, this process of grouping cannot alter its sum, and consequently (Art. 101) this integral will also converge to the sum.

But in general the value obtained depends on k; for example, $1-1+1-1+\ldots$ gives $0+0+0+\ldots(=0)$ if k is 2 or any even number, but $1-1+1-\ldots(=\frac{1}{2})$ if k is odd.

As another example, take $1-2+3-4+\ldots$, which has the sum $\frac{1}{4}$ by the foregoing methods.

With $k=2$ we get $-1-1-1-1-\ldots$, which diverges to $-\infty$; while $k=3$ gives $2-5+8-\ldots$, which is the same as

$$3(1-2+3-4+\ldots)-(1-1+1-1+\ldots)=\tfrac{3}{4}-\tfrac{1}{2}=\tfrac{1}{4}.$$

118. A further extension of the method of summation.

Suppose that $\phi(x)$ is a positive function, which steadily decreases to 0 as x tends to ∞, in such a way that all the integrals

$$c_n = \int_0^\infty \phi(x) x^n \, dx$$

are convergent. Then if we consider the equation

$$\int_0^\infty \phi(x)\left(\sum_0^\infty \frac{u_n}{c_n} x^n\right) dx = u_0 + u_1 + u_2 + \ldots,$$

it is easily seen to be correct when Σu_n is absolutely convergent.

For then
$$\int_0^\infty \left|\phi(x)\frac{u_n}{c_n} x^n\right| dx = |u_n|,$$

because $\phi(x)$ and x^n are positive; and accordingly the test B of Art. 176, Appendix, can be applied to justify the inversion of the order of summation and integration. Thus

$$\int_0^\infty \left(\sum_0^\infty \phi(x)\frac{u_n}{c_n} x^n\right) dx = \sum_0^\infty \left(\int_0^\infty \phi(x)\frac{u_n}{c_n} x^n \, dx\right) = \Sigma u_n.$$

Hence, if $\Sigma u_n t^n$ is convergent within the interval $(-1, +1)$, we have

$$\Sigma u_n t^n = \int_0^\infty \phi(x)\left[\sum_0^\infty \frac{u_n}{c_n}(xt)^n\right] dx = \frac{1}{t}\int_0^\infty \phi\left(\frac{x}{t}\right)\left(\sum_0^\infty \frac{u_n}{c_n} x^n\right) dx.$$

Now, because $\phi(x/t)$ steadily decreases as x increases, we have, by applying Abel's theorem for definite integrals (Appendix, Art. 171),

$$\lim_{t\to 1}\int_0^\infty \phi\left(\frac{x}{t}\right)\left(\sum_0^\infty \frac{u_n}{c_n}x^n\right)dx=\int_0^\infty \phi(x)\left(\sum_0^\infty \frac{u_n}{c_n}x^n\right)dx,$$

provided that the latter integral converges.

Thus, *provided that* $\Sigma u_n t^n$ *converges absolutely within the interval* $(-1, +1)$, *we have*

$$\lim_{t\to 1}\Sigma u_n t^n=\int_0^\infty \phi(x)\left(\Sigma\frac{u_n}{c_n}x^n\right)dx,$$

provided that the latter integral is convergent.

Thus, if for two different functions ϕ, ψ we can prove that the corresponding integrals are convergent, we can infer that their values are equal. In particular, *if we can shew that Borel's integral is convergent, we can obtain its value (when more convenient) by means of any integral of the form*

$$\int_0^\infty \phi(x)\left(\Sigma\frac{u_n}{c_n}x^n\right)dx,$$

which also converges.

Ex. Consider the series of Ex. 3, Art. 112 (with $b=1$).

$$1+at+a(a+1)\frac{t^2}{2!}+a(a+1)(a+2)\frac{t^3}{3!}+\dots.$$

We know that this series is summable by means of Borel's integral, if the real part of t is less than 1; so to find its value we could take

$$\phi(x)=e^{-x}x^{a-1},\quad c_n=\Gamma(a+n).$$

Then
$$\Sigma\frac{u_n(xt)^n}{c_n}=\frac{1}{\Gamma(a)}e^{xt}$$

and
$$\frac{1}{\Gamma(a)}\int_0^\infty e^{-x}x^{a-1}e^{xt}\,dx=\frac{1}{\Gamma(a)}\int_0^\infty e^{-(1-t)x}x^{a-1}\,dx=\frac{1}{(1-t)^a}.$$

119. Euler's method for summing oscillatory series.

Euler (*Inst. Calc. Diff.*, Pars II. cap. I.) employed his transformation (already given in Art. 24) for summing oscillatory series. This method is in many cases the most rapid in practice; and we shall apply it to some of the examples of Art. 103, before examining its theoretical foundation.

Ex. 1. $1-t+t^2-t^3+\ldots$.

From the coefficients $1, 1, 1, 1, \ldots$

we get the series of differences $0, 0, 0, \ldots,$

and so the transformation gives simply

$$\frac{1}{1+t}. \qquad \text{[Exs. 2, 4, Art. 103.]}$$

Ex. 2. $1-2+3-4+5-\ldots$.

From the coefficients $1, 2, 3, 4, 5, \ldots$

we get the differences $-1, -1, -1, -1, \ldots$

$0, 0, 0, \ldots,$

and so the transformation gives

$$\tfrac{1}{2}-\tfrac{1}{4}=\tfrac{1}{4}. \qquad \text{[Ex. 3, Art. 103.]}$$

Ex. 3. $1-2^2+3^2-4^2+5^2-\ldots$.

Here the coefficients are $1, 4, 9, 16, 25, \ldots$

and the differences are $-3, -5, -7, -9, \ldots$

$2, 2, 2, \ldots$

$0, 0, \ldots$.

Hence the sum is $\tfrac{1}{2}-\tfrac{3}{4}+\tfrac{2}{8}=0.$ [Ex. 2, Art. 110.]

Ex 4. As easy examples for practice we may give the following, taken from Euler for the most part:

$$1-3+5-7+9-\ldots = 0$$

$$1-3+6-10+15-\ldots = \tfrac{1}{8}$$

$$\left.\begin{aligned} 1-2^3+3^3-4^3+5^3-\ldots &= -\tfrac{1}{8} \\ 1-2^4+3^4-4^4+5^4-\ldots &= 0 \end{aligned}\right\} \quad \text{[Ex. 2, Art. 110.]}$$

It will be seen that in all these cases the results found from Borel's integral agree with those obtained by using Euler's transformation. This fact suggests the conjecture that a general relation can be obtained between the two methods; and we shall investigate this point in the following article.

120. Connexion between Euler's transformation and Borel's integral.

Suppose that $u_n=a_n t^n$, where a_n is real and positive, while t is real and less than 1. Then, as in Art. 24, we introduce the coefficients

$$\begin{aligned} b_0 &= a_0, \\ b_1 &= a_0-a_1=Da_0, \\ b_2 &= a_0-2a_1+a_2=D^2a_0, \\ b_3 &= a_0-3a_1+3a_2-a_3=D^3a_0, \text{ etc.} \end{aligned}$$

From these equations, we find at once

$$a_0 = b_0,$$
$$a_1 = b_0 - b_1 = Db_0,$$
$$a_2 = b_0 - 2b_1 + b_2 = D^2 b_0,$$
$$a_3 = b_0 - 3b_1 + 3b_2 - b_3 = D^3 b_0, \text{ etc.}$$

Thus, in Borel's integral we have

$$u(x) = \Sigma u_n \frac{x^n}{n!} = \Sigma a_n \frac{(xt)^n}{n!},$$

or, expressed in terms of $b_0, b_1, b_2, \ldots,$

$$\begin{aligned} u(x) = {} & b_0 \\ & + (b_0 - b_1)\, xt \\ & + (b_0 - 2b_1 + b_2) \frac{(xt)^2}{2!} \\ & + (b_0 - 3b_1 + 3b_2 - b_3) \frac{(xt)^3}{3!} \\ & + \ldots . \end{aligned}$$

We can obtain the value of $u(x)$ by summing this series according to columns, provided that the series converges *absolutely*.

Thus, provided that the series

$$|b_0| + |b_1|\, xt + |b_2| \frac{(xt)^2}{2!} + \ldots$$

converges, we find the equation

$$u(x) = e^{xt} \left[b_0 - b_1 xt + b_2 \frac{(xt)^2}{2!} - b_3 \frac{(xt)^3}{3!} + \ldots \right].$$

Hence, Borel's integral is

$$\int_0^\infty e^{-x} u(x)\, dx = \int_0^\infty e^{-x(1-t)} \left[b_0 - b_1 (xt) + \ldots \right] dx.$$

If we integrate this series term-by-term, we obtain Euler's series

$$\frac{b_0}{1-t} - \frac{b_1 t}{(1-t)^2} + \frac{b_2 t^2}{(1-t)^3} - \frac{b_3 t^3}{(1-t)^4} + \cdots$$

as equivalent to Borel's integral; although of course we have still to consider the validity of the transformations.

Now, if Euler's series converges, we can repeat the argument of Art. 100 (with a few obvious changes) to prove that then the series

$$b_0 - b_1 xt + b_2 \frac{(xt)^2}{2!} - \dots$$

is absolutely convergent, and that Borel's integral is equal to the sum of Euler's series. Thus we have obtained the theorem:

If Euler's series is convergent, the series $\Sigma a_n t^n$ is summable by Borel's method if t is real and less than 1, *and the two sums are equal.**

It is natural to enquire if this result holds for complex values of t; but it does not seem possible to apply the same method, owing to the fact that the lemma of Art. 80 leads to a difficulty here.

Thus we should have to take

$$v_n = (-1)^n b_n \frac{t^n}{(1-t)^{n+1}}, \qquad \lambda_n = \int_0^\lambda e^{-x(1-t)} \frac{x^n}{n!} (1-t)^{n+1} dx,$$

and to consider the continuity of $\Sigma v_n \lambda_n$ as $\lambda \to \infty$.

The lemma requires the convergence of Σv_n and of $\Sigma |\lambda_n - \lambda_{n+1}|$; now here we have

$$\lambda_n - \lambda_{n+1} = e^{-\lambda(1-t)} \frac{\lambda^{n+1}(1-t)^{n+1}}{(n+1)!}$$

by direct integration.

Now if $t = r + is$, $|e^{-\lambda(1-t)}| = e^{-\lambda(1-r)}$, and $|1-t|^2 = (1-r)^2 + s^2$,

thus

$$\Sigma |\lambda_n - \lambda_{n+1}| = e^{-\lambda(1-r)}(e^{\lambda|1-t|} - 1)$$
$$= e^{\lambda\delta} - e^{-\lambda(1-r)},$$

where $\delta = |1-t| - (1-r)$.

Now δ is *positive*, at least when s is not zero, and consequently the sum $\Sigma |\lambda_n - \lambda_{n+1}| \to \infty$ with λ. Thus we cannot use the lemma to infer the continuity of $\Sigma v_n \lambda_n$ from that of Σv_n.

On the other hand, if we assume that the series $\Sigma |b_n| \rho^n$ is convergent, where $\rho = |t|/(1-r)$ (r being the real part of t), we can apply test B of Art. 176, Appendix. For since $\Sigma |b_n| \rho^n$ is convergent, it follows that

$$\Sigma (-1)^n b_n \frac{(xt)^n}{n!}$$

* It should be noted that if t is *negative*, no restriction is implied on its magnitude.

converges absolutely and uniformly with respect to x in any finite interval. We can now infer from our test that

$$\int_0^\infty e^{-x(1-t)}\left[\Sigma(-1)^n b_n \frac{(xt)^n}{n!}\right]dx$$

converges to the same sum as $\Sigma(-1)^n \frac{b_n t^n}{(1-t)^{n+1}}$.

We have in fact

$$\int_0^\infty e^{-x(1-t)} b_n \frac{(xt)^n}{n!}\, dx < |b_n| \cdot |t|^n \int_0^\infty e^{-x(1-r)} \frac{x^n}{n!} dx,$$

and the last expression reduces to $|b_n| \rho^n/(1-r)$.

Thus the series $\displaystyle\sum_0^\infty \int_0^\infty e^{-x(1-t)} b_n x^n \frac{t^n}{n!}\, dx$

is convergent, because $\Sigma |b_n| \rho^n$ converges. Consequently, in virtue of the article quoted, we can write

$$\int_0^\infty e^{-x(1-t)}\left[\sum_0^\infty (-1)^n b_n \frac{(xt)^n}{n!}\right] dx = \sum_0^\infty (-1)^n b_n t^n \int_0^\infty e^{-x(1-t)} \frac{x^n}{n!} dx$$
$$= \sum_0^\infty (-1)^n \frac{b_n t^n}{(1-t)^{n+1}}.$$

From Art. 50, we see that the series $\Sigma |b_n| \rho^n$ converges if $\rho < l$, say; thus Borel's and Euler's sums are equal if $|t| < l(1-r)$, that is to say at all points* within a conic of eccentricity l, whose focus is at the origin and whose directrix is the line $r = 1$.

Euler's series will converge in an area bounded by the circle $|t| = l|1-t|$, which is the auxiliary circle of the conic; if $l < 1$, the area is within the circle because $t = 0$ is inside the circle; if $l > 1$, the area is outside the circle because $t = 0$ is so. If $l = 1$, the boundary is a straight line ($r = \frac{1}{2}$) and the area is to the left of the line in the ordinary form of diagram; and then of course the conic is a parabola.

By appealing to the Theory of Functions, we can now see that *Borel's integral and Euler's series must be equal at all points where both converge.* But there is no obvious means of determining the region of convergence of one from that of the other, as will appear from the two simple examples given below.

---

* In case $l > 1$, the conic is a hyperbola, and we must take only points within *one* branch; that is, the branch for which $r < 1$, and this is the *left-hand* branch in the ordinary way of drawing the diagram (with $r = 1$ to the right of the origin).

Ex. 1. Take $a_n = 1/2^n$: this gives $Da_n = a_n - a_{n+1} = 1/2^{n+1}$.

Thus $$a_0 = 1,\quad Da_0 = \tfrac{1}{2},\quad D^2a_0 = \tfrac{1}{4}, \text{ etc.},$$

and so Euler's series is

$$\frac{1}{1-t} - \frac{1}{2}\frac{t}{(1-t)^2} + \frac{1}{4}\frac{t^2}{(1-t)^3} - \dots,$$

which converges if $|t| < 2|1-t|$; that is, in the region outside the circle whose diameter is the line joining the two points $\frac{2}{3}$, 2.

But Borel's integral gives

$$u(x) = e^{xt/2},$$

and so the integral is convergent, when the real part of t is less than 2.

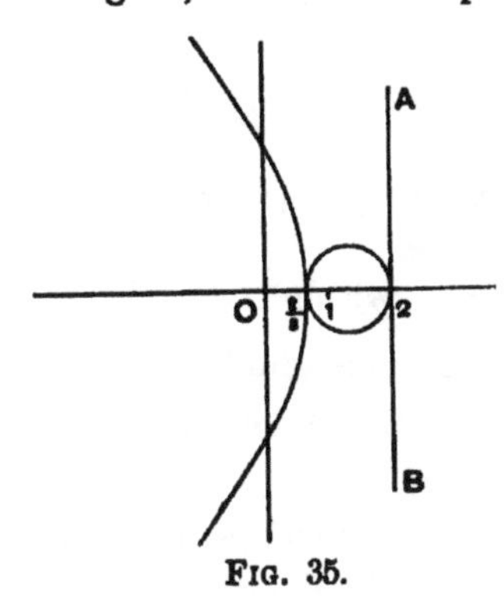

Fig. 35.

In the diagram, the region of convergence of Euler's series is the area outside the circle; and that of Borel's integral is the space to the left of the line AB. The area in which they are proved to be equal by the method given above is the area to the left of the hyperbola

$$|t| < 2(1-r).$$

The area in which the two are actually equal is the area which lies to the left of AB and is outside the circle. In this case one or other of the methods can be applied for *every* value of t, except $t=2$.

The original series $\Sigma a_n t^n$ converges within the circle $|t|=2$, which is not drawn.

Ex. 2. Consider next the case $a_n = 3^n$.

Then $$Da_n = a_n - a_{n+1} = -2 \times 3^n,\quad D^2a_n = 2^2 \times 3^n, \text{ etc.}$$

Thus $$a_0 = 1,\quad Da_0 = -2,\quad D^2a_0 = +4, \text{ etc.},$$

and so Euler's series is

$$\frac{1}{1-t} + \frac{2t}{(1-t)^2} + \frac{4t^2}{(1-t)^3} + \dots,$$

which converges if $|t| < \frac{1}{2}|1-t|$, that is in the area inside the circle whose diameter is the line joining the points $\frac{1}{3}$, -1.

But in Borel's integral we have

$$u(x) = e^{3tx},$$

and so the integral converges when the real part of t is less than $\frac{1}{3}$.

In the diagram, the interior of the circle represents the area of convergence for Euler's series; and Borel's integral converges in the space to the left of AB. The two expressions have been proved to be equal in

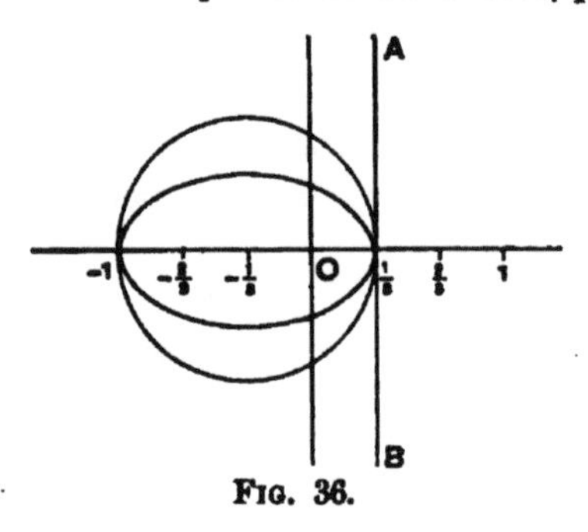

FIG. 36.

the area within the ellipse ($|t| < \frac{1}{2}(1-r)$); but they are actually equal at all points within the circle.

The original series converges within the circle $|t| = \frac{1}{3}$, which is not drawn.

Ex. 3. The reader may examine similarly any example such as

$$a_n = k^n \text{ or } nk^n,$$

where k is a positive constant.

In particular, if $a_n = 1$ or n, Euler's series terminates for any value of t: Borel's integral converges only if the real part of t is less than 1.

121. Numerical examples of Euler's transformation.

Euler's transformation lends itself very naturally to the numerical evaluation of non-convergent series.

Ex. 1. As an example let us take the series

$$-(t + \tfrac{1}{2}t^2 + \tfrac{1}{3}t^3 + \ldots) \quad [\text{which is equal to } \log(1-t), \text{ if } |t| < 1]$$

and write $t = -2$.

We obtain $2 - \frac{1}{2} \cdot 2^2 + \frac{1}{3} \cdot 2^3 - \frac{1}{4} \cdot 2^4 + \ldots$; and we shall utilise the calculations already made in Art. 24; thus we sum the first eight terms separately, which give $-19{\cdot}314286$, to six decimals. The remainder can be put in the form

$$\frac{2^9}{3}[b_0 + \tfrac{2}{3}b_1 + (\tfrac{2}{3})^2 b_2 + \ldots],$$

where $b_0, b_1, b_2, \ldots$ are the differences given in Ex. 2, Art. 24.

But we can obtain a general formula for these differences, and so establish the convergence of the transformed series.

For
$$Da_n = a_n - a_{n+1} = \frac{1}{n} - \frac{1}{n+1} = \frac{1}{n(n+1)},$$

$$D^2 a_n = Da_n - Da_{n+1} = \frac{1}{n(n+1)} - \frac{1}{(n+1)(n+2)} = \frac{1 \cdot 2}{n(n+1)(n+2)},$$

and so on; leading to

$$D^p a_n = p!/[n(n+1)(n+2)\ldots(n+p)]$$

Clearly this expression is always less than 1, and so the transformed series certainly converges, since $|t|/|1-t|=\frac{2}{3}$.

With the differences found previously we find the values

$$
\begin{aligned}
b_0 &= \cdot 111111 \\
\tfrac{2}{3}b_1 &= 7407 \\
(\tfrac{2}{3})^2 b_2 &= 898 \\
(\tfrac{2}{3})^3 b_3 &= 150 \\
(\tfrac{2}{3})^4 b_4 &= 31 \\
(\tfrac{2}{3})^5 b_5 &= 7 \\
(\tfrac{2}{3})^6 b_6 &= 2 \\
& \overline{\cdot 119606}
\end{aligned}
$$

If we apply the formula given above, we see that the remainder lies between $\frac{7}{24}$ and $\frac{7}{12}$ of the last term retained, so that we can take

$$b_0+\tfrac{2}{3}b_1+(\tfrac{2}{3})^2 b_2+\ldots=\cdot 119607.$$

Now $\frac{2^9}{3}(\cdot 119607)=20\cdot 4129$, and so the series should be

$$20\cdot 4129-19\cdot 3143=1\cdot 0986,$$

which agrees with $\log 3$ to the last figure.

Ex. 2. As a second example, for comparison with Borel's numerical work,* let us take

$$1-\tfrac{1}{2}t-\frac{1\,.\,1}{2\,.\,4}t^2-\frac{1\,.\,1\,.\,3}{2\,.\,4\,.\,6}t^3-\ldots\,[=(1-t)^{\frac{1}{2}} \text{ if } |t|<1],$$

and again write $t=-2$.

The sum of the first three terms is $1\cdot 5$; and we shall apply the transformation to the following terms.

We have

$$Da_n=a_n-a_{n+1}=\frac{1\,.\,3\ldots(2n-3)}{2\,.\,4\ldots 2n}-\frac{1\,.\,3\ldots(2n-1)}{2\,.\,4\ldots(2n+2)}=3\,.\,\frac{1\,.\,3\ldots(2n-3)}{2\,.\,4\ldots(2n+2)},$$

and proceeding thus we find

$$D^p a_n=\frac{1\,.\,3\ldots(2n-3)\,.\,1\,.\,3\ldots(2p+1)}{2\,.\,4\,.\,6\ldots(2n+2p)}.$$

Thus, putting $n=3$,

$$b_p=\frac{1}{16}\cdot\frac{3\,.\,5\ldots(2p+1)}{8\,.\,10\ldots(2p+6)}.$$

It follows that b_p decreases as p increases, so that Euler's series converges; and consequently the integral of Borel's method is also convergent; this deserves mention because Borel himself does not seem to have succeeded in proving this directly (*l.c.* p. 118). We have, however, already established the convergence in Art. 112, Ex. 4:

* *Liouville's Journal* (5), t. 2, 1896, p. 119.

In our case, we have

$$\begin{aligned} c_0 &= \quad b_0 = \quad a_3 = \tfrac{1}{16}, \\ c_1 &= \tfrac{2}{3} b_1 = \tfrac{2}{3} \cdot \tfrac{3}{8} \, c_0 = \tfrac{1}{4} c_0, \\ c_2 &= (\tfrac{2}{3})^2 b_2 = \tfrac{2}{3} \cdot \tfrac{5}{10} c_1 = \tfrac{1}{3} c_1, \\ c_3 &= (\tfrac{2}{3})^3 b_3 = \tfrac{2}{3} \cdot \tfrac{7}{12} c_2 = \tfrac{7}{18} c_2, \\ c_4 &= (\tfrac{2}{3})^4 b_4 = \tfrac{2}{3} \cdot \tfrac{9}{14} c_3 = \tfrac{3}{7} c_3, \end{aligned}$$

and so on.

Then the series is $1{\cdot}5 + \frac{1}{3} \cdot 2^3 (c_0 + c_1 + c_2 + \dots)$.

The numerical values given by these formulae have been checked by direct calculation of $a_3, a_4, a_5, \dots$ and their differences; we have then

$$\begin{array}{rr} c_1 = & {\cdot}062500 \\ c_2 = & 15625 \\ c_3 = & 5208 \\ c_4 = & 2025 \\ c_5 = & 868 \\ c_6 = & \underline{398} \\ & {\cdot}086624 \\ & \underline{378} \\ & {\cdot}087002 \end{array} \qquad \begin{array}{rr} c_7 = & {\cdot}000192 \\ c_8 = & 96 \\ c_9 = & 50 \\ c_{10} = & 26 \\ c_{11} = & \underline{14} \\ & {\cdot}000378 \end{array}$$

A rough estimate of the remainder gives $\frac{4}{3} c_{11}$, so that we get

$$c_1 + c_2 + c_3 + \dots = {\cdot}087021,$$

the approximation being probably in excess of the true value.

Thus we find for the sum of the original series

$$1{\cdot}5 + {\cdot}232056 = 1{\cdot}732056.$$

Now $$\sqrt{3} = 1{\cdot}7320508\dots,$$

so that our approximation is a very good one; it is closer than Borel's, although he works to 7 figures and uses 34 terms of the series.

Ex. 3. Euler calculates in this way the value of the series

$$\log_{10} 2 - \log_{10} 3 + \log_{10} 4 - \dots,$$

starting to take differences at $\log_{10} 10$. He obtains 0·0980601. (Compare Ex. 10, p. 351.)

Euler also attempts to evaluate $1! - 2! + 3! - 4! + \dots$ by this method, and he obtains ·4008 ..., but although the first and second figures agree with those found in Arts. 98 and 132 (1), yet it does not appear that his method rests on a satisfactory basis here.

122. Cesàro's method of summation.

It has been proved (Art. 34) that when two convergent series are multiplied together, the product-series is *at most simply indeterminate*. That is, if s_n denotes the sum to $(n+1)$ terms, the limit

(1) $$\lim_{n \to \infty} (s_0 + s_1 + s_2 + \dots + s_n)/(n+1)$$

exists and is finite. We have also seen that this mean-value appears in the theorem of Frobenius (Art. 51) on

$$\lim_{x \to 1} (\Sigma u_n x^n).$$

Cesàro has proposed to adopt the limit (1), if it exists, as the definition of the sum of a non-convergent series; or, more generally, he defines the sum as

$$(2) \qquad \lim_{n \to \infty} \{S_n^{(r)}/A_n^{(r)}\},$$

where

$$S_n^{(r)} = s_n + rs_{n-1} + \frac{r(r+1)}{2!} s_{n-2} + \dots \frac{r(r+1)\dots(r+n-1)}{n!} s_0$$

and $$A_n^{(r)} = \{(r+1)(r+2)\dots(r+n)\}/n!.$$

Of course the limit (1) is the special case of (2) which is given by putting $r=1$.

With these definitions, it is evident that*

$$\Sigma S_n^{(r)} x^n = (1-x)^{-r} (\Sigma s_n x^n)$$

and $$\Sigma A_n^{(r)} x^n = (1-x)^{-(r+1)}.$$

Now $$(1-x)^{-(r+1)} = (1+x+x^2+\dots)(1-x)^{-r},$$

and by equating coefficients of x^n we see that

$$A_n^{(r)} = 1 + r + \frac{r(r+1)}{2!} + \dots + \frac{r(r+1)\dots(r+n-1)}{n!}.$$

Thus the denominator is equal to the sum of the coefficients in the numerator of $S_n^{(r)}/A_n^{(r)}$; and so, in particular, if

$$s_0 = s_1 = s_2 = \dots = s_n = \dots,$$

we see that they are all equal to $S_n^{(r)}/A_n^{(r)}$.

Further, since

$$\Sigma s_n x^n = (1+x+x^2+\dots)(\Sigma u_n x^n),$$

we have $$\Sigma S_n^{(r)} x^n = (1-x)^{-(r+1)} (\Sigma u_n x^n);$$

and consequently

$$S_n^{(r)} = u_n + (r+1) u_{n-1} + \frac{(r+1)(r+2)}{2!} u_{n-2} + \dots$$

$$+ \frac{(r+1)(r+2)\dots(r+n)}{n!} u_0.$$

---

* For $$(1-x)^{-r} = 1 + rx + \frac{r(r+1)}{2!} x^2 + \frac{r(r+1)(r+2)}{3!} x^3 + \&c.$$

If this series is multiplied by $\Sigma s_n x^n$ (by the ordinary rule) the coefficient of x^n is seen to be $S_n^{(r)}$ by inspection.

Thus we can write also

$$S_n^{(r)}/A_n^{(r)} = u_0 + \frac{n}{r+n} u_1 + \frac{n(n-1)}{(r+n-1)(r+n)} u_2 + \dots$$
$$+ \frac{n!}{(r+1)(r+2)\dots(r+n)} u_n.$$

From the last expression it is evident that the effect of increasing r is to give more weight to the comparatively early terms of the series Σu_n; it is therefore obvious that by increasing r we may hope to counteract the oscillatory character of the series Σu_n, and so to replace the series by a convergent limit.

Now, since

$$\Sigma S_n^{(r)} x^n = (1-x)^{-(r+1)} \Sigma u_n x^n,$$

we have
$$\Sigma S_n^{(r+1)} x^n = (1+x+x^2+\dots) \Sigma S_n^{(r)} x^n,$$
and similarly,
$$\Sigma A_n^{(r+1)} x^n = (1+x+x^2+\dots) \Sigma A_n^{(r)} x^n.$$

By equating coefficients of x^n, we have

$$S_n^{(r+1)} = S_0^{(r)} + S_1^{(r)} + S_2^{(r)} + \dots + S_n^{(r)},$$
$$A_n^{(r+1)} = A_0^{(r)} + A_1^{(r)} + A_2^{(r)} + \dots + A_n^{(r)}.$$

Consequently, from Stolz's theorem (Appendix, Art. 152, II.), if $\lim\limits_{n\to\infty} [S_n^{(r)}/A_n^{(r)}]$ exists, so also does $\lim\limits_{n\to\infty} [S_n^{(r+1)}/A_n^{(r+1)}]$; and the second limit is equal to the first. Thus *if the limit* (2) *exists for any value of r, it exists also for any higher value of r.*

If $r=k$ is the *least* value of r for which the limit (2) exists, Cesàro calls the series Σu_n *k-ply indeterminate.*

123. Extension of Frobenius's theorem.

We have tacitly assumed that the power-series used in the last article would converge absolutely. This is, however, capable of easy proof whenever $S_n^{(r)}/A_n^{(r)}$ tends to a finite limit l; for then we can fix an upper limit C to the absolute value of this quotient. Thus $\Sigma S_n^{(r)} x^n$ will converge absolutely when $\Sigma A_n^{(r)} x^n = (1-x)^{-(r+1)}$ does so; that is to say, for values of x such that $|x| < 1$. Since $\Sigma u_n x^n$ and $\Sigma s_n x^n$ can be derived from $\Sigma S_n^{(r)} x^n$ by multiplying this series by $(1-x)^{r+1}$ and $(1-x)^r$ respectively, we infer that $\Sigma u_n x^n$ and $\Sigma s_n x^n$ are also convergent for $|x| < 1$.

Again, by Art. 51, we see that

$$\lim_{x \to 1} \frac{\Sigma S_n^{(r)} x^n}{\Sigma A_n^{(r)} x^n} = \lim_{n \to \infty} \frac{S_n^{(r)}}{A_n^{(r)}} = l.$$

But
$$\frac{\Sigma S_n^{(r)} x^n}{\Sigma A_n^{(r)} x^n} = \frac{(1-x)^{-(r+1)} \Sigma u_n x^n}{(1-x)^{-(r+1)}} = \Sigma u_n x^n,$$

so that
$$\lim_{x \to 1} \Sigma u_n x^n = l,$$

which is *the extension of Frobenius's theorem.*

Illustrations of this theorem are afforded by Exs. 3–5, Art. 126.

This result appears to be novel; but a closely related theorem has been given by Hölder (*Math. Annalen*, Bd. 20, p. 535). Let us write

$$T_n^{(1)} = \frac{1}{n+1}(s_0 + s_1 + \ldots + s_n) = S_n^{(1)}/A_n^{(1)},$$

$$T_n^{(2)} = \frac{1}{n+1}[T_0^{(1)} + T_1^{(1)} + \ldots + T_n^{(1)}],$$

$$\ldots\ldots\ldots\ldots\ldots\ldots\ldots\ldots\ldots\ldots\ldots\ldots$$

$$T_n^{(r+1)} = \frac{1}{n+1}[T_0^{(r)} + T_1^{(r)} + \ldots + T_n^{(r)}];$$

then if $\lim_{n \to \infty} T_n^{(r)} = l$, Hölder has proved that $\lim_{x \to 1} \Sigma u_n x^n = l$. It seems likely that the means found by Cesàro's method and by Hölder's process must be the same, and this has been proved up to $r=2$. For higher values of r, Knopp* has proved that whenever Hölder's method gives a limit, Cesàro's will give the same limit.

It is usually better to apply Cesàro's method than Hölder's to evaluate a given series on account of its greater simplicity.

124. Inferences from Cesàro's definition.

If Σu_n is r-ply indeterminate and has the sum l, in Cesàro's sense, it follows from the definition that

$$\lim \{S_n^{(r)}/A_n^{(r)}\} = l.$$

Thus we can write

$$\Sigma S_n^{(r)} x^n - l \Sigma A_n^{(r)} x^n = \rho(x),$$

where $\rho(x) = \Sigma \rho_n x^n$ is a power-series such that

(1) $$\lim \{\rho_n / A_n^{(r)}\} = 0.$$

Hence, remembering the identities of Art. 122, we have now the identity

(2) $$(1-x)^{-(r+1)} \Sigma u_n x^n = l(1-x)^{-(r+1)} + \rho(x),$$

* K. Knopp, *Inauguraldissertation* (Berlin, 1907), p. 19.

where the coefficients in $\rho(x)$ are subject to condition (1) given above. And, conversely, if (2) holds, the series Σu_n is *at most* r-ply indeterminate and has the sum l.

It is to be noticed that (2) does *not* allow us to write

$$(1-x)^{-r}\Sigma u_n x^n = l(1-x)^{-r} + \Sigma \sigma_n x^n,$$

where

$$\lim \{\sigma_n / A_n^{(r-1)}\} = 0.$$

For we have $\sigma_n = \rho_n - \rho_{n-1}$ and $A_n^{(r-1)} = A_n^{(r)} - A_{n-1}^{(r)}$,

and so if $|\rho_{n-1}| < \epsilon A_{n-1}^{(r)}$ and $|\rho_n| < \epsilon A_n^{(r)}$,

we can only deduce that $|\sigma_n| < \epsilon \{A_{n-1}^{(r)} + A_n^{(r)}\}$

or

$$|\sigma_n / A_n^{(r-1)}| < \epsilon \{1 + (2n/r)\}.$$

Thus, owing to the presence of n in the factor multiplying ϵ, it cannot be inferred that

$$\lim \{\sigma_n / A_n^{(r-1)}\} = 0,$$

if we know nothing more than that the condition (1) is satisfied.

A simple example is given by taking $r=2$ and

$$\rho(x) = 1 - 2x + 3x^2 - \dots.$$

Then

$$\sigma(x) = 1 - 3x + 5x^2 - \dots,$$

so that $\sigma_n / A_n^{(1)}$ oscillates between -2 and $+2$.

In particular, if we have an identity

(3) $$(1-x)^{-(r+1)}\Sigma u_n x^n = l(1-x)^{-(r+1)} + P(x)(1-x)^{-r},$$

where $P(x)$ is a polynomial not divisible by $1-x$, the series Σu_n is at most r-ply indeterminate, and has the sum l.

For if there are p terms in $P(x)$, and if M is the greatest of the absolute values of the coefficients of $P(x)$, it is clear that the coefficient of x^n in $P(x)(1-x)^{-r}$ is less than

$$MpA_n^{(r-1)},$$

and the quotient of this by $A_n^{(r)}$ is

$$Mpr/(r+n),$$

which tends to zero as n tends to ∞.

In practice the most common form of identity is

(4) $$(1-x)^{-(r+1)}\Sigma u_n x^n = l(1-x)^{-(r+1)} + P_1(x)(1-x^2)^{-r},$$

from which we can deduce the same results as from (3).

125. Cesàro's theorem on the multiplication of series.

The following lemma will be needed:

If Σu_n, Σv_n are two series such that $|u_n|/A_n^{(r)}$ has a finite upper limit, while $v_n/A_n^{(s)}$ tends to zero, then the quotient

$$(u_0 v_n + u_1 v_{n-1} + \dots + u_n v_0)/A_n^{(r+s+1)}$$

tends to zero. It may be noticed that either (or both) of r, s may be zero here.

For, under the given conditions, we can find C and m such that

$$|u_n| < CA_n^{(r)}, \quad |v_n| < CA_n^{(s)}, \quad \text{for all values of } n,$$

and

$$|v_n| < \epsilon A_n^{(s)}, \quad \text{if } n > m.$$

Then $$\left|\sum_{\nu=0}^{m} u_{n-\nu} v_\nu\right| < C^2 A_n^{(r)} \{A_0^{(s)} + A_1^{(s)} + \ldots + A_m^{(s)}\} = C^2 A_n^{(r)} A_m^{(s+1)}$$

and $$\left|\sum_{\nu=m+1}^{n} u_{n-\nu} v_\nu\right| < C\epsilon \{A_n^{(r)} A_0^{(s)} + A_{n-1}^{(r)} A_1^{(s)} + \ldots + A_0^{(r)} A_n^{(s)}\} = C\epsilon B_n, \text{ say.}$$

Now B_n is the coefficient of x^n in $(1-x)^{-(r+1)} \times (1-x)^{-(s+1)} = (1-x)^{-(r+s+2)}$, and so B_n is equal to $A_n^{(r+s+1)}$.

Hence $$\left|\sum_{\nu=0}^{n} u_{n-\nu} v_\nu\right| < C^2 A_n^{(r)} A_m^{(s+1)} + C\epsilon A_n^{(r+s+1)},$$

now $$A_n^{(r)} / A_n^{(r+s+1)} \leqq A_n^{(r)} / A_n^{(r+1)} = r/(n+r+1),$$

so that $$\lim_{n\to\infty} \{A_n^{(r)} / A_n^{(r+s+1)}\} = 0.$$

Consequently, $$\overline{\lim_{n\to\infty}} \left|\sum_{\nu=0}^{n} u_{n-\nu} v_\nu\right| / A_n^{(r+s+1)} \leqq C\epsilon,$$

and, since ϵ can be taken as small as we please, it follows that

$$\lim (u_0 v_n + u_1 v_{n-1} + \ldots + u_n v_0) / A_n^{(r+s+1)} = 0.$$

We proceed now to the proof of Cesàro's theorem.

Suppose that Σu_n and Σv_n are r-ply and s-ply indeterminate respectively, and that their sums are U, V, so that we can write

$$(1-x)^{-(r+1)} (\Sigma u_n x^n) = U(1-x)^{-(r+1)} + \rho(x),$$
$$(1-x)^{-(s+1)} (\Sigma v_n x^n) = V(1-x)^{-(s+1)} + \sigma(x),$$

where $\rho(x) = \Sigma \rho_n x^n$, $\sigma(x) = \Sigma \sigma_n x^n$ are such that

$$\lim \{\rho_n / A_n^{(r)}\} = 0, \quad \lim \{\sigma_n / A_n^{(s)}\} = 0.$$

Then $$(\Sigma u_n x^n) \times (\Sigma v_n x^n) = \Sigma w_n x^n,$$

where $$w_n = u_0 v_n + u_1 v_{n-1} + \ldots + u_n v_0.$$

Hence $(1-x)^{-(r+s+2)} \Sigma w_n x^n = UV(1-x)^{-(r+s+2)} + R(x)$,

where $R(x) = U(1-x)^{-(r+1)} \sigma(x) + V(1-x)^{-(s+1)} \rho(x) + \rho(x)\sigma(x)$.

To each of the terms in $R(x)$ we can apply the lemma given above; for instance, in $U(1-x)^{-(r+1)}$ the coefficient of x^n is $UA_n^{(r)}$, and so if

$$U(1-x)^{-(r+1)} \sigma(x) = \Sigma X_n x^n,$$

we have $$\lim \{X_n / A_n^{(r+s+1)}\} = 0.$$

Similarly for the two other terms in $R_n(x)$, and so if

$$R(x) = \Sigma R_n x^n,$$

we have $$\lim \{R_n / A_n^{(r+s+1)}\} = 0.$$

Thus we see from (1) and (2) of the last article that the degree of indeterminacy of Σw_n is not higher than $(r+s+1)$, and that its sum is equal to UV.*

This leads to the theorem: *The product of a series (whose degree of indeterminacy is r) by a series (whose degree is s) can be formed as if the series were convergent. The degree of indeterminacy of the resulting series cannot exceed* $(r+s+1)$.

126. Examples of Cesàro's method.

Ex. 1. The series $1-1+1-1+1-1+\ldots$ is *simply* indeterminate and has the sum $\frac{1}{2}$.

Ex. 2. The series $1-2+3-4+5-6+\ldots$ is *doubly* indeterminate and has the sum $\frac{1}{4}$. For here we find $s_0=1$, $s_1=-1$, $s_2=2$, $s_3=-2, \ldots$; and if we write $r=2$ in the formula of Art. 122, we have to evaluate

$$\lim_{n\to\infty}\frac{s_n+2s_{n-1}+3s_{n-2}+\ldots+(n+1)s_0}{\frac{1}{2}(n+1)(n+2)}.$$

Now $\quad (n+1)s_0+ns_1=1, \quad (n-1)s_2+(n-2)s_3=2, \quad$ etc.

Thus, when n is even $(=2m)$, we have

$$(n+1)s_0+ns_1+\ldots+s_n=1+2+3+\ldots+m+(m+1)=\tfrac{1}{2}(m+1)(m+2).$$

When n is odd $(=2m+1)$, we have again

$$(n+1)s_0+ns_1+\ldots+s_n=1+2+3+\ldots+m+(m+1)=\tfrac{1}{2}(m+1)(m+2).$$

Thus the fraction has the limiting value $\frac{1}{4}$.

It will be observed that

$$1-2+3-4+5-6+\ldots=(1-1+1-1+\ldots)^2,$$

and, according to Cesàro's theorem of the last article, $(1-1+1-1+\ldots)^2$ should be *at most trebly* indeterminate; and this of course is verified here.

Ex. 3. The series $1-2^2+3^2-4^2+\ldots$ is *trebly* indeterminate, for we have

$$1-2^2x+3^2x^2-4^2x^3+\ldots=\frac{d}{dx}[x(1+x)^{-2}]=(1-x)(1+x)^{-3}.$$

Hence $\quad (1-x)^{-4}(1-2^2x+3^2x^2-4^2x^3+\ldots)=(1-x^2)^{-3},$

which obviously satisfies the condition (3) of Art. 124 and gives

$$1-2^2+3^2-4^2+\ldots=0.$$

Ex. 4. In like manner, we find

$$1-2^3x+3^3x^2-4^3x^3+\ldots=(x^2-4x+1)(1+x)^{-4}.$$

Now $\dfrac{x^2-4x+1}{(1+x)^4}$ has the value $-\frac{1}{8}$ for $x=1$, so that we have

$$\frac{x^2-4x+1}{(1+x)^4}=-\tfrac{1}{8}+\frac{(1-x)P(x)}{(1+x)^4},$$

where $P(x)$ is a polynomial of degree 3.

* The reader will probably find it instructive to use this method to establish the theorem for $r=0$, $s=0$, already established in Art. 34. Cesàro's treatment of the general case is on the same lines as the proof given in Art. 34.

Hence we find

$$(1-x)^{-5}(1-2^3x+3^3x^2-4^3x^3+\ldots+\tfrac{1}{8})=P(x)(1-x^2)^{-4},$$

and so $$1-2^3+3^3-4^3+\ldots=-\tfrac{1}{8},\qquad \text{Art. 124, (4)}$$

this series being *quadruply* indeterminate.

Ex. 5. It is not difficult to see that in general the series

$$1-2^{r-1}+3^{r-1}-4^{r-1}+\ldots$$

is r-ply indeterminate, because

$$e^t-2^{r-1}e^{2t}+3^{r-1}e^{3t}-\ldots=\frac{d^{r-1}}{dt^{r-1}}\left(\frac{e^t}{1+e^t}\right).$$

Thus $1-2^{r-1}x+3^{r-1}x^2-\ldots$ is of the form $P(x)/(1+x)^r$, where $P(x)$ is a polynomial of degree $(r-2)$. Thus, proceeding as above, we find

$$(1-x)^{-(r+1)}[(1-2^{r-1}x+3^{r-1}x^2-\ldots)-l]=Q(x)(1-x^2)^{-r},$$

where $Q(x)$ is a polynomial of degree $(r-1)$; and

$$\left.\begin{aligned} l=\lim_{t\to 0}\left[\frac{d^{r-1}}{dt^{r-1}}\left(\frac{e^t}{1+e^t}\right)\right]&=(2^{2n}-1)\frac{B_n}{2n}, \text{ if } r=2n,\\ \text{or } &=0, \qquad\qquad \text{if } r \text{ is } odd. \end{aligned}\right\}\ \text{Art. 93.}$$

Hence the sum of the series is l by (4) of Art. 124. [CESÀRO.]

Ex. 6. The reader will have little difficulty in verifying Exs. 3–5 by multiplication. For instance we find

$$(1-1+1-1+\ldots)(1-2+3-4+\ldots)=1-3+6-10+\ldots$$

or $$1-3+6-10+\ldots=\tfrac{1}{8}.$$

But $1-2^2+3^2-4^2+\ldots=2(1-3+6-10+\ldots)-(1-2+3-4+\ldots)$,

and so $$1-2^2+3^2-4^2+\ldots=\tfrac{2}{8}-\tfrac{1}{4}=0.$$

Ex. 7. The following series are all *simply* indeterminate:

$$\begin{aligned} 1+\cos\theta+\cos 2\theta+\cos 3\theta+\ldots&=\tfrac{1}{2},\\ \sin\theta+\sin 2\theta+\sin 3\theta+\ldots&=\tfrac{1}{2}\cot\tfrac{1}{2}\theta,\\ \cos\theta+\cos 3\theta+\cos 5\theta+\ldots&=0,\\ \sin\theta+\sin 3\theta+\sin 5\theta+\ldots&=\tfrac{1}{2}\operatorname{cosec}\theta. \end{aligned}$$

These results agree with those found by Borel's method (Art. 103).

It is of some historical interest to note the first of these series was evaluated by d'Alembert (using the mean-value process), *Opuscula Math.*, t. 4, Paris, 1768.

127. Limitations of Cesàro's method.

The mean-value process has rather narrow limits of usefulness; to see this, let us determine the general type of series which can be summed in this way. We have seen that

$$\Sigma S_n^{(r)}x^n=(1-x)^{-(r+1)}(\Sigma u_n x^n),$$

so that $$\Sigma u_n x^n=(1-x)^{r+1}[\Sigma S_n^{(r)}x^n],$$

which gives

$$u_n=S_n^{(r)}-(r+1)S_{n-1}^{(r)}+\frac{(r+1)r}{2!}S_{n-2}^{(r)}-\ldots \text{ to } (r+2) \text{ terms.}$$

Now if $S_n^{(r)}/A_n^{(r)}$ approaches a definite limit l, we can find a value p such that when $\nu > p-r$, all the terms $S_\nu^{(r)}/A_\nu^{(r)}$ lie between $(l-\epsilon)$ and $(l+\epsilon)$.

Thus $$|u_n| < (l+\epsilon)\left[A_n^{(r)} + \frac{(r+1)r}{2!}A_{n-2}^{(r)} + \ldots\right]$$

$$-(l-\epsilon)\left[(r+1)A_{n-1}^{(r)} + \frac{(r+1)r(r-1)}{3!}A_{n-3}^{(r)} + \ldots\right], \quad (n>p).$$

The coefficient of l in this expression is easily seen to be zero in virtue of the fact that

$$(1-x)^{r+1}[\Sigma A_n^{(r)} x^n] = 1,$$

so that $$A_n^{(r)} - (r+1)A_{n-1}^{(r)} + \frac{(r+1)r}{2!}A_{n-2}^{(r)} - \ldots = 0.$$

Thus $$|u_n| < \epsilon A_n^{(r)}\left[1 + (r+1) + \frac{(r+1)r}{2!} + \ldots\right], \quad (n>p).$$

That is, $$|u_n| < \epsilon . 2^{r+1} A_n^{(r)} \quad (n>p),$$

and so $$\lim_{n\to\infty} u_n/A_n^{(r)} = 0.$$

But $$A_n^{(r)} = (n+r)!/n!r!,$$

so that $$\lim_{n\to\infty} A_n^{(r)}/n^r = 1/r!.$$

Thus, we have the result

$$\lim_{n\to\infty} (u_n/n^r) = 0;$$

which gives a necessary condition that the series Σu_n may be r-ply indeterminate.

There are many series summable by the former methods which do not satisfy this condition; in fact the simplest of all,

$$1 - t + t^2 - t^3 + \ldots, \quad (t>1)$$

does not satisfy it, and is therefore not summable in Cesàro's sense.

128. Borel's original method contrasted with Cesàro's.

We may regard the most general form of mean amongst $s_0, s_1, s_2, \ldots$ as given by

$$\frac{x_0 s_0 + x_1 s_1 + \ldots + x_n s_n + \ldots}{x_0 + x_1 + \ldots + x_n + \ldots}$$

where x_n is *never negative*, and the series Σx_n, $\Sigma x_n s_n$ are convergent.

Here we may either (i) fix every factor x_n once for all, or (ii) fix x_n as a function of a variable x which is then made to tend to some definite limit. For a variety of reasons, the first alternative may be ruled out; and so we are left with the second.

Cesàro's first and second means (for series of simple and double indeterminacy) come under case (ii) by taking

$$x_n = 1, \qquad \text{if } n \leqq x, \qquad x_n = 0, \quad \text{if } n > x;$$

or

$$x_n = x + 1 - n, \quad \text{if } n \leqq x, \qquad x_n = 0, \quad \text{if } n > x;$$

where x is afterwards made to tend to ∞ (through integral values). And the other means are given by similar, but more complicated, formulae. In these cases the oscillation of the sequence (s_n) is converted into convergence by the decreasing character of the factors (x_n); so that little weight is attached to the terms with high indices, which are the very terms that are of importance in studying the series.

To meet this difficulty, Borel chooses the factors x_n so as to increase at first to a maximum (whose position varies with x), and afterwards to steadily decrease. The position of the maximum recedes as x increases, which ensures that the terms with high indices play an important part in determining the mean.

The most natural type of function x_n satisfying these conditions is $x^n/n!$, which increases so long as $n < x$, and finally decreases very rapidly.

This gives as the mean

$$\lim_{x \to \infty} \frac{s_0 + s_1 x + s_2 x^2/2! + \dots}{1 + x + x^2/2! + \dots} = \lim_{x \to \infty} e^{-x}\left(s_0 + s_1 x + s_2 \frac{x^2}{2!} + \dots\right),$$

which is the definition discussed in Art. 114.

129. Connexion between Borel's sum and Cesàro's mean for a given series Σu_n.

We have proved (Art. 123) that if Cesàro's mean gives a limit l, then

$$\lim_{t \to 1} (\Sigma u_n t^n) = l.$$

Further, we have proved (Art. 111) that

$$\lim_{t\to 1}(\Sigma u_n t^n)=\int_0^\infty e^{-x}u(x)\,dx,$$

provided that the latter converges. Thus, *when Cesàro's mean exists, it gives the value of Borel's integral, provided that the latter is convergent*; the integral may, however, oscillate between extreme limits which include Cesàro's mean.* Obviously the same result is true for any mode of summation given by the integral of Art. 118,

$$\int_0^\infty \phi(x)\left(\Sigma\frac{u_n}{c_n}x^n\right)dx.$$

Hardy has proved in his second paper (see also the small type below) that if $r=1$, and if a further condition is satisfied, the convergence of Borel's integral will follow from the existence of Cesàro's mean. In general, with higher values for r, it seems likely that the same condition will suffice to deduce the convergence of the integral from the existence of the mean. However, the algebraical difficulties involved seem at present too formidable to make it worth while to write out a rigorous proof, except for the case $r=1$.

We have seen (Art. 114) that the existence of the limit

$$\lim_{x\to\infty}\left[e^{-x}\left(\Sigma s_n\frac{x^n}{n!}\right)\right]$$

implies the existence of Borel's integral, and that the two expressions are equal; and we shall prove now that if

$$\lim_{n\to\infty}(s_0+s_1+\ldots+s_n)/(n+1)=l,$$

then

$$\lim_{x\to\infty}\left[e^{-x}\left(\Sigma s_n\frac{x^n}{n!}\right)\right]=l,$$

subject to Hardy's condition. It is then obvious from Art. 114 that Borel's integral must also be equal to l.

Now suppose that H, h are the upper and lower limits of

$$\sigma_n=(s_0+s_1+\ldots+s_n)/(n+1)$$

as n ranges from 0 to ∞, while H_m, h_m are those of $\sigma_m, \sigma_{m+1}, \sigma_{m+2}, \ldots$: then, from Art. 153 of the Appendix, we see that

$$\frac{s_0v_0+s_1v_1+s_2v_2+\ldots+s_nv_n}{v_0+v_1+v_2+\ldots+v_n}$$

---

* It is of course understood that Cesàro's mean exists. If we are dealing with series in general, the mean may oscillate though the integral converges (see Art. 127).

lies between
$$H_m + \frac{(h_m - h) m v_m + (H_m - h_m) p v_p}{v_0 + v_1 + v_2 + \dots + v_n}$$
and
$$h_m - \frac{(H - H_m) m v_m + (H_m - h_m) p v_p}{v_0 + v_1 + v_2 + \dots + v_n},$$
where $v_r = x^r/r!$ and v_p is the maximum term. Here p must be equal to the integral part of x; and on making n tend to ∞, $v_0 + v_1 + \dots + v_n$ has the limit e^x, and so we get

$$e^{-x}\left(\Sigma s_n \frac{x^n}{n!}\right) < H_m + e^{-x}\left[(h_m - h)\frac{x^m}{(m-1)!} + (H_m - h_m)\frac{x^p}{(p-1)!}\right]$$
$$\text{and } > h_m - e^{-x}\left[(H - H_m)\frac{x^m}{(m-1)!} + (H_m - h_m)\frac{x^p}{(p-1)!}\right].$$

Now, corresponding to any value of m, we can write
$$h_m \geqq l - \epsilon_m, \quad H_m \leqq l + \epsilon_m,$$
where $\epsilon_m \to 0$ as m increases. And there is a constant A such that
$$h_m - h < A, \quad H - H_m < A;$$
thus we find
$$\left| e^{-x}\left(\Sigma s_n \frac{x^n}{n!}\right) - l \right| < \epsilon_m + e^{-x}\left[A\frac{x^m}{(m-1)!} + 2\epsilon_m \frac{x^p}{(p-1)!}\right].$$

Next, if we use Stirling's formula (Appendix, Art. 179), we see that
$$\log\left[e^{-x}\frac{x^p}{(p-1)!}\right] = -x + p\log x - (p - \tfrac{1}{2})\log p + p - \tfrac{1}{2}\log(2\pi) + P,$$
where P tends to 0 as $p \to \infty$.

Now $$p(\log x - \log p) = p\log\left(1 + \frac{x-p}{p}\right) = x - p + P',$$
where P' also tends to 0, because $x - p < 1$.

Thus $$\log\left[e^{-x}\frac{x^p}{(p-1)!}\right] = \tfrac{1}{2}\log\left(\frac{p}{2\pi}\right) + P_1, \quad \text{where } P_1 \to 0,$$
or
$$e^{-x}\frac{x^p}{(p-1)!} < K\sqrt{p},$$
where K denotes a number not necessarily the same in all inequalities but always contained between certain fixed limits, such as ·0001 and 10000.

Similarly, we can prove that if we take m as the integral part of $\frac{1}{2}p$,
$$e^{-x}\frac{x^m}{(m-1)!} < K\left(\frac{2}{e}\right)^m \sqrt{m}.$$

Hence, if m is the integral part of $\frac{1}{2}x$, we have
$$\left| e^{-x}\left(\Sigma s_n \frac{x^n}{n!}\right) - l \right| < \epsilon_m + K\sqrt{m}\left\{\left(\frac{2}{e}\right)^m + \epsilon_m\right\}.$$

Thus we have $$\lim_{x\to\infty}\left[e^{-x}\left(\Sigma s_n \frac{x^n}{n!}\right)\right] = l,$$
provided that the condition $$\lim_{m\to\infty}(\epsilon_m \sqrt{m}) = 0$$
is satisfied.

Hardy remarks that the condition $\lim (\epsilon_m \sqrt{m}) = 0$ is by no means necessary for the truth of the theorem; in fact, it is not satisfied by the following *convergent* series (which has the sum 0 and is therefore summable in virtue of Art. 100).

Let
$$s_0 = 0, \quad s_n = n^{-a}, \quad (0 < a < \tfrac{1}{2})$$
and then
$$\epsilon_n = \left(\sum_1^n r^{-a}\right)/(n+1).$$

Thus by Art 11, we see that
$$\epsilon_n > \frac{1}{1-a}\left[\frac{1}{(n+1)^a} - \frac{1}{n+1}\right],$$
so that
$$\lim (\epsilon_n \sqrt{n}) = \infty, \text{ although } \lim \epsilon_n = 0.$$

But Hardy shews by another example (*l.c.* p. 41) that if the condition $\lim (\epsilon_n \sqrt{n}) = 0$ is broken, the series Σu_n *may* not be summable; so that the condition is not merely a consequence of the special presentation of the argument.

ASYMPTOTIC SERIES.

130. Euler's use of asymptotic series.

One of the most instructive examples of the application of non-convergent series was given by Euler in using his formula of summation (Art. 95) for the calculation of certain finite sums.*

Thus, taking $f(x) = 1/x$, $a = 1$, $b = n$, Euler finds
$$1 + \frac{1}{2} + \frac{1}{3} + \ldots + \frac{1}{n} = \log n + \frac{1}{2n} - \frac{B_1}{2n^2} + \frac{B_2}{4n^4} - \frac{B_3}{6n^6} + \ldots + \text{const.}$$

Now this series, if continued to infinity, does not converge, because we have (Art. 92)
$$\frac{B_r}{B_{r-1}} = \frac{2r(2r-1)}{4\pi^2} \frac{\Sigma \dfrac{1}{n^{2r}}}{\Sigma \dfrac{1}{n^{2r-2}}};$$

but, if $r > 3$, $\Sigma \dfrac{1}{n^{2r-2}} < 1 \Big/ \left(1 - \dfrac{1}{2^4}\right)$ (see Art. 7), and $\Sigma \dfrac{1}{n^{2r}} > 1$,

so that
$$\frac{B_r}{2rn^{2r}} \Big/ \frac{B_{r-1}}{2(r-1)n^{2r-2}} > \frac{15(r-1)^2}{16n^2\pi^2};$$

hence the terms in the series steadily increase in numerical value after a certain value of r (depending on n). It does not appear whether Euler realised that the series could never converge; but he was certainly aware of the fact that it does not

* *Inst. Calc. Diff.*, 1755 (Pars Posterior), cap. vi.

converge for $n=1$. He employed the series for $n=10$ to calculate the constant (*Euler's constant*,* Art. 11),

$$C=0{\cdot}5772156649015325\ldots,$$

which he regarded as the "sum" of the series

$$\frac{1}{2}+\frac{B_1}{2}-\frac{B_2}{4}+\frac{B_3}{6}-\ldots.$$

The reason why this series can be used, although not convergent, is that *the error in the value obtained by stopping at any particular stage in the series, is less than the next term in the series.* The truth of this statement follows at once from the general theorem proved below (see Art. 131) by observing that $f^r(x)$ is of constant sign and has the *same* sign as $f^{r+2}(x)$.

This fact enables us to see at once that all of Euler's results are correct, after making a few unimportant changes.

We quote a few of Euler's results for verification:

Ex. 1.
$$1+\frac{1}{2}+\ldots+\frac{1}{n}=7{\cdot}48547, \quad \text{if } n=1000,$$
$$=14{\cdot}39273, \quad \text{if } n=1000000.$$

Euler gives the values to 13 decimals.

Ex. 2. Shew that

$$1+\frac{1}{3}+\frac{1}{5}+\ldots+\frac{1}{2n-1}=\frac{1}{2}(C+\log n)+\log 2+\frac{B_1}{8n^2}-\frac{(2^3-1)B_2}{64n^4}+\ldots,$$

and find formulae for

$$1-\frac{1}{2}+\frac{1}{3}-\frac{1}{4}+\ldots+\frac{1}{2n-1}-\frac{1}{2n},$$

and
$$\frac{1}{a+b}+\frac{1}{2a+b}+\frac{1}{3a+b}+\ldots+\frac{1}{na+b}.$$

Ex. 3. Prove that

$$\log 3=1+\tfrac{1}{2}-\tfrac{2}{3}+\tfrac{1}{4}+\tfrac{1}{5}-\tfrac{2}{6}+\tfrac{1}{7}+\tfrac{1}{8}-\tfrac{2}{9}+\ldots,$$
$$\log 4=1+\tfrac{1}{2}+\tfrac{1}{3}-\tfrac{3}{4}+\tfrac{1}{5}+\tfrac{1}{6}+\tfrac{1}{7}-\tfrac{3}{8}+\tfrac{1}{9}+\ldots,$$

and so on.

In fact the sum to $3n$ terms of the first series is

$$\left(1+\frac{1}{2}+\frac{1}{3}+\ldots+\frac{1}{3n}\right)-3\left(\frac{1}{3}+\frac{1}{6}+\ldots+\frac{1}{3n}\right)$$
$$=(C+\log 3n+R_{3n})-(C+\log n+R_n),$$

where R_{3n}, R_n tend to zero.

---

* Writing $n=500$ and 1000 in this series, J. C. Adams has calculated C to 260 places (*Proc. Roy. Soc.*, vol. 27, 1878; and *Math. Papers*, vol. 1, p. 459). This requires a knowledge of $B_1, B_2, \ldots, B_{62}$ which had been previously tabulated by Adams (*Math. Papers*, vol. 1, pp. 453, 455).

Ex. 4. Taking $f(x)=1/x^2$, prove similarly that

$$\frac{1}{10^2}+\frac{1}{11^2}+\frac{1}{12^2}+\dots \text{ to } \infty = \cdot 0951663357,$$

and deduce that

$$\frac{\pi^2}{6}=1\cdot 6449340668.$$

Ex. 5. Shew similarly that

$$1+\frac{1}{2^3}+\frac{1}{3^3}+\frac{1}{4^3}+\dots=1\cdot 2020569032.$$

Euler obtained in this manner the numerical values of $\Sigma 1/n^r$ from $r=2$ to 16, each calculated to 18 decimals (*l.c.* p. 456); Stieltjes has carried on the calculations to 32 decimals from $r=2$ to 70 (*Acta Math.*, Bd. 10, p. 299). The values to 10 decimals (for $r=2$ to 9) are given in Chrystal's *Algebra*, vol. 2, p. 367.

Ex. 6. If $f(x)=1/(l^2+x^2)$, prove that

$$\frac{1}{l^2+1}+\frac{1}{l^2+2^2}+\dots+\frac{1}{l^2+n^2}$$

$$=\frac{1}{l}\left(\frac{\pi}{2}-\theta\right)-\frac{1}{2}\left(\frac{1}{l^2}-\frac{1}{l^2+n^2}\right)+\frac{\pi}{l(e^{2l\pi}-1)}-\frac{B_1}{2}\frac{\sin^2\theta\sin 2\theta}{l^3}+\frac{B_2}{4}\frac{\sin^4\theta\sin 4\theta}{l^5}-\dots,$$

where $\tan\theta=l/n$; the constant is determined by allowing n to tend to ∞ and using the series found at the end of Art. 92.

Ex. 7. In particular, by writing $l=n$ (in Ex. 6), we find

$$\pi=4n\left(\frac{1}{n^2+1}+\frac{1}{n^2+2^2}+\dots+\frac{1}{n^2+n^2}\right)+\frac{1}{n}-\frac{4\pi}{e^{2n\pi}-1}$$

$$+\frac{B_1}{1\,.\,n^2}-\frac{B_3}{3\,.\,2^2\,.\,n^6}+\frac{B_5}{5\,.\,2^4\,.\,n^{10}}-\frac{B_7}{7\,.\,2^6\,.\,n^{14}}+\dots.$$

By writing $n=5$, Euler calculates the value of π to 15 decimals.

Ex. 8. If $f(x)=\log x$, we obtain Stirling's series

$$\log(n!)=(n+\tfrac{1}{2})\log n-n+\tfrac{1}{2}\log(2\pi)+\frac{B_1}{1\,.\,2n}-\frac{B_2}{3\,.\,4n^3}+\dots,$$

which is found differently in Art. 132 below.

131. The remainder in Euler's formula.

In virtue of the results of Art. 95 (small type), we can write

$$(1)\quad f(a)+f(a+1)+\dots+f(b)=\int_a^b f(t)\,dt+\tfrac{1}{2}[f(b)+f(a)]$$

$$+\frac{B_1}{2!}[f'(b)-f'(a)]-\frac{B_2}{4!}[f'''(b)-f'''(a)]$$

$$+\dots+(-1)^n\frac{B_{n-1}}{(2n-2)!}[f^{2n-3}(b)-f^{2n-3}(a)]+R_n,$$

where

$$R_n=-\frac{1}{(2n)!}\int_0^1\phi_{2n}(t)[f^{2n}(a+t)+f^{2n}(a+t+1)+\dots+f^{2n}(b+t-1)]\,dt.$$

Now $$R_n - R_{n+1} = (-1)^{n+1} \frac{B_n}{(2n)!} [f^{2n-1}(b) - f^{2n-1}(a)]$$

in virtue of the results found in that article; and in the integrals R_{n+1}, R_n, the polynomials $\phi_{2n}(t)$, $\phi_{2n+2}(t)$ are both of constant sign (Art. 94), but their signs are *opposite*. Thus *if we assume that the signs of* $f^{2n}(x)$, $f^{2n+2}(x)$ *are the same and that their common sign remains constant for all values of* x *from* a *to* b, the integrals R_{n+1}, R_n have *opposite* signs.

Hence $$|R_n| < |R_n - R_{n+1}| < \frac{B_n}{(2n)!} |f^{2n-1}(b) - f^{2n-1}(a)|.$$

Thus *the error involved in omitting* R_n *from the equation* (1) *is numerically less than the next term, and has the same sign*; that is, the series so obtained has the same property as a convergent series of decreasing terms which have alternate signs. Theoretically, however, the convergent series can be pushed to an *arbitrary* degree of approximation, while (1) cannot; but in practice the series (1) usually gives quite as good an approximation as is necessary for ordinary calculations.

132. Further examples of asymptotic series.

(1) *The logarithmic integral.*

The integral* $$\int_x^\infty e^{-t} \frac{dt}{t}$$

is often denoted by the symbol $-\text{li}(e^{-x})$.

In many problems, it is important to calculate the value of this integral for *large* values of x. Now we can write†

$$\int_x^\infty \frac{e^{-t}}{t} dt = \int_1^\infty e^{-t} \frac{dt}{t} - \int_0^1 (1-e^{-t}) \frac{dt}{t} - \int_1^x \frac{dt}{t} + \int_0^x (1-e^{-t}) \frac{dt}{t}$$

$$= -C - \log|x| + x - \frac{1}{2}\frac{x^2}{2!} + \frac{1}{3}\frac{x^3}{3!} - \frac{1}{4}\frac{x^4}{4!} + \dots,$$

where C is Euler's constant (see Appendix, Art. 178).

---

*If x is *negative*, the principal value of the integral is to be taken (see Art. 164). The symbol "li" denotes *logarithmic integral*; the meaning of this terminology is evident on writing $u = e^{-t}$, $y = e^{-x}$, and then $\text{li}(y) = \int_0^y du/\log u$.

†When x is *negative* all these integrals are convergent except $\int_1^x \frac{dt}{t}$, of which we must take the principal value; that is

$$\lim_{\epsilon \to 0} \left(\int_1^\epsilon \frac{dt}{t} + \int_{-\epsilon}^x \frac{dt}{t} \right) = \lim_{\epsilon \to 0} \left[\log \epsilon + \log\left(-\frac{x}{\epsilon}\right) \right]$$
$$= \log(-x) = \log|x|.$$

But this expansion, although convergent for all values of x, is unsuitable for calculation when $|x|$ is large, just as the exponential series is not convenient for calculating high powers of e. To meet this difficulty we proceed as follows:

If x is positive, we write $t=x(1+u)$, and then we have

$$\int_x^\infty \frac{e^{-t}}{t}\,dt = e^{-x}\int_0^\infty \frac{e^{-xu}}{1+u}\,du$$

$$= e^{-x}\left[\int_0^\infty \{1-u+u^2-\ldots+(-1)^{n-1}u^{n-1}\}e^{-xu}\,du + (-1)^n\int_0^\infty \frac{u^n}{1+u}e^{-xu}\,du\right].$$

This gives $\quad e^{-x}\left[\frac{1}{x}-\frac{1}{x^2}+\frac{2!}{x^3}-\frac{3!}{x^4}+\ldots+(-1)^{n-1}\frac{(n-1)!}{x^n}+R_n\right],$

where $\quad |R_n| < n!\,x^{-(n+1)}.$

This result can also be found by integration by parts.

When x is large, the terms of this series at first decrease very rapidly. Thus, *up to a certain degree of accuracy*, this series is very convenient for numerical work when x is large; but we cannot get beyond a certain approximation, because the terms finally increase beyond all limits.

For example, with $x=10$, the estimated limits for R_9, R_{10} are equal and are less than any other remainder. And the ratio of their common value to the first term in the series is about 1 : 2500. To get this degree of accuracy from the first series we should need 35 terms. Again, with $x=20$, the ratio of R_{10} to the first term is less than $1 : 10^6$; but 80 terms of the ascending series do not suffice to obtain this degree of approximation.

When x is negative, we write

$$x=-\xi \text{ and } t=x(1-u)=\xi(u-1);$$

then we find

$$\mathrm{li}(e^\xi) = e^\xi P\int_0^\infty \frac{e^{-u\xi}}{1-u}\,du$$

$$= e^\xi\left[\int_0^\infty (1+u+u^2+\ldots+u^{n-1})e^{-u\xi}\,du + P\int_0^\infty \frac{u^n e^{-u\xi}}{1-u}\,du\right],$$

where P denotes the principal value of the integral (Appendix, Art. 164). Thus

$$\mathrm{li}(e^\xi) = e^\xi\left[\frac{1}{\xi}+\frac{1}{\xi^2}+\frac{2!}{\xi^3}+\ldots+\frac{(n-1)!}{\xi^n}+R_n\right],$$

where $\quad R_n = P\int_0^\infty \frac{u^n e^{-u\xi}}{1-u}\,du.$

Stieltjes* has proved by an elaborate discussion, which is too lengthy to be given here, that in this case also we get the best approximation by taking n equal to the integral part of ξ, and that the value of R_n is then of the order $e^{-\xi}(2\pi/\xi)^{\frac{1}{2}}$.

It is not without interest to note that according to Lacroix (*Calcul. Diff. et Int.*, Paris, 1819, t. 3, p. 517) these two expansions were utilised by Mascheroni to find Euler's constant C.

Another application is to be found in the "summation" of

$$1!-2!+3!-4!+\ldots,$$

taking the value of C as known (Art. 130 above).

If we write $x=1$ and equate the series on pp. 325, 326, we have

$$-C+\left(1-\frac{1}{2.2!}+\frac{1}{3.3!}-\ldots\right)=e^{-1}(1-1!+2!-3!+\ldots),$$

which gives

$$1!-2!+3!-4!+\ldots=1+e\left[C-\left(1-\frac{1}{2.2!}+\frac{1}{3.3!}-\ldots\right)\right].$$

Lacroix gives the value 0·7965996 as the value of the series in round brackets, which yields the "sum"

$$1!-2!+3!-4!+\ldots=\cdot 4036526,$$

agreeing with the result found from Euler's continued fraction in Art. 98.

Lacroix (*l.c.* p. 389) gives another calculation of this oscillatory series by using the method of approximate quadrature to evaluate the integral

$$\int_1^\infty \frac{e^{-t}}{t}\,dt=\int_0^1 \frac{e^{-1/y}}{y}\,dy=\frac{1}{e}(1-2!+3!-\ldots),$$

which gives the sum ·403628.... Lacroix attributes the calculation to Euler, but without a reference; and he also suggests the application of approximate quadrature to the integral

$$\frac{1}{e}\int_0^1 \frac{dv}{1-\log v},$$

which is found by writing $v=e^{1-t}$, but he gives no numerical results.

(2) *Fresnel's integrals.*

Consider the two integrals

$$U=\int_x^\infty \frac{\cos t}{\sqrt{t}}\,dt, \qquad V=\int_x^\infty \frac{\sin t}{\sqrt{t}}\,dt, \qquad (x>0),$$

which are met with in the theory of Physical Optics, and also in the theory of deep-water waves.†

* *Annales de l'École Normale Supérieure* (3), t. 3, 1886, p. 201.

† Historically the hydrodynamical application seems to have occurred first (see Lamb, *Proc. Lond. Math. Soc.* (2), vol. 2, 1904, p. 371); and the chief properties of the integrals were worked out by Poisson and Cauchy in connection with this problem (for references and details see Lamb's paper).

We have $$U+iV=\int_x^\infty \frac{e^{it}}{\sqrt{t}}\,dt,$$

so that if $t=x(1+u)$ we have

$$U+iV=e^{ix}\sqrt{x}\int_0^\infty \frac{e^{ixu}}{(1+u)^{\frac{1}{2}}}\,du.$$

Now $$\int_0^\infty \frac{e^{ixu}}{(1+u)^{\frac{1}{2}}}\,du=-\frac{1}{ix}+\int_0^\infty \frac{1}{2ix}\,\frac{e^{ixu}}{(1+u)^{\frac{3}{2}}}\,du$$

by applying the process of integration by parts. Continuing thus, we have

$$\int_0^\infty \frac{e^{ixu}}{(1+u)^{\frac{1}{2}}}\,du=\frac{i}{x}+\frac{1}{2x^2}-\frac{1.3.i}{2^2x^3}-\frac{1.3.5}{2^3x^4}+\frac{1.3.5.7}{2^4x^4}\int_0^\infty \frac{e^{ixu}\,du}{(1+u)^{\frac{9}{2}}},$$

and the process can be continued as far as we please. A moment's consideration shews that the remainder integral at any stage is less in absolute value than the last term of the series; and thus for any value of x we can determine the stage at which it is best to stop in numerical work.

We are thus led to the asymptotic equation (see Art. 133)

$$U+iV\sim\frac{ie^{ix}}{\sqrt{x}}\left[1+\frac{1}{2ix}+\frac{1.3}{(2ix)^2}+\frac{1.3.5}{(2ix)^3}+\dots\right]$$

$$\sim\frac{ie^{ix}}{\sqrt{x}}(X-iY), \text{ say.}$$

The series for $X-iY$ can be "summed" by observing that

$$1.3.5\dots(2n-1)=2^n\Gamma(n+\tfrac{1}{2})/\Gamma(\tfrac{1}{2}).$$

Thus $$\frac{1.3.5\dots(2n-1)}{2^n}=\frac{1}{\sqrt{\pi}}\int_0^\infty e^{-v}v^{n-\frac{1}{2}}\,dv,$$

and so, applying Art. 118, we see that if the series $X-iY$ can be summed in Borel's way, its sum is given by

$$X-iY=\frac{x}{\sqrt{\pi}}\int_0^\infty \frac{e^{-v}}{x+iv}\,\frac{dv}{\sqrt{v}}, \quad \text{(see Art. 136).}$$

Thus $$X=\frac{x^2}{\sqrt{\pi}}\int_0^\infty \frac{e^{-v}}{v^2+x^2}\,\frac{dv}{\sqrt{v}},\quad Y=\frac{x}{\sqrt{\pi}}\int_0^\infty \frac{e^{-v}v}{v^2+x^2}\,\frac{dv}{\sqrt{v}},$$

and $$U=\frac{1}{\sqrt{x}}(-X\sin x+Y\cos x),\quad V=\frac{1}{\sqrt{x}}(X\cos x+Y\sin x).$$

Of course we have not given a complete proof that these expressions are equal to the original integrals; but it is easy to complete the proof by differentiating with respect to x. We have, in fact,

$$\frac{d}{dx}(U+iV)=-\frac{e^{ix}}{\sqrt{x}}.$$

Thus we must have $$\frac{d}{dx}\left[\frac{ie^{ix}}{\sqrt{x}}(X-iY)\right]=-\frac{e^{ix}}{\sqrt{x}}.$$

Hence we find the condition

$$\frac{d}{dx}(X-iY)+\left(i-\frac{1}{2x}\right)(X-iY)=i$$

or $$\frac{dX}{dx}-\frac{X}{2x}+Y=0,\quad \frac{dY}{dx}-\frac{Y}{2x}-X=-1.$$

It is easy to verify that these equations are satisfied by the last pair of integrals for X, Y and that these integrals tend to 1, 0 respectively as $x\to\infty$; thus we may infer that U, V and X, Y are actually related in the manner suggested by the foregoing work. The integrals X, Y seem to be due to Cauchy, and the asymptotic expansion to Poisson (see Lamb's paper, already quoted).

It is perhaps worth while to make the additional remark that the relations between X, Y and U, V are most naturally suggested by the use of the asymptotic expansion.

(3) *Stirling's series.*

It can be proved (Appendix, Art. 180) that

$$\log\Gamma(1+x)=\psi(x)+2\int_0^\infty\frac{\operatorname{arc\,tan}(v/x)}{e^{2\pi v}-1}dv,$$

where $$\psi(x)=(x+\tfrac{1}{2})\log x-x+\tfrac{1}{2}\log(2\pi).$$

Now (Art. 64), we have

$$\operatorname{arc\,tan}(v/x)=(v/x)-\tfrac{1}{3}(v/x)^3+\tfrac{1}{5}(v/x)^5-\dots$$
$$+(-1)^{n-1}\frac{1}{2n-1}(v/x)^{2n-1}+R_n,$$

where $$|R_n|<\frac{1}{2n+1}(v/x)^{2n+1}.$$

Hence (Appendix, Art. 176, Ex. 3), we have

$$\int_0^\infty\frac{\operatorname{arc\,tan}(v/x)}{e^{2\pi v}-1}dv=\frac{B_1}{1\,.\,2x}-\frac{B_2}{3\,.\,4x^3}+\frac{B_3}{5\,.\,6x^5}-\dots$$
$$+(-1)^{n-1}\frac{B_n}{(2n-1)\,.\,2n\,.\,x^{2n-1}}+R_n',$$

where R_n' is numerically less than the first term omitted from the series.

If we take the quotient of two consecutive terms and remark that (compare Art. 130)

$$B_{n+1}/B_n = (2n+1)(2n+2)Q/4\pi^2,$$

where Q is a factor slightly less than 1, we see that the least value for the remainder is given by taking n equal to the integral part of πx; but the first two terms give a degree of accuracy which is ample for ordinary calculations.*

(4) The reader may discuss, by methods analogous to those of (1) and (2) above, the following integrals:

$$\int_x^\infty \frac{e^{-t}}{\sqrt{t}}\,dt, \quad \int_x^\infty \frac{\cos t}{t}\,dt, \quad \int_x^\infty \frac{\sin t}{t}\,dt,$$

the first of which is related to the error-function, while the second and third are the cosine- and sine-integrals.

133. Poincaré's theory of asymptotic series.

All the investigations of Arts. 130–132 resemble one another to the following extent:

Starting from some function $J(x)$, we develop it formally in a series

$$a_0 + \frac{a_1}{x} + \frac{a_2}{x^2} + \frac{a_3}{x^3} + \dots.$$

This series is not convergent, but yet the sum of the first $(n+1)$ terms gives an approximation to $J(x)$ which differs from $J(x)$ by less than K_n/x^{n+1}, where K_n depends only on n and not on x.

Thus, if S_n denotes the sum of the first $(n+1)$ terms, we have

$$\lim_{x\to\infty} x^n(J - S_n) = 0.$$

In all such cases, we say that *the series is asymptotic to the function*; and the relation may be denoted by the symbol

$$J(x) \sim a_0 + \frac{a_1}{x} + \frac{a_2}{x^2} + \dots.$$

Such series were called *semiconvergent* by the older writers.

---

* An elementary treatment of this approximation will be found (for the case when x is an integer) in a paper by the author (*Messenger of Maths.*, vol. 36, 1906, p. 81).

It is to be noticed, however, that the same series may be asymptotic to more than one function; for example, since

$$\lim_{x\to\infty}(x^n e^{-x})=0,$$

the same series will represent $J(x)$ and $J(x)+Ae^{-x}$.

It follows from the definition that *we can add and subtract asymptotic series as if they were convergent.*

Next, take the product of two asymptotic series, assuming that the rule of Art. 34 still applies. We then find, if

$$J(x)\sim a_0+\frac{a_1}{x}+\frac{a_2}{x^2}+\dots \text{ and } K(x)\sim b_0+\frac{b_1}{x}+\frac{b_2}{x^2}+\dots,$$

the formal product $\Pi(x)=c_0+\dfrac{c_1}{x}+\dfrac{c_2}{x^2}+\dots,$

where $$c_n=a_0b_n+a_1b_{n-1}+\dots+a_nb_0.$$

Let S_n, T_n, Σ_n denote the sums of the first $(n+1)$ terms in these three series, then we have, say

$$J(x)=S_n+\rho/x^n, \quad K(x)=T_n+\sigma/x^n,$$

where ρ, σ are functions of x which tend to zero as $x\to\infty$.

Now, by definition Σ_n coincides with the product S_nT_n up to and including the terms in $1/x^n$; thus $S_nT_n-\Sigma_n$ contains terms from $1/x^{n+1}$ to $1/x^{2n}$. We can therefore write

$$S_nT_n=\Sigma_n+P_n/x^{2n},$$

where P_n is a polynomial in x, whose highest term is of degree $(n-1)$.

Thus $$\left[J(x)-\frac{\rho}{x^n}\right]\left[K(x)-\frac{\sigma}{x^n}\right]=\Sigma_n+P_n/x^{2n}$$

or $$x^n[J(x).K(x)-\Sigma_n]=\rho K(x)+\sigma J(x)+(P_n-\rho\sigma)/x^n.$$

Now, as $x\to\infty$, $\quad J(x)\to a_0,\ K(x)\to b_0,\ \rho\to 0,\ \sigma\to 0,$

and accordingly

$$\lim_{x\to\infty} x^n[J(x).K(x)-\Sigma_n]=\lim_{x\to\infty} P_n/x^n=0.$$

Thus the product $J(x).K(x)$ is represented asymptotically by $\Pi(x)$; or *asymptotic series can be multiplied together as if they were convergent*; and in particular we can obtain any power of an asymptotic series by the ordinary rules.

Let us now consider the possibility of substituting an asymptotic series in a power-series. In the first place, we may evidently write $J(x)=a_0+J_1(x)$ and substitute a_0+J_1 for J in the series*

$$f(J)=c_0+c_1J+c_2J^2+c_3J^3+\dots,$$

*Of course c_n no longer represents $a_0b_n+\dots+a_nb_0$.

and rearrange in powers of J_1, *provided that* $|a_0|$ *is less than the radius of convergence* (Art. 84); because $\lim J_1 = 0$, and we can therefore take x large enough to satisfy the restriction that $a_0 + J_1$ is to be less than the radius of convergence.

This having been done, we may consider the substitution of the asymptotic series

$$\frac{a_1}{x} + \frac{a_2}{x^2} + \frac{a_3}{x^3} + \dots$$

for J_1 in the series

$$F(J_1) = C_0 + C_1 J_1 + C_2 J_1^2 + C_3 J_1^3 + \dots.$$

Let us make a formal substitution, as if the series for J_1 were convergent; then we obtain some new series

$$\Sigma = D_0 + \frac{D_1}{x} + \frac{D_2}{x^2} + \frac{D_3}{x^3} + \dots,$$

where

$D_0 = C_0$, $D_1 = C_1 a_1$, $D_2 = C_1 a_2 + C_2 a_1^2$, $D_3 = C_1 a_3 + 2C_2 a_1 a_2 + C_3 a_1^3$, etc.

Let us denote by S_n and Σ_n the sum of the terms up to $1/x^n$ in J_1 and Σ respectively.

Now, if $$\Sigma_n' = C_0 + C_1 S_n + C_2 S_n^2 + \dots + C_n S_n^n,$$
Σ_n' and Σ_n agree up to terms in $1/x^n$, and consequently $\Sigma_n' - \Sigma_n$ is a polynomial in $1/x$, ranging from terms in $(1/x)^{n+1}$ to $(1/x)^{n^2}$; thus

(1) $$\lim_{x\to\infty} x^n(\Sigma_n' - \Sigma_n) = 0.$$

Next, if $$T_n = C_0 + C_1 J_1 + C_2 J_1^2 + \dots + C_n J_1^n,$$
we have, since S_n represents J_1 asymptotically, $\lim_{x\to\infty} x^n(J_1^r - S_n^r) = 0$, and therefore

(2) $$\lim_{x\to\infty} x^n(T_n - \Sigma_n') = 0.$$

Finally, $$F - T_n = C_{n+1} J_1^{n+1} + C_{n+2} J_1^{n+2} + \dots;$$
thus, since $F(J_1)$ is convergent,

$$|F - T_n| < M J_1^{n+1},$$

where M is a constant.

Hence, we find

(3) $$\lim_{x\to\infty} x^n(F - T_n) = 0,$$

because $$\lim_{x\to\infty} (x^n J_1^{n+1}) = \lim_{x\to\infty} (a_1^{n+1}/x) = 0.$$

By combining (1), (2) and (3), we see now that

$$\lim_{x\to\infty} x^n(F - \Sigma_n) = 0.$$

Thus the series Σ represents $F(J_1)$ asymptotically; and consequently *an asymptotic series may be substituted in a power-series and rearranged (just as if convergent), provided that its first term is numerically less than the radius of convergence.*

Further, a reference to the foregoing proof shews that we use the convergence of the series $f(J)$ in two places only, first in order to rearrange in powers of J_1, and secondly to establish the inequality $|F-T_n|<MJ_1^{n+1}$.

Now this inequality is satisfied if the series

$$C_1J_1+C_2J_1^2+C_3J_1^3+\dots$$

is *asymptotic* to $F(J_1)$; and then we must suppose that a_0 is zero in order to get any result at all, so that $J=J_1$ and we can entirely avoid the restriction that $f(J)$ is convergent. Thus, *an asymptotic series* **whose first term is zero** *may be substituted in another asymptotic series, and the result may be rearranged just as if both series were convergent.*

An application of the former result is to establish the *rule for division* (assuming that a_0 is not zero). For we can write

$$J(x)=a_0(1+K),$$

where

$$K\sim\frac{a_1}{a_0x}+\frac{a_2}{a_0x^2}+\dots.$$

Then $\quad [J(x)]^{-1}=a_0^{-1}(1-K+K^2-K^3+\dots),$

and we can thus construct an asymptotic series for $[J(x)]^{-1}$ by exactly the same rule as if the series for $J(x)$ were convergent. Thus, applying the rule for multiplication, we see that *we can divide any asymptotic series by any other asymptotic series, just as if they were convergent.*

Finally, let us consider *the integration of an asymptotic series* (in which $a_0=0$, $a_1=0$).

If

$$J(x)\sim\frac{a_2}{x^2}+\frac{a_3}{x^3}+\frac{a_4}{x^4}+\dots,$$

we have

$$|J-S_n|<\epsilon/x^n, \quad \text{if } x>x_0.$$

Thus

$$\left|\int_x^\infty J\,dx-\int_x^\infty S_n dx\right|<\frac{\epsilon}{(n-1)x^{n-1}}, \quad \text{if } x>x_0,$$

so that $\int_x^\infty J\,dx$ is represented asymptotically by

$$\frac{a_2}{x}+\frac{a_3}{2x^2}+\frac{a_4}{3x^3}+\dots.$$

But, on the other hand, *an asymptotic series cannot safely be differentiated without additional investigation*, for the existence of an asymptotic series for $J(x)$ does not imply the existence of one for $J'(x)$.

Thus $e^{-x}\sin(e^x)$ has an asymptotic series

$$0+\frac{0}{x}+\frac{0}{x^2}+\ldots.$$

But its differential coefficient is $-e^{-x}\sin(e^x)+\cos(e^x)$, which oscillates as x tends to ∞; and consequently the differential coefficient has no asymptotic expansion.

On the other hand, if we know that $J'(x)$ has an asymptotic expansion, it must be the series obtained by the ordinary rule for term-by-term differentiation.

This follows by applying the theorem on integration to $J'(x)$; but a direct proof is quite as simple, and perhaps more instructive. We make use of the theorem that *if $\phi(x)$ has a definite finite limit as x tends to ∞, then $\phi'(x)$ either oscillates or tends to zero as a limit.**

If $J(x) \sim a_0+a_1/x+a_2/x^2+\ldots$, we have

$$\lim x^{n+1}[J(x)-S_n(x)]=a_{n+1}.$$

Thus the differential coefficient

$$x^{n+1}[J'(x)-S_n'(x)]+(n+1)x^n[J(x)-S_n(x)],$$

if it has a definite limit, must tend to zero. But $x^n[J(x)-S_n(x)]$ does tend to zero, so that $\lim x^{n+1}[J'(x)-S_n'(x)]$, if it exists, is zero.

That is, if $J'(x)$ has an asymptotic series, it is

$$-a_1/x^2-2a_2/x^3-3a_3/x^4-\ldots.$$

It is instructive to contrast the rules for transforming and combining asymptotic series with those previously established for convergent series. Thus *any* two asymptotic series can be multiplied together: whereas the product of two convergent series need not give a convergent series (see Arts. 34, 35). Similarly *any* asymptotic series may be integrated term-by-term, although not every convergent series can be integrated (Art. 45).

---

* In fact if $\phi(x)$ tends to a definite limit we can find x_0 so that

$$|\phi(x)-\phi(x_0)|<\epsilon, \quad \text{if } x>x_0.$$

Thus, since $$\frac{\phi(x)-\phi(x_0)}{x-x_0}=\phi'(\xi), \quad \text{where } x>\xi>x_0,$$

we find $$|\phi'(\xi)|<\epsilon/(x-x_0).$$

So $\phi'(x)$ cannot approach any *definite* limit other than zero; but the last inequality does not exclude *oscillation*, since ξ may not take *all* values greater than x_0 as x tends to ∞.

On the other hand, as we have just explained, we cannot differentiate *any* asymptotic series unless we know from independent reasoning that the corresponding derivate has an asymptotic expansion; although, in dealing with a convergent series, we can apply the test for uniform convergence directly to the differentiated series, and so *infer* that the derived function has an expansion (Art. 46).

These contrasts, however, are not to be regarded as surprising. In a convergent series, the parameter with respect to which we differentiate or integrate is strictly an *auxiliary* variable, and in no way enters into the definition of the convergence of the series; but in an asymptotic series, the very definition depends on the parameter x. The contrast may be illustrated in an even more fundamental way; *any* coefficients whatever may define a perfectly good asymptotic series. Indeed an asymptotic series is not a completed whole in the same sense as a convergent series.

It is sometimes convenient to extend our definition and say that J is represented asymptotically by the series

$$\Phi+\left(a_0+\frac{a_1}{x}+\frac{a_2}{x^2}+\dots\right)\Psi$$

when $\dfrac{J-\Phi}{\Psi}$ is represented by $a_0+\dfrac{a_1}{x}+\dfrac{a_2}{x^2}+\dots$, where Φ, Ψ are two suitably chosen functions of x.

Thus, for example, we can deduce from Stirling's series the asymptotic formula

$$\Gamma(x+1)\sim e^{-x}x^x(2\pi x)^{\frac{1}{2}}\left(1+\frac{C_1}{x}+\frac{C_2}{x^2}+\dots\right),$$

where
$$C_1=\frac{B_1}{2}=\frac{1}{12},$$

$$C_2=-\frac{B_2}{12}+\frac{B_1^2}{8}=-\frac{11}{5760}, \text{ etc.}$$

Hitherto x has been supposed to tend to ∞ through *real, positive* values; but the theory remains unaltered if x is complex and tends to ∞ in any other definite direction. But *a non-convergent series cannot represent asymptotically the same one-valued analytic function J for all arguments of x.*

In fact, if we can determine constants M, R, such that

$$\left|J - a_0 - \frac{a_1}{x}\right| < \frac{M}{|x|^2}, \text{ when } |x| > R,$$

it follows from elementary theorems in the theory of functions that $J(x)$ is a regular function of $1/x$, and consequently the asymptotic series is convergent.

For different ranges of variation of the argument of x, we may have different asymptotic representations of the same function which between them give complete information as to its nature. A good illustration of this phenomenon is afforded by the Bessel functions which have been discussed at length by Stokes.*

134. Applications of Poincaré's theory.

An important application of Poincaré's theory is to the solution of differential equations.† The method may be summed up in the following steps:

1st. A formal solution is obtained by means of a non-convergent series.

2nd. It is shewn, by independent reasoning, that a solution exists which is capable of asymptotic representation. Thus we may either, as has been done in Arts. 131, 132 above, deduce a definite integral from the series first calculated; or we may find a solution as a definite integral directly, and then identify it with the series.

3rd. The region is determined in which the asymptotic representation is valid.

Poincaré has in fact proved‡ that every linear differential equation which has *polynomial* coefficients may be solved by asymptotic series; but his work is restricted to the case in which the independent variable tends to ∞ along a specified

* *Camb. Phil. Trans.*, vol. 9, 1850, vol. 10, 1857, p. 105, and vol. 11, 1868, p. 412; *Math. and Phys. Papers*, vol. 2, p. 350, vol. 4, pp. 77, 283. See also *Acta Mathematica*, vol. 26, 1902, p. 393, and *Papers*, vol. 5, p. 283. Stokes remarks that in the asymptotic series examined by him, the change in representation occurs at a value of the argument which gives the same sign to all the terms of the divergent series.

† Some interesting remarks on the sense in which an asymptotic series gives a solution of a differential equation, have been made by Stokes (*Papers*, vol. 2, p. 337).

‡ *Acta Mathematica*, t. 8, 1886, p. 303.

direction, and the regions are not determined. This gap has been filled by Horn in a number of special cases;* and we may refer to a detailed discussion of a special differential equation by Jacobsthal in the paper quoted below in Art. 139.

Other applications of Poincaré's theory have been made by Barnes and Hardy in constructing the asymptotic representation of functions given by power-series. A convenient summary with very full references is given by Barnes;† the method adopted by Barnes is beyond the limits of this book, as it depends on the theory of contour integration. We give, however, in the following article a simple method due to Stokes, which can be used in dealing with certain types of *real* series.

135. Stokes' asymptotic expression for the series

$$\Sigma \frac{\Gamma(n+a_1+1)\ldots\Gamma(n+a_r+1)}{\Gamma(n+b_1+1)\ldots\Gamma(n+b_s+1)} x^n = \Sigma X_n,$$

where x is real and $s > r$.‡

Write $s-r=\mu$, $\Sigma b - \Sigma a = \lambda$, and consider the term X_{t+p}, where t is large, and p is not of higher order than $\sqrt{t}$.

We find, neglecting terms of order $1/\sqrt{t}$,

$$\log X_{t+p} = (t+p)\log x - \mu[(t+\tfrac{1}{2})\log t + \tfrac{1}{2}\log(2\pi) - t]$$
$$- (p\mu + \lambda)\log t - \tfrac{1}{2}\mu p^2/t$$

(see Art. 180, Appendix).

It is convenient to suppose that x is of the form t^μ where t is an integer (a restriction which can be removed by using more elaborate methods); and then X_t is the greatest term because $\log x = \mu \log t$, and so

$$\log X_{t+p} = \mu t - \tfrac{1}{2}\mu\log(2\pi t) - \lambda\log t - \tfrac{1}{2}\mu p^2/t,$$

or

$$X_{t+p} = \frac{e^{t\mu} t^{-\lambda}}{(2\pi t)^{\frac{1}{2}\mu}} \exp(-\tfrac{1}{2}\mu p^2/t).$$

This gives the asymptotic expression

$$\frac{e^{t\mu} t^{-\lambda}}{(2\pi t)^{\frac{1}{2}\mu}}(1+2q+2q^4+2q^9+\ldots),$$

---

* See a series of papers in the *Mathematische Annalen*, from Bd. 49 onwards, and some papers in *Crelle's Journal*. A good summary of the theory with many references is given in Horn's *Gewöhnliche Differentialgleichungen*, Abschnitt VII.

† *Phil. Trans.*, series A, vol. 206, 1906, p. 249; see also *Quarterly Journal*, vol. 38, 1907, pp. 108, 116.

‡ *Proc. Camb. Phil. Soc.*, vol. 6, 1889; *Math. and Phys. Papers*, vol. 5, p. 221.

where $q=e^{-\frac{1}{2}\mu/t}$. Making use of Art. 51, Ex. 3, we see that (since q approaches the limit 1) the series in brackets is represented approximately by $\pi^{\frac{1}{2}}(1-q)^{-\frac{1}{2}}$, or by $(2\pi t/\mu)^{\frac{1}{2}}$.

Thus the asymptotic expression is

$$\frac{e^{t\mu}t^{-\lambda}}{\mu^{\frac{1}{2}}(2\pi t)^{\frac{1}{2}(\mu-1)}}, \text{ where } t=x^{\frac{1}{\mu}}.$$

Hardy* has proved, somewhat in the same way, that if

$$f(x)=\Sigma\frac{x^{n^2}}{n^2!},$$

$f(x)$ is represented asymptotically by $Ae^x/(2\pi x)^{\frac{1}{4}}$, where

$$A=1+2(q+q^4+q^9+\ldots), \text{ and } q=e^{-2}.$$

SUMMATION OF ASYMPTOTIC SERIES.

136. Extension of Borel's definition of summable series.

In the foregoing work (Arts. 130–135) we have shewn how to obtain asymptotic expansions of certain given functions, and we have established rules of calculation for these expansions. We are now led to consider the converse problem of summing a given asymptotic series; and the natural method is to apply Borel's integral. But on trial, it appears that Borel's method is not sufficiently powerful to sum even the simplest asymptotic series, such as (Arts. 98, 132)

$$1-\frac{1!}{x}+\frac{2!}{x^2}-\frac{3!}{x^3}+\ldots, \qquad (x>0).$$

If we apply Borel's method of summation to this series we are led to use the associated function

$$u(t, x)=1-\frac{t}{x}+\frac{t^2}{x^2}-\ldots$$

which converges only if $t<x$. Thus the results of Arts. 99-113 no longer apply, but we shall now proceed to modify the definition so as to obtain corresponding results for such series.

We have $\qquad u(t, x)=x/(x+t), \quad$ if $t<x$,

and this defines a function which is regular, as t ranges from 0 to ∞. Thus we can agree to take the integral

$$\int_0^\infty \frac{xe^{-t}}{x+t}dt$$

---

* *Proc. Lond. Math. Soc.* (2), vol. 2, 1904, p. 339.

as defining the sum of the series above. But we can also write $t=xv$ in this integral, and so obtain a second definition

$$x\int_0^\infty \frac{e^{-xv}\,dv}{1+v}.$$

The reader will recognize that the former integral agrees with Euler's, which was given in Art. 98.

The two integrals are exactly equivalent so long as x is real and positive; but if x is complex the former is convergent except when x is real and negative, whereas the latter converges only if the real part of x is positive.

It is not difficult to shew that the former integral defines an analytic function in the part of the plane for which the least distance from the negative real axis is any positive number l. This part of the plane is bounded by two parallel lines joined by a semicircle, as in the figure.

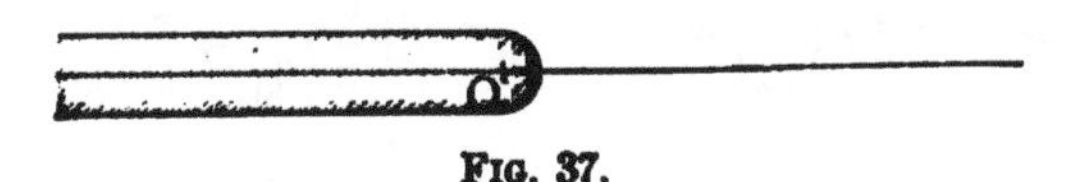

FIG. 37.

To establish this property, we note that

$$\left|\frac{d}{dx}\left(\frac{xe^{-t}}{x+t}\right)\right| = \left|\frac{te^{-t}}{(x+t)^2}\right| < \frac{t}{l^2}e^{-t},$$

and that the integral $\int_0^\infty \frac{t}{l^2}e^{-t}\,dt$ converges, so that the test of Art. 174 (Appendix) can be used.

In like manner the second integral can be proved to define an analytic function in the part of the plane for which the real part of x is greater than l; this region is bounded by a line parallel to the imaginary axis.

The method applied in this example suggests a general process for summing an asymptotic series.

Suppose that

$$a_0+\frac{a_1}{x}+\frac{a_2}{x^2}+\frac{a_3}{x^3}+\dots$$

is an asymptotic series such that the associated series

$$f(v)=a_0+a_1v+a_2\frac{v^2}{2!}+a_3\frac{v^3}{3!}+\dots$$

converges for certain positive values of v, and that the function $f(v)$ defined by the series for these values of v is continuous from $v=0$ to ∞.

Then we define the sum of the series by the integral

$$\int_0^\infty e^{-t}f\left(\frac{t}{x}\right)dt, \quad (x>0).$$

To make any practical use of this definition *we must, however, assume further that a positive number l can be found such that*

$$\lim_{v\to\infty} e^{-lv} f^n(v)=0,$$

where n is any index of differentiation.

It is evident that then the integral defines a function which is analytic within the region indicated in the figure.

137. Rules for calculation.

It is easy to see that *the series of the last article represents the integral asymptotically*, x being real and positive.

For, if we integrate by parts we have

$$J(x)=\int_0^\infty e^{-t} f\left(\frac{t}{x}\right) dt=\left[-e^{-t} f\left(\frac{t}{x}\right)\right]_0^\infty+\frac{1}{x}\int_0^\infty e^{-t} f'\left(\frac{t}{x}\right) dt$$

$$=a_0+\frac{1}{x}\int_0^\infty e^{-t} f'\left(\frac{t}{x}\right) dt,$$

because $f(0)=a_0$. We continue this process, and we obtain

$$J(x)=a_0+\frac{a_1}{x}+\frac{a_2}{x^2}+\dots+\frac{a_n}{x^n}+\frac{1}{x^{n+1}}\int_0^\infty e^{-t} f^{n+1}\left(\frac{t}{x}\right) dt.$$

Now, by hypothesis, we can find the positive constants l, A so that

$$|f^{n+1}(v)|<Ae^{lv},$$

and therefore $\left|\frac{1}{x^{n+1}}\int_0^\infty e^{-t} f^{n+1}\left(\frac{t}{x}\right) dt\right|<\frac{A}{x^n(x-l)}.$

Consequently,

$$\lim_{x\to\infty}\left[x^n\left\{J(x)-\left(a_0+\frac{a_1}{x}+\dots+\frac{a_n}{x^n}\right)\right\}\right]=0,$$

which establishes the asymptotic property.

It will be remembered (see Art. 133) that the differentiation of an asymptotic series requires special consideration; but *this special type of asymptotic series may be differentiated any number of times.* For

$$J'(x)=\frac{d}{dx}\int_0^\infty e^{-t} f\left(\frac{t}{x}\right) dt=-\frac{1}{x}\int_0^\infty e^{-t}\frac{t}{x} f'\left(\frac{t}{x}\right) dt,$$

the differentiation under the integral sign being permissible (if $x>l$), because

$$|te^{-t} f'(t/x)|<Ate^{-(x-l)t},$$

and the integral $\int_0^\infty te^{-(x-l)t}\,dt$ is convergent.

Now, within the interval of convergence of $f(v)$ we have

$$vf'(v)=a_1v+2a_2\frac{v^2}{2!}+3a_3\frac{v^3}{3!}+\ldots,$$

so that $J'(x)$ is represented asymptotically by the series

$$-\frac{1}{x}\left(\frac{a_1}{x}+\frac{2a_2}{x^2}+\frac{3a_3}{x^3}+\ldots\right),$$

or by
$$-\frac{a_1}{x^2}-\frac{2a_2}{x^3}-\frac{3a_3}{x^4}-\ldots,$$

which is the series derived from the original series for $J(x)$ by formal differentiation, as could be anticipated from Art. 133. Clearly the operation may be repeated any number of times.

The reader who is acquainted with the elements of function-theory will have no difficulty in proving that these results are valid within the part of the x-plane indicated by the diagram of Art. 136.

Let us consider next the *algebraic* operations; it follows from Poincaré's theory that two asymptotic series may be added and multiplied as if convergent. But we must also prove that the product can be represented by an integral of the type considered above.

We have proved that

$$c_0+\frac{c_1}{x}+\frac{c_2}{x^2}+\ldots$$

is the asymptotic series corresponding to the product, if

$$c_n=a_0b_n+a_1b_{n-1}+\ldots+a_nb_1.$$

Now if both $\quad f(t)=\Sigma a_n\frac{t^n}{n!},\quad g(t)=\Sigma b_n\frac{t^n}{n!}$

converge under the condition $|t|\leqq r$, we can find a constant M such that $\quad |a_n|r^n<M.n!,\quad |b_n|r^n<M.n!.$

Thus $$|c_nr^n|<M^2[n!+(n-1)!+(n-2)!.2!+\ldots+n!]$$
$$<M^2(n+1)!,$$

and consequently $\Sigma\frac{c_nv^n}{n!}$ will converge if $|v|<r$. Next, as in Art. 106 above, we have

$$\Sigma c_n\frac{v^{n+1}}{(n+1)!}=\int_0^v f(v-t)g(t)\,dt,\quad \text{if } |v|<r,$$

and consequently by differentiation we find

$$\Sigma c_n \frac{v^n}{n!} = a_0 g(v) + \int_0^v f'(v-t) g(t)\, dt.$$

Thus Borel's associated function for the product-series is

$$F(v) = a_0 g(v) + \int_0^v f'(v-t) g(t)\, dt$$

and
$$|F(v)| \leqq |a_0| . |g(v)| + \int_0^v |f'(v-t)| . |g(t)|\, dt.$$

But, by hypothesis, we can find the positive constants A and l such that

$$|f'(t)| < Ae^{lt}, \quad |g(t)| < Ae^{lt}.$$

Hence
$$|F(v)| < |a_0| Ae^{lv} + A^2 v e^{lv},$$

and consequently we find

$$\lim_{v \to \infty} [e^{-2lv} F(v)] = 0.$$

Similarly, we can prove that

$$\lim_{v \to \infty} [e^{-2lv} F^n(v)] = 0.$$

Thus $F(v)$ satisfies all the necessary conditions.

By combining these results, we have the following rule: *If a polynomial expression contains a certain number of asymptotic series* (of the present type) and their derivates, the value of the polynomial is expressible asymptotically by a series (of the same type), obtained by applying the ordinary rules of calculation, as if the series were all convergent. And the result cannot be identically zero, unless the terms of the resulting series are zero.*

The reader who is familiar with elementary function-theory will have no difficulty in extending these results to the complex variable, in a suitably restricted area.

---

* Of course some of the functions may have ordinary convergent expansions in $1/x$.

138. Another integral for an asymptotic series.

There are a number of interesting examples which do not come immediately under the foregoing method, but can be treated very easily by making a slight change in the definition. If we suppose $\phi(x)$ to be a function such as is specified in Art. 118, it is readily seen that the previous results are still true for the integral

$$J(x)=\int_0^\infty \phi(t) f\left(\frac{t}{x}\right) dt, \quad \text{where } f(v)=\sum_0^\infty \frac{a_n}{c_n} v^n,$$

provided that the series $f(v)$ has a non-zero radius of convergence ρ, and that certain restrictions are satisfied as to the rate of increase of $|f^n(v)|$ with v.

In particular, if we can find a function $\psi(\xi)$ such that

$$\frac{a_n}{c_n}=(-1)^n \int_0^1 \psi(\xi) \xi^n \, d\xi,$$

we shall have the equation

$$f(v)=\int_0^1 \frac{\psi(\xi)}{1+v\xi} d\xi.$$

For this is easily proved to be true when $|v|<\rho$, and we can take this integral as the definition of $f(v)$ for larger values of v.

Then, $$|f(v)|<\int_0^1 |\psi(\xi)| \, d\xi$$

and $$|f^n(v)|<\int_0^1 \xi^n |\psi(\xi)| \, d\xi.$$

That the integral is in fact represented asymptotically by the series follows at once by observing that

$$f(v)=$$
$$\int_0^1 \psi(\xi)\left[1-v\xi+v^2\xi^2+\ldots+(-1)^{n-1}v^{n-1}\xi^{n-1}+(-1)^n \frac{v^n \xi^n}{1+v\xi}\right]$$
$$=\frac{a_0}{c_0}+\frac{a_1}{c_1}v+\frac{a_2}{c_2}v^2+\ldots+\frac{a_{n-1}}{c_{n-1}}v^{n-1}+R_n,$$

where $$|R_n|<\left|\frac{a_n}{c_n}\right| v^n.$$

Hence

$$J(x)=\int_0^\infty \phi(t) f\left(\frac{t}{x}\right) dt = a_0+\frac{a_1}{x}+\ldots+\frac{a_{n-1}}{x^{n-1}}+R_n',$$

where $$|R_n'|<|a_n|/x^n.$$

Again*
$$J'(x) = -\int_0^\infty \frac{t}{x^2}\phi(t) f'\left(\frac{t}{x}\right) dt,$$
and, as previously, this can be shewn to give the asymptotic series obtained by the ordinary rule for differentiation.

A special type is given by taking
$$\phi(t) = e^{-t} t^{\lambda-1}, \quad (\lambda > 0),$$
and then
$$c_n = \Gamma(n+\lambda).$$

Thus our definition becomes
$$J(x) = \int_0^\infty e^{-t} f\left(\frac{t}{x}\right) t^{\lambda-1}\, dt$$
where
$$f(v) = \Sigma \frac{a_n}{\Gamma(n+\lambda)} v^n.$$

It is easy to repeat the previous argument (with small changes) to establish the corresponding theorems for operating with this integral.

As an example of the last method, we may point out that in Ex. 2, Art. 132, we could infer the differential equations for X, Y by simply operating on the asymptotic series.

Another example is given by the series
$$\Gamma(\lambda) - \Gamma(1+\lambda)\frac{1}{x} + \Gamma(2+\lambda)\frac{1}{x^2} - \dots$$
which leads to $f(v) = 1/(1+v)$, so that the series represents asymptotically the integral
$$\int_0^\infty \frac{x}{t+x} e^{-t} t^{\lambda-1}\, dt.$$

139. Differential equations.

We shall conclude by giving a few examples of the way in which asymptotic series present themselves in the solution of differential equations; and we shall illustrate the methods of Arts. 136–138 by summing these series.

1. Let us try to solve the differential equation†
$$\frac{dy}{dx} = \frac{a}{x} + by, \quad (b > 0),$$

---

* Differentiation under the integral sign is justified, because we suppose
$$|f'(t/x)| < \int_0^1 \xi |\psi(\xi)|\, d\xi$$
and $\int_0^\infty t\phi(t)\,dt$ converges.

Thus the test of Art. 172 (3) applies.

† This is the simplest case of a general type of equations examined by Borel (*Annales de l'École Normale Supérieure* (3), t. 16, 1899, p. 95).

by means of an asymptotic series

$$y=A_0+\frac{A_1}{x}+\frac{A_2}{x^2}+\dots.$$

On substitution, we find

$$-\frac{A_1}{x^2}-\frac{2A_2}{x^3}-\frac{3A_3}{x^4}-\dots=\frac{a}{x}+b\left(A_0+\frac{A_1}{x}+\frac{A_2}{x^2}+\dots\right).$$

This gives

$$A_0=0,\quad A_1=-\frac{a}{b},\qquad A_2=-\frac{A_1}{b}=\frac{a}{b^2},$$

$$A_3=-\frac{2A_2}{b}=-\frac{1\,.\,2a}{b^3},\quad A_4=-\frac{3A_3}{b}=\frac{1\,.\,2\,.\,3\,.\,a}{b^4},\ \text{etc.}$$

Thus we find the formal solution

$$y=-\frac{a}{bx}\left[1-\frac{1}{bx}+2!\left(\frac{1}{bx}\right)^2-3!\left(\frac{1}{bx}\right)^3+\dots\right],$$

and by Art. 136 this gives the equivalent integral

$$-a\int_0^\infty\frac{e^{-t}dt}{t+bx},$$

and it is easy to verify directly that this integral does satisfy the given equation, as of course must be the case in virtue of Art. 137.

2. Consider the modified Bessel's equation

$$\frac{d^2y}{dx^2}+\frac{1}{x}\frac{dy}{dx}-y=0.$$

Write $y=e^{-x}x^{-\frac{1}{2}}\eta$, and then we find

$$x^2\left(\frac{d^2\eta}{dx^2}-2\frac{d\eta}{dx}\right)+\frac{1}{4}\eta=0.$$

If we substitute $\qquad \eta=1+\frac{a_1}{x}+\frac{a_2}{x^2}+\frac{a_3}{x^3}+\dots,$

we obtain

$$x^2\left(\frac{1\,.\,2a_1}{x^3}+\frac{2\,.\,3a_2}{x^4}+\frac{3\,.\,4a_3}{x^5}+\dots\right)$$
$$+2x^2\left(\frac{a_1}{x^2}+\frac{2a_2}{x^3}+\frac{3a_3}{x^4}+\frac{4a_4}{x^5}+\dots\right)$$
$$+\frac{1}{4}\left(1+\frac{a_1}{x}+\frac{a_2}{x^2}+\frac{a_3}{x^3}+\dots\right)=0.$$

Hence we find

$$2a_1=-(\tfrac{1}{2})^2,\ 2\,.\,2a_2=-(\tfrac{3}{2})^2a_1,\ 2\,.\,3\,.\,a_3=-(\tfrac{5}{2})^2a_2,\dots,$$

and so

$$a_1=-\frac{1}{2}\cdot\frac{1}{4},\quad a_2=\frac{1^2\,.\,3^2}{2\,.\,4}\frac{1}{4^2},\quad a_3=-\frac{1^2\,.\,3^2\,.\,5^2}{2\,.\,4\,.\,6}\cdot\frac{1}{4^3},\ \text{etc.}$$

Thus η is represented by the series

$$1-\frac{1}{2}\frac{1}{4x}+\frac{1^2.3^2}{2.4}\frac{1}{(4x)^2}-\frac{1^2.3^2.5^2}{2.4.6}\frac{1}{(4x)^3}+\ldots.$$

That is,
$$\frac{a_n\Gamma(\frac{1}{2})}{\Gamma(n+\frac{1}{2})}=(-1)^n\frac{1.3.5\ldots(2n-1)}{2.4.6\ldots 2n}\frac{1}{2^n},$$

and so if we put
$$c_n=\frac{\Gamma(n+\frac{1}{2})}{\Gamma(\frac{1}{2})}=\frac{1}{\sqrt{\pi}}\int_0^\infty e^{-t}t^{n-\frac{1}{2}}\,dt,$$

we have
$$\Sigma\frac{a_n}{c^n}v^n=(1+\tfrac{1}{2}v)^{-\frac{1}{2}}.$$

This leads to the integral (Art. 138)

$$\eta=\frac{1}{\sqrt{\pi}}\int_0^\infty e^{-t}t^{-\frac{1}{2}}\left(1+\tfrac{1}{2}\frac{t}{x}\right)^{-\frac{1}{2}}dt.$$

Hence
$$y=\frac{e^{-x}}{\sqrt{\pi}}\int_0^\infty\frac{e^{-t}dt}{[t(x+\frac{1}{2}t)]^{\frac{1}{2}}},$$

which can be proved directly to be a solution of the given equation. Similarly, we can obtain the solution

$$e^{-x}x^{-k}\int_0^\infty e^{-t}[t(x+\tfrac{1}{2}t)]^{k-\frac{1}{2}}dt$$

for the equation

$$\frac{d^2y}{dx^2}+\frac{1}{x}\frac{dy}{dx}-\left(1+\frac{k^2}{x^2}\right)y=0.$$

3. In like manner the differential equation

$$x\frac{d^2y}{dx^2}+(a+\beta-x)\frac{dy}{dx}-ay=0,$$

is satisfied formally by the asymptotic power-series

$$y=\frac{1}{x^a}\left[1-\frac{a(1-\beta)}{1.x}+\frac{a(a+1)(1-\beta)(2-\beta)}{1.2.x^2}-\ldots\right],$$

which leads to the integral

$$x^{1-a-\beta}\int_0^\infty e^{-t}t^{a-1}(x+t)^{\beta-1}dt.$$

This equation has been very fully discussed by W. Jacobsthal (*Math. Annalen*, Bd. 56, 1903, p. 129), to whose paper we may refer for details as to the various solutions.

We have now completed a tolerably full account of the theory of non-convergent series, so far as it has been yet developed on the "arithmetic" side. Its applications to the Theory of Functions lie beyond the scope of this book, and we have made no attempt to give more than a few theorems whose proofs can be readily supplied by any reader interested in such developments.

EXAMPLES.

Integration of Summable Series.

1. We have seen (Art. 103) that the Borel sum of the series

$$\sin(x+a)+\sin 2(x+a)+\sin 3(x+a)+\ldots$$

is $\frac{1}{2}\cot\frac{1}{2}(x+a)$; and so we find the sum of the series

$$2(\cos x\sin a+\cos 2x\sin 2a+\cos 3x\sin 3a+\ldots)$$

to be $$\tfrac{1}{2}\cot\tfrac{1}{2}(x+a)-\tfrac{1}{2}\cot\tfrac{1}{2}(x-a)=\sin a/(\cos x-\cos a).$$

This suggests* $$P\int_0^\pi \frac{\cos nx\,.\,dx}{\cos x-\cos a}=\pi\frac{\sin na}{\sin a},\qquad (0<a<\pi),$$

which is easily verified by using the trigonometrical identity

$$\frac{\cos nx-\cos na}{\cos x-\cos a}=\frac{1}{\sin a}[\sin na+2\cos x\sin(n-1)a+\ldots+2\cos(n-1)x\sin a].$$

2. Just as in the last example we get

$$2(\sin x\cos a+\sin 2x\cos 2a+\ldots)=-\sin x/(\cos x-\cos a);$$

and so integrating we find

$$\log 4(\cos x-\cos a)^2=-4\Sigma\frac{1}{n}\cos nx\cos na,$$

the constant being found by using the value of the series for $x=\frac{1}{2}\pi$ (Art. 65).

Thus $$\int_0^\pi \log(\cos x-\cos a)^2 dx=-2\pi\log 2,$$

$$\int_0^\pi \log(\cos x-\cos a)^2\cos nx\,dx=-(2\pi/n)\cos na.$$

Another form of the first integral is

$$\int_0^{\frac{1}{2}\pi}\log(a^2\cos^2\theta-b^2\sin^2\theta)^2 d\theta=\pi\log\{\tfrac{1}{4}\,|\,a^2-b^2\,|\,\}.$$

Deduce that if $r^2<1$,

$$\int_0^\pi \frac{\log 4(\cos x-\cos a)^2}{1-2r\cos x+r^2}dx=\frac{2\pi}{1-r^2}\log(1-2r\cos a+r^2).$$

3. Since we have

$$\cot mx=2\sin 2mx+2\sin 4mx+2\sin 6mx+\ldots,$$

it is suggested* that if μ is a multiple of m,

$$\int_0^\pi \sin 2\mu x\cot mx\,dx=\pi.$$

This result is easily verified independently.

---

*The results of Art. 109 do not allow us to integrate over the value $x=a$ in Ex. 1, nor over the values $x=n\pi/2m$ in Ex. 3.

4. From Art. 103 (4) we see that

$$\mathop{\mathscr{S}}_0^\infty e^{i(x+n)\theta} = \tfrac{1}{2}i\, e^{i(x-\frac{1}{2})\theta}/\sin\tfrac{1}{2}\theta.$$

Thus
$$\mathop{\mathscr{S}}_0^\infty \sin(x+n)\theta = \tfrac{1}{2}\cos(x-\tfrac{1}{2})\theta/\sin\tfrac{1}{2}\theta,$$

and
$$\mathop{\mathscr{S}}_0^\infty \cos(x+n)\theta = -\tfrac{1}{2}\sin(x-\tfrac{1}{2})\theta/\sin\tfrac{1}{2}\theta.$$

Hence, integrating with respect to θ, we find

$$\sum_0^\infty \frac{\cos(x+n)\theta}{x+n} = \frac{1}{2}\int_\theta^\pi \frac{\cos(x-\frac{1}{2})\theta}{\sin\frac{1}{2}\theta}\,d\theta + \cos(\pi x)\sum_0^\infty \frac{(-1)^n}{x+n},$$

$$\sum_0^\infty \frac{\sin(x+n)\theta}{x+n} = \frac{1}{2}\int_\theta^\pi \frac{\sin(x-\frac{1}{2})\theta}{\sin\frac{1}{2}\theta}\,d\theta + \sin(\pi x)\sum_0^\infty \frac{(-1)^n}{x+n}.$$

The series $\sum_0^\infty (-1)^n/(x+n)$ is sometimes denoted by $\beta(x)$, and can be expressed by means of the ψ-function (see Exs. 42–44, p. 475), since

$$\frac{1}{2}\left[\psi\left(\frac{x+1}{2}\right) - \psi\left(\frac{x}{2}\right)\right] = \sum_0^\infty \frac{(-1)^n}{x+n} = \beta(x).$$

Thus, *if x is rational*, the series can be summed in finite terms; the case $x=1$ has occurred in Arts. 65, 90, so take now $x=\frac{1}{2}$ as a further example.

We get
$$\sum_0^\infty \frac{\cos(n+\frac{1}{2})\theta}{n+\frac{1}{2}} = \frac{1}{2}\int_\theta^\pi \operatorname{cosec}\tfrac{1}{2}\theta\, d\theta = \log\cot\tfrac{1}{4}\theta,$$

$$\sum_0^\infty \frac{\sin(n+\frac{1}{2})\theta}{n+\frac{1}{2}} = \beta(\tfrac{1}{2}) = 2(1-\tfrac{1}{3}+\tfrac{1}{5}-\ldots) = \tfrac{1}{2}\pi,$$

where $0<\theta<\pi$.

If θ/π is rational the series may be expressed by means of ψ-functions; and so the integrals are then expressible in the same form.

If we allow θ to tend to 0 in the sine-series, we get

$$\frac{1}{2}\int_0^\pi \frac{\sin(x-\frac{1}{2})\theta}{\sin\frac{1}{2}\theta}\,d\theta = \tfrac{1}{2}\pi - \sin(\pi x)\,\beta(x). \qquad \text{[HARDY.]}$$

[For
$$\lim_{\theta\to 0}\sum_1^\infty \left(\frac{1}{x+n} - \frac{1}{n}\right)\sin(x+n)\theta = 0$$

by Weierstrass's M-test, and

$$\sum_1^\infty \frac{\sin(x+n)\theta}{n} = \tfrac{1}{2}(\pi-\theta)\cos x\theta - \tfrac{1}{2}\sin x\theta\log(4\sin^2\tfrac{1}{2}\theta).]$$

Cesàro's Mean-value Process.

5. A specially interesting example of Cesàro's process was given recently by Fejér, who applied it to the Fourier-series for a function $f(x)$ which does not satisfy Dirichlet's conditions (Art. 174, App. III.), so that the series need not converge.

In fact, if
$$a_n = \frac{1}{\pi}\int_0^{2\pi} f(\theta)\cos n\theta\, d\theta,\quad b_n = \frac{1}{\pi}\int_0^{2\pi} f(\theta)\sin n\theta\, d\theta,$$

and
$$a_0 = \frac{1}{2\pi}\int_0^{2\pi} f(\theta)\, d\theta,$$

and we write
$$u_0 = a_0,\quad u_n = a_n\cos nx + b_n\sin nx,$$

we find that

$$s_n = \frac{1}{2\pi}\int_0^{2\pi} f(\theta)\frac{\sin(n+\frac{1}{2})(\theta-x)}{\sin\frac{1}{2}(\theta-x)}\,d\theta,$$

and that the arithmetic mean of $s_0, s_1, \ldots, s_{n-1}$ is

$$\frac{1}{2n\pi}\int_0^{2\pi} f(\theta)\left\{\frac{\sin\frac{1}{2}n(\theta-x)}{\sin\frac{1}{2}(\theta-x)}\right\}^2 d\theta.$$

By dividing this integral into two, we find from Ex. 7, Art. 173 (App. III.), that it has the limit $f(x)$ if the function $f(x)$ is continuous. The extension to cases when $f(x)$ has a finite number of ordinary discontinuities presents no fresh difficulty; but the proof under the single restriction that $f(x)$ must be integrable is beyond our range.*

6. Let us write σ_n for the arithmetic mean used in the last example, then it is easily seen that if $f(x)$ is continuous the convergence of σ_n to its limit $f(x)$ is *uniform* for all values of x from 0 to 2π. Thus

$$\lim_{n\to\infty}\int_0^{2\pi}[f(x)-\sigma_n]^2\,dx = 0.$$

Since
$$\sigma_{n-1} = a_0 + \sum_{r=1}^{n-1}\frac{n-r}{n}(a_r\cos rx + b_r\sin rx),$$

we find that (paying attention to the definitions of a_r, b_r)

$$J_n = \int_0^{2\pi}[f(x)-\sigma_{n-1}]^2\,dx = \int_0^{2\pi}[f(x)]^2\,dx - 2\pi\left[a_0^2 + \tfrac{1}{2}\sum_{r=1}^{n}\left(1-\frac{r^2}{n^2}\right)(a_r^2+b_r^2)\right].$$

Thus
$$a_0^2 + \tfrac{1}{2}\sum_{r=1}^{m}(a_r^2+b_r^2)\left(1-\frac{r^2}{n^2}\right) < \frac{1}{2\pi}\left\{\int_0^{2\pi}[f(x)]^2\,dx - J_n\right\},$$

where m is any number less than n; and so, taking the limit as n tends to ∞, we find

$$a_0^2 + \tfrac{1}{2}\sum_{r=1}^{m}(a_r^2+b_r^2) \leqq \frac{1}{2\pi}\int_0^{2\pi}[f(x)]^2\,dx, \text{ because } \lim J_n = 0.$$

Thus the series $\sum_{r=1}^{\infty}(a_r^2+b_r^2)$ is convergent (Art. 7), and so we may apply Tannery's theorem (Art. 49) to J_n, which gives

$$\frac{1}{2\pi}\int_0^{2\pi}[f(x)]^2\,dx = a_0^2 + \tfrac{1}{2}\sum_{r=1}^{\infty}(a_r^2+b_r^2).$$

[This result is due to de la Vallée Poussin; see also a paper by Hurwitz (*Math. Annalen*, Bd. 57, 1903, p. 425).]

7. We have seen that Σa_n^2, Σb_n^2 are convergent, using the notation of Exs. 5, 6; thus (Ex. 15, Ch. II.), we see that

$$\frac{1}{n}\Sigma|a_n| \text{ and } \frac{1}{n}\Sigma|b_n|$$

are convergent.

* L. Fejér, *Math. Annalen*, Bd. 58, 1904, p. 51; Lebesgue, *Séries Trigonométriques*, Paris, 1906, pp. 92–104.

Hence the series

$$a_0x+\Sigma\frac{1}{n}\{a_n\sin nx+b_n(1-\cos nx)\}$$

is convergent; and its sum is therefore equal to the sum found by taking the arithmetic mean. But this is equal to

$$\lim_{n\to\infty}\int_0^x \sigma_n\,dx=\int_0^x f(x)\,dx,$$

because σ_n converges *uniformly* to the value $f(x)$.

[This result is also due to de la Vallée Poussin.]

8. Consider the application of Frobenius's theorem to the series

$$1-xt+x^4t^2-x^9t^3+\ldots,$$

where t is a complex number of absolute value 1, but is not equal to -1, and x is real. It is easily proved that

$$s_0=1,\quad s_1=s_2=s_3=\frac{1-t^2}{1+t},\quad s_4=s_5=\ldots=s_8=\frac{1+t^3}{1+t},\ \text{etc.},$$

and generally $s_n=\{1-(-t)^\nu\}/(1+t)$, if $(\nu-1)^2\leqq n<\nu^2$.

Hence the arithmetic mean is found to be $1/(1+t)$,

and thus
$$\lim_{x\to1}(1-xt+x^4t^2-x^9t^3+\ldots)=1/(1+t)$$

or
$$\lim_{x\to1}(1-x\cos\theta+x^4\cos2\theta-x^9\cos3\theta+\ldots)=\tfrac12,$$

$$\lim_{x\to1}(x\sin\theta-x^4\sin2\theta+x^9\sin3\theta-\ldots)=\tfrac12\tan(\tfrac12\theta).$$

[See also Appendix, Art. 155.]

9. Apply Cesàro's mean-value process to the series

$$f_1-(f_1+f_2)+(f_1+f_2+f_3)-(f_1+f_2+f_3+f_4)+\ldots,$$

where f_n is positive and decreases steadily to zero, but Σf_n diverges.

The sum is

$$\tfrac12(f_1-f_2+f_3-f_4+\ldots)$$
$$=\lim_{x\to1}[f_1-(f_1+f_2)x+(f_1+f_2+f_3)x^2-\ldots].\qquad\text{[Hardy.]}$$

[We have, in fact,

$$s_1=f_1,\ s_2=-f_2,\ s_{2n-1}=f_1+f_3+\ldots+f_{2n-1},\ s_{2n}=-(f_2+f_4+\ldots+f_{2n}),$$

and so the arithmetic mean of $s_1, s_2, \ldots, s_{2n}$ is

$$(t_2+t_4+\ldots+t_{2n})/2n,\ \text{where } t_{2n}=f_1-f_2+f_3-\ldots-f_{2n}.$$

Apply Stolz's theorem (Art. 152, II.), and we find that the limit of the arithmetic mean is equal to $\frac12\lim t_{2n}$. Again, from Stolz's theorem we see that $\lim(s_{2n+1}/n)=\lim f_{2n+1}=0$, and so the arithmetic mean of $s_1, s_2, \ldots, s_{2n+1}$ tends to the same limit as that of $s_1, s_2, \ldots, s_{2n}$. The result can also be found by applying Art. 124 to the product $(1-1+1-\ldots)(f_0-f_1+f_2-\ldots)$.]

10. Illustrations of the last example are given by taking

$$1-(1+\tfrac{1}{2})+(1+\tfrac{1}{2}+\tfrac{1}{3})-\dots=\tfrac{1}{2}\log 2,$$
$$\log 2-\log 3+\log 4-\dots=\tfrac{1}{2}\log(\tfrac{1}{2}\pi).$$

[In the second we use Wallis's product (Art. 70): it is instructive to notice also that, to the base 10, $\frac{1}{2}\log(\frac{1}{2}\pi)=\cdot 098060$ to 6 decimal places, which verifies Euler's calculation quoted in Ex. 3, Art. 121.]

11. It follows from Art. 11 that

$$\left(1+\frac{1}{2}+\dots+\frac{1}{n}\right)-\log n$$

steadily decreases, and that its limit is Euler's constant C.

Thus the series $\Sigma(-1)^{n-1}\left[\left(1+\frac{1}{2}+\dots+\frac{1}{n}\right)-C-\log n\right]$

is convergent and, from the last example, its sum is seen to be

$$\tfrac{1}{2}\log 2-\tfrac{1}{2}C+\tfrac{1}{2}\log(\tfrac{1}{2}\pi)=\tfrac{1}{2}(\log\pi-C)=\cdot 28376. \qquad \text{[HARDY.]}$$

[The value of the sum can be verified by observing that the sum to $2n$ terms is

$$\left(\log\frac{2}{1}-\frac{1}{2}\right)+\left(\log\frac{4}{3}-\frac{1}{4}\right)+\dots+\left(\log\frac{2n}{2n-1}-\frac{1}{2n}\right).]$$

Asymptotic Series.

12. Establish the asymptotic formula

$$v_n=\int_{n\pi}^{\infty}\frac{\sin x}{x}\,dx\sim\frac{(-1)^n}{n\pi}\left[1-\frac{1.2}{(n\pi)^2}+\frac{1.2.3.4}{(n\pi)^4}-\dots\right],$$

which leads very easily to the numerical results

$$v_3=-\cdot 1040,\quad v_4=+\cdot 0787,\quad v_5=-\cdot 0631,\quad v_6=+\cdot 0527,$$

and so on.

[Compare Art. 132.] [GLAISHER.]

13. Use the integral of Art. 180 to shew that

$$\log\frac{\Gamma(x+a)}{\Gamma(x)}=\int_0^{\infty}\left[ae^{-t}-e^{-xt}\left(\frac{1-e^{-at}}{1-e^{-t}}\right)\right]\frac{dt}{t},$$

and deduce the asymptotic expansion

$$a\log x+\sum_1^{\infty}\frac{(-1)^{n-1}}{x^n}\,\frac{\phi_{n+1}(a)}{n(n+1)},$$

where $\phi_n(a)$ is the Bernoullian function of Art. 94. [SONINE.]

14. $$\int_0^{\infty}\frac{e^{-kx}}{1+x^2}dx=\int_k^{\infty}\frac{dt}{t}\sin(t-k)\sim\frac{1}{k}-\frac{2!}{k^3}+\frac{4!}{k^5}-\dots,$$

the error obtained by stopping at any stage being less than the following term in the series. [CAUCHY and DIRICHLET.]

[For the first integral use the identity

$$\frac{1}{1+x^2}=1-x^2+x^4-\dots\pm\frac{x^{2n}}{1+x^2}.$$

For the second, integrate by parts; the equality between the integrals is suggested by the series and can be established directly.]

15. $\int_0^\infty \frac{e^{-kx^2}}{1+x^2}dx = \sqrt{\pi}e^k \int_{\sqrt{k}}^\infty e^{-t^2}dt \sim \frac{\sqrt{\pi}}{2\sqrt{k}}\left[1-\frac{1}{2k}+\frac{1.3}{(2k)^2}-\frac{1.3.5}{(2k)^3}+\dots\right]$,

the error being again less than the following term.

[Apply the same methods again.]

By writing
$$\int_{\sqrt{k}}^\infty e^{-t^2}dt = \int_0^\infty e^{-t^2}dt - \int_0^{\sqrt{k}} e^{-t^2}dt$$

and integrating the latter by parts, we find that the first integral is equal to

$$\frac{\pi}{2}\left[e^k - \sum_0^\infty \frac{k^{n+\frac{1}{2}}}{\Gamma(n+\frac{3}{2})}\right].$$

16. Generally, if $0<s<1$, we find

$$\int_0^\infty \frac{e^{-kx}x^{-s}}{1+x}dx = \Gamma(1-s)e^k\int_k^\infty e^{-t}t^{s-1}dt = \frac{\pi}{\sin(s\pi)}\left[e^k - \sum_0^\infty \frac{k^{n+s}}{\Gamma(n+s+1)}\right]$$
$$\sim k^{s-1}\Gamma(1-s)\left[1-\frac{1-s}{k}+\frac{(1-s)(2-s)}{k^2}-\frac{(1-s)(2-s)(3-s)}{k^3}+\dots\right].$$

And similar expressions can be given for

$$\int_0^\infty \frac{e^{-kx}x^{-s}}{1+x^t}dx.$$

17. Apply Art. 131 to the function $f(x)=x^{-(1+\lambda)}$, $\lambda>0$, and deduce in particular the formula

$$\frac{1}{10^{1+\lambda}}+\frac{1}{11^{1+\lambda}}+\dots \text{ to } \infty \sim \frac{1}{10^\lambda}\left[\frac{1}{\lambda}+\frac{1}{20}+\frac{\lambda+1}{1200}-\frac{(\lambda+1)(\lambda+2)(\lambda+3)}{72\times 10^5}+\dots\right].$$

Hence evaluate the sum

$$1+\frac{1}{2^{\frac{4}{3}}}+\frac{1}{3^{\frac{4}{3}}}+\dots$$

to five decimal places.

18. If
$$u_n = P\int_0^\infty \frac{x^{2n}e^{a^2(1-x)^2}}{1-x^2}dx,$$

shew that
$$u_n - u_{n+1} = \sqrt{\pi}e^{a^2}\frac{1.3\dots(2n-1)}{2^{n+1}a^{2n+1}},$$

and that
$$u_0 = \int_0^\infty \frac{dx}{1-x^2}\left[e^{a^2(1-x^2)}-1\right] = \sqrt{\pi}\int_0^a e^{t^2}dt.$$

Hence prove that

$$\int_0^a e^{t^2}dt = e^{a^2}\left[\frac{1}{2a}+\frac{1}{4a^3}+\dots+\frac{1.3\dots(2n-3)}{2^n a^{2n-1}}\right]+\frac{u_n}{\sqrt{\pi}},$$

and that the value of the remainder is approximately $(a^2-n+\frac{1}{6})/\sqrt{(2n)}$.

In particular, for $a=4$, by taking 16 terms we can infer that the value of the integral $\int_0^4 e^{t^2}dt$ lies between 1149400·6 and 1149400·8.

[STIELTJES, *Acta Math.*, Bd. 9, p. 167.]

19. Obtain from Art. 131 the asymptotic expansion

$$\frac{1}{e^t-1}+\frac{1}{e^{2t}-1}+\frac{1}{e^{3t}-1}+\ldots$$

$$\sim\frac{1}{t}(C-\log t)+\frac{1}{4}+\frac{(B_1)^2}{2!}\frac{t}{2}-\frac{(B_2)^2}{4!}\frac{t^3}{4}-\frac{(B_3)^2}{6!}\frac{t^5}{6}-\ldots,$$

and deduce the behaviour of the series

$$\frac{x}{1-x}+\frac{x^2}{1-x^2}+\frac{x^3}{1-x^3}+\ldots$$

as x approaches 1, by writing $t=\log(1/x)$.

[SCHLÖMILCH, *Compendium der höheren Analysis*, II., p. 238.]

20. From the equation

$$\frac{1}{n^2}=\int_0^\infty te^{-nt}dt=\int_0^1\log\left(\frac{1}{1-x}\right)(1-x)^{n-1}dx,$$

we obtain

$$1+\frac{1}{2^2}+\frac{1}{3^2}+\ldots+\frac{1}{n^2}=\int_0^1\frac{1-(1-x)^n}{x}\log\left(\frac{1}{1-x}\right)dx$$

$$=\int_0^1\left[\sum_{r=1}^\infty\frac{1}{r}x^{r-1}\{1-(1-x)^n\}\right]dx$$

$$=\sum_{r=1}^\infty\left[\frac{1}{r^2}-\frac{1.2\ldots(r-1)}{r(n+1)(n+2)\ldots(n+r)}\right].$$

Thus

$$1+\frac{1}{2^2}+\ldots+\frac{1}{n^2}=\frac{\pi^2}{6}-\frac{1}{n+1}-\frac{\frac{1}{2}}{(n+1)(n+2)}-\frac{\frac{2}{3}}{(n+1)(n+2)(n+3)}-\ldots.$$

[SCHLÖMILCH.]

21. Similarly prove that

$$1+\frac{1}{2}+\frac{1}{3}+\ldots+\frac{1}{n}=C+\log n+\frac{1}{2n}-\frac{1/12}{n(n+1)}-\frac{1/12}{n(n+1)(n+2)}-\ldots,$$

$$\log(n!)=\tfrac{1}{2}\log(2\pi)+(n+\tfrac{1}{2})\log n-n-\frac{1/12}{n}+\frac{1/360}{n(n+1)(n+2)}-\ldots.$$

[See SCHLÖMILCH, *Compendium der höheren Analysis*, II., pp. 97–100; *Uebungsbuch der höheren Analysis*, II., pp. 192–195.]

22. Prove also that

$$\int_x^\infty\frac{1}{t}e^{-t}dt=\frac{1}{x}e^{-x}\left[1-\frac{1}{x+1}+\frac{1}{(x+1)(x+2)}-\ldots\right],$$

the following numerators being 2, 4, 14, 38, ...

And $$\int_x^\infty\frac{1}{\sqrt{t}}e^{-t}dt=\frac{1}{\sqrt{x}}e^{-x}\left[1-\frac{\frac{1}{2}}{x+1}+\frac{\frac{1}{4}}{(x+1)(x+2)}-\ldots\right],$$

the following numerators being $\frac{5}{8}$, $\frac{9}{16}$, $\frac{129}{32}$, ...

[SCHLÖMILCH, *Compendium* ..., II., p. 270.]

23. More generally prove that

$$\int_x^\infty e^{-t}t^{-\lambda}dt = e^{-x}x^{-\lambda}\left[1-\frac{a_1}{x+1}+\frac{a_2}{(x+1)(x+2)}-\dots\right],$$

where
$$t(t-1)\dots(t-m+1)=t^m - C_1 t^{m-1} + C_2 t^{m-2} - \dots,$$
$$\lambda_n = \lambda(\lambda+1)\dots(\lambda+n-1)$$
and
$$a_m = \lambda_m - C_1\lambda_{m-1} + C_2\lambda_{m-2} - \dots.$$

[SCHLÖMILCH, *l.c.*]

24. It must not be assumed that if $\lim(a_n/b_n)=1$ and $\Sigma a_n x^n$ is convergent for all values of x, that the two functions $\Sigma a_n x^n$, $\Sigma b_n x^n$ have the same asymptotic representations.

For example, consider the two series

$$\cos x = 1 - \frac{x^2}{2!} + \frac{x^4}{4!} - \frac{x^6}{6!} + \dots,$$

$$\cos x + \frac{\sinh x}{x} = \left(1+\frac{1}{1}\right) - \frac{x^2}{2!}\left(1-\frac{1}{3}\right) + \frac{x^4}{4!}\left(1+\frac{1}{5}\right) - \frac{x^6}{6!}\left(1-\frac{1}{7}\right) + \dots.$$

Miscellaneous Applications.

25. Shew that if we attempt to find a Fourier sine-series for $\cot x$ from $x=0$ to $x=\pi$, we obtain the series

$$2(\sin 2x + \sin 4x + \sin 6x + \dots),$$

which can be proved to have the sum $\cot x$ by using either Borel's integral or Cesàro's mean value (Arts. 103, 126).

26. Hardy has extended the result of the last example as follows:

If $f(x)$ satisfies the conditions of Fourier's expansion from $x=0$ to 2π, except that near $x=a$, $f(x)-\left[\frac{A_1}{x-a}+\frac{A_2}{(x-a)^2}+\dots+\frac{A_m}{(x-a)^m}\right]$ satisfies these conditions; then $f(x)$ can be expanded as a Fourier series which is summable except at $x=a$.

In particular, if $m=1$, we have the simple result

$$f(x) = a_0 + \mathscr{S}_1^\infty (a_n \cos nx + b_n \sin nx),$$

where
$$a_0 = \frac{1}{2\pi} P\int_0^{2\pi} f(x)\,dx,\quad a_n = \frac{1}{\pi}P\int_0^{2\pi} f(x)\cos nx\,dx,$$
$$b_n = \frac{1}{\pi}P\int_0^{2\pi} f(x)\sin nx\,dx;$$

and this is easily extended to the case in which $f(x)$ has any finite number of such singular points. [HARDY, *Messenger of Maths.*, vol. 33, 1903, p. 137.]

27. Applying a method similar to that of Ex. 2, p. 42, prove that

$$1 - \frac{1}{2m} + \frac{1.2}{2m(2m-1)} - \dots = \frac{2m+1}{m+1}$$

or
$$(2m)! - (2m-1)!\,.\,1! + (2m-2)!\,.\,2! - \dots = \frac{(2m+1)!}{m+1},$$

while
$$(2m+1)! - (2m)!\,.\,1! + (2m-1)!\,.\,2! - \dots = 0.$$

Deduce that the product of the two asymptotic series

$$1+\frac{1!}{x}+\frac{2!}{x^2}+\frac{3!}{x^3}+\dots, \quad 1-\frac{1!}{x}+\frac{2!}{x^2}-\frac{3!}{x^3}+\dots$$

is represented by

$$1+\frac{3!}{2x^2}+\frac{5!}{3x^4}+\frac{7!}{4x^6}+\dots.$$

As a verification, observe that if the first two series are denoted by xu, xv respectively, then term-by-term differentiation gives (compare Art. 137)

$$\frac{du}{dx}=-u+\frac{1}{x}, \quad \frac{dv}{dx}=v-\frac{1}{x}.$$

Thus
$$\frac{d}{dx}(uv)=\frac{v-u}{x}\sim -2\left(\frac{1!}{x^3}+\frac{3!}{x^5}+\frac{5!}{x^7}+\dots\right),$$

and so
$$uv\sim\frac{1!}{x^2}+\frac{3!}{2x^4}+\frac{5!}{3x^6}+\dots,$$

which agrees with the result deduced above from direct multiplication.

28. Le Roy's method described in Art. 115 can be applied to Stirling's series (Art. 132),

$$\sum_{n=0}^{\infty}\frac{(-1)^n B_{n+1}}{(2n+1)(2n+2)x^{2n+1}},$$

taking $p=2$ in Le Roy's formula.

[We find in fact
$$U_2(\xi)=\sum_0^{\infty}\frac{(-1)^n B_{n+1}\xi^{2n}}{(2n+2)!\,x^{2n+1}},$$

a series which converges absolutely if $|\xi|<2\pi|x|$. Also (Art. 176, Ex. 2)

$$B_{n+1}=4(n+1)\int_0^{\infty}\frac{t^{2n+1}dt}{e^{2\pi t}-1},$$

so that
$$U_2(\xi)=2\sum_0^{\infty}\frac{(-1)^n\xi^{2n}}{(2n+1)!\,x^{2n+1}}\int_0^{\infty}\frac{t^{2n+1}dt}{e^{2\pi t}-1}$$
$$=2\int_0^{\infty}\frac{\sin(\xi t/x)dt}{\xi(e^{2\pi t}-1)},$$

the interchange of summation and integration being justifiable by Art. 176, A, so long as $|\xi|<2\pi|x|$. Thus, in agreement with the principle explained in Art. 136, we can agree to define $U_2(\xi)$ as always given by the last integral. Then if it is permissible to invert the order of integration with respect to ξ and t, we have

$$\int_0^{\infty}U_2(\xi)e^{-\xi}d\xi=2\int_0^{\infty}\frac{dt}{e^{2\pi t}-1}\int_0^{\infty}\frac{\sin(\xi t/x)}{\xi}e^{-\xi}d\xi,$$

and we get the sum
$$2\int_0^{\infty}\frac{\arctan(t/x)}{e^{2\pi t}-1}dt,$$

but this last inversion is not particularly easy to justify.]

29. As examples of series which can be readily summed by means of various integrals similar to those of Art. 136, we give the following simple cases:

$$1-2!x+4!x^2-6!x^3+\ldots,$$
$$1-3!x+5!x^2-7!x^3+\ldots,$$
$$\Gamma(\lambda)-\Gamma(1+\lambda)x+\Gamma(2+\lambda)x^2-\ldots,$$
$$\Gamma(\lambda)-\Gamma(2+\lambda)x+\Gamma(4+\lambda)x^2-\ldots,$$
$$1-\tfrac{1}{2}(2!)x+\frac{1.3}{2.4}(4!)x^2-\ldots.$$

30. Determine a formal solution of the equation

$$x\frac{d^2y}{dx^2}+(p+q+x)\frac{dy}{dx}+py=0$$

in the form $$e^{-x}x^{-q}\left(1+\frac{A_1}{x}+\frac{A_2}{x^2}+\ldots\right),$$

and express the result as a definite integral

[The integral is $\dfrac{e^{-x}}{x^q\Gamma(q)}\displaystyle\int_0^\infty t^{q-1}\left(1+\frac{t}{x}\right)^{p-1}e^{-t}dt,$ if $q>0$.]

31. Obtain a formal solution of the differential equation

$$\frac{d^2y}{dx^2}+(2n+1-x^2)y=0$$

in the form $$e^{-\frac{1}{2}x^2}x^n\left[1-\frac{n(n-1)}{4x^2}+\frac{n(n-1)(n-2)(n-3)}{4.8x^4}-\ldots\right],$$

and express the sum of this series as a definite integral.

References.

For Exs. 1, 2, see Hardy, *Proc. Lond. Math. Soc.* (1), vol. 34, p. 55, and vol. 35, p. 81. For a justification of the integration used in Ex. 1, see Hardy, *Trans. Camb. Phil. Soc.*, 1904, vol. 19, p. 310.

For Exs. 14–16, see Dirichlet's *Bestimmte Integrale* (ed. Arendt), pp. 208–215; Nielsen, *Mathematische Annalen*, Bd. 59, p. 94; Hardy, *Proc. Lond. Math. Soc.* (2), vol. 3, p. 453. The equality of the integrals of Ex. 14 was proved by Cauchy, *Œuvres*, t. 1, p. 377.

In Exs. 20–23 the series are *not asymptotic but convergent*; they may be used instead of some of the asymptotic series found in Arts. 130, 132. But the law of the coefficients is much simpler in the asymptotic series; and for this reason they are usually preferable.

APPENDIX I.

ARITHMETIC THEORY OF IRRATIONAL NUMBERS AND LIMITS.

140. Infinite decimals.

If we apply the ordinary process of division to convert a rational fraction a/b to a decimal, it is evident that either the process must terminate or else the quotients must recur after $(b-1)$ divisions at most; for in dividing by b, there are not more than $(b-1)$ different remainders possible, (namely 1, 2, 3, ... $b-1$).

For instance, $\frac{1}{8} = \cdot 125$, terminating after three divisions.
Again $\frac{5}{7} = \cdot\dot{7}1428\dot{5}$, recurring after six $(=7-1)$ divisions.
Also $\frac{5}{14} = \cdot 3\dot{5}7142\dot{8}$, recurring after seven divisions.
And $\frac{2}{13} = \cdot\dot{1}5384\dot{6}$, recurring after six $(=\frac{1}{2}(13-1))$ divisions.

If the decimal part is purely periodic and contains p figures, the decimal can be expressed in the form $P/(10^p-1)$, by means of the formula for summing a Geometrical Progression (Art. 6). Thus b must be a factor of, or equal to, 10^p-1; and so b is not divisible by either 2 or 5. Conversely, it follows from Euler's extension of Fermat's theorem that when b is not divisible by either 2 or 5, an index p can be found so that 10^p-1 is divisible by b; thus a/b is of the form $P/(10^p-1)$ and so can be expanded as a periodic decimal with p figures in the period.

But if the decimal part is mixed, containing n non-periodic and p periodic figures (as for $\frac{5}{14}$), b must contain either 2^n or 5^n as a factor; because the decimal part of $(a\times 10^n)/b$ will be purely periodic. The relation between p and the other prime factors of b cannot be discussed so simply. But it is proved in the Theory of Numbers (see, for instance, Gauss, *Disq. Arithm.*, §§ 83–92, 308–318) that if $b = 2^\alpha 5^\beta r^\rho s^\sigma t^\tau \ldots$, where $r, s, t, \ldots$ are primes (not 2 or 5), then n is the greater of α, β, while p is a factor of the L.C.M. of $r^{\rho-1}(r-1)$, $s^{\sigma-1}(s-1)$, $t^{\tau-1}(t-1)$, ...

If now we agree to replace a terminated (say ·125) by an infinite decimal ·125000 ..., it will be evident that *any rational fraction can be expressed as an infinite decimal.**

* According to the rules of arithmetic, we could also replace ·125 by $\cdot 124\dot{9}$, but it is more convenient to have a unique form, and we shall adhere to ·125000 ....

But we can easily see that the rational numbers do not exhaust *all* infinite decimals.

Thus consider the decimal

$$\cdot 1010010001000010\ldots,$$

which consists of unities separated by zeros, the number of zeros increasing by one at each stage. Clearly this decimal neither terminates nor recurs: and it is therefore *not rational.*

Similarly, we may take a decimal

$$\cdot 11101010001010001\ldots$$

formed by writing unity when the order of the decimal place is prime (1, 2, 3, 5, 7, ...), and zero when it is composite (4, 6, 8, 9, 10, ...). This cannot be rational, since the primes do not form a sequence which recurs (in rank), and their number is infinite, as appears from the Theory of Numbers.

If the primes recurred in rank after a certain stage, it would be possible to find the integers a, b, such that all the numbers

$$a,\ a+b,\ a+2b,\ \ldots$$

would be prime. Now this is impossible, since $a+ab$ is divisible by a; and therefore the primes do not recur.

If the primes were finite in number, we could denote them by p_1, p_2, p_3, ..., p_n; and then the number

$$(p_1 p_2 p_3 \ldots p_n)+1$$

would not be divisible by any prime, which is absurd. Thus the number of primes is infinite; this theorem and proof are Euclid's (Bk. ix, Prop. 20).

As an example of a different type, consider the infinite decimal obtained by applying the regular arithmetic process for extracting the square-root to a non-square such as 2; this process gives a sequence of digits*

$$1\cdot 414213562373\ldots.$$

This decimal has the property that, if A_n denotes the value of the first n digits after the point,

$$\left(A_n+\frac{1}{10^n}\right)^2 > 2 > A_n^2,$$

which gives $2-A_n^2 < 3/10^n$, since $2A_n+1/10^n < 3$.

* A rapid way of finding the decimal is to use the series of Euler, Ex. B. 12, Ch. VIII.

To see that this decimal cannot terminate or recur, we have only to prove that *there is no rational fraction whose square is equal to* 2.

This is nearly obvious, but we can give a formal proof thus: Suppose, if possible, that $(a/b)^2=2$; we may assume that a, b are positive integers which are mutually prime, and therefore at least one of them is *odd*. Now since $a^2=2b^2$, a cannot be *odd*; so that b must be odd. But if we write $a=2c$, we get $2c^2=b^2$, so that b cannot be odd; we thus arrive at a contradiction.

141. The order of the system of infinite decimals.

It is possible, and in many ways it is distinctly best, to build up the whole theory of rational numbers on the basis of *order*, regarding the numbers as marks distinguishing certain objects arranged in a definite order. If, as usual, we place the larger numbers to the right of the smaller, along a straight line, we shall then regard the inequalities $A>B$, $B>C$ simply as meaning that the mark A is to the right of B and the mark C to the left of B, so that B falls between A and C.

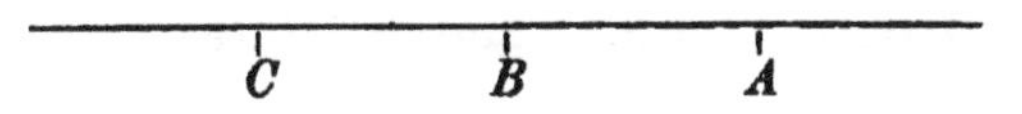

We shall now prove that *we can obtain the same arrangement by reference to the corresponding infinite decimals, without comparing the rational numbers directly.*

Suppose that we find

$$A=a_0+\frac{a_1}{10}+\frac{a_2}{10^2}+\dots+\frac{a_n}{10^n}+\dots,$$

$$B=b_0+\frac{b_1}{10}+\frac{b_2}{10^2}+\dots+\frac{b_n}{10^n}+\dots,$$

and that the integers a_r, b_r are the same up to a certain stage;* say that we find

$$a_0=b_0,\ a_1=b_1,\dots,\ a_{n-1}=b_{n-1},\ \text{but } a_n>b_n.$$

Write $$A_r=a_0+\frac{a_1}{10}+\frac{a_2}{10^2}+\dots+\frac{a_r}{10^r},$$

with a corresponding interpretation for B_r. Then we have

$$A_n-B_n=(a_n-b_n)/10^n \geqq 1/10^n.$$

* They cannot be always the same, or A would be equal to B. Note that a_0, b_0 may be negative, but that $a_1, a_2, \dots, b_1, b_2, \dots$ are all positive and less than 10.

Also $10^n(B-B_n)$ is a rational number, and is less than 1, in virtue of the method of finding B_n from B.

Thus $$B < B_n + 1/10^n,$$
while $$A \geqq A_n$$
and $$A_n \geqq B_n + 1/10^n.$$
Hence $$A > B.$$

Thus, *in order to determine the relative position of two infinite decimals (derived from rational fractions), we need only compare their digits, until we arrive at a stage where the corresponding digits are different; the relative value of these digits determines the relative position of the two decimals.*

By extending this rule to *all* infinite decimals (whether derived from rational numbers or not) we can assign a perfectly definite order to the whole system: for example, the decimal ·1010010001 ... given in Art. 140 would be placed between the two decimals

$$\cdot 1010000 \ldots \text{(zeros) and } \cdot 1011000 \ldots \text{(zeros)},$$

and also between

$$\cdot 101001000 \ldots \text{(zeros) and } \cdot 101001100 \ldots \text{(zeros)},$$

and so on.

Similarly, we may shew that the infinite decimal derived from extracting the square root of 2 must be placed between $\frac{24}{17}$ and $\frac{58}{41}$. For, by division, we find

$$\tfrac{24}{17} = 1{\cdot}411 \ldots, \quad \tfrac{58}{41} = 1{\cdot}4146 \ldots,$$

so that, in agreement with the rule,

$$\tfrac{24}{17} < 1{\cdot}41421 \ldots < \tfrac{58}{41}.$$

It must not be forgotten that *at present the new infinite decimals are purely formal expressions,* although, as we have explained, they fall in perfectly definite order into the scheme of infinite decimals derived from rational fractions.

142. Additional arithmetical examples of infinite decimals which are not rational.

Consider first the sequence of fractions (a_n), where

$$a_n = 1 + \frac{1}{2!} + \frac{1}{3!} + \ldots + \frac{1}{n!}.$$

If m is any integer, and $n > m$, we find

$$a_n - a_m = \frac{1}{(m+1)!} + \frac{1}{(m+2)!} + \ldots + \frac{1}{n!},$$

which is less than

$$\frac{1}{m!}\left[\frac{1}{m+1}+\frac{1}{(m+1)^2}+\dots+\frac{1}{(m+1)^{n-m}}\right]<\frac{1}{m!}\left(\frac{1}{m+1}\right)\Big/\left(1-\frac{1}{m+1}\right).$$

Thus, if $n>m+1$, the decimal for a_n lies between the decimals given by

$$a_{m+1}=a_m+\frac{1}{(m+1)!} \quad\text{and}\quad a_m+\frac{1}{m(m!)}.$$

In this way we find successively the limits

1·66	and 1·75,	$m=2$
1·708	and 1·720,	$m=3$
1·7166	and 1·7188,	$m=4$
1·71805	and 1·71834,	$m=5$
1·71825	and 1·71829,	$m=6$

and so on.

As m increases, these two decimals become more and more nearly equal; and we are thus led to construct an infinite decimal (1·718281828...), which we regard as equivalent to the expression

$$(1) \qquad 1+\frac{1}{2!}+\frac{1}{3!}+\frac{1}{4!}+\dots \text{ to } \infty.$$

It will now be proved that this infinite decimal cannot agree with the one which corresponds to any rational number.

For the decimal corresponding to

$$\frac{1}{2!}+\frac{1}{3!}+\frac{1}{4!}+\dots+\frac{1}{n!}$$

must be less than the decimal derived from

$$\frac{1}{2}+\frac{1}{2\,.\,3}+\frac{1}{2\,.\,3^2}+\dots+\frac{1}{2\,.\,3^{n-2}}.$$

And the last expression is

$$\frac{1}{2}\left(1-\frac{1}{3^{n-1}}\right)\Big/\left(1-\frac{1}{3}\right)=\frac{3}{4}\left(1-\frac{1}{3^{n-1}}\right).$$

Hence, no matter how many terms we take from $\frac{1}{2!}+\frac{1}{3!}+\dots$, the decimal derived from their sum will be less than the decimal ·75.

But if c is any integer greater than 1,

$$\frac{1}{2!}>\frac{1}{c+1},\quad \frac{1}{3!}>\frac{1}{(c+1)(c+2)},\quad \frac{1}{4!}>\frac{1}{(c+1)(c+2)(c+3)},$$

and so on. Thus the decimal representing any number of terms from

$$\frac{1}{c+1}+\frac{1}{(c+1)(c+2)}+\frac{1}{(c+1)(c+2)(c+3)}+\dots$$

must be less than ·75.

Suppose now, if possible, that (1) could lead to an infinite decimal agreeing with the decimal derived from a/c, where a and c are positive integers. Multiply by $c!$, and (1) becomes

$$[(2.3\dots c)+(3.4\dots c)+(4.5\dots c)+\dots+c+1]$$

$$+\frac{1}{c+1}+\frac{1}{(c+1)(c+2)}+\dots.$$

The terms in brackets give some integer I, say, and so we find that

$$a(c-1)!-I=\frac{1}{c+1}+\frac{1}{(c+1)(c+2)}+\dots,$$

that is, an integer equal to a decimal which is less than ·75, which is absurd. Thus no fraction such as a/c can give the same infinite decimal as (1) does.

Consider next the continued fractions

$$b_n=1+\frac{1}{1+}\,\frac{1}{1+}\,\frac{1}{1+}\dots \text{ to } n \text{ terms.}$$

Here, we recall the facts that if

$$b_m=p/q, \quad b_{m+1}=r/s,$$

then

$$|ps-qr|=1,$$

while b_n lies between b_m and b_{m+1} if $n>m+1$.

Thus we find for the successive values of b_n,

$$1,\ 2,\ \tfrac{3}{2},\ \tfrac{5}{3},\ \tfrac{8}{5},\ \tfrac{13}{8},\ \dots,$$

and so, converting to decimals, we see that b_n lies between

1	
	2
1·5	
	1·67
1·6	
	1·625
1·615	
	1·6191

As m increases, b_m and b_{m+1} become more and more nearly equal, and lead to an infinite decimal 1·6180340..., which can be considered as equivalent to the infinite continued fraction

$$(2) \qquad 1+\frac{1}{1+}\frac{1}{1+}\frac{1}{1+}\frac{1}{1+}\dots \text{ to } \infty.$$

This infinite decimal cannot be derived from a rational fraction a/c; for if it were we should have the inequality

$$\left|\frac{p}{q}-\frac{a}{c}\right| < \left|\frac{p}{q}-\frac{r}{s}\right| = \frac{1}{qs},$$

so that
$$|pc-aq| < c/s.$$

Since $(pc-aq)$ is an integer, the last condition gives $c > s$; but this is obviously absurd, because the denominator of the cth convergent is greater than c (if $c > 5$).

Similarly, the continued fraction

$$(3) \qquad 1+\frac{1}{2+}\frac{1}{2+}\frac{1}{2+}\dots$$

can be proved to lead to an infinite decimal which is not rational.

The reader who is familar with the theory of continued fractions will see that the square of (3) converges to the value 2; while (2) can be interpreted in connexion with the first geometrical example of Art. 143.

As a somewhat different example, it is easy to see that the infinite decimal

$$(4) \qquad \log_{10}2 = \cdot 301029995663981\dots$$

cannot be rational. For if it were equal to a/c, we should have $10^a = 2^c$; but 10^a must end with 0, whereas 2^c ends with 2, 4, 6 or 8. Thus $10^a = 2^c$ is impossible.

Similarly, we can see that 3, 5, 6, 7, 11,... cannot have rational logarithms.

143. Geometrical examples.

From the examples given in Arts. 140, 142 it is evident that the system of rational numbers is by no means sufficient to fulfil all the needs of algebra. We shall now give an example to shew that it does not suffice for geometry.

Let a straight line AB be divided at C in "golden section" (as in Euclid, Book II., prop. 11), so that $AC:CB = AB:AC$.

A E D C B

It is then easy to see that AC must be greater than CB, but less than twice CB.* Cut off AD equal to CB; it follows at once that AC is divided at D in the same ratio as AB is divided at C. For we have

$$AC : AD = AC : CB = AB : AC,$$

and consequently

$$AD : DC = AD : (AC - AD) = AC : (AB - AC) = AC : CB.$$

Also AD is less than half of AB.

Thus if we repeat this process $2n$ times, we arrive at a line AN, which is less than the 2^nth part of AB.

Now suppose, if possible, that AC/AB can be expressed as a rational fraction r/s; then $AD/AB = CB/AB$ is $(s-r)/s$, and DC/AB is $(2r-s)/s$. Hence AE/AB is $(2r-s)/s$ and ED/AB is $(2s-3r)/s$. Continuing this argument, we see that AN/AB must be some multiple of $1/s$; and so cannot be less than $1/s$. But we have seen than AN/AB is less than $1/2^n$, so that we are led to a contradiction, because we can choose n so that 2^n exceeds s. Thus the ratio $AC : AB$ cannot be rational.

It is not difficult to prove similarly that the ratio of the side to the diagonal of a square is not expressible as a rational fraction. In fact, let ABC in the figure represent half a square of which AB is a diagonal; it is at once evident that AB is greater than AC and less than $2AC$. Cut off $BD = BC$, and erect

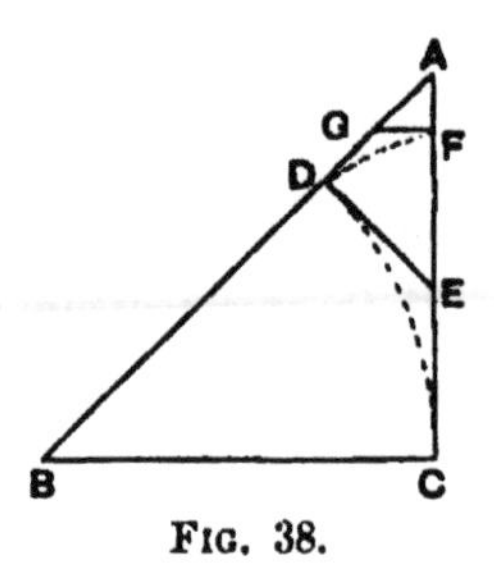

Fig. 38.

DE perpendicular to AB at D; then we have $ED = DA$, and $EC = ED$, because BE is a line of symmetry for the quadrilateral $BCED$. Thus $EC = DA$. If we repeat the same construction on the triangle ADE, we see in the same way that

$$AF = FG = GD.$$

* The first follows from the definition; and so we see that $AB = AC + CB$ is less than twice AC. Now, since $AC : CB = AB : AC$, it follows that AC is less than twice CB.

Thus $AD(=\frac{1}{2}FC)$ is less than half AC; and similarly, AF is less than half AD. Thus, by continuing the construction, we arrive at an isosceles triangle ANP, such that AN is less than the 2^nth part of AC.

But if AC/AB is a rational fraction r/s, then AD/AB is $(s-r)/s$, so that AF/AB is $(3r-2s)/s$; and continuing the process, we see that AN/AB is not less than $1/s$, or that AN/AC is not less than $1/r$, which leads to a contradiction, as before.

Ex. The reader may shew geometrically that the continued fraction (2) of the last article converges to the ratio $AB : AC$ in the golden section; while (3) converges to the ratio of the diagonal to the side of a square.

144. A special classification of rational numbers.

The examples of Arts. 140–3 indicate the need for developing some theory of *irrational numbers*. But before proceeding to a formal definition, which will be found in the next article, we shall give some considerations which shew how infinite decimals which do not recur lead up to Dedekind's definition.

The infinite decimal 1·41421 ... discussed in Art. 140 enables us to divide *all* rational numbers into two classes:

(A) *The lower class*, which contains all rational fractions (such as $\frac{24}{17}$) less than or equal to some term of the sequence of terminated decimals

$$1{\cdot}4,\quad 1{\cdot}41,\quad 1{\cdot}414,\quad 1{\cdot}4142, \text{ etc.}$$

(B) *The upper class*, which contains all rational fractions (such as $\frac{58}{41}$) greater than *every* term of the sequence.

It is then clear that

(i) Any number in the upper class is greater than every number in the lower class.

(ii) There is no greatest number in the lower class; and no least number in the upper class.

To see the truth of the second statement, we may observe that, if

$$l=(4+3k)/(3+2k),$$

we have $\quad l-k=2(2-k^2)/(3+2k),\quad 2-l^2=(2-k^2)/(3+2k)^2.$

Hence, if k is any rational number of the lower class, we have $l>k$, because $k^2<2$; and, for the same reason, $l^2<2$, so that l will also belong to the lower class. There is therefore no greatest number in the lower class.

If now we suppose k to be a rational number of the upper class, we prove by a similar argument that l is also a number of the upper class, but is less than k.

Ex. 1. Prove similarly that if $k^2 < N$, then $l > k$, $l^2 < N$, where

$$l = (Na + bk)/(b + ak) \text{ and } b^2 > Na^2.$$

Ex. 2. Establish inequalities similar to those of Ex. 1, taking

$$l = k(3N + k^2)/(N + 3k^2).$$ [DEDEKIND.]

Ex. 3. The formula, corresponding to Ex. 2, for the nth root of N is

$$l = k[(n+1)N + (n-1)k^n]/[(n-1)N + (n+1)k^n].$$

Ex. 4. Utilise the last example to find approximations to $2^{\frac{1}{3}}$; the first two may be taken as 1, $\frac{5}{4}$.

The classification of rational numbers which has been just described can, however, be obtained by a different process. From the arithmetical process of extracting the square-root of 2, it is evident that

$$(1{\cdot}4)^2,\ (1{\cdot}41)^2,\ (1{\cdot}414)^2,\ (1{\cdot}4142)^2, \ldots$$

are all less than 2; but the sequence contains numbers which are as close to 2 as we please. Thus the lower class contains every positive rational number whose square is less than 2; and it also contains all negative rational numbers. Since the two classes together contain *all* rational numbers, it follows that the upper class must contain every positive rational number whose square is greater than 2.

Thus the same classification is made by putting,

(A) In the *lower class*, all negative numbers and all positive numbers whose square is less than 2.

(B) In the *upper class*, all positive numbers whose square is greater than 2.

145. Dedekind's definition of irrational numbers.

Suppose that some rule has been chosen which separates *all* rational numbers into two classes, such that any number in the upper class is greater than every number in the lower class. Thus, if a number k belongs to the upper class, so also does every rational number greater than k.

There are then three mutually exclusive possibilities:

(1) There may be a number g in the lower class which is greater than every other number in that class.

(2) There may be a number l in the upper class which is less than every other number in that class.

(3) Neither g nor l may exist.

The cases (1), (2) lend themselves very readily to geometrical interpretation, by representing any rational number by a point on a line. Thus OP will

represent the fraction m/n, if the length OP is m times the nth part of the unit of length.

In case (1), the upper class consists of all rational points to the right of g

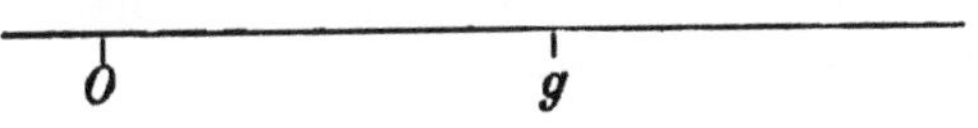

on the line; and the lower class consists of g and all rational points to the left of g.

Similarly, we can illustrate case (2). It might be thought at first sight that g and l might exist simultaneously; but this is excluded by the hypothesis

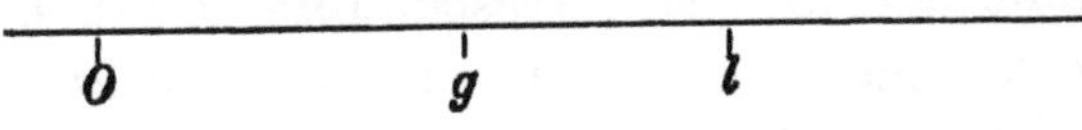

that *all* rational numbers are to be classed. Now $\frac{1}{2}(g+l)$ is rational and falls between g and l; and this would escape classification.

That there are cases in which neither g nor l can exist is clear from the example given in the last article, where it was proved that there could be no greatest number in the lower class, and no least number in the upper class.

For example, let us illustrate on a straight line the approximations to $\sqrt{2}$, which are derived from the convergents to the continued fraction

$$1+\frac{1}{2+}\,\frac{1}{2+}\,\frac{1}{2+}\cdots.$$

The convergents of the lower class are seen to be

$$1,\quad \tfrac{7}{5},\ \tfrac{41}{29},\ \ldots,$$

while

$$\tfrac{3}{2},\ \tfrac{17}{12},\ \tfrac{99}{70},\ \ldots$$

are those of the upper class.

It may be observed that if $\dfrac{p}{q}$ is a convergent of either class, the next convergent of the same class is $\dfrac{3p+4q}{2p+3q}$, while the intermediate convergent of the other class is $\dfrac{p+2q}{p+q}$.

Their representative points are as shewn in the diagrams, the second figure being a large-scale reproduction of the segment of the first which falls between $\frac{7}{5}$ and $\frac{3}{2}$.

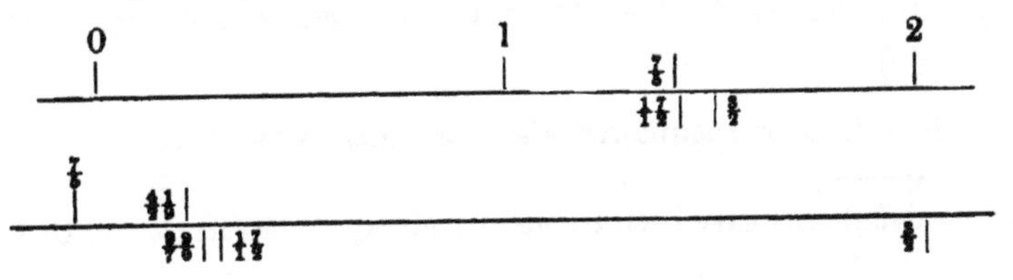

It is clear that in case (3) the rule gives a cleavage or *section* in the rational numbers; and to fill up the gap so caused in our number-system, we agree to *regard every such section as defining a new number.* This constitutes *Dedekind's definition of irrational numbers.** For it is clear from what has been said that these new numbers cannot be rational.

On the other hand, in cases (1), (2) there is no section, and so no new number is introduced.

146. Definitions of equal, greater, less; deductions.

For the present we use the following notation:

An irrational number is denoted by a Greek letter, such as α, β; the numbers of the corresponding lower class by small italics, as a, b; those of the upper class by capital italics, as A, B. The classes themselves may be denoted by adding brackets, as (a), (A).

These definitions may be indicated graphically thus

a α A

It is an obvious extension of the ordinary use of the symbols $<$, $>$, to write

$$a<\alpha<A, \quad b<\beta<B.$$

In particular, we say that *α is positive* when 0 belongs to (a); *α is negative* when 0 belongs to (A).

Two irrational numbers are equal, if their classes are the same; in symbols we write $\alpha=\beta$ if $(a)=(b)$ and $(A)=(B)$.

The reader who is acquainted with Euclid's theory of ratio will recognise that this definition of equality is exactly the same as that which he adopts in his *Elements.* Euclid in fact says that $A:B=C:D$, provided that the inequalities $mA \gtreqless nB$ are accompanied by $mC \gtreqless nD$, for any values of m, n whatever. In Dedekind's theory, the inequality $mA>nB$ implies that n/m is in the lower class defining $A:B$; thus Euclid's definition implies that $A:B$ and $C:D$ have the same lower class and the same upper class.

On the other hand, *the number α is less than the number β*, when part of the upper class (A) belongs to the lower class (b), so that at least one rational number r belongs both to (A) and to (b).

This definition of inequality also coincides with Euclid's.

* Other definitions have been framed by Méray, Weierstrass and G. Cantor.

It follows at once from the definition that if $\alpha < \beta$ and $\beta < \gamma$, then $\alpha < \gamma$.

Again, *if* $\alpha < \beta$, *an infinite number of rational numbers fall between* α *and* β. For at least one rational number r exists such that $\alpha < r < \beta$. Now there is no greatest number in the class (b), so that we can find another rational number s which belongs to (b) and is greater than r. Thus

$$\alpha < r < s < \beta.$$

Then if x, y are any two positive integers, we have

$$r < \frac{rx+sy}{x+y} < s,$$

so that all these rational numbers lie between α and β.

147. Deductions from the definitions.

Any irrational number (α) *can be expressed as an infinite decimal.*

For there will be some integer n_0 (positive or negative) such that n_0 belongs to the lower class (a) and n_0+1 belongs to the upper class (A). Then consider the rational fractions

$$n_0,\ n_0+\tfrac{1}{10},\ n_0+\tfrac{2}{10},\ n_0+\tfrac{3}{10}, \ldots,\ n_0+\tfrac{9}{10},\ n_0+1;$$

some of these belong to class (a), the rest to class (A). Suppose that $n_0+n_1/10$ is the greatest in class (a), so that we have

$$n_0+\frac{n_1}{10} < \alpha < n_0+\frac{n_1+1}{10}.$$

Continuing this process, we arrive at the result

$$n_0+\frac{n_1}{10}+\frac{n_2}{10^2}+\ldots+\frac{n_r}{10^r} < \alpha < n_0+\frac{n_1}{10}+\frac{n_2}{10^2}+\ldots+\frac{n_r+1}{10^r}.$$

If we call these two decimal fractions a_r, A_r, it is plain that $A_r-a_r(=1/10^r)$ can be made less than any prescribed rational fraction merely by taking r to be sufficiently great; and if we continue the process indefinitely we see that α is the number defined by the infinite decimal $n_0 \cdot n_1 n_2 n_3 \ldots$.

The argument just given shews also that *we can always determine numbers* A, a *belonging to the two classes such that* $A-a$ *is less than an arbitrarily small rational fraction.*

It is useful to note further that a_r can be chosen so as to exceed any prescribed number a' of the lower class. For let

a'' be another number of the lower class which is greater than a'; and then choose r so that $10^r > 1/(a''-a')$. Then

$$A_r - a_r < a'' - a', \text{ or } a_r - a' > A_r - a''.$$

But $A_r > a''$, so that $a_r > a'$.

Modified form of Dedekind's definition.

Suppose that a classification of the rational numbers has the following properties:

(1) if a belongs to the lower class, so does every rational number less than a;

(2) if A belongs to the upper class, so does every rational number greater than A;

(3) every number a is less than any number A;

(4) numbers A, a can be found in the two classes such that $A-a$ is less than an arbitrary rational fraction.

Such a classification defines a single number, rational or irrational.

For any rational number r which does not belong to either class must lie between the two classes, since any number less than a number of the lower class must also belong to the lower class; and therefore r must exceed every number of the lower class: similarly, r must be less than every number of the upper class. Hence, if a, A are any two numbers of the two classes $a < r < A$.

Suppose now that s is a second rational number which belongs to neither class; then $a < s < A$. Hence $|r-s|$ must be less than $A-a$; but this is impossible, since by hypothesis A, a can be chosen so that $A-a$ is less than any assigned rational fraction.

Consequently, not more than one rational number can escape classification; if there is one such number, the classification may be regarded as defining that number; but if there is no rational number which escapes classification, we have obtained a Dedekind section, and have therefore defined an irrational number.

148. Algebraic operations with irrational numbers.

The negative of an irrational number a is defined by means of the lower class $-A$ and the upper class $-a$; and it is denoted by $-a$.

The reciprocal of an irrational number a is defined most easily by restricting the classes at first to contain only terms of *one* sign; and then the reciprocal $1/a$ is defined by the lower class $1/A$ and the upper class $1/a$. Thus if the number a is positive, the complete specification of the classes for $1/a$ will be given by putting the whole of $1/A$ in the lower class, together with all negative numbers, while the upper class will contain only the positive part of $1/a$; and a corresponding definition is easily framed for $1/a$ when a is negative.

The absolute value of an irrational number a is always positive and is equal to a or $-a$, according as a is positive or negative; it is denoted by $|a|$.

Addition of two irrationals.

Suppose α, β to be the two given irrationals, so that $a < \alpha < A, b < \beta < B$. Then classify the rational numbers by making $a+b$ a typical member of the lower class and $A+B$ of the upper class. This rule obviously satisfies conditions (1)–(3) of Art. 147. To prove that it satisfies condition (4) and so defines a *single* number, we note that

$$(A+B)-(a+b)=(A-a)+(B-b),$$

and, as explained at the beginning of Art. 147, we can find a, A and b, B so as to make $A-a$ and $B-b$ each less than $\frac{1}{2}\epsilon$; and then $(A+B)-(a+b)$ is less than ϵ. Hence our classification defines a number which may be rational or irrational; this number is called the sum $\alpha+\beta$.

It follows at once that $\alpha+(-\alpha)=0$; for here the lower class is represented by the type $a-A$, and the upper class by $A-a$. That is, the lower class consists of all negative rational numbers and the upper class of all positive rational numbers; hence, zero is the only rational number not classed, and therefore is the number defined by the classification.

Subtraction.

In virtue of the relation $\beta+(-\beta)=0$, we may define $\alpha-\beta$ as equal to the sum $\alpha+(-\beta)$.

Multiplication (positive irrational numbers).

For simplicity of statement, we omit the *negative* numbers from the lower classes; and then we define the product $\alpha\beta$ by using the type ab for the lower class and AB for the corresponding upper class. To prove that this defines a single number, let ϵ denote an arbitrary positive rational fraction less than 1;

then choose any rational number P which is greater than $\alpha+\beta+1$. Next find numbers A, a and B, b such that $A-a<\epsilon_1$, $B-b<\epsilon_1$, where

$$\epsilon_1=\epsilon/P.$$

The determination of A, a, and B, b is possible in virtue of Art. 147. Now we have

$$AB-ab<(a+\epsilon_1)(b+\epsilon_1)-ab$$

or
$$AB-ab<\epsilon_1(a+b+\epsilon_1)<\epsilon_1(a+b+1)<\epsilon_1 P.$$

That is,
$$AB-ab<\epsilon.$$

Thus the classification by means of ab and AB defines a single number which may be rational or irrational; and this number is called $\alpha\beta$.

In particular, if $\beta=1/\alpha$, the product is equal to 1; for the lower class is represented by a/A and the upper class by A/a. That is, the lower class contains all rational numbers less than 1 and the upper class all rational numbers greater than 1. Consequently the product is equal to 1, the single rational number which escapes classification.

Multiplication of negative irrational numbers is reduced at once to that of positive numbers by agreeing to accept the "rule of signs" as established for rational numbers.

Division.

In consequence of the relation $(1/\beta)\times\beta=1$, we may define the quotient α/β as equal to the product $\alpha\times(1/\beta)$.

It is at once evident that any of the fundamental laws of algebra which have been established for rational numbers remain true for irrational numbers.

Thus, we have the following laws:

$$\alpha+0=\alpha,\quad \alpha+\beta=\beta+\alpha,\quad \alpha+(\beta+\gamma)=(\alpha+\beta)+\gamma,$$

$$\alpha\times 1=\alpha,\quad \alpha\beta=\beta\alpha,\quad \alpha(\beta\gamma)=(\alpha\beta)\gamma,$$

$$\alpha(\beta+\gamma)=\alpha\beta+\alpha\gamma,$$

$$|\alpha|+|\beta|\geqq|\alpha+\beta|\geqq|\alpha|-|\beta|.$$

For example, let us prove the theorem $\alpha+\beta=\beta+\alpha$.

By definition we have

$$(a+b)<\alpha+\beta<(A+B)$$

and
$$(b+a)<\beta+\alpha<(B+A).$$

But $a+b=b+a$ and $A+B=B+A$, so that $\alpha+\beta$ and $\beta+\alpha$ are defined by the same two classes and are accordingly equal.

The reader will find it a good exercise to write out proofs of the other laws in a similar way. After this he may attempt to construct a theory of irrational indices and of logarithms, on the foundation of Dedekind's theory. It is necessary to first define a^λ and then to prove that $a^\lambda . a^\mu = a^{\lambda+\mu}$, and so on, finally shewing that the equation in λ, $a^\lambda = \beta$, has a root, where a, β are positive and λ, μ may be either positive or negative.

149. The principle of convergence for monotonic sequences whose terms may be either rational or irrational.

A monotonic sequence (a_n) leads to a section in the system of rational numbers as follows:

Suppose for definiteness that the sequence is an increasing one, in which the terms remain less than a fixed number A, so that

$$a_1 \leqq a_2 \leqq a_3 \leqq \ldots \leqq a_n \leqq \ldots < A.$$

Now if k is any rational number, one of two alternatives must occur; either some term in the sequence (a_n) will be equal to or greater than k, or else every term of the sequence will be less than k. We define the class (b) as the class of all rational numbers k which satisfy the first condition, the class (B) as the class of all which satisfy the second condition. Typical numbers of the class (b) are the rational numbers which belong to the sequence; while (B) contains every rational number greater than A, and possibly some rational numbers less than A.

It is clear that the classes (b), (B) together contain *all* rational numbers, and therefore give a *section* which defines some number β, rational or irrational. We may call (B), (b) the upper and lower classes respectively, defined by the sequence (a_n).

Now every rational number greater than β belongs to the upper class (B), and is therefore greater than any term a_n. And the same is true of every irrational number γ greater than β; for there will be rational numbers between γ and β, and these rational numbers are greater than any term a_n: thus γ is also greater than any term a_n. Consequently *no term in the sequence (a_n), whether rational or irrational, can exceed β.*

On the other hand, every rational number less than β must belong to the lower class (b). Now if ϵ is any positive number, there will be rational numbers between β and $\beta-\epsilon$; and, since these numbers are less than β, they must belong to the lower class (b). That is, *there must be some term of the sequence, say a_m, which is greater than or equal to $\beta-\epsilon$.*

Hence, since $a_n \geqq a_m$, if $n > m$,

we have $\beta \geqq a_n \geqq \beta - \epsilon$, if $n > m$.

That is, $\lim a_n = \beta$.

A good example of such a sequence is afforded by the terminated decimals derived from an infinite decimal; and it will be seen at once that the section described here is an obvious extension of the method used in Art. 144 above.

Suppose next that the terms of the sequence, while still increasing, do not remain less than any fixed number A. It is then evident that if $a_m > A$, we have $a_n > A$ if $n > m$.

Thus
$$\lim_{n\to\infty} a_n = \infty.$$

Exactly similar arguments can be applied to a decreasing sequence.

As an example we shall give a proof of the theorem that *any continuous monotonic function attains just once every value between its greatest and least values.* Suppose that $f(x)$ steadily increases from $x=a$ to $x=c$, so that

$$b < d, \text{ if } f(a)=b \text{ and } f(c)=d.$$

Then if l is any number between b and d, we consider $f\{\frac{1}{2}(a+c)\}$, which is also between b and d; suppose that this is found to be less than l, write then

$$a_1 = \tfrac{1}{2}(a+c), \quad b_1 = f(a_1) < l,$$
$$c_1 = c, \quad d_1 = f(c_1) > l.$$

On the other hand, when $f\{\frac{1}{2}(a+c)\}$ is greater than l, we write

$$a_1 = a, \quad b_1 = f(a_1) < l,$$
$$c_1 = \tfrac{1}{2}(a+c), \quad d_1 = f(c_1) > l.$$

Continuing the process we construct two sequences (a_n), (c_n), the first never decreasing and the second never increasing; and $c_n - a_n = (c-a)/2^n$, so that (a_n), (c_n) have a common limit k. Also by the method of construction it is evident that $f(a_n) < l < f(c_n)$; unless it happens that at some stage we find $f(a_n) = l$, in which case the theorem requires no further discussion.

Now since $f(x)$ is *continuous* we can find an integer ν such that $f(c_n) - f(a_n) < \epsilon$, if $n > \nu$; and both $f(k)$ and l are contained between $f(a_n)$ and $f(c_n)$. Thus we can find ν so that

$$|f(k) - l| < \epsilon, \quad \text{if } n > \nu,$$

and therefore, as in Art. 1 (6), $f(k) = l$. From the method of construction it is clear that there is only one value such as k; and this is also evident from the monotonic nature of $f(x)$.

150. Maximum and minimum limiting values of a sequence of rational or irrational terms.

Suppose first that all the terms of the sequence are less than some fixed rational number P, and let p be a smaller rational number, such that an infinity of terms a_n are greater than p. Then if we bisect the interval (p, P) by $\frac{1}{2}(p+P)$, it is evident that either an infinity or a finite number of terms a_n fall between $\frac{1}{2}(p+P)$ and P; in the former case we write $p_1=\frac{1}{2}(p+P)$, $P_1=P$; in the latter we write $p_1=p$, $P_1=\frac{1}{2}(p+P)$. We have thus constructed a smaller interval (p_1, P_1) which contains an infinity of terms a_n; and we can repeat the process as often as we wish. A few stages are indicated in the diagram.

FIG. 39.

Then the sequence $p, p_1, p_2, p_3 \ldots$ never decreases, and remains less than P; and so the sequence (p_n) determines a number G (which may of course be either rational or irrational, as in the last article). Then the sequence (P_n) has the same limit G, because $P_n - p_n = (P-p)/2^n \to 0$ as n increases. Thus, if ϵ is an arbitrarily small positive number, we can find n so that $p_n \geqq G-\epsilon$, $P_n \leqq G+\epsilon$.

Consequently *an infinite number of terms a_n lie between $G-\epsilon$ and $G+\epsilon$, but only a finite number are greater than $G+\epsilon$.*

Thus we can determine a sub-sequence from (a_n) which has G as its limit; and we can find a certain stage after which all the terms of the sequence are less than $G+\epsilon$; thus no convergent sub-sequence can have a limit greater than G. *These properties shew that G is the maximum limit of the sequence (a_n).* (See Art. 5.)

If no such number as P can be found in the foregoing argument, there are numbers of the sequence (a_n) greater than any assignable number, *so that the maximum limit is then ∞.* On the other hand, if no such number as p can be found, there will be only a *finite* number of terms greater than $-N$, however large N may be, and consequently

$$\lim a_n = -\infty.$$

For the sake of uniformity we may say even then that the sequence has a maximum limit, which is, of course, $-\infty$.

All the foregoing discussion can be at once modified to establish the existence of a *minimum limit* (g or $-\infty$).

151. The general principle of convergence stated in Art. 3 is both necessary and sufficient; the terms being rational or irrational.

In the first place, the condition is obviously *necessary*; for if $\lim_{n\to\infty} a_n = l$, we know that an index m can be found to correspond to ϵ, in such a way that

$$|l - a_n| < \tfrac{1}{2}\epsilon, \quad \text{if } n \geqq m.$$

Thus $\quad |a_n - a_m| \leqq |l - a_n| + |l - a_m| < \epsilon, \quad \text{if } n > m.$

In the second place, the condition is *sufficient*; for let m be fixed so that

$$|a_n - a_m| < \tfrac{1}{3}\epsilon, \quad \text{if } n > m,$$

or

$$a_m - \tfrac{1}{3}\epsilon < a_n < a_m + \tfrac{1}{3}\epsilon, \quad \text{if } n > m.$$

Then it follows from the last article that the sequence (a_n) has a finite maximum limit G; so that an infinity of terms fall between $G - \frac{1}{3}\epsilon$ and $G + \frac{1}{3}\epsilon$. Choose one of these, say a_p, whose index p is greater than m. Thus we have

$$G - \tfrac{1}{3}\epsilon < a_p < G + \tfrac{1}{3}\epsilon.$$

Also $\quad a_p - \frac{1}{3}\epsilon < a_m < a_p + \frac{1}{3}\epsilon, \quad$ since $p > m$.

Thus $\quad G - \frac{2}{3}\epsilon < a_m < G + \frac{2}{3}\epsilon,$

and since $\quad a_m - \frac{1}{3}\epsilon < a_n < a_m + \frac{1}{3}\epsilon, \quad$ if $n > m$,

it follows that $\quad G - \epsilon < a_n < G + \epsilon, \quad$ if $n > m$.

Thus $a_n \to G$; and consequently the sequence is convergent. Of course in this case $g = G$, the extreme limits being equal in a convergent sequence.

Various proofs of this general theorem have been published, some being *apparently* much shorter than the foregoing series of articles. But on examining the foundations of the shorter investigations it will be seen that in all cases the apparent brevity is obtained by avoiding the definition of an irrational number. This virtually implies a shirking of the whole difficulty; for this difficulty consists essentially* in proving that (under the condition of Art. 3) a sequence may be used to define a "number."

* Pringsheim (*Encyklopädie*, I. A. 3, 14) says: "As the truth of this theorem rests essentially and exclusively *on an exact definition of irrational numbers*, naturally the first accurate proofs are connected with the *arithmetical* theories of irrational numbers, and with the associated revision and improvement of the older *geometrical* views."

152. Theorems on limits of quotients.

I. *If* $\lim a_n = 0$ *and* $\lim b_n = 0$; *and if, in addition, the sequence* (b_n) **steadily decreases**, then

$$\lim \frac{a_n}{b_n} = \lim \frac{a_n - a_{n+1}}{b_n - b_{n+1}},$$

provided that the **second** *quotient has a definite limit, finite or infinite.*

Suppose first that the limit is finite and equal to l; then if ϵ is an arbitrarily small positive fraction, m can be found so that

$$l - \epsilon < \frac{a_n - a_{n+1}}{b_n - b_{n+1}} < l + \epsilon, \quad \text{if } n > m;$$

or, since $(b_n - b_{n+1})$ is *positive*, we have

$$(l-\epsilon)(b_n - b_{n+1}) < a_n - a_{n+1} < (l+\epsilon)(b_n - b_{n+1}).$$

Change n to $n+1, n+2, \dots n+p-1$ and add the results; then we find

$$(l-\epsilon)(b_n - b_{n+p}) < a_n - a_{n+p} < (l+\epsilon)(b_n - b_{n+p}).$$

Now take the limit of this result as $p \to \infty$; we obtain

$$(l-\epsilon)b_n \leqq a_n \leqq (l+\epsilon)b_n,$$

because $a_{n+p} \to 0$, $b_{n+p} \to 0$ by hypothesis.

Hence, since b_n is positive, we have

$$|(a_n/b_n) - l| \leqq \epsilon, \quad \text{if } n > m,$$

or

$$\lim (a_n/b_n) = l.$$

On the other hand, if the limit is ∞, we can find m so that

$$\frac{a_n - a_{n+1}}{b_n - b_{n+1}} > N, \quad \text{if } n > m,$$

however large N may be. By exactly the same argument as before, we find

$$a_n - a_{n+p} > N(b_n - b_{n+p}),$$

which leads to

$$a_n \geqq N b_n,$$

or

$$a_n/b_n \geqq N, \quad \text{if } n > m.$$

Thus

$$\lim (a_n/b_n) = \infty.$$

There is no difficulty in extending the argument to prove that, with the same restriction on the sequence (b_n),

$$\underline{\lim} \frac{a_n - a_{n+1}}{b_n - b_{n+1}} \leqq \overline{\lim} \frac{a_n}{b_n} \leqq \overline{\lim} \frac{a_n - a_{n+1}}{b_n - b_{n+1}}.$$

This theorem should be compared with the theorem (L'Hospital's) of the Differential Calculus, that:

If $$\lim_{x\to\infty}\phi(x)=0,\quad \lim_{x\to\infty}\psi(x)=0,$$

then $$\lim_{x\to\infty}\phi(x)/\psi(x)=\lim_{x\to\infty}\phi'(x)/\psi'(x),$$

provided that the **second** *limit exists and that* $\psi'(x)$ *has a constant sign for values of* x *greater than some fixed value.*

II. *If* b_n **steadily increases** *to* ∞, *then*

$$\lim\frac{a_n}{b_n}=\lim\frac{a_{n+1}-a_n}{b_{n+1}-b_n},$$

provided that the **second** *limit exists.**

For if the second limit is finite and equal to l, as in I. above, we see by a similar argument that it is possible to choose m so that

$$(l-\epsilon)(b_n-b_m)<a_n-a_m<(l+\epsilon)(b_n-b_m),\quad \text{if } n>m.$$

Thus, since b_n is positive,

$$(l-\epsilon)\left(1-\frac{b_m}{b_n}\right)+\frac{a_m}{b_n}<\frac{a_n}{b_n}<(l+\epsilon)\left(1-\frac{b_m}{b_n}\right)+\frac{a_m}{b_n}.$$

Now, since $b_n\to\infty$, we have

$$\lim_{n\to\infty}\left[(l-\epsilon)\left(1-\frac{b_m}{b_n}\right)+\frac{a_m}{b_n}\right]=l-\epsilon$$

and $$\lim_{n\to\infty}\left[(l+\epsilon)\left(1-\frac{b_m}{b_n}\right)+\frac{a_m}{b_n}\right]=l+\epsilon.$$

And so we find

$$l-\epsilon\leqq\underline{\lim}\frac{a_n}{b_n}\leqq\overline{\lim}\frac{a_n}{b_n}\leqq l+\epsilon.$$

But these extreme limits are independent of m, and therefore also of ϵ; and so the inequalities can only be true if each of the extreme limits is equal to l. Hence

$$\lim(a_n/b_n)=l.$$

Similarly, if the limit is ∞, we can find m, so that

$$a_n-a_m>N(b_n-b_m),\quad \text{if } n>m,$$

however great N may be.

Thus $$\frac{a_n}{b_n}>\frac{a_m}{b_n}+N\left(1-\frac{b_m}{b_n}\right),$$

---

* Extended by Stolz from a theorem given by Cauchy for the case $b_n=n$.

and the limit of the expression on the right is N, as $n \to \infty$, so that $\underline{\lim}\,(a_n/b_n) \geqq N$; now this minimum limit is independent of m, and therefore of N.

Thus $\quad \underline{\lim}\,(a_n/b_n) = \infty \quad$ or $\quad \lim\,(a_n/b_n) = \infty$.

There is no difficulty in proving similarly that with the same restriction on the sequence (b_n)

$$\underline{\lim}\,\frac{a_{n+1}-a_n}{b_{n+1}-b_n} \leq \underline{\lim}\,\frac{a_n}{b_n} \leq \overline{\lim}\,\frac{a_n}{b_n} \leq \overline{\lim}\,\frac{a_{n+1}-a_n}{b_{n+1}-b_n}.$$

Theorem II. should be compared with the following theorem (L'Hospital's) of the Differential Calculus:

If $\psi(x)$ **increases steadily** *to* ∞ *with* x, *then*

$$\lim \frac{\phi(x)}{\psi(x)} = \lim \frac{\phi'(x)}{\psi'(x)},$$

provided that the **second** *limit exists.*

It is probably not out of place to say a word on this important theorem, since only one of the commoner English text-books contains a correct proof.* By the extended form of the mean-value theorem we have

$$\frac{\phi(x)-\phi(a)}{\psi(x)-\psi(a)} = \frac{\phi'(\xi)}{\psi'(\xi)}, \text{ where } x > \xi > a,$$

and $\psi(x)-\psi(a)$ is positive. Thus, if $\phi'(x)/\psi'(x)$ tends to a limit l, we can choose a so that

$$(l-\epsilon)[\psi(x)-\psi(a)] < \phi(x)-\phi(a) < (l+\epsilon)[\psi(x)-\psi(a)].$$

And from here onwards the argument proceeds as for sequences.

Similarly, if $\phi'(x)/\psi'(x) \to \infty$, we can find a so that

$$\phi(x)-\phi(a) > N[\psi(x)-\psi(a)],$$

and again the same argument can be used.

Ex. 1. If $\quad \left.\begin{aligned} a_n &= 1^p + 2^p + \ldots + n^p \\ b_n &= n^{p+1} \end{aligned}\right\} \quad p+1 > 0,$

we have $$\frac{a_{n+1}-a_n}{b_{n+1}-b_n} = \frac{(n+1)^p}{(n+1)^{p+1}-n^{p+1}} = \frac{1/n}{(1+1/n)^{p+1}-1}\left(1+\frac{1}{n}\right)^p.$$

Now $$\lim_{h\to 0} \frac{(1+h)^{p+1}-1}{h} = p+1,$$

by a fundamental limit of the Differential Calculus.

Hence, $$\lim \frac{a_{n+1}-a_n}{b_{n+1}-b_n} = \frac{1}{p+1},$$

and so $$\lim \frac{a_n}{b_n} = \frac{1}{p+1}.$$

---

* Gibson's *Elementary Treatise on the Calculus.*

Ex. 2. If $$a_n = \log n,\ b_n = n,$$
we have $$\frac{a_{n+1} - a_n}{b_{n+1} - b_n} = \log\left(1 + \frac{1}{n}\right),$$
so that $$\lim\left(\frac{\log n}{n}\right) = 0. \quad \text{(Compare Art. 161.)}$$

Similarly, if $a_n = (\log n)^2$, $b_n = n$, we find
$$\frac{a_{n+1} - a_n}{b_{n+1} - b_n} = [\log n + \log(n+1)] \log\left(1 + \frac{1}{n}\right) < \frac{2}{n} \log(n+1),$$
which tends to 0 by the previous result.

Thus $$\lim [(\log n)^2/n] = 0.$$

Similarly we can prove that $\lim [(\log n)^k/n] = 0$.

The reader may also verify this result by using L'Hospital's theorem.

Ex. 3. If $$a_n = p^n,\ b_n = n,$$
we have $$a_{n+1} - a_n = p^n(p-1),\ b_{n+1} - b_n = 1,$$
and hence
$$\lim(a_{n+1} - a_n) = 0, \quad \text{if } p \leqq 1,$$
$$\text{or} \quad = \infty, \quad \text{if } p > 1.$$

Thus
$$\lim(p^n/n) = 0, \quad \text{if } p \leqq 1,$$
$$\text{or} \quad = \infty, \quad \text{if } p > 1, \quad \text{(Art. 161).}$$

Of course Ex. 3 is only another form of Ex. 2.

Ex. 4. Even when (a_n) and (b_n) are both monotonic, $\lim a_n/b_n$ need not exist. In this case, the theorem shews that $\lim (a_{n+1} - a_n)/(b_{n+1} - b_n)$ does not exist. An example is given by
$$a_n = p^n[q + (-1)^n],\ b_n = p^n \quad (p > 1).$$

Here
$$a_{n+1} - a_n = p^n[pq + p(-1)^{n+1} - q - (-1)^n]$$
$$= p^n[(p-1)q - (p+1)(-1)^n],$$
and so a_n steadily increases if $q > (p+1)/(p-1)$.

Then we have
$$\underline{\lim}\, \frac{a_{n+1} - a_n}{b_{n+1} - b_n} = q - \frac{p+1}{p-1}, \quad \overline{\lim}\, \frac{a_{n+1} - a_n}{b_{n+1} - b_n} = q + \frac{p+1}{p-1},$$
while
$$\underline{\lim}(a_n/b_n) = q - 1, \qquad \overline{\lim}(a_n/b_n) = q + 1.$$

Since $$(p+1)/(p-1) > 1,$$
these results agree with the extended form of Theorem II.

Ex. 5. If $$a_n = 3n + (-1)^n,\ b_n = 3n + (-1)^{n+1},$$
we see that $$a_{n+1} - a_n = 3 + 2(-1)^{n+1},\ b_{n+1} - b_n = 3 + 2(-1)^n.$$

Thus $(a_{n+1} - a_n)/(b_{n+1} - b_n)$ oscillates between $\frac{1}{5}$ and 5, although
$$\lim (a_n/b_n) = 1.$$

Again, if $$a_n = (n+1)^2 + (-1)^n n,\ b_n = (n+1)^2 + (-1)^{n+1} n,$$
we find
$$a_{n+1} - a_n = 2n + 3 + (-1)^{n+1}(2n+1),\ b_{n+1} - b_n = 2n + 3 + (-1)^n(2n+1),$$

so that $(a_{n+1}-a_n)/(b_{n+1}-b_n)$ is alternately $2(n+1)$ and $1/2(n+1)$; and so

$$\underline{\lim}\frac{a_{n+1}-a_n}{b_{n+1}-b_n}=0, \quad \overline{\lim}\frac{a_{n+1}-a_n}{b_{n+1}-b_n}=\infty.$$

But $$\lim(a_n/b_n)=1.$$

Thus, even when $(a_{n+1}-a_n)/(b_{n+1}-b_n)$ oscillates infinitely, a_n/b_n may have a definite limit.

Ex. 6. If b_n does not **steadily** increase, the Theorem II. is not necessarily true.

Thus take $$a_n=n+1, \quad b_n=[2+(-1)^n]n,$$
so that $$a_{n+1}-a_n=1, \quad b_{n+1}-b_n=2+(-1)^{n+1}(2n+1).$$

Consequently $$\lim\frac{a_{n+1}-a_n}{b_{n+1}-b_n}=0,$$
but yet $$\underline{\lim}(a_n/b_n)=\tfrac{1}{3}, \quad \overline{\lim}(a_n/b_n)=1.$$

In the same way L'Hospital's theorem may fail when $\psi'(x)$ changes sign infinitely often.

Thus consider $$\phi(x)=x+1+\sin x\cos x,$$
$$\psi(x)=e^{\sin x}(x+\sin x\cos x),$$
for which we find that $\phi'(x)/\psi'(x)\to 0$, while $\phi(x)/\psi(x)$ oscillates between $1/e$ and e, as $x\to\infty$.

Ex. 7. Consider the case,
$$\phi(x)=x+a\sin x, \quad \psi(x)=x, \quad (a>0),$$
and prove that $$\lim_{x\to\infty}\phi(x)/\psi(x)=1,$$
while $$\underline{\lim}\,\phi'(x)/\psi'(x)=1-a, \quad \overline{\lim}\,\phi'(x)/\psi'(x)=1+a.$$

153. An extension of Abel's Lemma.

To determine limits for the fraction

$$X_n=\frac{b_0v_0+b_1v_1+\ldots+b_nv_n}{a_0v_0+a_1v_1+\ldots+a_nv_n},$$

where a_r, v_r are positive and the sequence (v_r) is monotonic.

Write $A_0=a_0, \; A_1=a_0+a_1, \; \ldots, \; A_n=a_0+a_1+\ldots+a_n,$

and $B_0=b_0, \; B_1=b_0+b_1, \; \ldots, \; B_n=b_0+b_1+\ldots+b_n.$

Then

$$X_n=\frac{B_0(v_0-v_1)+B_1(v_1-v_2)+\ldots+B_{n-1}(v_{n-1}-v_n)+B_nv_n}{A_0(v_0-v_1)+A_1(v_1-v_2)+\ldots+A_{n-1}(v_{n-1}-v_n)+A_nv_n}.$$

If the sequence (v_n) steadily decreases, since v_r, A_r are all positive, we can obtain an upper limit to X_n by writing

H_mA_r in place of $B_r \quad (r=m, \; m+1, \; \ldots, \; n)$

and HA_r in place of $B_r, \quad (r=0, \; 1, \; \ldots, \; m-1),$

where H, H_m are the upper limits of

$$\left(\frac{B_0}{A_0}, \frac{B_1}{A_1}, \dots, \frac{B_n}{A_n}\right), \text{ and of } \left(\frac{B_m}{A_m}, \frac{B_{m+1}}{A_{m+1}}, \dots, \frac{B_n}{A_n}\right),$$

respectively. This follows because the factors (v_0-v_1), (v_1-v_2), $\dots$, $(v_{n-1}-v_n)$ and v_n are all positive.

Thus

$$X_n < H_m + (H-H_m)\frac{A_0(v_0-v_1)+A_1(v_1-v_2)+\dots+A_{m-1}(v_{m-1}-v_m)}{A_0(v_0-v_1)+A_1(v_1-v_2)+\dots+A_nv_n},$$

and, if we replace A_r by its value in terms of $a_0+a_1+\dots+a_r$, we obtain,* since $H \geqq H_m$ by definition,

$$X_n < H_m + (H-H_m)\frac{a_0v_0+a_1v_1+\dots+a_mv_m}{a_0v_0+a_1v_1+\dots+a_nv_n}.$$

In like manner we prove that

$$X_n > h_m - (h_m-h)\frac{a_0v_0+a_1v_1+\dots+a_mv_m}{a_0v_0+a_1v_1+\dots+a_nv_n},$$

where h, h_m are the corresponding lower limits of B_r/A_r.

Secondly, suppose that the sequence (v_n) *steadily increases.* In the numerator of X_n the factors (v_0-v_1), (v_1-v_2), $\dots$, $(v_{n-1}-v_n)$ are all negative, while v_n is positive, so that the value of the numerator is increased by writing

$$hA_r \text{ in place of } B_r, \quad (r=0, 1, \dots, m-1)$$

and $\quad h_mA_r$ in place of B_r, $\quad (r=m, m+1, \dots, n-1)$,

while in the last term we must put H_mA_n in place of B_n. This changes the numerator to

$$\begin{aligned} h[A_0(v_0-v_1)+\dots+A_{m-1}(v_{m-1}-v_m)]+h_m[A_m(v_m-v_{m+1})+\dots \\ +A_{n-1}(v_{n-1}-v_n)]+H_mA_nv_n \\ =h_m(a_0v_0+\dots+a_nv_n)+(H_m-h_m)A_nv_n \\ +(h_m-h)(A_mv_m-a_0v_0-\dots-a_mv_m), \end{aligned}$$

and, since $h_m \geqq h$, this again will not be decreased by omitting the negative terms in the last bracket. Hence, since the denominator is positive,

$$X_n < h_m + \frac{(H_m-h_m)A_nv_n+(h_m-h)A_mv_m}{a_0v_0+a_1v_1+\dots+a_nv_n}.$$

---

*The numerator is actually

$$a_0v_0+a_1v_1+\dots+a_mv_m-(a_0+a_1+\dots+a_m)v_m$$

which is less than the value given above.

Similarly, we find

$$X_n > H_m - \frac{(H_m - h_m)A_n v_n + (H - H_m)A_m v_m}{a_0 v_0 + a_1 v_1 + \ldots + a_n v_n}.$$

Thirdly, if the sequence (v_n) *first increases and afterwards decreases,** we may suppose that the term v_p is the greatest in the sequence, and let m be less than p. Here the factors $(v_0 - v_1)$, ..., $(v_{p-1} - v_p)$ are negative, while $(v_p - v_{p+1})$, ..., $(v_{n-1} - v_n)$, v_n are positive. Thus the numerator of X_n is not greater than

$$\begin{aligned} h[A_0(v_0 - v_1) + \ldots + A_{m-1}(v_{m-1} - v_m)] \\ + h_m[A_m(v_m - v_{m+1}) + \ldots + A_{p-1}(v_{p-1} - v_p)] \\ + H_m[A_p(v_p - v_{p+1}) + \ldots + A_{n-1}(v_{n-1} - v_n) + A_n v_n] \\ = H_m(a_0 v_0 + \ldots + a_n v_n) \\ + (H_m - h_m)(A_p v_p - a_0 v_0 - \ldots - a_p v_p) \\ + (h_m - h)(A_m v_m - a_0 v_0 - \ldots - a_m v_m). \end{aligned}$$

Hence we deduce, by an argument similar to the last,

$$X_n < H_m + \frac{(h_m - h)A_m v_m + (H_m - h_m)A_p v_p}{a_0 v_0 + a_1 v_1 + \ldots + a_n v_n}.$$

Similarly, we find

$$X_n > h_m - \frac{(H - H_m)A_m v_m + (H_m - h_m)A_p v_p}{a_0 v_0 + a_1 v_1 + \ldots + a_n v_n}.$$

Ex. Prove that if $a_n \to a$ and Σb_n is divergent, *although* b_n *need not be positive,*

$$\lim \frac{a_0 b_0 + a_1 b_1 + \ldots + a_n b_n}{b_0 + b_1 + \ldots + b_n} = a,$$

provided that

$$\frac{|b_0| + |b_1| + \ldots + |b_n|}{b_0 + b_1 + \ldots + b_n} < K.$$

[JENSEN.]

154. Other theorems on limits.

It follows at once from Theorem II. of Art. 152, that if $(s_{n+1} - s_n)$ has a definite limit, s_n/n tends to the same limit. Thus by writing $s_n = a_1 + a_2 + \ldots + a_n$, we see that *if a sequence* (a_n) *has a limit, this limit is also equal to*

$$\lim \frac{1}{n}(a_1 + a_2 + \ldots + a_n).$$

---

* In this case the sequence (v_n) is not, strictly speaking, monotonic; but it saves trouble to examine it here.

Of course the second limit may exist, when the first does not; thus with $a_{2n-1}=1$, $a_{2n}=0$, the second limit becomes $\frac{1}{2}$, although the sequence (a_n) oscillates.

THEOREM III. *If all the terms of a sequence (a_n) are positive, and if $\lim (a_{n+1}/a_n)$ is definite, so also is $\lim a_n^{\frac{1}{n}}$; and the two limits are the same.* [CAUCHY.]

For, if $\lim (a_{n+1}/a_n)$ is finite and equal to l (not zero), we can write $a_n = c_n l^n$, so that

$$\lim (c_{n+1}/c_n) = 1.$$

Thus we can find m, so that

$$1-\epsilon < c_{n+1}/c_n < 1+\epsilon, \quad \text{if } n \geqq m.$$

By multiplication, we obtain

$$(1-\epsilon)^{n-m} < c_n/c_m < (1+\epsilon)^{n-m},$$

so that
$$c_m(1-\epsilon)^n < \quad c_n \quad < c_m(1+\epsilon)^n.$$

Hence
$$c_m^{\frac{1}{n}}(1-\epsilon) < \quad c_n^{\frac{1}{n}} \quad < c_m^{\frac{1}{n}}(1+\epsilon).$$

Now
$$\lim_{n\to\infty} c_m^{\frac{1}{n}} = 1 \quad \text{(see Ex. 4, p. 17)},$$

so that
$$1-\epsilon \leqq \underline{\lim}\, c_n^{\frac{1}{n}} \leqq \overline{\lim}\, c_n^{\frac{1}{n}} \leqq 1+\epsilon.$$

Now these extreme limits are independent of m and therefore of ϵ; and ϵ is arbitrarily small, so that the inequalities can only be true if each of these limits is equal to 1, or if

$$\lim c_n^{\frac{1}{n}} = 1.$$

Thus
$$\lim a_n^{\frac{1}{n}} = l = \lim (a_{n+1}/a_n).$$

But if $\lim (a_{n+1}/a_n) = \infty$, we can find m, so that

$$a_{n+1}/a_n > N, \quad \text{if } n \geqq m,$$

however great N may be.

Hence, as above,
$$a_n/a_m > N^{n-m}$$

or
$$a_n^{\frac{1}{n}} > N[a_m/N^m]^{\frac{1}{n}}.$$

But
$$\lim_{n\to\infty} [a_m/N^m]^{\frac{1}{n}} = 1 \qquad \text{(Ex. 4, p. 17)},$$

so that
$$\underline{\lim}\, a_n^{\frac{1}{n}} \geqq N.$$

This minimum limit is independent of m and therefore of N; and N is arbitrarily great, so that we must have

$$\lim a_n^{\frac{1}{n}} = \infty.$$

The case when $\lim(a_{n+1}/a_n)=0$ can be reduced to the last by writing $a_n=1/b_n$, because a_n is positive.

It is not difficult to extend the previous argument to prove that in general

$$\underline{\lim}\,(a_{n+1}/a_n)\leqq\underline{\lim}\,a_n^{\frac{1}{n}}\leqq\overline{\lim}\,(a_{n+1}/a_n)\,;$$

and also that

$$\underline{\lim}\,(a_{n+1}/a_n)^{\frac{1}{c_n}}\leqq\underline{\lim}\,a_n^{\frac{1}{b_n}}\leqq\overline{\lim}\,(a_{n+1}/a_n)^{\frac{1}{c_n}},$$

if $c_n=b_{n+1}-b_n$ and b_n steadily increases to ∞.

Ex. 1. To find $$\lim\frac{1}{n}(n!)^{\frac{1}{n}},$$

we write $$a_n=(n!)/n^n,$$

so that $$a_{n+1}/a_n=n^n/(n+1)^n=(1+1/n)^{-n}.$$

Thus $$\lim\frac{1}{n}(n!)^{\frac{1}{n}}=\lim a_n^{\frac{1}{n}}=\lim\frac{a_{n+1}}{a_n}=\frac{1}{e},$$

by Art. 158. This result can be verified at once by reference to Stirling's formula for $n!$ (see Art. 179).

Ex. 2. To find $\lim\frac{1}{n}[(m+1)(m+2)\dots(m+n)]^{\frac{1}{n}}$, where m is fixed,

we write $$a_n=(m+1)(m+2)\dots(m+n)/n^n,$$

and then $$\frac{a_{n+1}}{a_n}=\frac{n+m+1}{n+1}\left(1+\frac{1}{n}\right)^{-n};$$

so that the limit again is $1/e$.

Ex. 3. Similarly we find that

$$\lim\frac{1}{n}[(n+1)(n+2)\dots 2n]^{\frac{1}{n}}=\frac{4}{e},$$

because $$\frac{a_{n+1}}{a_n}=\frac{2(2n+1)}{n+1}\left(1+\frac{1}{n}\right)^{-n}.$$

THEOREM IV. *If the sequences* (a_n), (b_n) *converge to the limits* a, b, *then*

$$\lim\frac{1}{n}(a_1b_n+a_2b_{n-1}+\dots+a_nb_1)=ab. \qquad \text{[CESÀRO.]}$$

For, let us write $a_n=a+p_n$, $b_n=b+q_n$, so that $p_n\to 0$, $q_n\to 0$. Then the given expression takes the form

$$ab+\frac{a}{n}(q_1+\dots+q_n)+\frac{b}{n}(p_1+\dots+p_n)+\frac{1}{n}(p_1q_n+\dots+p_nq_1).$$

Now, by the result at the beginning of this article, we see that

$$\lim\frac{1}{n}(p_1+p_2+\dots+p_n)=0,\quad \lim\frac{1}{n}(q_1+q_2+\dots+q_n)=0,$$

and consequently we have only to prove that

$$\lim \frac{1}{n}(p_1 q_n + p_2 q_{n-1} + \ldots + p_n q_1) = 0.$$

Now, since p_n, q_n both tend to 0, we can find a constant A which is greater than $|p_r|$ and $|q_r|$; and further, we can find μ so that

$$|p_r| < \epsilon/A, \quad |q_r| < \epsilon/A, \quad \text{if } r > \mu.$$

Then, if $n > 2\mu$, we have

$$|p_r q_{n-r+1}| < A\,|q_{n-r+1}| < \epsilon, \quad \text{if } r \leqq \tfrac{1}{2}n,\ n - r \geqq \tfrac{1}{2}n > \mu,$$

and $$|p_r q_{n-r+1}| < A\,|p_r| < \epsilon, \quad \text{if } r \geqq \tfrac{1}{2}n > \mu.$$

Consequently

$$\frac{1}{n}|(p_1 q_n + p_2 q_{n-1} + \ldots + p_n q_1)| < \epsilon, \quad \text{if } n > 2\mu,$$

and so tends to zero, as n tends to ∞; thus the theorem is established.

155. THEOREM V. *If Σb_n, Σc_n are two divergent series of positive terms, then*

$$\lim \frac{c_0 s_0 + c_1 s_1 + \ldots + c_n s_n}{c_0 + c_1 + \ldots + c_n} = \lim \frac{b_0 s_0 + b_1 s_1 + \ldots + b_n s_n}{b_0 + b_1 + \ldots + b_n},$$

provided that the second limit exists, and that either (1) c_n/b_n *steadily decreases, or* (2) c_n/b_n *steadily increases subject to the condition**

$$\frac{c_n}{c_0 + c_1 + \ldots + c_n} < K \frac{b_n}{b_0 + b_1 + \ldots + b_n},$$

where K is fixed.

Let us write for brevity

$$B_n = b_0 + b_1 + \ldots + b_n, \qquad C_n = c_0 + c_1 + \ldots + c_n,$$
$$P_n = b_0 s_0 + b_1 s_1 + \ldots + b_n s_n, \quad Q_n = c_0 s_0 + c_1 s_1 + \ldots + c_n s_n.$$

Let us also write $c_n/b_n = v_n$; then

$$\frac{Q_n}{C_n} = \frac{(b_0 s_0) v_0 + \ldots + (b_n s_n) v_n}{b_0 v_0 + \ldots + b_n v_n}.$$

* The second part of the theorem is due to Hardy, *Quarterly Journal*, vol. 38, 1907, p. 269; the first is given by Cesàro, *Bulletin des Sciences Mathématiques* (2), t. 13, 1889, p. 51. Of course $K > 1$ in case (2), in virtue of the fact that c_n/b_n increases.

To this fraction we can apply the result of Art. 153 above, and we find, in the first case, when (v_n) *decreases*,

$$(1) \qquad h_m-(h_m-h)\frac{C_m}{C_n}<\frac{Q_n}{C_n}<H_m+(H-H_m)\frac{C_m}{C_n},$$

where H, h are the upper and lower limits of P_0/B_0, P_1/B_1, ..., to ∞, and H_m, h_m are those of P_m/B_m, P_{m+1}/B_{m+1}, ..., to ∞.

If $\lim P_n/B_n=l$, we can find m so that $h_m\geqq l-\epsilon$, $H_m\leqq l+\epsilon$, and so we have from (1)

$$(l-\epsilon)-(l-h)\frac{C_m}{C_n}<\frac{Q_n}{C_n}<(l+\epsilon)+(H-l)\frac{C_m}{C_n}.$$

Now $C_n\to\infty$, so that if we take the limit of the last inequality, we find

$$(2) \qquad l-\epsilon\leqq\underline{\lim}\,\frac{Q_n}{C_n}\leqq\overline{\lim}\,\frac{Q_n}{C_n}\leqq l+\epsilon.$$

Hence, since ϵ is arbitrarily small, each of the extreme limits in (2) must be equal to l; or

$$\lim(Q_n/C_n)=l.$$

In like manner, if $P_n/B_n\to\infty$, we can find m so that $h_m\geqq N$, and then

$$Q_n/C_n>N-(N-h)C_m/C_n.$$

Since $C_n\to\infty$, we find from the last inequality that

$$\underline{\lim}\,(Q_n/C_n)\geqq N,$$

and so

$$\lim(Q_n/C_n)=\infty.$$

In the second case, when c_n/b_n *increases*, we have

$$H_m-(H_m-h_m)\frac{B_nc_n}{C_nb_n}-(H-H_m)\frac{B_mc_m}{C_nb_m}$$
$$<\frac{Q_n}{C_n}<h_m+(H_m-h_m)\frac{B_nc_n}{C_nb_n}+(h_m-h)\frac{B_mc_m}{C_nb_m}.$$

Now, by hypothesis, $\dfrac{c_n}{C_n}<K\dfrac{b_n}{B_n}$,

where K is a constant greater than unity. Hence

$$(3) \qquad H_m-K(H_m-h_m)-K(H-H_m)\frac{C_m}{C_n}$$
$$<\frac{Q_n}{C_n}<h_m+K(H_m-h_m)+K(h_m-h)\frac{C_m}{C_n};$$

then, just as (2) is deduced from (1), we find from (3) that

$$l-(2K-1)\epsilon\leqq\underline{\overline{\lim}}\,\frac{Q_n}{C_n}\leqq l+(2K-1)\epsilon,$$

from which the same result follows as before.

It is instructive to note that *in the first case the series* Σc_n *diverges more slowly than* Σb_n, *while in the second case* Σc_n *diverges more rapidly than* Σb_n, *but the final condition excludes series which diverge too fast.*

It should be noticed that if s_n tends to a definite limit, Theorem V. is an immediate corollary from Theorem II. of Art. 152; for then both fractions have the same limit as s_n.

The applications of most interest arise when

$$b_0 = b_1 = \dots = b_n = 1,$$

and then we have the result:

If Σc_n *is a divergent series of positive terms, then*

$$\lim \frac{c_0 s_0 + c_1 s_1 + \dots + c_n s_n}{c_0 + c_1 + \dots + c_n} = \lim \frac{s_0 + s_1 + \dots + s_n}{n+1},$$

provided that the second limit exists and either (1) *that* c_n *steadily decreases, or* (2) *that* c_n *steadily increases, subject to the restriction*

$$nc_n < K(c_0 + c_1 + \dots + c_n),$$

where K is a fixed number.

Ex. 1. A specially interesting application arises from applying the theorem of Frobenius (Art. 51) to the series

$$a_0 + a_1 x^{c_0} + a_2 x^{c_0+c_1} + a_3 x^{c_0+c_1+c_2} + \dots,$$

where $c_0, c_1, c_2, \dots$ form an increasing sequence of positive integers, satisfying the condition just given.

Here it is evident that the series must be written in the form

$$a_0 + (0)x + (0)x^2 + \dots + a_1 x^{c_0} + (0)x^{c_0+1} + \dots + a_2 x^{c_0+c_1} + \dots,$$

so that
$$A_0 = a_0,\quad A_1 = a_0,\quad \dots,\quad A_{c_0-1} = a_0,$$
$$A_{c_0} = a_0 + a_1,\quad A_{c_0+1} = a_0 + a_1,\quad \dots,\ \text{and so on.}$$

Generally we have $A_\nu = a_0 + a_1 + \dots + a_n$,

if
$$c_0 + c_1 + \dots + c_{n-1} \leqq \nu < c_0 + c_1 + \dots + c_n.$$

Thus, if $s_n = a_0 + a_1 + \dots + a_n$, we have

$$A_0 + A_1 + \dots + A_\nu = s_0 c_0 + s_1 c_1 + \dots + s_{n-1} c_{n-1} + s_n(\nu - c_0 - c_1 - \dots - c_{n-1}),$$

and therefore Frobenius's mean, if it exists, is given by

$$\lim \frac{s_0 c_0 + s_1 c_1 + \dots + s_n c_n}{c_0 + c_1 + \dots + c_n},$$

which we have proved to be the same as

$$\lim (s_0 + s_1 + \dots + s_n)/(n+1),$$

provided that the last limit exists.

Ex. 2. Interesting special cases of Ex. 1 are given by taking

$$c_0 + c_1 + \dots + c_n = (n+1)^2,\quad (n+1)^3,\quad \text{etc.},$$

for which K may be taken as 2, 3 respectively.

Thus we have the results

$$\lim_{x\to 1}(a_0+a_1x+a_2x^4+a_3x^9+\dots)=\lim_{n\to\infty}\frac{s_0+s_1+\dots+s_n}{n+1}$$

and $$\lim_{x\to 1}(a_0+a_1x+a_2x^8+a_3x^{27}+\dots)=\lim_{n\to\infty}\frac{s_0+s_1+\dots+s_n}{n+1}.$$

Ex. 3. But if we write* $c_0+c_1+\dots+c_n=2^{n+1}$, we have $c_0=2$, $c_n=2^n$, and so

$$nc_n/(c_0+c_1+\dots+c_n)=\tfrac{1}{2}n.$$

That is, our condition is broken, so that *we have no right to anticipate the existence of the limit,*

$$\lim_{x\to 1}(a_0+a_1x^2+a_2x^4+a_3x^8+a_4x^{16}+\dots)$$

when the limit of $(s_0+s_1+\dots+s_n)/(n+1)$ *exists*; and as a matter of fact the particular series $1-x^2+x^4-x^8+x^{16}-\dots$ can be proved to oscillate as x tends to 1 (compare Ex. 30, p. 489). [HARDY.]

Ex. 4. We can use this theorem to establish Cesàro's theorem of Art. 22, by taking $s_n=\pm 1$. Then, in the notation of that article,

$$s_1+s_2+\dots+s_n=p_n-q_n$$

and $$\lim\left[\left(\sum_1^n c_r s_r\right)\Big/\left(\sum_1^n c_r\right)\right]=0,$$

because $\sum_1^\infty c_r s_r$ converges, while $\sum_1^\infty c_r$ diverges.

Thus $(p_n-q_n)/n$ cannot approach any limit other than zero. [CESÀRO.]

Ex. 5. Write $$B_n\left(\frac{c_n}{b_n}-\frac{c_{n+1}}{b_{n+1}}\right)=f_n$$

and $$F_n=f_0+f_1+\dots+f_n=C_n-\frac{c_{n+1}}{b_{n+1}}B_n.$$

Also let $\lambda_n=P_n/B_n$; then prove that

$$Q_n=\sum_0^n\lambda_r f_r+\lambda_n(C_n-F_n)$$

or $$\frac{Q_n}{C_n}=\lambda_n+\frac{F_n}{C_n}\left(\frac{\Sigma\lambda_r f_r}{F_n}-\lambda_n\right).$$

Prove that, in case (1) of the theorem,† $F_n\to\infty$, but that $F_n/C_n<1$; and by applying Art. 152, II. deduce that

$$(Q_n/C_n-\lambda_n)\to 0.$$ [CESÀRO.]

---

* It is perhaps worth while to call attention to the fact that the sequence

2, 4, 8, 16, 32, ...

does actually increase faster than

1, 8, 27, 64, 125 ....

In fact the 10th and 11th terms are 1024, 2048 in the first sequence, and 1000, 1331 in the second.

† This is the only point in the proof at which care is necessary.

EXAMPLES.

Irrational Numbers.

1. (1) If A is a rational number lying between a^2 and $(a+1)^2$, prove that

$$a+\frac{A-a^2}{2a+1}<\sqrt{A}<a+\frac{A-a^2+\frac{1}{4}}{2a+1}.$$

(2) If $$(a+\sqrt{A})^n=p_n+q_n\sqrt{A},$$
where a, p_n, q_n are rational numbers, prove that

$$p_{n+1}=ap_n+Aq_n, \quad q_{n+1}=aq_n+p_n,$$

and that $$p_n^2-Aq_n^2=(a^2-A)^n.$$

Thus if a is an approximation to $\sqrt{A}$, p_n/q_n is a closer approximation.

[The approximation p_3/q_3 is the same as that used by Dedekind (see Art. 145).]

2. (1) If a, b, x, y are rational numbers such that

$$(bx-ay)^2+4(x-a)(y-b)=0,$$

prove that either (i) $x=a$ and $y=b$, or (ii) $\sqrt{(1-ab)}$ and $\sqrt{(1-xy)}$ are rational numbers. [*Math. Trip.* 1903.]

(2) If the equations in x, y,

$$ax^2+2bxy+cy^2=1=lx^2+2mxy+ny^2,$$

have only rational solutions, then

$$\sqrt{[(b-m)^2-(a-l)(c-n)]} \text{ and } \sqrt{[(an-cl)^2+4(am-bl)(cm-bn)]}$$

are both rational. [*Math. Trip.* 1899.]

3. If α is irrational and a, b, c, d are rational (but such that ad is not equal to bc), then

$$a\alpha+b \text{ and } (a\alpha+b)/(c\alpha+d)$$

are irrational numbers, except when $a=0$ in the former.

4. Any irrational number α can be expressed in the form

$$\alpha=c_0+\frac{c_1}{a}+\frac{c_2}{a^2}+\frac{c_3}{a^3}+\dots,$$

where a is any assigned positive integer and c_1, c_2, c_3, ... are positive integers less than a. Thus, in the scale of notation to base a, we may write α as a decimal

$$c_0 \cdot c_1c_2c_3\dots.$$

For example, with $a=2$, that is, in the binary scale, we find

$$\sqrt{2}=1\cdot0110101000001\dots.$$

5. If a_1, a_2, a_3, ... is an infinite sequence of positive integers such that n can be found to make $(a_1a_2a_3\dots a_n)$ divisible by N, whatever the integer N may be, then any number α can be expressed in the form

$$\alpha=c_0+\frac{c_1}{a_1}+\frac{c_2}{a_1a_2}+\frac{c_3}{a_1a_2a_3}+\dots, \quad 0\leqq c_n<a_n.$$

In order that a may be rational, c_n must be equal to $a_n - 1$ after a certain value of n. [CANTOR.]

For instance, $\sqrt{2} = 1 + \frac{0}{2!} + \frac{2}{3!} + \frac{1}{4!} + \frac{4}{5!} + \frac{4+\theta}{6!}$, $0 < \theta < \frac{1}{4}$.

The restriction that $(a_1 a_2 \dots a_n)$ must be divisible by N is essential; thus, if $c_n = n$, $a_n = 2n+1$, we find

$$\frac{1}{3} + \frac{2}{3 \cdot 5} + \frac{3}{3 \cdot 5 \cdot 7} + \dots \text{ to } n \text{ terms} = \frac{1}{2}\left[1 - \frac{1}{3 \cdot 5 \cdot 7 \dots (2n+1)}\right],$$

and so the sum to infinity is $\frac{1}{2}$, which is *rational*, although c_n is not equal to $a_n - 1$.

6. If we can determine a divergent sequence of integers (q_n) such that

$$\lim (p_n - q_n a) = 0,$$

where p_n is the integer nearest to $q_n a$, then a must be irrational. Apply this (a special case of Ex. 8) to the series in Ex. 7.

Establish also the converse theorem, and deduce that when a is irrational we can find an integer N such that $Na - M$ is as near to any assigned number β $(0 < \beta < 1)$ as we please, where M is the integral part of Na.

[For the first part, note that if a were equal to r/s,

$$|p_n - q_n a| \geqq 1/s.$$

For the second part, express a as a continued fraction.]

7. The sums of the series

$$\sum_1^\infty \left(\frac{1}{p}\right)^{n^2}, \quad \sum_1^\infty \left(-\frac{1}{p}\right)^{n^2}, \quad \sum_1^\infty \frac{q^n}{p^{n^2}},$$

where p, q are any positive integers, are irrational numbers. The same holds for the product $\Pi(1 - p^{-n})$. [EISENSTEIN.]

[For simple proofs and extensions, see Glaisher, *Phil. Mag.* (4), vol. 45, 1873, p. 191.]

8. If a is the root of an algebraic equation of degree k (with integral coefficients), we can find a constant K such that

$$\left|\frac{p}{q} - a\right| > \frac{1}{Kq^{k-1}},$$

where p, q are any positive integers. Thus if we can find a divergent sequence of integers (q_n) such that

$$|p_n - q_n a| < q_n^{-k},$$

where p_n is the nearest integer to aq_n, *then a is not an algebraic number of degree k.*

Consequently, if $\quad a = c_0 + \frac{c_1}{10} + \frac{c_2}{10^{1 \cdot 2}} + \dots + \frac{c_n}{10^{n!}} + \dots,$

where $c_1, c_2, c_3, \dots$ are less than 10, by taking the sequence $q_n = 10^{n!}$, we can prove that *a is transcendental.* [LIOUVILLE.]

9. Suppose that a is an irrational number which is converted into a continued fraction

$$\frac{1}{a_0+}\,\frac{1}{a_1+}\,\frac{1}{a_2+\ldots},$$

and p_n/q_n is the convergent which precedes the quotient a_n; write further

$$Q_{n+1}=q_nA_n+q_{n-1},$$

where
$$A_n=a_n+\frac{1}{a_{n+1}}+\frac{1}{a_{n+2}}+\ldots.$$

Then shew that
$$|\sin ma\pi|>K/Q_{n+1},$$

if
$$q_n<m<q_{n+1};$$

and also that
$$|\sin q_n a\pi|=\sin(\pi/Q_{n+1})=(1+\epsilon_n)\pi/Q_{n+1},$$

where ϵ_n tends to zero as n tends to ∞.

[Hardy, *Proc. Lond. Math. Soc.* (2), vol. 3, p. 444.]

Monotonic Sequences.

10. (1) If in a sequence (a_n) each term lies between the two preceding terms, shew that it is compounded of two monotonic sequences.

(2) If a sequence of positive numbers (a_n) is monotonic, prove that the sequence (b_n) of its geometric means is also monotonic, where

$$b_n^n=a_1a_2\ldots a_n.$$

(3) If $c_1, c_2, \ldots, c_p$ are real positive numbers, and if

$$\mu_n=(c_1^n+c_2^n+\ldots+c_p^n)/p,$$

prove that the sequence (μ_{n+1}/μ_n) steadily increases; and deduce that the same is true of $\mu_n^{\frac{1}{n}}$.

11. If
$$S_n=\sum_{r=1}^{n}\left[1-\left(\frac{r}{n}\right)^k\right]a_r,$$

where a_r is positive and independent of n, shew that if Σa_r is convergent, its sum gives the value of $\lim S_n$ (see Art. 49).

Conversely, if $\lim S_n$ exists, shew that Σa_r converges, and that its sum is equal to the limit of S_n.

Apply to Ex. 12, taking $k=1$; and to Ex. 6, Ch. XI., taking $k=2$.

12. Apply Cauchy's theorem (Art. 152, II.) to prove that

$$\lim\frac{1}{n}\left[\frac{n}{1}+\frac{n-1}{2}+\frac{n-2}{3}+\ldots+\frac{1}{n}-\log(n!)\right]=C,$$

where C is Euler's constant (Art. 11).

Prove also that for all values of n, the expression lies between 0 and 1.

[*Math. Trip.* 1907.]

13. Apply Stolz's theorem (Art. 152, II.) to prove that if

$$\lim\left(a_{n+1}-a_n+\lambda\frac{a_n}{n}\right)=l,\quad\text{where }\lambda>-1,$$

then
$$\lim(a_{n+1}-a_n)=\lim\frac{a_n}{n}=\frac{l}{1+\lambda}.$$

[Mercer.]

[If $P_n=(1-\lambda)(1-\frac{1}{2}\lambda)(1-\frac{1}{3}\lambda)\dots(1-\lambda/n)$ and $c_n=a_n/P_{n-1}$, then the given expression can be written

$$(c_{n+1}-c_n)P_n=(1+\lambda)(c_{n+1}-c_n)/(b_{n+1}-b_n),$$

where $b_n=n/P_{n-1}$. Then b_n increases with n, if $1+\lambda$ is positive; and Art. 42 shews that $b_n \sim \Gamma(1-\lambda)n^{1+\lambda}$, which tends to infinity with n. Thus $\lim(c_n/b_n)=l/(1+\lambda)$, from which we get the given results.]

Quasi-monotonic Sequences.

Given a sequence (a_n), let ν be the greatest index for which

$$a_1,\ a_2,\ a_3,\ \dots,\ a_\nu$$

are *all* less than a_n; then ν is a function of n, say $\phi(n)$. As a general rule $\phi(n)$ is not monotonic and does not tend to infinity.

When $\phi(n)$ tends to infinity with n (not necessarily steadily) the sequence (a_n) is called quasi-monotonic (increasing).

It is easy to frame a corresponding definition for the *decreasing* case.

14. Prove that:

(1) A quasi-monotonic sequence (increasing) either diverges to ∞ or converges to a limit l which is greater than any term of the sequence.

(2) Conversely, if a sequence diverges to ∞ or converges to a limit l greater than any term of the sequence, then the sequence is quasi-monotonic and increasing.

(3) Any convergent sequence can be sub-divided into two quasi-monotonic sequences having a common limit.

Illustrations of quasi-monotonic sequences are given by

$$a_n=n+(-1)^n p,\quad a_n=n^2+(-1)^n np,\quad a_n=1-\frac{1}{n}\{2+(-1)^n\},$$

where, in the first two, p is a constant. [HARDY.]

Infinite Sets of Numbers.

15. For some purposes of analysis we need to use *infinite sets of numbers which cannot be arranged as a sequence*; when a set can be arranged as a sequence, it is often called *countable* or *enumerable.*

Prove that the set of all real numbers lying between 0 and 1 is not countable. [CANTOR.]

[For if $a_1, a_2, a_3, \dots$ is any sequence of such numbers, we can write each term as an infinite decimal

$$a_1=\cdot a_{1,1}a_{1,2}a_{1,3}\dots,\quad a_2=\cdot a_{2,1}a_{2,2}a_{2,3}\dots,\ \text{etc.}$$

Then consider the infinite decimal

$$\gamma=\cdot c_1c_2c_3\dots,$$

in which c_n is subject only to the condition of not being the same as $a_{n,n}$. It is clear that γ does not belong to the sequence (a_n); and so (a_n) cannot exhaust all the real numbers between 0 and 1.]

16. Given any infinite set of numbers (k) we can construct a Dedekind section by placing in the upper class all rational numbers greater than *any* number k, and in the lower class all rational numbers less than *some* number k.

This section defines *the upper limit of the set*; prove that this upper limit has the properties stated on p. 12 for the upper limit of a sequence. Frame also a corresponding definition for the lower limit of the set k; and define both upper and lower limits by using the method of continued bisection (as in Art. 150).

17. The *limiting values of an infinite set of numbers* consist of numbers λ such that an infinity of terms of the set fall between $\lambda-\epsilon$ and $\lambda+\epsilon$, however small ϵ may be.

Given an infinite set of numbers (k) we can construct a Dedekind section by placing in the upper class all rational numbers which are greater than all but a finite number of the terms k, and in the lower class all rational numbers less than an *infinite* number of terms k.

This section defines *the maximum limit of the set*; prove that the maximum limit is a limiting value of the set, in accordance with the definition given above; and further that no limiting value of the set can exceed the maximum limit (compare Art. 5, p. 13). Frame a corresponding definition for *the minimum limit* and state the analogous properties.

Goursat's Lemma.

18. Suppose that an interval has the property that round every point P of the interval we can mark off a sub-interval such that a certain inequality denoted by $\{Q, P\}$ is satisfied for every point Q of the sub-interval. Then we can divide the whole interval into a *finite* number of parts, such that each part contains at least one point (P') for which the inequality $\{Q', P'\}$ is satisfied at every point Q' of the part in which P' lies.

[Bisect the original interval; if either half does not satisfy the condition, bisect it again; and so on. If we continue the process, one of two alternatives must occur: either we shall obtain a set of sub-divisions which satisfy the condition, or else, however small the divisions may be, there is always at least one part which does not satisfy the condition.

In the former case, the lemma is proved; in the latter, we have an infinite sequence of intervals (a_n, b_n), each half the preceding, so that

$$b_{n+1}-a_{n+1}=\tfrac{1}{2}(b_n-a_n) \text{ and } a_n \leqq a_{n+1} < b_{n+1} \leqq b_n,$$

and the condition is not satisfied in any interval (a_n, b_n).

Now the sequences (a_n), (b_n) have a common limit l; mark off round l the interval (f, g) within which the inequality $\{Q, l\}$ holds. Then we have $a_n \leqq l \leqq b_n$, and we can choose n so that $b_n - a_n < l-f = g-l$.

Thus $$a_n > f+(b_n-l) \geqq f \text{ and } b_n < g+(l-a_n) \leqq g,$$

so that the interval (a_n, b_n) is contained within the interval (f, g); and therefore the inequality $\{Q, l\}$ is satisfied at all points of (a_n, b_n). That is, if we take l'' at l, the required condition *is* satisfied in $(a_n b_n)$; and so we are led to a contradiction by assuming that the lemma is not true.]*

19. Suppose that an interval has the property that round every point P of the interval we can mark off a sub-interval such that the inequality $\{Q, R\}$ is satisfied for any two points Q, R of the sub-interval. Then we can divide the whole interval into a *finite* number of parts, such that the inequality $\{Q', R'\}$ is satisfied for any two points which lie in the *same* part.

[Prove as in Ex. 18.]

20. State and prove Exs. 18, 19 in forms applicable to regions in a plane; or to volumes in space.

Continuous Functions.

21. If $f(x)$ is continuous in the interval (a, b), prove that it assumes, at least once in the interval,

(1) every value between $f(a)$ and $f(b)$,

(2) the upper and lower limits (H and h) of $f(x)$ in the interval.

[Apply the method of continued bisection.

In case (2) we get an infinite sequence of intervals (a_n, b_n) such that H is the upper limit of $f(x)$ in the interval (a_n, b_n); let (a_n), (b_n) tend to the common limit l. Then if $f(l) < H$, choose δ so that

$$f(x) - f(l) < \tfrac{1}{2}[H - f(l)], \quad \text{if } |x - l| < \delta.$$

Then choose n so that $b_n - a_n \leqq \delta$; and we find that $f(x) < \tfrac{1}{2}[H + f(l)]$ at all points of (a_n, b_n), contrary to hypothesis.]

22. Use Ex. 18 to prove that if $f(x)$ is continuous in an interval (a, b), then the interval can be divided into a finite number of parts (the number depending on ϵ) such that

$$|f(x_2) - f(x_1)| < \epsilon,$$

where x_1, x_2 are any two points in the same sub-division. [HEINE.]

23. A function is said to be *finite in an interval* if its absolute value has a finite upper limit in the interval.

Deduce from Ex. 22 that if $f(x)$ is continuous in an interval, it is also finite in the interval; and also that δ can be found so that

$$|f(\xi_2) - f(\xi_1)| < \epsilon,$$

where ξ_1, ξ_2 are *any* two points of the interval satisfying $|\xi_2 - \xi_1| < \delta$.

* The argument developed here was given first by Goursat in the special form suitable for his proof of Cauchy's theorem (a reproduction of this proof will be found in his *Cours d'Analyse*, t. 2, p. 85); the general statements of Exs. 18, 19 seem to have been given first by Dr. Baker.

APPENDIX II.

DEFINITIONS OF THE LOGARITHMIC AND EXPONENTIAL FUNCTIONS.

156. In the text it has been assumed that

$$\frac{d}{dx}(\log x) = \frac{1}{x},$$

and a number of allied properties of the logarithm have been used. It is customary in English books on the Calculus to deduce the differential coefficient of $\log x$ from the exponential limit (Art. 57) or else from the exponential series (Art. 58). It would, therefore, seem illogical to assume these properties of logarithms in the earlier part of the theory; although, no doubt, we could have obtained these limits quite at the beginning of the book. But from the point of view adopted it seemed more natural to place all special limits after the general theorems on convergence. It is, therefore, desirable to indicate an independent treatment of the logarithmic function; and it seems desirable to use this way of introducing the function in a first course on the Calculus.*

157. Definition of the logarithmic function.

There appears to be no real need for the logarithm at the beginning of the Differential Calculus, but we require it in the Integral Calculus as soon as fractions have to be integrated. At first it is probably best to denote $\int dx/x$ by $L(x)$, and postpone the discussion of the nature of the function $L(x)$ until

* See a paper by Bradshaw (*Annals of Mathematics* (2), vol. 4, 1903, p. 51) and Osgood's *Lehrbuch der Funktionentheorie*, Bd. 1, pp. 487-500.

after the definite integral has been introduced. We shall assume, for the present, the theorem that $\int_a^b y\,dx$ represents the area between a curve, the axis of x and the two ordinates $x=a$, $x=b$; an arithmetic treatment of the theorem will be given below (Art. 163).

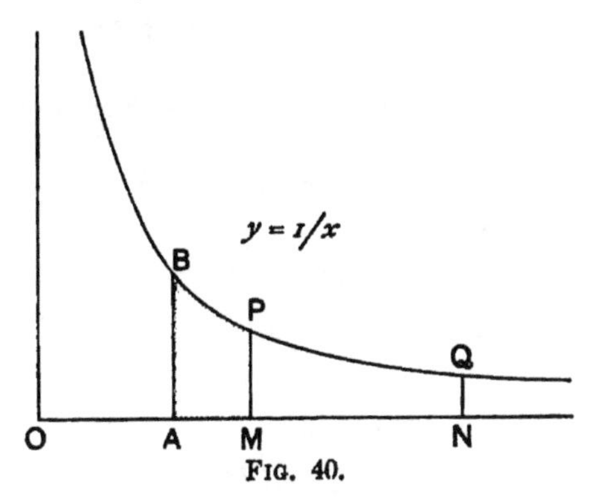

FIG. 40.

Let the rectangular hyperbola $y=1/x$ be drawn, then we shall denote by $L(x)$ the area $ABPM$ bounded by the curve, the fixed ordinate AB $(x=1)$, the axis of x and the variable ordinate PM; or, in the notation of the Calculus, we write

$$(1) \qquad L(x)=\int_1^x dx/x,$$

where, as will be evident from the figure, x is supposed *positive.*

It is obvious from the definition that

$$(2) \qquad L(1)=0.$$

Further, if parallels are drawn through B and P to the axis of x, we obtain two rectangles, one enclosing the area $ABPM$ and the other entirely within $ABPM$.

Thus we have

$$(3) \qquad x-1>L(x)>(x-1)/x,$$

or, with a slight change of notation,

$$(3a) \qquad x>L(1+x)>x/(1+x).$$

Although (3) has only been proved when $x>1$, yet it is easy to shew similarly that the inequalities (3) hold good *algebraically*, when $x<1$. But care must be taken to notice that when x is less than 1, in (3), or negative, in (3a), all the members of the inequalities are *negative*; thus, for the numerical values the inequalities would have to be reversed.

For instance, we get from (3),

$$-\tfrac{1}{2} > L(\tfrac{1}{2}) > -1.$$

But in numerical value $\tfrac{1}{2} < |L(\tfrac{1}{2})| < 1.$

Again, if we take an ordinate QN, such that $ON = 2 . OM$, we have

$$L(2x) - L(x) = PMNQ > \text{rect. } MN . NQ,$$

and

$$< \text{rect. } MN . MP.$$

That is, $x\left(\frac{1}{x}\right) > L(2x) - L(x) > x\left(\frac{1}{2x}\right)$, or

(4) $$1 > L(2x) - L(x) > \tfrac{1}{2}.$$

Thus, we get $1 > L(2) > \tfrac{1}{2}$, since $L(1) = 0$,

$$1 > L(4) - L(2) > \tfrac{1}{2},$$

$$1 > L(8) - L(4) > \tfrac{1}{2},$$

and so on.

It follows by addition that

$$n > L(2^n) > \tfrac{1}{2}n.$$

Now, if $x > x_0$, it is evident from the figure that

$$L(x) > L(x_0).$$

Hence $(n+1) > L(2^{n+1}) > L(x) > L(2^n) > \tfrac{1}{2}n,$

if $2^{n+1} > x > 2^n$, and it is evident that $L(x)$ *tends to infinity with* x (Art. 1, Note 2), or

(5) $$\lim_{x \to \infty} L(x) = \infty .$$

Again, if we write $x = 1/t$, we have

$$\frac{dx}{x} = -\frac{dt}{t},$$

so that $$L(x) = \int_1^x \frac{dx}{x} = -\int_1^t \frac{dt}{t} = -L(t), \text{ or}$$

(6) $$L(1/x) = -L(x).$$

Hence, as x approaches zero, since $1/x$ tends to infinity, $L(x)$ tends towards negative infinity, or

(7) $$\lim_{x \to 0} L(x) = -\infty .$$

Again, *the function* $L(x)$ *is continuous for all positive values of* x. For we see at once that $|L(x+h) - L(x)|$ lies between two rectangles, one of which is equal to $|h|/x$ and the other to $|h|/(x+h)$. Thus

$$|L(x+h) - L(x)| < \epsilon, \quad \text{if } |h| < \epsilon x/(1+\epsilon),$$

which proves the continuity of $L(x)$.

It follows from the fundamental relation between differentiation and integration that

$$(8) \qquad \frac{d}{dx}[L(x)]=\frac{1}{x};$$

but without appealing to this general fact we can obtain the result by noticing that $[L(x+h)-L(x)]/h$ is contained between $1/x$ and $1/(x+h)$. Thus

$$\frac{d}{dx}[L(x)]=\lim_{h\to 0}\frac{1}{h}[L(x+h)-L(x)]=\frac{1}{x}.$$

If $x=a$, b are two ordinates such that $b>a>1$, we have

$$L(b)-L(a)<(b-a)/a$$

by exactly the same argument as we used to establish (4). Further, from (3), we have

$$L(a)>(a-1)/a.$$

Hence
$$\frac{L(a)}{a-1}>\frac{1}{a}>\frac{L(b)-L(a)}{b-a},$$

or
$$(b-a)L(a)>(a-1)[L(b)-L(a)].$$

Thus
$$(b-1)L(a)>(a-1)L(b),$$

or
$$\frac{L(a)}{a-1}>\frac{L(b)}{b-1}, \text{ where } b>a>1.$$

Consequently, *the function $L(x)/(x-1)$ decreases as x increases*; which corresponds to the nearly obvious geometrical fact that *the mean ordinate between AB and PM decreases as x increases.*

As an example, the reader may prove the last result by differentiation.

The figure below will give a general idea of the course of the logarithmic function.

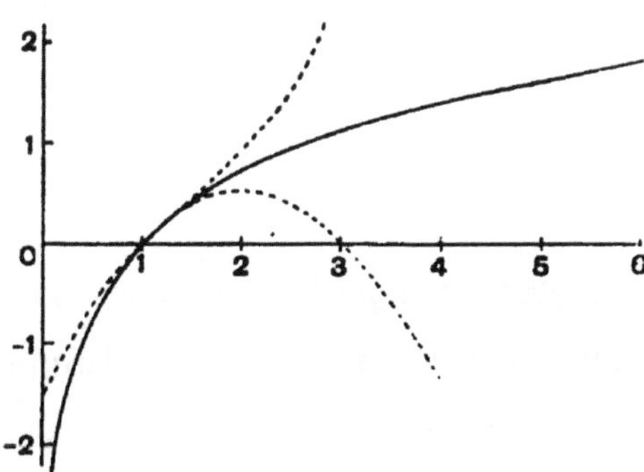

FIG. 41.

The dotted lines represent the curves

$$y=(x-1)-\tfrac{1}{2}(x-1)^2,$$
$$y=(x-1)-\tfrac{1}{2}(x-1)^2+\tfrac{1}{3}(x-1)^3.$$

158. Fundamental properties of the logarithmic function.

In the formula
$$L(u)=\int_1^u \frac{dx}{x},$$
change the independent variable from x to ξ by writing $x=\xi/v$; we find then
$$L(u)=\int_v^{uv}\frac{d\xi}{\xi}=\int_1^{uv}\frac{d\xi}{\xi}-\int_1^{v}\frac{d\xi}{\xi},$$
or, going back to the definition,
$$L(u)=L(uv)-L(v).$$

Thus

(1) $$L(uv)=L(u)+L(v).$$

From equation (1) it follows at once that

(2) $$L(u^n)=nL(u),$$

where n is any rational number.*

Suppose now that e is a number such that $L(e)=1$; the existence of e follows from the fact that $L(x)$ is a continuous function, which steadily increases from $-\infty$ to $+\infty$ as x varies from 0 to ∞ (see Art. 149).

Then equation (2) gives, for rational values of n,

(3) $$L(e^n)=n,$$

which proves that $L(x)$ must agree with the logarithm to base e, as ordinarily defined; we shall, therefore, write for the future $\log x$ in place of $L(x)$.

We can obtain an approximation to the value of e by observing that, when n is positive, (3) of Art. 157 gives, on writing $x=1+1/n$,
$$\frac{1}{n}>\log\left(1+\frac{1}{n}\right)>\frac{1}{n+1}, \text{ or}$$

(4) $$1>\log\left(1+\frac{1}{n}\right)^n>\frac{n}{n+1}.$$

Thus, as n increases, $\log\left(1+\frac{1}{n}\right)^n$ tends to 1 as its limit; and so, since the logarithmic function is continuous and monotonic, $\left(1+\frac{1}{n}\right)^n$ must tend to e.

* Equation (2) may be used (see Bradshaw's paper, §4) to establish the existence of roots which are not evident on geometrical grounds; for example, the fifth root. Of course, from the point of view adopted in this book, it is more natural to establish the existence of such roots by using Dedekind's section.

Similarly, we have

(5) $$1 < \log\left(1-\frac{1}{n}\right)^{-n} < \frac{n}{n-1}.$$

Thus, we find

(6) $$\lim_{n\to\infty}\left(1+\frac{1}{n}\right)^n = e = \lim_{n\to\infty}\left(1-\frac{1}{n}\right)^{-n}.$$

It is easy to give a direct proof that the expressions (6) have a definite limit. For we have proved (Art. 157, end) that $(\log x)/(x-1)$ decreases as x increases.

Thus, if $x=1+1/n$, we see that $\log(1+1/n)^n$ increases with n; and therefore $(1+1/n)^n$ does so. In the same way we prove that $(1-1/n)^{-n}$ decreases as n increases.

But from (4) and (5) we see that $(1+1/n)^n$ is less than $(1-1/n)^{-n}$, and is therefore less than $(1-1/2)^{-2}$ if $n>2$.

Thus $(1+1/n)^n < 4$, and consequently $(1+1/n)^n$ converges to a definite limit e (by Art. 149).

As a matter of fact, however, these limits are not very convenient for numerical computation. Their geometric mean gives a better approximation; for it will be seen that

$$\log\frac{n+1}{n-1} = \int_0^{\frac{1}{n}}\frac{2dt}{1-t^2} = \int_0^{\frac{1}{n}}2\left(1+\frac{t^2}{1-t^2}\right)dt.$$

Hence

$$\int_0^{\frac{1}{n}}2(1+t^2)dt < \log\left(\frac{n+1}{n-1}\right) < \int_0^{\frac{1}{n}}2\left(1+\frac{t^2}{1-1/n^2}\right)dt,$$

or $$\frac{2}{n}\left(1+\frac{1}{3n^2}\right) < \log\left(\frac{n+1}{n-1}\right) < \frac{2}{n}\left\{1+\frac{1}{3(n^2-1)}\right\}.$$

Thus

(7) $$1+\frac{1}{3n^2} < \log\left(\frac{n+1}{n-1}\right)^{\frac{n}{2}} < 1+\frac{1}{3(n^2-1)}.$$

With $n=100$ it will be found that

$$\left(1+\frac{1}{n}\right)^n = 2{\cdot}7048,\quad \left(1-\frac{1}{n}\right)^{-n} = 2{\cdot}7320,\quad \left(\frac{n+1}{n-1}\right)^{\frac{n}{2}} = 2{\cdot}7184,$$

the third of which is only wrong by a unit in the last place.

It is perhaps of historical interest to note that Napier calculated the first table of logarithms by means of an approximate formula allied to (7), namely,

(8) $$\log\frac{a}{b} = \frac{a-b}{2}\left(\frac{1}{a}+\frac{1}{b}\right).$$

The error in the approximation (8) can easily be shewn to be about $\frac{1}{6}[(a-b)/a]^3$, and in Napier's work $(a-b)/a$ does not exceed $5/10^5$; so that Napier's approximation is right to the 13th decimal place.

Napier's *definition* of a logarithm is exactly equivalent to the definite integral which we have employed; he supposes that the velocity of a moving point P is proportional to the distance of P from a fixed point O, and is directed towards O; so that the time represents the logarithm of the distance OP, if the initial distance OP is taken as unit.

159. The exponential function.

Since the logarithmic function $\log y$ steadily increases as y increases from 0 to $+\infty$, it follows from Art. 149 that, corresponding to any assigned real value of x, there is a real positive solution of the equation

$$\log y = x.$$

We call y the *exponential function* when x is the independent variable and write $y = \exp x$; the graph of the function can be obtained by turning over Fig. 41, p. 399 and interchanging x and y. The figure obtained is shewn below:

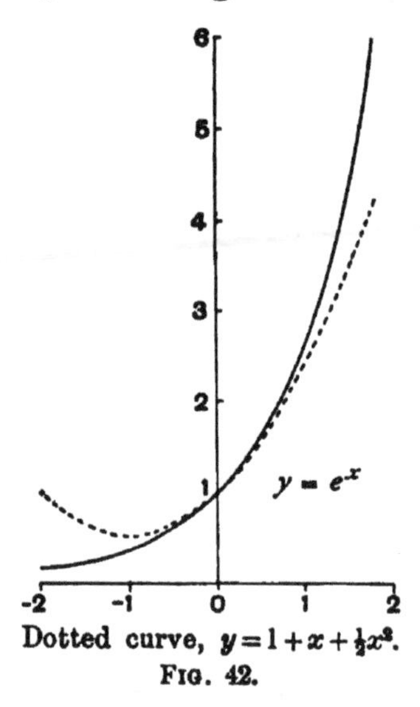

Dotted curve, $y = 1 + x + \frac{1}{2}x^2$.

FIG. 42.

It is evident that *the exponential function is single-valued,**

* Generally, the function inverse to a given function is single-valued in any interval for which the given function steadily increases (or steadily decreases).

because if $y_2 > y_1$, $\log y_2 > \log y_1$. Thus two different values of y cannot correspond to the same value of x in the equation $\log y = x$, so that y is a single-valued function of x.

Suppose now that $\log(y+k) = x+h$,

so that $h = \log(y+k) - \log y = \log(1+k/y)$.

Thus, from Art. 157 (3), we have

(1) $$k/(y+k) < h < k/y,$$

or, since y and $y+k$ are *positive*,

(2) $$hy < k < h(y+k).$$

Hence, *the function* $\exp x$ *increases with* x; and if

$$|h| < \epsilon/(y+\epsilon),$$

we see that $|k| < \epsilon$, and consequently *the exponential function is continuous.*

We have proved in Art. 158 (3) that, when x is *rational*, the exponential function is the *positive* value of e^x. If now x is irrational, defined by the upper and lower classes (A), (a), we have

$$e^a = \exp a < \exp x < \exp A = e^A,$$

because the exponential function increases with x. Also since $\exp x$ is continuous, $e^A - e^a$ can be made as small as we please; and consequently (compare p. 370) $\exp x$ is the single number defined by the classes (e^a), (e^A). Thus $\exp x$ coincides with Dedekind's definition of e^x, when x is irrational; and so *for all values of* x, *the exponential function is the positive value of* e^x.

Since $\log 1 = 0$, it follows that $e^0 = 1$, and so we have, from the continuity of the exponential function,

(3) $$\lim_{x \to 0} e^x = e^0 = 1.$$

Again, because $\log y + \log y' = \log(yy')$, we have

(4) $$e^{x+x'} = e^x . e^{x'}.$$

Of course (3) and (4) agree with the ordinary laws of indices, as established for rational numbers in books on algebra.

From the definitions of the logarithm and exponential functions it follows at once that

$$\frac{1}{y}\frac{dy}{dx} = 1, \text{ or } \frac{dy}{dx} = y.$$

Thus *the exponential function has a derivate equal to itself*, that is,

(5) $$\frac{d}{dx}(e^x) = e^x.$$

This result can also be deduced at once from (2) above.

The following inequalities are often useful:*

(6) $e^x > 1 + x$, for any value of x.

(7) $e^x < 1/(1-x)$, if $0 < x < 1$.

These follow from (3a) of Art. 157.

160. Some miscellaneous inequalities.

Since, when $x > 1$, we have from (3) of Art. 157,

(1) $$\log x < x - 1 < x,$$

it follows that if n is any positive index,

(2) $$\log x^n < x^n \text{ or } \log x < x^n/n.$$

Again, from the same article, we see that if x and n are positive,

$$x/(n+x) < \log(1 + x/n) < x/n.$$

Thus, we find

(3) $$e^\xi < \left(1 + \frac{x}{n}\right)^n < e^x, \quad \text{if } \xi = \frac{xn}{n+x}.$$

Since $\lim_{n\to\infty} \xi = x$, and since the exponential is a continuous function, it follows from (3) that

(4) $$e^x = \lim_{n\to\infty}\left(1 + \frac{x}{n}\right)^n.$$

Similarly, we can prove that if $n > x > 0$,

(3a) $$e^{x_1} > \left(1 - \frac{x}{n}\right)^{-n} > e^x, \quad \text{if } x_1 = \frac{xn}{n-x},$$

so that $$e^x = \lim_{n\to\infty}\left(1 - \frac{x}{n}\right)^{-n}.$$

When n is a positive integer, we have

$$\left(1 + \frac{x}{n}\right)^n = 1 + x + \frac{x^2}{2}\left(1 - \frac{1}{n}\right) + \dots + \left(\frac{x}{n}\right)^n,$$

and since all the terms are positive, this gives, from (3),

$$e^x > 1 + x + \frac{x^2}{2}\left(1 - \frac{1}{n}\right),$$

and consequently, by taking the limit as $n \to \infty$,

(5) $$e^x \geqq 1 + x + \tfrac{1}{2}x^2, \quad \text{if } x > 0.$$

Similarly, we can prove that

$$e^x \geqq 1 + x + \tfrac{1}{2}x^2 + \tfrac{1}{6}x^3, \quad \text{if } x > 0.$$

* The geometrical meaning of (6) is simply that the exponential curve lies entirely above any of its tangents.

161. Some limits; the logarithmic scale of infinity.

We have seen in (2) of the last article that

$$\log x < x^n/n, \quad \text{if } x > 1,\ n > 0,$$

and so

$$\log x/x^{2n} < 1/nx^n.$$

Hence

$$\lim_{x\to\infty} (\log x/x^{2n}) = 0,$$

or simply

(1) $$\lim_{x\to\infty} (\log x/x^n) = 0, \quad \text{if } n > 0.$$

Since $\log x = -\log(1/x)$, the last equation is the same as

(2) $$\lim_{x\to 0} (x^n \log x) = 0, \quad \text{if } n > 0.$$

From (3) of the last article we see that

$$(x/n)^n < e^x,$$

or

$$x^n < n^n e^x, \quad \text{if } x > 0,\ n > 0.$$

Hence

$$x^n e^{-2x} < n^n e^{-x},$$

and therefore

$$\lim_{x\to\infty} (x^n e^{-2x}) = 0,$$

or by writing x in place of $2x$, we have

(3) $$\lim_{x\to\infty} (x^n e^{-x}) = 0, \quad \text{if } n > 0.$$

Of course we can also write (3) in the form

(4) $$\lim_{x\to\infty} (e^x/x^n) = \infty, \quad \text{if } n > 0.$$

The results (3) and (4) are of course true if $n \leqq 0$; but no proof is then required.

We can also obtain (1) and (4) by appealing to L'Hospital's rule, Art. 152 above.

Thus

$$\lim_{x\to\infty} \frac{\log x}{x} = \lim_{x\to\infty} \frac{1/x}{1} = 0,$$

and by changing x to x^n we get (1).

Similarly,

$$\lim_{x\to\infty} \frac{e^{ax}}{x} = \lim_{x\to\infty} \frac{ae^{ax}}{1} = \infty, \quad a > 0.$$

If we write $a = 1/n$ and raise the last to the nth power, we get (4).

The limits (1)–(4) form the basis of *the logarithmic scale of infinity*. It follows from (1) that $\log x$ tends to ∞ more slowly than any positive power of x, however small its index may be; hence, *a fortiori*, $\log(\log x)$ tends to ∞ still more slowly, and so on. On the other hand, we see from (4) that e^x tends to ∞ faster than any power of x, however large its index may be; and hence *a fortiori* e^{e^x} tends to ∞ still faster, and so on. Thus we can construct a succession of functions, all tending to ∞, say,

$$\ldots < \log(\log x) < \log x < x < e^x < e^{e^x} < \ldots,$$

and *each function tends to ∞ faster than any power of the preceding function, but more slowly than any power of the following function.*

It is easy to see, however, that this scale by no means exhausts all types of increase to infinity. Thus, for instance, the function

$$e^{(\log x)^2} = x^{\log x}$$

tends to ∞ more slowly than e^x, but more rapidly than any (fixed) power of x.

Similarly, $$x^x = e^{x \log x}$$

tends to infinity more rapidly than e^x, but more slowly than e^{x^2} or than e^{e^x}.

Other examples will be found at the end of this Appendix (Exs. 11, 12, p. 412).

162. The exponential series.

If we write

$$X_n = 1 + x + \frac{x^2}{2!} + \ldots + \frac{x^n}{n!}$$

we have

$$\frac{dX_n}{dx} = X_{n-1}.$$

Thus $$\frac{d}{dx}(1 - e^{-x}X_n) = e^{-x}(X_n - X_{n-1}) = e^{-x}\frac{x^n}{n!},$$

and so, since $1 - e^{-x}X_n$ is zero for $x = 0$, we have

$$1 - e^{-x}X_n = \int_0^x e^{-t}\frac{t^n}{n!}dt,$$

a result which can also be easily obtained by applying the method of repeated integration by parts to the integral

$$\int_0^x e^{-t}dt.$$

Multiplying the last equation by e^x, we find

$$e^x = 1 + x + \frac{x^2}{2!} + \frac{x^3}{3!} + \ldots + \frac{x^n}{n!} + e^x\int_0^x \frac{t^n}{n!}e^{-t}dt.$$

But when x is positive, e^{-t} in the last integral is less than 1, and so

$$\int_0^x \frac{t^n}{n!} e^{-t} dt < \int_0^x \frac{t^n}{n!} dt = \frac{x^{n+1}}{(n+1)!}.$$

When x is negative ($= -\xi$, say), e^{-t} is less than e^ξ, and so

$$\left| \int_0^{-\xi} \frac{t^n}{n!} e^{-t} dt \right| < \int_0^{\xi} \frac{t^n}{n!} e^{\xi} dt = \frac{\xi^{n+1}}{(n+1)!} e^{\xi}.$$

We have therefore

$$e^x = 1 + x + \frac{x^2}{2!} + \ldots + \frac{x^n}{n!} + R_n,$$

where $$R_n < \frac{x^{n+1} e^x}{(n+1)!}, \quad \text{when } x > 0,$$

or $$|R_n| < \frac{|x|^{n+1}}{(n+1)!}, \quad \text{when } x < 0.$$

In either case $\lim_{n\to\infty} R_n = 0$ (Art. 2, Ex. 4), and so

$$e^x = 1 + x + \frac{x^2}{2!} + \frac{x^3}{3!} + \ldots \text{ to } \infty.$$

163. The existence of an area for the rectangular hyperbola.

We give here a proof that the rectangular hyperbola has an area which can be found by a definite limiting process; this seems essential, since few English books give any adequate arithmetic discussion of the area of a curve. The method applies at once to any curve which can be divided into a *finite* number of parts, in each of which the ordinate steadily increases or steadily decreases; although the actual proof refers only to a curve like the rectangular hyperbola in which the ordinate constantly *decreases*.

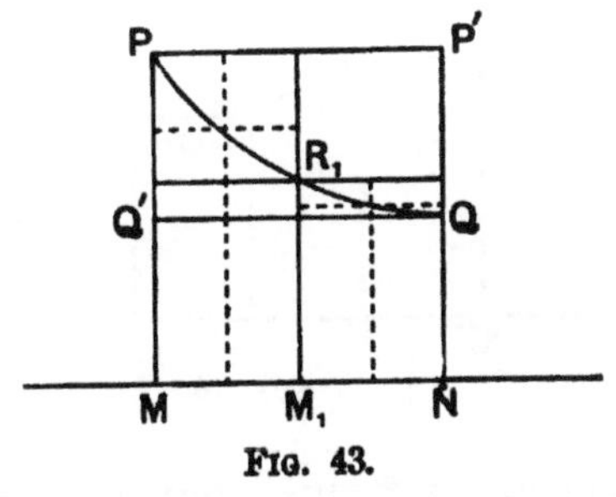

FIG. 43.

Consider any strip of the figure, bounded by the curve (PQ), the axis of x (MN), and two ordinates PM, QN. We can

associate with this strip an *outer* rectangle $PMNP'$ and an *inner* rectangle $Q'MNQ$; now bisect MN at M_1 and draw the ordinate R_1M_1. This gives two outer and two inner rectangles; namely, PM_1, R_1N outside the curve, R_1M, QM_1 inside. But the sum PM_1+R_1N is obviously less than the original outer rectangle; and R_1M+QM_1 is greater than the original inner rectangle.

If we again bisect MM_1 and M_1N, we obtain four outer and four inner rectangles; and the sum of the outer rectangles has again been diminished while the sum of the inner has been increased by the bisection.

When MN is divided into 2^n equal parts, let us denote the sum of the outer rectangles by S_n and the sum of the inner rectangles by s_n. Then

$$S_0 > S_1 > S_2 > \dots > S_n > \dots,$$

and

$$s_0 < s_1 < s_2 < \dots < s_n < \dots.$$

Also $\quad S_n > s_n, \quad (n=0, 1, 2, 3, \dots).$

Now, from the figure, we see that the difference $S_1 - s_1$ is the sum of the two rectangles PR_1, R_1Q, which is equal to

$$\tfrac{1}{2}MN(PM-QN) = \tfrac{1}{2}(S_0 - s_0).$$

Similarly, $\quad S_2 - s_2 = \tfrac{1}{2}(S_1 - s_1) = \dfrac{1}{2^2}(S_0 - s_0),$

and generally $\quad S_n - s_n = \tfrac{1}{2}(S_{n-1} - s_{n-1}) = \dots = \dfrac{1}{2^n}(S_0 - s_0).$

It is therefore clear from Art. 149 that S_n and s_n approach a common limit as n increases; this limit, say A, is called the area of the figure $PMNQ$.

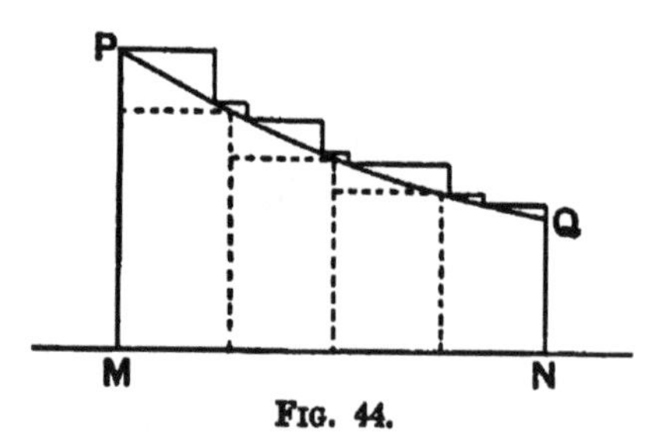

FIG. 44.

But it is essential to prove that *we find the* **same** *area A in whatever way the base MN is supposed divided to form the rectangles.*

Let Σ denote the sum of the outer rectangles when MN is divided up in any manner regular or irregular; and let σ denote the sum of the corresponding inner rectangles. Then a glance at the figure* will shew that for any value of n we have

$$\Sigma > s_n, \quad \sigma < S_n,$$

where of course Σ, σ are quite independent of n.

Thus, since $$\lim S_n = A = \lim s_n,$$

we have $$\Sigma \geqq A, \quad \sigma \leqq A.$$

But $$\Sigma - \sigma \leqq \beta(PM - QN),$$

where β is the breadth of the widest rectangle contained in the sums Σ, σ.

Hence we can choose a value δ such that

$$\Sigma - \sigma < \epsilon, \quad \text{if } \beta < \delta,$$

and therefore, since $\Sigma \geqq A \geqq \sigma$, we have

$$\Sigma - A < \epsilon, \quad A - \sigma < \epsilon, \quad \text{if } \beta < \delta.$$

Thus, $$\lim_{\beta \to 0} \Sigma = A = \lim_{\beta \to 0} \sigma.$$

That is, we obtain the *same* area A, in whatever way the base MN is divided, *provided that the largest sub-division tends to zero.*

Extensions of the definition of integration.

If the function to be integrated is not monotonic, but finite in the interval (a, b) (for definition, see Ex. 23, p. 395), we construct S_n, s_n by taking H_r, h_r, the upper and lower limits of the function in the interval γ_r, and writing $$S_n = \Sigma H_r \gamma_r, \quad s_n = \Sigma h_r \gamma_r,$$

where γ_r is obtained by dividing (a, b) into 2^n equal parts.

Then it is easy to see that S_n and s_n are each monotonic, and so have definite limits as n tends to infinity. These limits need not be equal, so far as we have gone at present; but we now introduce *Riemann's condition of integrability.*

Suppose that in *any* sub-division of (a, b) into sub-intervals $\eta_1, \eta_2, \ldots, \eta_\nu$, we denote by ω_r the difference between the upper and lower limits of the function in the interval η_r, then it must be possible to find δ so as to make

$$\sum_{r=1}^{\nu} \eta_r \omega_r < \epsilon,$$

for all modes of division of the interval such that $\eta_1, \eta_2, \ldots, \eta_\nu$ are *all* less than δ.

* To avoid confusion we have only indicated the rectangles Σ and s_n, the latter being dotted; the reader will have no difficulty in constructing a similar figure for σ and S_n.

Under this condition we have obviously

$$\lim (S_n - s_n) = 0,$$

so that $$\lim S_n = \lim s_n = A, \text{ say.}$$

Then, just as above, we prove that for *any* mode of division

$$\Sigma \geqq A, \quad \sigma \leqq A,$$

and by Riemann's condition,

$$\Sigma - \sigma < \epsilon, \quad \text{if } \eta_r < \delta.$$

Thus $$\lim \Sigma = A = \lim \sigma.$$

It is easy to shew that *a continuous function is integrable*; for (see Ex. 22, App. I.) we can find δ so that $\omega_r < \epsilon/(b-a)$, if $\eta_r < \delta$. Thus we find $\Sigma\eta_r\omega_r < \epsilon$, because $\Sigma\eta_r = b - a$.

It is also easy to extend the definition of integration to a function of two variables, say x, y; let us consider the meaning of

$$\iint f(x, y)\,dx\,dy,$$

where x ranges from a to b, and y from a' to b'.

If we divide (a, b) into 2^m equal parts and (a', b') into 2^n equal parts, we obtain two sums

$$S_{m,n} = \Sigma H_{\mu,\nu}\gamma_{\mu,\nu}, \quad s_{m,n} = \Sigma h_{\mu,\nu}\gamma_{\mu,\nu},$$

where $H_{\mu,\nu}$ and $h_{\mu,\nu}$ are the upper and lower limits in a sub-rectangle $\gamma_{\mu,\nu}$. Then, just as above, we see that $S_{m,n}$ decreases if either m or n is increased, while $s_{m,n}$ increases; thus $S_{m,n}$ and $s_{m,n}$ have each a limit when m, n tend to infinity *in any manner* (see Ch. V., Art. 31).

Further, if $f(x, y)$ is *continuous* we prove as above that

$$\lim S_{m,n} = \lim s_{m,n} = V, \text{ say.}$$

Now we have, from the definition of single integrals,

$$\lim_{m\to\infty}\left(\lim_{n\to\infty} S_{m,n}\right) = \int_a^b dx \int_{a'}^{b'} f(x, y)\,dy$$

and $$\lim_{n\to\infty}\left(\lim_{m\to\infty} S_{m,n}\right) = \int_{a'}^{b'} dy \int_a^b f(x, y)\,dx,$$

so that these *two repeated integrals are each equal to V, and therefore to each other.*

EXAMPLES.

1. Prove directly from the integral for $\log x$ that $2\frac{1}{2} < e < 3$.

[For we have $$\log(2\tfrac{1}{2}) = \int_1^{2\frac{1}{2}} \frac{dt}{t}, \quad \log 3 = \int_1^3 \frac{dt}{t}.$$

If we take these integrals from 1 to $1\frac{1}{4}$, from $1\frac{1}{4}$ to $1\frac{1}{2}$, etc., we find that

$$\log(2\tfrac{1}{2}) < \tfrac{1}{4} + \tfrac{1}{5} + \tfrac{1}{6} + \ldots + \tfrac{1}{9} < 1.$$

$$\log 3 > \tfrac{1}{5} + \tfrac{1}{6} + \tfrac{1}{7} + \ldots + \tfrac{1}{12} > 1.]$$

2. When x is positive, shew that the functions

$$\frac{\log(1+x)}{x} \quad \text{and} \quad (1+x)\frac{\log(1+x)}{x}$$

are both monotonic; and sketch their graphs.

3. If $a_n \to l$ as n tends to ∞, prove that

$$\lim n(a_n^{\frac{1}{n}} - 1) = \log l, \ \lim (1 + a_n/n)^n = e^l.$$

Deduce that

$$\lim \left[\frac{1}{p}(a_1^{\frac{1}{n}} + a_2^{\frac{1}{n}} + \ldots + a_p^{\frac{1}{n}})\right]^n = (a_1 a_2 \ldots a_p)^{\frac{1}{p}}.$$

4. Determine which of the two expressions

$$(\tfrac{1}{2}e)^{\sqrt{3}}, \quad (\sqrt{2})^{\frac{1}{2}\pi}$$

is the greater. [*Oxford Sen. Schol.*]

[Take logarithms and note that

$$\sqrt{3}/(\sqrt{3} + \tfrac{1}{4}\pi) < \tfrac{2}{5}\sqrt{3} < \cdot 6929 < \log 2,$$

since (Art. 64) $\log 2 > \cdot 6931$.]

5. If ρ_n is numerically less than a fixed number A, independent of n, and if

$$\log\left(1 + \frac{\rho_n}{n}\right) = \frac{\sigma_n}{n},$$

then $$\underline{\overline{\lim}}\, \sigma_n = \underline{\overline{\lim}}\, \rho_n.$$

Also if $$\log\left(1 + \frac{1}{n} + \frac{\rho_n}{n \log n}\right) = \frac{1}{n} + \frac{\sigma_n}{n \log n},$$

then $$\underline{\overline{\lim}}\, \sigma_n = \underline{\overline{\lim}}\, \rho_n.$$

[Compare Art 12 (4).]

6. Use the last example to shew that if

$$\log \frac{a_n}{a_{n+1}} = \frac{\mu}{n} + \frac{\omega_n}{n^\lambda}, \quad (\lambda > 1, \ |\omega_n| < A)$$

the series of positive terms Σa_n converges if $\mu > 1$, and otherwise diverges.

Deduce that the series

$$\Sigma (n!)^2 n^{n^2-1} e^n (n-1)^{-(n+1)^2}$$

is divergent. Compare (4) and (5) of Art. 12.

7. If $1 + x > 0$, prove that $x^2 > (1+x)\{\log(1+x)\}^2$.

[*Math. Trip.* 1906.]

[Write $\log(1+x) = 2\xi$, and use the fact that $e^\xi - e^{-\xi} > 2\xi$ if ξ is positive.]

8. Prove that as x ranges from -1 to ∞, the function

$$\frac{1}{\log(1+x)} - \frac{1}{x}$$

remains continuous and steadily decreases from 1 to 0.

[*Math. Trip.* 1894.]

[From the last example, we see that the derivate is negative; discontinuity is only possible at $x=0$, and when $|x|$ is small we find that

$$\frac{1}{\log(1+x)}-\frac{1}{x}=\frac{1}{2}-\frac{x}{12}+\dots,$$

the series converging if $|x|<1$.]

9. Shew how to determine X so that

$$e^x>Mx^N, \quad \text{if } x>X,$$

where M, N are any assigned large numbers.

[We have to make

$$x>\log M+N\log x,$$

which (since $\log x<2\sqrt{x}$) can be satisfied by taking $x>2\log M$ and $16N^2$. But as a rule these determinations of X are unnecessarily large.]

10. The logarithmic function $\log x$ is not a rational function of x. [Apply Art. 161.]

11. What is the largest number which can be expressed algebraically by means of three 9's? Estimate the number of digits in this number when written in the ordinary system of numeration.

12. Arrange the following functions in the order of the rapidity with which they tend to infinity with x:

$$x^x,\ x^{\log x},\ (\log x)^x,\ (\log x)^{(\log x)^2},\ (\log x)^{\log\log x},\ (\log\log x)^{\sqrt{\log x}}.$$

Indicate the position of each of these functions in between the members of the standard logarithmic scale.

13. If we assume the binomial series for any integral exponent, and suppose n to be an integer greater than $|x|$, we find

$$\left(1+\frac{x}{n}\right)^n=1+x+\left(1-\frac{1}{n}\right)\frac{x^2}{2!}+\left(1-\frac{1}{n}\right)\left(1-\frac{2}{n}\right)\frac{x^3}{3!}+\dots \text{ to } (n+1) \text{ terms},$$

$$\left(1-\frac{x}{n}\right)^{-n}=1+x+\left(1+\frac{1}{n}\right)\frac{x^2}{2!}+\left(1+\frac{1}{n}\right)\left(1+\frac{2}{n}\right)\frac{x^3}{3!}+\dots \text{ to } \infty.$$

Deduce that, if x is positive,

$$\left(1+\frac{x}{n}\right)^n<1+x+\frac{x^2}{2!}+\frac{x^3}{3!}+\dots \text{ to } \infty<\left(1-\frac{x}{n}\right)^{-n},$$

and so obtain the exponential series.

14. Shew that

$$e^x\sum_1^\infty\frac{(-1)^{n-1}x^n}{n(n!)}=\sum_1^\infty\left(1+\frac{1}{2}+\dots+\frac{1}{n}\right)\frac{x^n}{n!}.$$

[If the product on the left is called v, we get

$$\frac{dv}{dx}-v=\frac{e^x-1}{x}=1+\frac{x}{2!}+\frac{x^2}{3!}+\dots.$$

Taking $v=\Sigma a_n x^n/n!$, we get at once, since $a_0=0$,

$$a_1=1, \quad a_n-a_{n-1}=1/n.$$

If we obtain the series for v by means of the rule for multiplication, we find the identity

$$n-\frac{1}{2}\frac{n(n-1)}{2!}+\frac{1}{3}\frac{n(n-1)(n-2)}{3!}-\dots \text{ to } n \text{ terms}=1+\frac{1}{2}+\frac{1}{3}+\dots+\frac{1}{n},$$

which is easily verified directly.]

15. Shew that, as $x \to 0$,

$$\left(\frac{e^x}{e^x-1}-\frac{1}{x}\right)\to\frac{1}{2}, \qquad \frac{1}{x^2}\{e^x-1-\log(1+x)\}\to 1.$$

[EULER.]

16. If $\chi_n=\left(1+\frac{1}{2}+\frac{1}{3}+\dots+\frac{1}{n}\right)-\frac{1}{2}\log\{n(n+1)\}-C$, where C is Euler's constant (Art. 11), prove that

$$0<\chi_n<\tfrac{1}{6}\{n(n+1)\}^{-1}.$$

[CESÀRO.]

[It will be seen that

$$\chi_{n-1}-\chi_n=\int_0^1\frac{x^2dx}{n(n^2-x^2)}<\frac{1}{3}\,\frac{1}{n(n^2-1)},$$

which gives the result.]

17. A good approximation to the function of Ex. 16 is given by taking

$$\chi_n=\frac{1}{6[n(n+1)+\frac{1}{5}]},$$

the error in which is of the order $1/(150n^6)$. [A. LODGE.]

[Apply Euler's series (Art. 130).]

18. Prove that

$$e^{-x}\left[\frac{x^{n+1}}{(n+1)!}+\frac{x^{n+2}}{(n+2)!}+\frac{x^{n+3}}{(n+3)!}+\dots\right]=\frac{x^n}{n!}\left(\frac{x}{n+1}-\frac{x^2}{n+2}+\frac{1}{2!}\frac{x^3}{n+3}-\dots\right).$$

[Differentiate, and both sides reduce to $e^{-x}(x^n/n!)$.]

19. Shew that the sequence

$$a_1=1,\ a_2=e^2,\ a_3=e^{e^3},\ a_4=e^{e^{e^4}},\ \dots$$

tends to infinity more rapidly than any member of the exponential scale.

20. Prove that the series

$$\Sigma(\log n)^p n^{-q}$$

converges if $q>1$ or if $q=1$ and $p<-1$; and otherwise diverges.

APPENDIX III.

SOME THEOREMS ON INFINITE INTEGRALS AND GAMMA-FUNCTIONS.

164. Infinite integrals: definitions.

If either the range is infinite or the subject of integration tends to infinity at some point of the range, an integral may be conveniently called *infinite*,* as differing from an ordinary integral very much in the same way as an infinite series differs from a finite series.

In the case of an infinite integral, the method commonly used to establish the existence of a finite integral will not apply, as will be seen if we attempt to modify the proof of Art. 163. We must accordingly frame a new definition:

First, *if the range is infinite, we define the integral* $\int_a^\infty f(x)dx$ *as equal to the limit* $\lim\limits_{\lambda\to\infty}\int_a^\lambda f(x)dx$ *when this limit exists.*

Secondly, *if the integrand tends to infinity at either limit (say that* $f(x)\to\infty$ *as* $x\to a$*), we define the integral* $\int_a^b f(x)dx$ *as equal to the limit*

$$\lim_{\delta\to 0}\int_{a+\delta}^{b} f(x)dx \qquad (\delta > 0)$$

when this limit exists.

---

* Following German writers (who use *uneigentlich*), some English authors have used the adjective *improper* to distinguish such integrals as we propose to call *infinite*. The term used here was introduced by Hardy (*Proc. Lond. Math. Soc.* (ser. 1), vol. 34, p. 16, footnote), and has several advantages, not the least of which is the implied analogy with the theory of *infinite* series.

If the integrand tends to infinity at a point c within the range of integration, it is usually best to divide the integral into two, and then we should define the integral by the equation

$$\int_a^b f(x)\,dx = \lim_{\delta\to 0}\int_a^{c-\delta} f(x)\,dx + \lim_{\delta_1\to 0}\int_{c+\delta_1}^b f(x)\,dx.$$

But in certain problems the two limits in the last equation are both infinite, while the *sum* of the two integrals tends to a finite limit if δ_1/δ tends to a finite limit; we then define *the principal value of the integral* by the equation

$$P\int_a^b f(x)\,dx = \lim_{\delta\to 0}\left[\int_a^{c-\delta} f(x)\,dx + \int_{c+\delta}^b f(x)\,dx\right].$$

It is at once evident that we can extend the use of the terms *converge, diverge,* and *oscillate** so as to apply to these definitions.

Exs. (of convergence).

1. $$\int_0^\infty \frac{dx}{1+x^2} = \lim_{\lambda\to\infty}\int_0^\lambda \frac{dx}{1+x^2} = \lim_{\lambda\to\infty}(\arctan\lambda) = \frac{\pi}{2},$$

2. $$\int_0^\infty e^{-ax}\,dx = \lim_{\lambda\to\infty}\int_0^\lambda e^{-ax}\,dx = \lim_{\lambda\to\infty}\frac{1}{a}(1-e^{-a\lambda}) = \frac{1}{a}, \quad \text{if } a>0,$$

3. $$\int_0^1 \frac{dx}{x^k} = \lim_{\delta\to 0}\int_\delta^1 \frac{dx}{x^k} = \lim \frac{1-\delta^{1-k}}{1-k} = \frac{1}{1-k}, \quad \text{if } 0<k<1,$$

4. $$\int_0^1 \frac{dx}{\sqrt{(1-x^2)}} = \lim_{\delta\to 0}\int_0^{1-\delta}\frac{dx}{\sqrt{(1-x^2)}} = \lim_{\delta\to 0}[\arcsin(1-\delta)] = \frac{\pi}{2}$$

5. $$\int_{-a}^b \frac{dx}{x^{\frac13}} = \lim_{\delta\to 0}\int_{-a}^{-\delta}\frac{dx}{x^{\frac13}} + \lim_{\delta_1\to 0}\int_{\delta_1}^b \frac{dx}{x^{\frac13}}$$
$$= \lim_{\delta\to 0}\frac{3}{2}(\delta^{\frac23} - a^{\frac23}) + \lim_{\delta_1\to 0}\frac{3}{2}(b^{\frac23} - \delta_1^{\frac23})$$
$$= \frac{3}{2}(b^{\frac23} - a^{\frac23}),$$

6. $$P\int_{-a}^b \frac{dx}{x} = \lim_{\delta\to 0}\left(\int_{-a}^{-\delta}\frac{dx}{x} + \int_\delta^b \frac{dx}{x}\right)$$
$$= \lim_{\delta\to 0}\left(-\log\frac{a}{\delta} + \log\frac{b}{\delta}\right) = \log\left(\frac{b}{a}\right),$$

where in the last two integrals we suppose a and b to be positive. It should be remarked that in the last case we should have

$$\int_{-a}^{-\delta}\frac{dx}{x} + \int_{\delta_1}^b \frac{dx}{x} = \log\frac{b}{a} + \log\frac{\delta}{\delta_1},$$

which, of course, does not tend to a definite limit unless δ/δ_1 does so.

---

* Stokes, *Math. and Phys. Papers*, vol. 1, p. 241

Exs. (of divergence and oscillation).

$$\int_0^\infty \frac{dx}{1+x} \textit{ diverges and } \int_0^\infty \sin x\,dx \textit{ oscillates,}$$

$$\int_0^1 \frac{dx}{x} \textit{ diverges and } \int_0^1 \frac{dx}{x^2}\sin\left(\frac{1}{x}\right) \textit{ oscillates.}$$

It must not be supposed that the two types of infinite integrals are fundamentally different. An infinite integral of one type can always be transformed so as to belong to the other type; thus, if $f(x)\to\infty$ as $x\to b$, but is continuous elsewhere in the interval (a, b), we can write

$$\xi = \frac{x-a}{b-x} \quad\text{or}\quad x = \frac{a+b\xi}{1+\xi}.$$

Then
$$\int_a^b f(x)\,dx = \int_0^\infty f\left(\frac{a+b\xi}{1+\xi}\right)\frac{(b-a)d\xi}{(1+\xi)^2}$$
and the integrand in ξ is everywhere finite.*

Ex. $\displaystyle\int_0^1 \frac{dx}{(1-x^2)^{\frac{1}{2}}} = \int_0^\infty \frac{d\xi}{(1+\xi)(1+2\xi)^{\frac{1}{2}}}.$

By reversing this transformation it may happen that an integral to ∞ can be expressed as a *finite* integral.

Ex. When $x=1/\xi$, $\int_1^\infty x^{-s}dx$ becomes $\int_0^1 \xi^{s-2}d\xi$, which is a finite integral if $s\geqq 2$ (both integrals still converge if $2>s>1$).

It is also possible in many cases to express a convergent infinite integral of the second type as a finite integral by a change of variable. Thus we have

$$\int_0^1 \frac{f(x)}{(1-x^2)^{\frac{1}{2}}}\,dx = \int_0^{\frac{1}{2}\pi} f(\sin\theta)\,d\theta$$

by writing $x=\sin\theta$, and the latter integral is finite if $f(x)$ is finite in the interval $(0, 1)$ (for definition see Ex. 23, p. 395). Kronecker in his lectures on definite integrals states that such a transformation is always possible, but although this is theoretically true, it is not effectively practicable† in all cases.

* Care must be taken in applying this kind of transformation when the infinity of $f(x)$ is *inside* the range of integration. Here it is usually safer to divide the integral into two, as already explained.

† If $f(x)\to\infty$ as $x\to a$, we can write $\int_x^b f(x)dx=\xi$, and introduce ξ as a new variable. Similarly in other cases; and in the same sense we can always express a divergent integral in the form $\int^\infty d\xi$.

165. Special case of monotonic functions.

Although, as we have pointed out in the last article, the definition of a definite integral requires in general a modification for the case of an infinite integral, yet we can obtain a direct definition of the integral as the limit of a sum, when the integrand steadily increases or steadily decreases.

Suppose first that in the integral $\int_a^\infty f(x)dx$ the function $f(x)$ steadily decreases to 0 for values of x greater than c; we may then consider only the integral $\int_c^\infty f(x)dx$, because the integral from a to c falls under the ordinary rules. Then let $x_0(=c)$, $x_1, x_2, x_3, \ldots$ be a sequence of values increasing to ∞; we have, as in Art. 11,

$$(x_{n+1}-x_n)f(x_n) > \int_{x_n}^{x_{n+1}} f(x)dx > (x_{n+1}-x_n)f(x_{n+1}).$$

Thus, if the integral $\int_c^\infty f(x)dx$ converges to the value I,

$$(1) \qquad \sum_0^\infty (x_{n+1}-x_n)f(x_n) \geqq I \geqq \sum_0^\infty (x_{n+1}-x_n)f(x_{n+1}).$$

Of the two series in (1), the second certainly converges, in virtue of the convergence of the integral and the fact that the series contains only positive terms. The first need not converge, if the rate of increase of (x_n) is sufficiently rapid; for instance, with $x_n = 2^{2^n}$ and $f(x) = 1/x^2$, it will be found that every term in the series is greater than $\frac{1}{2}$.

However, by taking x_n to be a properly chosen function of some parameter h (as well as of n), we can easily ensure the convergence of both series in (1); and we can also prove that the two series have a common limit as $x_{n+1}-x_n$ is made to tend to zero by varying h; this common limit must be equal to I, in virtue of the inequalities (1).

For example, suppose that $x_{n+1}-x_n$ is independent of n and equal to h, say; then $x_n = c+nh$ and we have

$$\sum_0^\infty (x_{n+1}-x_n)f(x_n) \quad = h[f(c)+f(c+h)+f(c+2h)+\ldots],$$

$$\sum_0^\infty (x_{n+1}-x_n)f(x_{n+1}) = h[f(c+h)+f(c+2h)+f(c+3h)+\ldots].$$

It follows that the difference between the two sums is $hf(c)$, so that both are convergent, and their difference tends to 0 with h; hence

$$I=\lim_{h\to 0} h[f(c)+f(c+h)+f(c+2h)+\ldots].$$

In like manner, if x_{n+1}/x_n is independent of n and equal to q, say, so that $x_n=cq^n$, we have

$$\sum_0^\infty (x_{n+1}-x_n)f(x_n) \quad =c(q-1)[f(c)+qf(cq)+q^2f(cq^2)+\ldots]$$

and $\sum_0^\infty (x_{n+1}-x_n)f(x_{n+1})=c[(q-1)/q][qf(cq)+q^2f(cq^2)+\ldots]$.

Thus we can again infer the convergence of the first series from that of the second, and we see that

$$I=\lim_{q\to 1} c(q-1)[f(c)+qf(cq)+q^2f(cq^2)+\ldots].$$

Ex. 1. Consider $\int_0^\infty xe^{-x}dx$, with $x_n=nh$.

We have then
$$\begin{aligned} I&=\lim_{h\to 0} h^2[e^{-h}+2e^{-2h}+3e^{-3h}+\ldots] \\ &=\lim_{h\to 0} h^2e^{-h}(1-e^{-h})^{-2} \\ &=\lim_{h\to 0} e^h\left(\frac{h}{e^h-1}\right)^2=1, \end{aligned}$$

a value which can be verified by integration by parts.

Ex. 2. Consider $\int_c^\infty x^{-s}dx$, (where $s>1$).

Here write $x_n=cq^n$, and we get

$$\begin{aligned} I&=\lim_{q\to 1} \frac{c(q-1)}{c^s}\left(1+\frac{q}{q^s}+\frac{q^2}{q^{2s}}+\ldots\right) \\ &=\lim_{q\to 1} \frac{q-1}{c^{s-1}}\Big/\left(1-\frac{1}{q^{s-1}}\right) \\ &=\lim_{q\to 1} \frac{q^{s-1}}{c^{s-1}}\left(\frac{q-1}{q^{s-1}-1}\right)=\frac{1}{(s-1)c^{s-1}} \end{aligned}$$

by applying one of the fundamental limits of the differential calculus.

Ex. 3. It can be proved by rather more elaborate reasoning that *if $f(x)$ steadily decreases to 0 as x tends to ∞, then*

$$\int_0^\infty \sin x\, f(x)\,dx=\lim_{h\to 0} h\sum_0^\infty f(nh)\sin nh, \quad \int_0^\infty \cos x\, f(x)\,dx=\lim_{h\to 0} h\sum_0^\infty f(nh)\cos nh.$$

Let us consider the simple example

$$\int_0^\infty \frac{\sin x}{x}dx,$$

the sum is then
$$h+(\sin h+\tfrac{1}{2}\sin 2h+\ldots).$$

Since h is positive (and less than 2π), the sum of the series in brackets is $\frac{1}{2}(\pi-h)$, by Art. 65, and so the whole sum is

$$\tfrac{1}{2}(\pi+h),$$

which gives the limit $\frac{1}{2}\pi$; that this gives the correct value for the integral can be verified by other methods (see Ex. 1, Art. 173).

Ex. 4. The reader may verify in the same way that

$$\int_0^\infty \frac{\sin^2 x}{x^2}\,dx = \tfrac{1}{2}\pi.$$

Ex. 5. By means of the integral $\int_0^\infty x^{k-1}e^{-x}dx$, we can prove that

$$\lim_{h\to 0} h^k(e^{-h}+2^{k-1}e^{-2h}+3^{k-1}e^{-3h}+\ldots)=\int_0^\infty x^{k-1}e^{-x}dx$$
$$=\Gamma(k),$$

a result which has already been found in Art. 51 by another method.

In like manner, if $f(x)\to\infty$ as $x\to 0$, but steadily decreases as x varies from 0 to b, we can prove that when $\int_0^b f(x)dx$ converges, we have

$$\int_0^b f(x)dx = \lim_{q\to 1} b(1-q)[f(b)+qf(bq)+q^2f(bq^2)+\ldots].$$

Ex. 6. Take $\int_0^b \log x\,dx$; we have to find

$$\lim_{q\to 1} b(1-q)[\log b + q\log(bq)+q^2\log(bq^2)+\ldots]$$
$$=\lim_{q\to 1} b(1-q)\left[\frac{\log b}{1-q}+\frac{q\log q}{(1-q)^2}\right]$$
$$=\lim_{q\to 1}[b(\log b)+bq(\log q)/(1-q)]$$
$$=b(\log b)-b,$$

as we may verify by direct integration.

In the previous work we have seen how to evaluate an infinite integral by calculating the limit of an *infinite* series; when the range is finite we can also obtain the result as the limit of a *finite* series; that is, we can replace a double limit by a single limit. (See also Ex. 50, p. 495.)

Thus, suppose that in the convergent integral $\int_a^b f(x)dx$ the integrand $f(x)\to\infty$ as $x\to a$, and that $f(x)$ steadily decreases from a to b. Then write $b-a=nh$, and an argument similar to that of Art. 11 will shew that $\int_{a+h}^b f(x)dx$ lies between the two sums

$$h[f(a+h)+f(a+2h)+\ldots+f(b-h)]$$

and
$$h[f(a+2h)+f(a+3h)+\ldots+f(b)].$$

Now, as $h \to 0$, the integral tends to a definite limit; and the difference between the two sums is $h[f(a+h)-f(b)]$, which tends to zero with h in virtue of the monotonic property of $f(x)$ (see pp. 423, 424 below). That is,

$$\int_a^b f(x)dx = \lim_{h \to 0} h\,[f(a+h)+f(a+2h)+\ldots+f(b)],$$

which gives the value of the integral as a *single* limit.

Ex. 7. Consider $\int_0^b x^{-s}dx$, where $0<s<1$.

Write $h=\frac{b}{n}$, and we have to find

$$\lim_{n\to\infty} \frac{b^{1-s}}{n^{1-s}}\left(1+\frac{1}{2^s}+\frac{1}{3^s}+\ldots+\frac{1}{n^s}\right)=\frac{b^{1-s}}{1-s}$$

(by Ex. 1, Art. 152, above).

Ex. 8. In the same way $\int_0^1 \log x\, dx$ is found as

$$\lim_{n\to\infty} \frac{1}{n}\sum_{r=1}^{n} \log\left(\frac{r}{n}\right)=\lim_{n\to\infty} \log\left(\frac{n!}{n^n}\right)^{\frac{1}{n}}=-1 \text{ (Ex. 1, Art. 154).}$$

Ex. 9. If we divide the last equation of Art. 69 by $\sin\theta$, and let θ tend to zero, we find, if $a=\pi/n$,

$$n=2^{n-1}\sin a \sin 2a \ldots \sin(n-1)a.$$

Now change from n to $2n$ and write h for a; we get, pairing the terms,

$$2n=2^{2n-1}\sin^2 h \sin^2 2h \ldots \sin^2(n-1)h.$$

Thus, extracting the square root,

$$\sin h \sin 2h \ldots \sin(n-1)h=n^{\frac{1}{2}}2^{1-n}, \quad (\text{if } h=\pi/2n),$$

and from this we can find $\int_0^{\frac{1}{2}\pi}\log\sin x$.

For this integral is equal to

$$\lim_{h\to 0} h[\log\sin h+\log\sin(2h)+\ldots+\log\sin(nh)]$$

$$=\lim_{n\to\infty}\frac{\pi}{2n}\left[\tfrac{1}{2}\log n-(n-1)\log 2\right]$$

$$=-\tfrac{1}{2}\pi\log 2.$$

166. Tests of convergence for infinite integrals with a positive integrand.

If the function $f(x)$ is positive, at least for sufficiently large values of x, it is clear that the integral $\int_a^\lambda f(x)\,dx$ steadily increases with λ; thus in virtue of the monotonic test for convergence, *the integral to* ∞ *cannot oscillate, and will converge if we can prove that*

$$\int_a^\lambda f(x)\,dx < A,$$

where A is independent of λ.

In practice the usual method of applying this test is to appeal to the *principle of comparison*, as in the case of series of positive terms; in fact, if $g(x)$ is a positive function for which $\int^\infty g(x)\,dx$ converges, then $\int^\infty f(x)\,dx$ also converges if $f(x) < g(x)$, at any rate for values of x greater than some fixed number c.

For then $$\int_c^\lambda f(x)\,dx < \int_c^\lambda g(x)\,dx < \int_c^\infty g(x)\,dx,$$

and this last expression is independent of λ.

Thus, suppose we consider $f(x) = x^\beta e^{-ax}$, where a is positive and β is either positive or negative; from Art. 161 above, we find that $x^\beta e^{-\frac{1}{2}ax} \to 0$ as $x \to \infty$, so that we can determine c to satisfy

$$x^\beta e^{-\frac{1}{2}ax} < 1, \qquad \text{if } x > c,$$

and then $$f(x) = x^\beta e^{-ax} < e^{-\frac{1}{2}ax}, \qquad \text{if } x > c.$$

Now (see Ex. 2, p. 415, Art. 164) $\int^\infty e^{-\frac{1}{2}ax}\,dx$ is convergent, and consequently $\int^\infty x^\beta e^{-ax}\,dx$ is also convergent.

If we write $X = e^x$, we find that

$$\int x^\beta e^{-ax}\,dx = \int (\log X)^\beta X^{-(1+a)}\,dX,$$

so that $\int^\infty (\log x)^\beta x^{-(a+1)}\,dx$ is convergent.

Examples of this type can be multiplied to any extent by the aid of the logarithmic scale of infinity (Art. 161).

Thus *if we can find a positive index a and a constant B, such that* **one** *of the conditions*

$$\left.\begin{array}{l}\text{(i) } f(x) < B(\log x)^\beta x^{-(1+a)},\\ \text{(ii) } f(x) < Bx^\beta e^{-ax},\end{array}\right\} \quad a > 0, \quad x > c,$$

is satisfied, the integral $\int_c^\infty f(x)\,dx$ *converges.*

The comparison test for divergence runs as follows:

If $G(x)$ is always positive and $\int^\infty G(x)\,dx$ is divergent, then so also is $\int^\infty f(x)\,dx$, if $f(x) > G(x)$, at any rate after a certain value of x.

We have proved (see the small type above) that if a is positive, c can be found so that

$$x^{\beta} e^{ax} > e^{\frac{1}{2}ax}, \quad \text{if } x > c,$$

whatever the index β may be.

Now $$\int_c^{\lambda} e^{\frac{1}{2}ax} dx = \frac{2}{a}\left(e^{\frac{1}{2}a\lambda} - e^{\frac{1}{2}ac}\right)$$

and this expression tends to ∞ with λ, so that the integral to ∞ is divergent.

Thus $\int^{\infty} x^{\beta} e^{ax} dx$ also diverges.

If $a=0$, it is easily seen that this integral diverges if $\beta \geqq -1$.

By changing the variable, we deduce as before that

$$\int^{\infty} (\log x)^{\beta} x^{-(1-a)} dx$$

diverges under the same conditions.

Accordingly *the integral* $\int_a^{\infty} f(x)\,dx$ *diverges if we can find an index* $a \geqq 0$, *such that one of the conditions*

$$\left.\begin{aligned} &\text{(i) } f(x) > B(\log x)^{\beta} x^{-(1-a)}, \\ &\text{(ii) } f(x) > Bx^{\beta} e^{ax}, \end{aligned}\right\} \quad \begin{aligned} &a > 0 \\ &\text{or } a=0,\ \beta \geqq -1, \end{aligned}$$

is satisfied.

These conditions are analogous to those of Art. 11 for testing the convergence of a series of positive terms; and, as there remarked, closer tests can be obtained by making use of other terms in the logarithmic scale (although such conditions are not of importance for our present purpose). But one striking feature presents itself in the theory of infinite integrals which has no counterpart in the theory of series. *An integral* $\int^{\infty} f(x)\,dx$ *may converge even though* $f(x)$ *does not tend to the limit zero.* Naturally, we must then have an oscillatory function, for $\underline{\lim} f(x)=0$ is obviously necessary in all cases of convergence; but we may even have $\overline{\lim} f(x)=\infty$. To see, in a general way, that this is possible, we may use a graphical method.

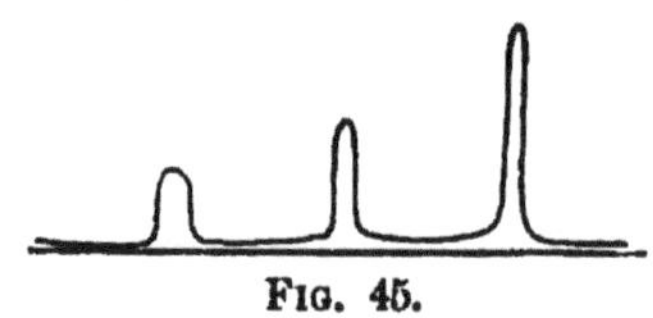

FIG. 45.

Consider a curve which has an infinite series of *peaks,* of steadily increasing height; then, it is quite possible to suppose that their widths are correspondingly decreased in such a way that the areas of the peaks form a convergent series; and consequently $\int^{\infty} f(x)\,dx$ may converge.

Let us consider in particular the function

$$f(x)=x^{\beta}/(1+x^{\alpha}\sin^2 x), \qquad (\alpha>\beta>0).$$

Here in general $f(x)$ is comparable with $x^{\beta-\alpha}$, but its graph comes up to the curve $y=x^{\beta}$, at every point for which x is a multiple of π.

In the interval from $n\pi$ to $(n+1)\pi$, we have

$$\frac{(n\pi)^{\beta}}{1+[(n+1)\pi]^{\alpha}\sin^2 x}<f(x)<\frac{[(n+1)\pi]^{\beta}}{1+(n\pi)^{\alpha}\sin^2 x}.$$

Now $$\int_{n\pi}^{(n+1)\pi}\frac{dx}{1+A\sin^2 x}=\frac{\pi}{(1+A)^{\frac{1}{2}}},$$

so that $$\frac{n^{\beta}\pi^{\beta+1}}{[1+(n+1)^{\alpha}\pi^{\alpha}]^{\frac{1}{2}}}<\int_{n\pi}^{(n+1)\pi}f(x)\,dx<\frac{(n+1)^{\beta}\pi^{\beta+1}}{(1+n^{\alpha}\pi^{\alpha})^{\frac{1}{2}}}.$$

From this it is evident that $\int^{\infty} f(x)\,dx$ converges or diverges with the series $\Sigma n^{\beta-\frac{1}{2}\alpha}$; that is, according as $\alpha>2(\beta+1)$ or $\alpha\leqq 2(\beta+1)$.

And generally, if $\phi(x)$, $\psi(x)$ steadily increase to ∞ with x, the integral

$$\int^{\infty}\frac{\phi(x)\,dx}{1+\psi(x)\sin^2(x\pi)}$$

converges if $\Sigma\phi(n+1)/[\psi(n)]^{\frac{1}{2}}$ converges and diverges if $\Sigma\phi(n)/[\psi(n+1)]^{\frac{1}{2}}$ diverges.

Ex. 1. $\int_0^{\infty}\frac{x^{\beta}dx}{1+x^{\alpha}|\sin x|}$ converges or diverges according as

$\alpha>\beta+1$ or $\alpha\leqq\beta+1$. [HARDY.*]

Ex. 2. $\int_0^{\infty}\phi(x)e^{-\psi(x)|\sin x|}dx$ converges with $\Sigma\frac{\phi(n\pi+\epsilon)}{\psi(n\pi-\epsilon)}$ and diverges with $\Sigma\frac{\phi(n\pi-\epsilon)}{\psi(n\pi+\epsilon)}$. [HARDY.]

Ex. 3. $\int_0^{\infty}\phi(x)e^{-\psi(x)\sin^2 x}dx$ converges with $\Sigma\frac{\phi(n\pi+\epsilon)}{\sqrt{\psi(n\pi-\epsilon)}}$ and diverges with $\Sigma\frac{\phi(n\pi-\epsilon)}{\sqrt{\psi(n\pi+\epsilon)}}$. [DU BOIS REYMOND.]

In spite of the last result, we can prove (as in Art. 9 for series) that *if $f(x)$ steadily decreases, the condition* $\lim xf(x)=0$ *is necessary for the convergence of* $\int^{\infty} f(x)dx$.

---

* *Messenger of Mathematics*, April 1902, Note VIII.

For here we have

$$\int_\lambda^\mu f(x)\,dx > (\mu-\lambda)f(\mu);$$

thus for convergence it is necessary to be able to find λ so that $(\mu-\lambda)f(\mu)$ is less than ϵ for *any* value of μ greater than λ.

Hence $\lim xf(x)=0$ is necessary for convergence. But even so, no such condition as $\lim(x\log x)f(x)=0$ is necessary in general (compare Art. 9); but it is easy to shew that if (for instance) $xf(x)$ is monotonic, then $x\log xf(x)$ must tend to 0. More generally, if $\phi(x)$ tends steadily to ∞ and $f(x)/\phi'(x)$ is monotonic, then $f(x)\phi(x)/\phi'(x)$ must tend to zero; this may be proved by changing the variable from x to $\phi(x)$. [PRINGSHEIM.]

It is perfectly easy to modify all the foregoing work* so as to apply to integrals in which the integrand tends to infinity, say at $x=0$.

The results are: *The integral* $\int_0^b f(x)dx$ *converges (if b is less than* 1), *provided that we can satisfy one of the conditions*

$$f(x) < Bx^{a-1}\left(\log\frac{1}{x}\right)^\beta,$$

where either (i) $a>0$ *or* (ii) $a=0,\ \beta<-1$.

On the other hand, the integral diverges when

$$f(x) > Bx^{a-1}\left(\log\frac{1}{x}\right)^\beta,$$

where (i) $a<0$ *or* (ii) $a=0,\ \beta\geqq-1$.

167. Examples.

To illustrate the last article, we consider two simple cases.

1. $$\int_0^\infty (e^{-x}-e^{-tx})\frac{dx}{x} \qquad (t>0).$$

It is easy to see that the integral converges, so far as concerns the upper limit, by applying the tests of the last article. There is an apparent difficulty at the lower limit, because of the factor $1/x$; but since

$$\frac{1}{x}(e^{-x}-e^{-tx})=\frac{1}{x}\left[\left(1-x+\frac{x^2}{2!}-\dots\right)-\left(1-xt+\frac{x^2t^2}{2!}-\dots\right)\right]$$

$$=(t-1)-\frac{x}{2!}(t^2-1)+\dots$$

the difficulty is apparent only.

* Or we may obtain the results directly by writing $1/x$ for x in the integral.

Now the integral is $\lim_{\delta\to 0}\int_\delta^\infty (e^{-x}-e^{-tx})\frac{dx}{x}$, and

$$\int_\delta^\infty e^{-tx}\frac{dx}{x}=\int_{t\delta}^\infty e^{-x}\frac{dx}{x},$$

by changing the variable of integration.

Hence our integral is

$$\lim_{\delta\to 0}\int_\delta^{t\delta} e^{-x}\frac{dx}{x}.$$

But $\int_\delta^{t\delta} e^{-x}\frac{dx}{x}$ lies between the values found by replacing e^{-x} by $e^{-\delta}$ and by $e^{-t\delta}$; these values are respectively

$$e^{-\delta}\log t \text{ and } e^{-t\delta}\log t,$$

both of which tend to $\log t$ as δ tends to 0.

Hence
$$\lim_{\delta\to 0}\int_\delta^{t\delta} e^{-x}\frac{dx}{x}=\log t,$$

and accordingly
$$\int_0^\infty (e^{-x}-e^{-tx})\frac{dx}{x}=\log t.$$

2. Consider
$$\int_0^\infty (Ae^{-ax}+Be^{-bx}+Ce^{-cx})\frac{dx}{x^2},$$

where a, b, c are positive and

$$A+B+C=0,\quad Aa+Bb+Cc=0.$$

It may be shewn as above that the integral converges when these conditions are satisfied.

Now consider

$$\int_\delta^\infty Ae^{-ax}\frac{dx}{x^2}=\int_{a\delta}^\infty Aae^{-x}\frac{dx}{x^2}=Aa\left(\int_\delta^\infty e^{-x}\frac{dx}{x^2}-\int_\delta^{a\delta} e^{-x}\frac{dx}{x^2}\right).$$

In virtue of the condition $\Sigma Aa=0$, it is now evident that

$$\int_\delta^\infty (\Sigma Ae^{-ax})\frac{dx}{x^2}=-\Sigma Aa\int_\delta^{a\delta} e^{-x}\frac{dx}{x^2}.$$

But

$$\int_\delta^{a\delta} e^{-x}\frac{dx}{x^2}=\int_\delta^{a\delta}\left(1-x+\frac{x^2}{2!}-\dots\right)\frac{dx}{x^2}=\frac{1}{\delta}\left(1-\frac{1}{a}\right)-\log a+\frac{1}{2!}\delta(a-1)+\dots.$$

Thus
$$\int_\delta^\infty (\Sigma Ae^{-ax})\frac{dx}{x^2}=\Sigma(Aa\log a)-\frac{\delta}{2!}(\Sigma Aa^2)+\dots,$$

and so
$$\int_0^\infty (\Sigma Ae^{-ax})\frac{dx}{x^2}=\Sigma(Aa\log a).$$

3. The reader can prove similarly that, if $p > n$,

$$\int_0^\infty \left(\sum_0^p A_r e^{-a_r x}\right)\frac{dx}{x^n} = -\sum_0^p A_r \frac{(-a_r)^{n-1}}{(n-1)!}\log a_r,$$

where $\quad \sum_0^p A_r a_r^k = 0. \quad (k=0, 1, 2, \ldots, n-1)$

168. Analogue of Abel's Lemma.

If the function $f(x)$ steadily decreases, but is always positive, in an interval (a, b), and if $|\phi(x)|$ is less than a fixed number A in the interval, then *

$$hf(a) < \int_a^b f(x)\phi(x)dx < Hf(a),$$

where H, h are the upper and lower limits of the integral

$$\chi(\xi) = \int_a^\xi \phi(x)dx,$$

as ξ ranges from a to b.

For, assuming that $f(x)$ is differentiable, we have

$$J = \int_a^b f(x)\phi(x)dx = f(b)\chi(b) - \int_a^b f'(x)\chi(x)dx.$$

Now, since $f(b)$ is *positive* and $f'(x)$ is everywhere *negative*, we obtain a value greater than J by replacing $\chi(b)$ and $\chi(x)$ in the last expression by H, and a value less than J by replacing them by h.

Thus we find

$$hf(b) - h\int_a^b f'(x)dx < J < Hf(b) - H\int_a^b f'(x)dx$$

or $$hf(a) < J < Hf(a).$$

Similarly, if H_1, h_1 are the limits of the integral $\chi(\xi)$ in the interval (a, c) and H_2, h_2 in the interval (c, b), we find

$$h_2 f(b) - h_1\int_a^c f'(x)dx - h_2\int_c^b f'(x)dx < J$$

$$< H_2 f(b) - H_1\int_a^c f'(x)dx - H_2\int_c^b f'(x)dx$$

or $$h_1[f(a) - f(c)] + h_2 f(c) < J < H_1[f(a) - f(c)] + H_2 f(c).$$

* If $f(x)$ should be discontinuous at $x = a$, $f(a)$ denotes the limit of $f(x)$ as x tends to a through larger values.

When $\phi(x)$ is a complex function of a real variable, it is easily seen that if u is any number, real or complex, and if η_1, η_2 are the upper limits of

$$\left|\int_a^\xi \phi(x)dx - u\right|,$$

as ξ ranges from a to c and from c to b, respectively, then

$$|J-uf(a)| < \eta_1[f(a)-f(c)]+\eta_2 f(c).$$

When $f(x)$ is complex, formulae corresponding to the lemma of Art. 80 can be obtained (see *Proc. Lond. Math. Soc.*, vol. 6, 1907, p. 65); but these results are not needed for our present purpose.

The first inequality on p. 426 is equivalent to the **Second Theorem of Mean Value**. To see this, note first that $\chi(\xi)$ is continuous, and so (Ex. 21, p. 395) assumes every value between h, H at least once in the interval (a, b). Thus the inequality leads to **Bonnet's theorem**

$$J=f(a)\chi(\xi_0), \text{ where } a \leqq \xi_0 \leqq b.$$

From this **du Bois Reymond's theorem,** which is true for any monotonic function $g(x)$, follows by writing $|g(x)-g(b)|$ for $f(x)$; thus we find the form commonly quoted

$$\int_a^b g(x)\phi(x)dx = g(a)\int_a^{\xi_0}\phi(x)dx + g(b)\int_{\xi_0}^b \phi(x)dx.$$

But, since the precise value of ξ_0 cannot be determined, *these equations contain no more information than the original inequality and not so much as the inequality at the foot of p.* 426.

Although the restriction that $f(x)$ is to be differentiable is of little importance here, yet it is theoretically desirable to establish such results as the foregoing with the greatest generality possible. We shall therefore give a second proof, based on one due to Pringsheim*, in which we assume nothing about the existence of $f'(x)$.

Divide the interval into n equal parts by inserting points $x_1, x_2, \ldots, x_{n-1}$, and write $x_0=a$, $x_n=b$; then we have

$$J=\int_a^b f(x)\phi(x)dx = \sum_{r=0}^{n-1} J_r,$$

where $$J_r = \int_{x_r}^{x_{r+1}} f(x)\phi(x)dx.$$

* *Münchener Sitzungsberichte*, Bd. 30, 1900, p. 209; this paper contains a more general form of the theorem, which is also deducible from the first inequality on p. 426. Another proof has been given by Hardy, *Messenger of Maths.*, vol. 36, 1906, p. 10.

Hence,* if $f_{r+1}=f(x_{r+1})$,

(1) $$J_r - f_{r+1}\int_{x_r}^{x_{r+1}}\phi(x)dx = \int_{x_r}^{x_{r+1}}[f(x)-f_{r+1}]\phi(x)dx;$$

but in virtue of the decreasing property of $f(x)$, $[f(x)-f_{r+1}]$ is positive in the last integral and is less than $(f_r - f_{r+1})$, so that

(2) $$\left|\int_{x_r}^{x_{r+1}}[f(x)-f_{r+1}]\phi(x)dx\right| < (f_r - f_{r+1})A(b-a)/n,$$

because $|\phi(x)|<A$ and $x_{r+1}-x_r=(b-a)/n$.

By adding up the equations (1), bearing in mind the inequality (2), we see that

(3) $$J - \sum_{r=0}^{n-1} f_{r+1}\int_{x_r}^{x_{r+1}}\phi(x)dx = R_n,$$

where $$|R_n| < \frac{A}{n}(b-a)(f_0 - f_n) < \frac{A}{n}(b-a)f_0,$$

because $f_n = f(b)$ is positive.

If now we apply Abel's Lemma (Art. 23) to the sum

$$\sum_{r=0}^{n-1} f_{r+1}\int_{x_r}^{x_{r+1}}\phi(x)dx,$$

we obtain the limits hf_1 and Hf_1 for it, because

$$\sum_{r=0}^{m-1}\int_{x_r}^{x_{r+1}}\phi(x)dx = \int_a^{x_m}\phi(x)dx,$$

and the sequence $f_1, f_2, \ldots, f_n$ is decreasing.

Thus, from (3), we find

(4) $$hf_1 - \frac{A}{n}(b-a)f_0 < J < Hf_1 + \frac{A}{n}(b-a)f_0,$$

where $$f_1 = f[a+(b-a)/n].$$

If now we take the limit of (4) as n tends to infinity, we obtain the desired result.†

In exactly the same way we can make the further inference that *if c lies between a, b, and if H_1, h_1 are the upper and lower limits of $\int_a^{\xi}\phi(x)dx$ as ξ ranges from a to c, while H_2, h_2 are those as ξ ranges from c to b, then*

$$h_1[f(a)-f(c)] + h_2 f(c) < \int_a^b f(x)\phi(x)dx < H_1[f(a)-f(c)] + H_2 f(c).$$

---

* In case $f(x)$ should be discontinuous at x_{r+1}, we define f_{r+1} as the limit of $f(x)$ when x approaches x_{r+1} through *smaller* values of x; this limit will exist in virtue of the monotonic property of $f(x)$.

† It will be seen that the condition $|\phi(x)|<A$ is by no means essential, and that it may be broken at an infinity of points, provided that $\int_a^b|\phi(x)|dx$ converges; for we can then make a division into sub-intervals, for each of which $\int_{x_r}^{x_{r+1}}|\phi(x)|dx$ is less than any assigned number. But Pringsheim has proved that it is only necessary to assume that $\phi(x)$ and $f(x)\times\phi(x)$ are integrable in the interval (a, b); compare *Proc. Lond. Math. Soc.*, vol. 6, 1907, p. 62.

169. Tests of convergence in general.

Applying the general test for convergence (Art. 3), we see that *the necessary and sufficient condition for the convergence of the integral* $\int_a^\infty f(x)dx$ *is that we can find* ξ *such that*

$$\left|\int_\xi^{\xi'} f(x)dx\right| < \epsilon,$$

where ξ' *may have* **any** *value greater than* ξ *and* ϵ *is arbitrarily small.*

However, just as for infinite series, the general test for convergence is usually replaced in practice by some narrower test which can be applied more quickly. The three chief tests are the following:

1. Absolute convergence.*

The integral $\int_a^\infty f(x)dx$ will certainly converge if $\int_a^\infty |f(x)| dx$ converges, because

$$\left|\int_\xi^{\xi'} f(x)dx\right| \leqq \int_\xi^{\xi'} |f(x)|\, dx.$$

But naturally the analogy between such integrals and absolutely convergent series is not quite complete, since there is no *order* in the values of a function.

In particular, if $|f(x)| < -g'(x)$, where $g(x)$ steadily decreases to zero as x increases, the integral $\int_a^\infty f(x)dx$ will converge.

2. Abel's test.

An infinite integral which converges (although not **absolutely***) will remain convergent after the insertion of a factor which is monotonic and less than a fixed number (in numerical value).*

Suppose that $\int_a^\infty \phi(x)dx$ converges, and that $\psi(x)$ is a monotonic function, such that $|\psi(x)| < A$: it is then evident that $\psi(x)$ tends to some limit l as $x \to \infty$.

Thus, if we write $f(x) = l - \psi(x)$ when $\psi(x)$ increases, or $\psi(x) - l$ if $\psi(x)$ decreases, we see that $f(x)$ is positive and

* The distinction between absolute and non-absolute convergence is clearly pointed out by Stokes (*Math. and Phys. Papers*, vol. 1, p. 241).

decreases to 0; and it is obviously sufficient to prove the convergence of $\int_a^\infty f(x)\phi(x)dx$.

Now, by the analogue of Abel's Lemma (Art. 168), we see that

$$\int_\xi^{\xi'} f(x)\phi(x)dx < Hf(\xi) < Hf(a),$$

where H is the upper limit to

$$\left|\int_\xi^X \phi(x)dx\right|$$

when X ranges from ξ to ξ'. Now, in virtue of the convergence of $\int_a^\infty \phi(x)dx$, we can determine ξ so as to make $H < \epsilon/f(a)$, and then

$$\left|\int_\xi^{\xi'} f(x)\phi(x)dx\right| < \epsilon,$$

so that the integral $\int_a^\infty f(x)\phi(x)\,dx$ converges.

Hence also $\int_a^\infty \psi(x)\phi(x)dx$ converges.

3. Dirichlet's test.

An infinite integral which oscillates finitely becomes convergent after the insertion of a monotonic factor which tends to zero as a limit.

Here again we have

$$\left|\int_\xi^{\xi'} f(x)\phi(x)dx\right| < Hf(\xi),$$

and H will be less than some fixed constant independent of ξ;* thus, since $f(x)\to 0$, we can find ξ so that $Hf(\xi) < \epsilon$, and consequently the integral $\int_a^\infty f(x)\phi(x)dx$ is convergent.

Although the tests (2), (3) are almost immediately suggested by the tests of Arts. 19, 20, yet it is not clear that they were ever given, in a complete form, until recently. Stokes (in 1847) was certainly aware of the theorem (3) in the case $\phi(x)=\sin x$ (*Math. and Phys. Papers*, vol. 1, p. 275), but he makes no reference to any extension, nor does he indicate his method of proof. The first general statements and proofs seem to be due to Hardy (*Messenger of Maths.*, vol. 30, 1901, p. 187); his argument is somewhat different from the foregoing, and is on the lines of the following treatment of the special case $\phi(x)=\sin x$.

---

* Since the integral $\int_a^\xi \phi(x)dx$ oscillates *finitely* it remains less than some fixed number C for all values of ξ; thus we have $H \leqq 2C$.

In this case, the curve $y=f(x)\sin x$ oscillates between the two curves $y=f(x)$, $y=-f(x)$, as indicated roughly in the figure.

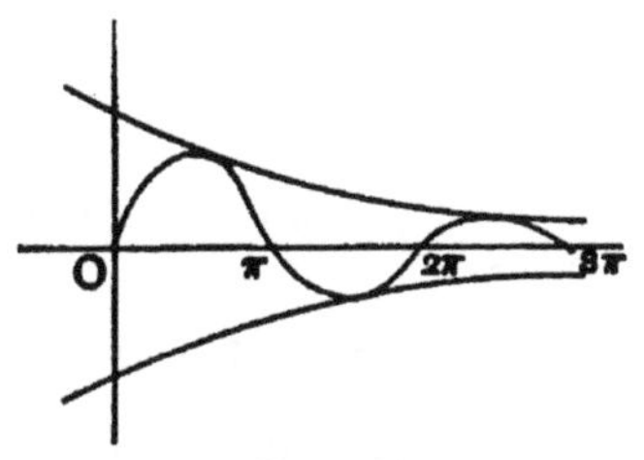

FIG. 46.

It is almost intuitively evident that the areas of the waves steadily decrease in value, and have alternate signs. In fact

$$\int_{2n\pi}^{(2n+1)\pi} f(x)\sin x\,dx = \int_0^\pi f(x+2n\pi)\sin x\,dx,$$

and since $\sin x$ is positive in the integral, this lies between $2f(2n\pi)$ and $2f(\overline{2n+1}\,\pi)$; so that it tends to zero as n increases to ∞. Further,

$$\int_{(2n+1)\pi}^{(2n+2)\pi} f(x)\sin x\,dx = -\int_0^\pi f(x+\overline{2n+1}\,\pi)\sin x\,dx,$$

which is obviously negative and numerically less than the area of the previous wave. It follows that $\int_a^\infty \sin x f(x)dx$ is convergent, by applying the theorem of Art. 21.

In general, if $\phi(x)$ changes sign infinitely often we can apply a similar method, using Dirichlet's test (Art. 20) to establish the convergence of the series.

If the integrand tends to ∞, say as $x\rightarrow a$, the general test for convergence and the test of absolute convergence run as follows:

The necessary and sufficient condition for the convergence of the integral $\int_a^b f(x)dx$ *is that we can find* δ *such that*

$$\left|\int_{a+\delta'}^{a+\delta} f(x)dx\right| < \epsilon,$$

where δ' *has* **any** *positive value less than* δ.

This condition is certainly satisfied if the integral $\int_a^b |f(x)|\,dx$ *converges; and the original integral is then said to converge absolutely.*

It is possible to write out corresponding modifications of Abel's and Dirichlet's tests; but such tests are not often needed in practice and are better left to the ingenuity of the reader.

Ex. 1. $\int_1^\infty \frac{\sin x}{x^p}dx$, $\int_1^\infty \frac{\cos x}{x^p}dx$ converge absolutely if $p>1$, and so does

$$\int_0^1 \frac{\cos x}{x^q}dx \text{ if } 0<q<1\text{; because } |\sin x| \leqq 1,\ |\cos x| \leqq 1.$$

Ex. 2. $\int_1^\infty \frac{\sin x}{x^p}dx$, $\int_1^\infty \frac{\cos x}{x^p}dx$ converge if $0<p<1$; and generally

$$\int^\infty \phi(x)\sin x\,dx,\quad \int^\infty \phi(x)\cos x\,dx$$

converge in virtue of Dirichlet's test, if $\phi(x)$ tends steadily to zero.

For $\left|\int_a^b \sin x\,dx\right| = |\cos a - \cos b| \leqq 2$, $\left|\int_a^b \cos x\,dx\right| = |\sin b - \sin a| \leqq 2$.

Ex. 3. Further examples of Dirichlet's test are given by

$f(x)=e^{-px}$, $(\log x)^{-p}$; $\phi(x)=(\sin x)^{-\frac{1}{2}}$, $\log(4\cos^2 x)$. [HARDY, *l.c.*]

170. Frullani's integrals.

As a simple and interesting example of the tests of the last article, let us consider the value of

$$\int_0^\infty \frac{\phi(ax)-\phi(bx)}{x}dx,$$

where $\phi(x)$ is such that $\int^\infty \phi(x)dx$ oscillates between finite limits (or converges).

Then, by applying Dirichlet's or Abel's test, we see that

$$\int_\delta^\infty \frac{\phi(ax)}{x}dx$$

is convergent and is equal to

$$\int_{a\delta}^\infty \frac{\phi(x)}{x}dx, \quad \text{if } a,\ \delta>0.$$

Thus $$\int_\delta^\infty \frac{\phi(ax)-\phi(bx)}{x}dx = \int_{a\delta}^{b\delta}\frac{\phi(x)}{x}dx = \int_a^b \frac{\phi(x\delta)}{x}dx,$$

and if $\phi(x)$ tends to a definite finite limit ϕ_0, as x tends to 0, the last integral has a finite range and a finite integrand: thus we have

$$\int_0^\infty \frac{\phi(ax)-\phi(bx)}{x}dx = \phi_0 \log\left(\frac{b}{a}\right).$$

In the same way we can prove that if

$$A+B+C=0,\quad Aa+Bb+Cc=0,$$

then $$\int_0^\infty \frac{dx}{x^2}[A\phi(ax)+B\phi(bx)+C\phi(cx)] = -(\Sigma A a \log a)\phi'(0).$$

For examples, take $\phi(x)=\cos x$ or $\sin x$.

The former integral can also be evaluated if $\phi(x)$ tends to a definite value ϕ_1 as x tends to ∞. For we have the identity

$$\int_\delta^\lambda \frac{\phi(ax)-\phi(bx)}{x}\,dx=\int_a^b \frac{\phi(x\delta)-\phi(x\lambda)}{x}\,dx,$$

by means of which the value of Frullani's integral may be proved to be

$$(\phi_0-\phi_1)\log(b/a).$$

The integral found in Art. 167 (1) is a particular case of this formula and also of that on p. 432.

Extensions of these integrals have been considered by Lerch and by Hardy.*

171. Uniform convergence of an infinite integral.†

If we consider the integral (supposed convergent)

$$\int_a^\infty f(x, y)dx,$$

the *least* value of ξ, for which the inequality

$$\left|\int_\lambda^\infty f(x, y)\,dx\right|<\epsilon, \quad (\lambda>\xi),$$

holds, is a function of y as well as of ϵ. In correspondence with Art. 43, we say that the integral **converges uniformly** in an interval (α, β), if for **all** values of y in the interval ξ remains less than a function $X(\epsilon)$, which depends on ϵ but is **independent of** y. But, if this condition cannot be satisfied for any interval which contains a particular value y_0, then y_0 is said to be a **point of non-uniform convergence** of the integral.

Ex. 1. If $f(x, y)=1/(x+y)^2$, where $y\geqq 0$, we find that

$$\int_\lambda^\infty f(x, y)dx=1/(\lambda+y).$$

Thus $\xi=(1/\epsilon)-y$ (if $y<1/\epsilon$), or $\xi=0$ (if $y>1/\epsilon$),
and so the integral is uniformly convergent for all *positive* values of y, since we may take $X(\epsilon)=1/\epsilon$.

Ex. 2. If $f(x, y)=y/(1+x^2y^2)$, we find that

$$\int_\lambda^\infty f(x, y)\,dx=\cot^{-1}(\lambda y), \quad \text{if } y\gtrless 0, \quad \text{or}=0, \quad \text{if } y=0.$$

Thus $\xi=\cot\epsilon/|y|$, if $y\gtrless 0$, or $\xi=0$, if $y=0$.

Hence $y=0$ is a point of non-uniform convergence for the integral.

---

* Lerch, *Sitzungsberichte d. k. Böhmischen Gesellschaft der Wiss.*, June 2, 1893; Hardy, *Messenger of Maths.*, vol. 34, 1904, pp. 11, 102, and *Quarterly Journal*, vol. 33, 1901, p. 113. See also Art. 173, Exs. 1–4.

† Stokes, *Math. and Phys. Papers*, vol. 1, p. 283.

But, just as in the case of series, we have usually in practice to introduce a test for uniform convergence which is similar to the general test for convergence (Art. 169): *The necessary and sufficient condition for the* **uniform** *convergence of the integral* $\int_a^\infty f(x, y)dx$, *in an interval of values of* y, *is that we can find a value of* ξ, **independent of** y, *such that*

$$\left|\int_\xi^{\xi'} f(x, y)dx\right| < \epsilon,$$

where ξ' *has* **any** *value greater than* ξ, *and* ϵ *is arbitrarily small.*

The only fresh point introduced is seen to be the fact that ξ must be *independent of* y. The proof that this condition is both necessary and sufficient follows precisely on the lines of Art. 43, with mere verbal alterations.

But in practical work we need more special tests which can be applied more quickly; the three most useful of these tests are:

1. **Weierstrass's test.**

Suppose that for all values of y *in the interval* (α, β), *the function* $f(x, y)$ *satisfies the condition*

$$|f(x, y)| < M(x),$$

where $M(x)$ *is a positive function,* **independent of** y. *Then, if the integral* $\int_a^\infty M(x)\,dx$ *converges, the integral* $\int_a^\infty f(x, y)\,dx$ *is absolutely and uniformly convergent for all values of* y *in the interval* (α, β).

For then we can choose ξ independently of y, so that $\int_\xi^\infty M(x)dx$ is less than ϵ; and therefore

$$\left|\int_\xi^{\xi'} f(x, y)dx\right| < \int_\xi^{\xi'} M(x)dx < \int_\xi^\infty M(x)dx < \epsilon.$$

Thus the integral converges uniformly; and it converges absolutely in virtue of Art. 169, (1).

2. **Abel's test.***

The integral $\int_a^\infty f(x, y)\phi(x)dx$ *is uniformly convergent in an interval* (α, β), *provided that* $\int_a^\infty \phi(x)dx$ *converges, and that,*

---

* Bromwich, *Proc. Lond. Math. Soc.* (2), vol. 1, 1903, p. 201.

for every fixed value of y in the interval (α, β), the function $f(x, y)$ is positive and steadily decreases as x increases, while $f(a, y)$ is less than a constant K (independent of y).

For then, in virtue of the analogue to Abel's Lemma, we have

$$\left|\int_{\xi}^{\xi'} f(x, y)\phi(x)dx\right| < Hf(\xi, y) < Hf(a, y) < HK,$$

where H is the upper limit to the expression $\left|\int_{\xi}^{\xi_1}\phi(x)dx\right|$, when ξ_1 ranges from ξ to ξ'. Now, since $\int_a^\infty \phi(x)dx$ is convergent, we can find ξ independently of y, so that $H < \epsilon/K$; and consequently the given integral converges uniformly.

It is evident that $\phi(x)$ may be replaced by $\phi(x, y)$, provided that $\int_a^\infty \phi(x, y)dx$ is *uniformly* convergent in the interval (α, β).

3. **Dirichlet's test.**

The integral $\int_a^\infty f(x, y)\phi(x)dx$ is uniformly convergent in an interval (α, β) if $\int_a^\infty \phi(x)dx$ oscillates between finite limits, and the function $f(x, y)$ is positive and steadily decreases as x increases (y being kept constant), provided that $f(x, y)$ tends to zero uniformly with respect to y in the interval (α, β).

For then $$\left|\int_{\xi}^{\xi'} f(x, y)\phi(x)dx\right| < Hf(\xi, y),$$

where H is less than some constant independent of y; we can then fix ξ, independently of y, to satisfy $f(\xi, y) < \epsilon/H$.

Again, $\phi(x)$ may contain y, provided that the extreme limits of $\int_a^\infty \phi(x)dx$ remain *finite* throughout the interval (α, β).

Ex. 3. *Weierstrass's test.*

$$\int_1^\infty \frac{\cos(xy)}{x^{1+a}}dx,\ \int_1^\infty \frac{\sin(xy)}{x^{1+a}}dx,\ \int_0^\infty \frac{\cos(xy)}{1+x^2}dx,\ \int_0^\infty \frac{\sin(xy)}{1+x^2}dx,\quad (a>0),$$

converge uniformly throughout any interval of variation of y.

Ex. 4. *Abel's test.*

$$\int_a^\infty e^{-xy}\frac{\cos x}{x}dx,\ \int_a^\infty e^{-xy}\frac{\sin x}{x}dx,\quad (a>0),$$

converge uniformly in any interval $(0 \leqq y \leqq A)$, because the integrals

$$\int_a^\infty \frac{\cos x}{x}dx,\ \int_a^\infty \frac{\sin x}{x}dx$$

converge in virtue of Art. 169, Ex. 2. [STOKES, *l.c.*, p. 284.]

Generally $\int_a^\infty e^{-xy}\phi(x)dx$ converges uniformly in any similar interval, provided that $\int_a^\infty \phi(x)dx$ converges.

Ex. 5. *Dirichlet's test.*

$$\int_1^\infty \frac{\cos x}{(x^2+y^2)^{\frac{1}{2}}}dx,\quad \int_1^\infty \frac{\sin x}{(x^2+y^2)^{\frac{1}{2}}}dx$$

converge uniformly throughout any interval of variation of y.

And $$\int_1^\infty \frac{x\cos(xy)}{1+x^2}dx,\quad \int_1^\infty \frac{x\sin(xy)}{1+x^2}dx,\quad \int_1^\infty \frac{\sin(xy)}{x}dx$$

converge uniformly in any interval which does not include $y=0$.

Of course the definition of, and the tests for, uniform convergence can be modified at once so as to refer to the second type of infinite integrals.

172. Applications of uniform convergence.

An integral $\int_a^\infty f(x, y)dx$ which converges uniformly in an interval (α, β) has properties strictly analogous to those of uniformly convergent series (Arts. 45, 46); and the proofs can be carried out on exactly the same lines. Thus we find:

1. *If $f(x, y)$ is a continuous function of y in the interval (α, β), the integral is also a continuous function of y, provided that it converges uniformly in the interval (α, β).**

Only verbal alterations are needed in the proof of the corresponding theorem for series (see Art. 45).

Ex. 1. Thus (see Ex. 3, Art. 171)

$$\int_1^\infty \frac{\cos(xy)}{x^{1+a}}dx,\quad \int_1^\infty \frac{\sin(xy)}{x^{1+a}}dx,\quad \int_0^\infty \frac{\cos(xy)}{1+x^2}dx,\quad \int_0^\infty \frac{\sin(xy)}{1+x^2}dx,\qquad (a>0),$$

are continuous functions of y in any interval.

Ex. 2. If $\int_a^\infty \phi(x)dx$ is convergent, then (see Ex. 4, Art. 171)

$$\lim_{y\to 0}\int_a^\infty e^{-xy}\phi(x)\,dx = \int_a^\infty \phi(x)\,dx.$$

[DIRICHLET.]

Ex. 3. But we must not anticipate the continuity at $y=0$ of

$$\int_0^\infty \frac{x\sin(xy)}{1+x^2}dx,\quad \int_0^\infty \frac{\sin(xy)}{x}dx,$$

and it is not hard to see that they are actually discontinuous.

(See Ex. 5, Art. 171, and Ex. 6, Art. 173.)

---

* Stokes, *l.c.*, p. 283.

2. *Under the same conditions as in* (1), *we may integrate with respect to y under the sign of integration, provided that the range falls within the interval* (α, β).

Again, the proof for series needs only verbal changes.

Ex. 4. If $u=\int_0^\infty \frac{\cos(xy)}{1+x^2}dx$, $\int_0^y u\,dy=\int_0^\infty \frac{\sin xy}{x(1+x^2)}dx$,

and if $v=\int_0^\infty \frac{\sin(xy)}{1+x^2}dx$, $\int_0^y v\,dy=\int_0^\infty \frac{1-\cos(xy)}{x(1+x^2)}dx$.

3. *The equation*

$$\frac{d}{dy}\int_a^\infty f(x, y)dx=\int_a^\infty \frac{\partial f}{\partial y}dx$$

*is valid, provided that the integral on the right converges uniformly and that the integral on the left is convergent.**

Write $\quad \phi(x, h)=\int_a^x \left[\frac{f(x, y+h)-f(x, y)}{h}-\frac{\partial f}{\partial y}\right]dx,$

and let us find ξ so that

$$\left|\int_\xi^X \frac{\partial f}{\partial y}dx\right|<\epsilon, \quad \text{if } X>\xi,$$

where ξ will be independent of y, and the inequality is correct for all values of y in the interval (α, β). Then, if $X>\xi$,

$$\int_\xi^X \frac{1}{h}[f(x, y+h)-f(x, y)]dx=\int_\xi^X dx\int_y^{y+h}\frac{1}{h}\frac{\partial f}{\partial y}dy$$

$$=\frac{1}{h}\int_y^{y+h}dy\int_\xi^X \frac{\partial f}{\partial y}dx,$$

because the value of the double integral of a continuous function, taken over a *finite* area, is independent of the order of integration (see Art. 163, p. 410).

Thus, $\quad \left|\int_\xi^X \frac{1}{h}[f(x, y+h)-f(x, y)]dx\right|<\epsilon,$

in virtue of our choice of ξ.

Now $\phi(X, h)-\phi(\xi, h)=\int_\xi^X \left[\frac{f(x, y+h)-f(x, y)}{h}-\frac{\partial f}{\partial y}\right]dx,$

so that $\quad |\phi(X, h)-\phi(\xi, h)|<2\epsilon.$

* The last condition is partly superfluous; compare the note on p. 119 for the case of series.

The last inequality holds for all values of X greater than ξ, and all values of $|h|$ under a certain limit. If we make X tend to ∞, we obtain

$$|\phi(\infty, h)| \leqq |\phi(\xi, h)| + 2\epsilon.$$

Since our choice of ξ is independent of h, we can now allow h to tend to zero without changing ξ; and, by definition,

$$\lim_{h\to 0} \phi(\xi, h) = 0;$$

thus we have

$$\overline{\lim_{h\to 0}} |\phi(\infty, h)| \leqq 2\epsilon.$$

Since ϵ is arbitrarily small, and $\phi(\infty, h)$ is independent of ϵ, this inequality can only be true if

$$\lim_{h\to 0} \phi(\infty, h) = 0,$$

or

$$\lim_{h\to 0} \int_a^\infty \frac{f(x, y+h) - f(x, y)}{h} dx = \int_a^\infty \frac{\partial f}{\partial y} dx.$$

Another theorem may be mentioned here, although the ideas involved are a little beyond our scope.

If, in the integral $F(z) = \int_a^\infty f(x, z)dx$, *the function* $f(x, z)$ *is an analytic function of the complex variable* z *at all points of a certain region* T *of the* z-*plane, then* $F(z)$ *is analytic within* T, *provided that a real positive function* $M(x)$ *can be found which makes the integral* $\int_a^\infty M(x)dx$ *convergent and satisfies the condition* $\left|\frac{\partial f}{\partial z}\right| < M(x)$ *at all points of* T.

Ex. 5. To shew the need for some condition such as that of uniform convergence, we may consider the integral $\int_0^\infty \frac{\sin(xy)}{x} dx$; if this is differentiated with respect to y under the integral sign, we find $\int_0^\infty \cos(xy)\, dx$, which does not converge.

Ex. 6. On the other hand, the equations

$$\frac{d}{dy}\int_0^\infty \frac{\cos(xy)}{1+x^2} dx = -\int_0^\infty \frac{x \sin(xy)}{1+x^2} dx,$$

$$\frac{d}{dy}\int_0^\infty e^{-xy} \frac{\sin x}{x} dx = -\int_0^\infty e^{-xy} \sin x\, dx$$

are quite correct. (See Exs. 1, 6, Art. 173.)

4. The analogue of Tannery's theorem (Art. 49).

If $\lim_{n\to\infty} f(x, n) = g(x), \quad \lim_{n\to\infty} \lambda_n = \infty,$

then

$$\lim_{n\to\infty} \int_a^{\lambda_n} f(x, n)dx = \int_a^\infty g(x)dx,$$

provided that $f(x, n)$ tends to its limit $g(x)$ uniformly in any fixed interval, and that we can determine a positive function $M(x)$ to satisfy $|f(x, n)| \leqq M(x)$, for all values of n, while $\int_a^\infty M(x)dx$ converges.

For, let ξ be chosen so that $\int_\xi^\infty M(x)dx$ is less than ϵ, then, if n is large enough to make $\lambda_n > \xi$, we have (as in Art. 49)

$$\left| \int_a^{\lambda_n} f dx - \int_a^\infty g\,dx \right| < \int_a^\xi |f-g|\,dx + 2\epsilon.$$

Since ξ is fixed, $|f-g|$ will tend to zero uniformly (as n tends to ∞) in the integral on the right; and so this integral tends to zero (Art. 45 (2)). The proof can now be completed in exactly the same way as in Art. 49.

Ex. 7. To see that some test such as Tannery's is necessary, consider the integral

$$\int_0^n \left(\frac{1}{1+x^2} - \frac{n}{n^2+x^2}\right) dx = \text{arc tan}\,\frac{n-1}{n+1}.$$

Since $n/(n^2+x^2) \leqq 1/n$, we have $\lim n/(n^2+x^2) = 0$, and so if we apply the rule, without reference to the existence of $M(x)$, we find the limit

$$\int_0^\infty \frac{dx}{1+x^2} = \frac{\pi}{2}.$$

But $(n-1)/(n+1)$ tends to 1, so that the integral approaches the limit $\frac{1}{4}\pi$ and not $\frac{1}{2}\pi$.

The second type of infinite integrals.

The reader should find little difficulty in stating and proving results, corresponding to (1)–(4) above, for the second type of integrals.

There is only one case of practical interest which may be found to offer some difficulty; this is the problem of differentiating an integral of the type*

$$F(y) = \int_a^b f(x, y)dx, \quad b = b(y) > a,$$

in which the upper limit varies with y, and is a point of discontinuity for $\frac{\partial f}{\partial y}$, although f is continuous there.

We assume that the integral $\int_a^b \frac{\partial f}{\partial y} dx$ is uniformly convergent for all values of y belonging to the interval with which we are concerned, and that $|b'(y)|$ remains less than a constant B for these values of y.

Then we can find a constant δ such that, if $0 < \xi < \delta$, we have

$$\left| \int_{b-\delta}^{b-\xi} \frac{\partial f}{\partial y} dx \right| < \frac{\epsilon}{1+B}, \quad \text{and} \quad |f(b-\delta, y) - f(b-\xi, y)| < \frac{\epsilon}{1+B}.$$

Now, if we write

$$\phi(\xi, y) = \int_a^{b-\xi} f dx, \quad Y = \int_a^b \frac{\partial f}{\partial y} dx + b'(y)f(b, y)$$

---

* A very simple example is given by taking, say,

$$F(y) = \int_0^y \sqrt{\{x(y-x)\}}dx, \quad y > 0.$$

we have, by the ordinary theorem for differentiating an integral,

$$\frac{d\phi}{dy}=\int_a^{b-\xi}\frac{\partial f}{\partial y}dx+b'(y)f(b-\xi, y),$$

and so, using the inequalities which define δ, we find

$$\left|\frac{d\phi}{dy}-Y\right|<\epsilon, \quad \text{if } 0<\xi\leqq\delta.$$

Also,

$$\frac{d}{dy}\{\phi(\xi, y)-\phi(\delta, y)\}=\int_{b-\delta}^{b-\xi}\frac{\partial f}{\partial y}dx+b'(y)\{f(b-\xi, y)-f(b-\delta, y)\},$$

so that

$$\left|\frac{d}{dy}\{\phi(\xi, y)-\phi(\delta, y)\}\right|<\epsilon;$$

and so, using a double integral as on p. 437, we see that

$$\left|\frac{1}{h}\{\phi(\xi, y+h)-\phi(\xi, y)\}-\frac{1}{h}\{\phi(\delta, y+h)-\phi(\delta, y)\}\right|<\epsilon.$$

In the last inequality, let ξ tend to 0, and $\phi(\xi, y)$ then tends to $F(y)$, so that

$$\left|\frac{1}{h}\{F(y+h)-F(y)\}-\frac{1}{h}\{\phi(\delta, y+h)-\phi(\delta, y)\}\right|\leqq\epsilon.$$

Thus we see that

$$\left|\frac{1}{h}\{F(y+h)-F(y)\}-Y\right|<2\epsilon+\left|\frac{1}{h}\{\phi(\delta, y+h)-\phi(\delta, y)\}-\frac{d\phi}{dy}\right|.$$

Take the limit of the last inequality as h tends to zero; then the right-hand tends to 2ϵ, because δ is independent of h; thus we find

$$\overline{\lim_{h\to 0}}\left|\frac{1}{h}\{F(y+h)-F(y)\}-Y\right|\leqq 2\epsilon,$$

and so by the same argument as before, we have

$$F'(y)=Y.$$

173. Applications of Art. 172.

Ex. 1. Consider first the integral

$$J=\int_0^\infty e^{-xy}(e^{-ax}-e^{-bx})\frac{dx}{x},$$

where a, b may be complex, provided that they have their real parts positive or zero.

Then J is uniformly convergent for all positive or zero values of y.* Now differentiate with respect to y. We obtain

$$-\int_0^\infty e^{-xy}(e^{-ax}-e^{-bx})dx=-\frac{1}{a+y}+\frac{1}{b+y},$$

and this integral converges uniformly so long as $y\geqq l>0$. Its value is therefore equal to dJ/dy, in virtue of Art. 172 (3).

* It is understood that neither a nor b is zero.

Now, $\lim_{y\to\infty} J=0$, by Art. 172 (1), so that

$$J=\int_y^\infty\left(\frac{1}{a+y}-\frac{1}{b+y}\right)dy.$$

Thus, using Ex. 2, Art. 172, we find

$$\int_0^\infty (e^{-ax}-e^{-bx})\frac{dx}{x}=\lim_{y\to 0} J=\int_0^\infty\left(\frac{1}{a+y}-\frac{1}{b+y}\right)dy.$$

The last integral can be written as $\log(b/a)$, which has the advantage of being of the same form as if a and b were real; but owing to the many-valued nature of the logarithm of a complex number, it is often safer to appeal directly to the integral.

In particular, if we write $a=1$, $b=i$, we have

$$\int_0^\infty (e^{-x}-e^{-ix})\frac{dx}{x}=\int_0^\infty\left(\frac{1}{1+y}-\frac{1}{y+i}\right)dy=\left[\log\frac{1+y}{\sqrt{(1+y^2)}}+i\tan^{-1}y\right]_0^\infty=\tfrac{1}{2}\pi i$$

or

$$\int_0^\infty (e^{-x}-\cos x)\frac{dx}{x}=0,\quad \int_0^\infty \sin x\frac{dx}{x}=\tfrac{1}{2}\pi.$$

Ex. 2. Generally, we can prove in the same way that

$$\int_0^\infty (\Sigma A e^{-ax})\frac{dx}{x}=\Sigma\int_0^\infty \frac{A}{a+y}dy=-\Sigma A\log a,$$

where $\Sigma A=0$ and the real parts of a, b, c, ... are positive or zero.

Ex. 3. By direct integration combined with (1) it will be found that

$$\int_0^\infty [e^{-ax}\{1+(a+c)x\}-e^{-bx}\{1+(b+c)x\}]\frac{dx}{x^2}=b-a+c\log\left(\frac{b}{a}\right),$$

where the logarithm is determined as before, and the real parts of a, b are not negative.

For example, if we take

$$a=1,\quad b=-i,\quad c=i,$$

we get

$$\int_0^\infty [e^{-x}\{1+(1+i)x\}-e^{ix}]\frac{dx}{x^2}=-i-1+\frac{\pi}{2},$$

or

$$\int_0^\infty [e^{-x}(1+x)-\cos x]\frac{dx}{x^2}=\frac{\pi}{2}-1,\quad \int_0^\infty (xe^{-x}-\sin x)\frac{dx}{x^2}=-1.$$

As another illustration take

$$b=1,\quad c=-(a+\tfrac{1}{2}).$$

Then we find

$$\int_0^\infty\left[\left(a-1\right)e^{-x}+\left(\frac{1}{x}-\frac{1}{2}\right)(e^{-ax}-e^{-x})\right]\frac{dx}{x}=\left(a+\frac{1}{2}\right)\log a-(a-1).$$

Ex. 4. It is easy to prove similarly that

$$\int_0^\infty [\Sigma\{A_1(1+ax)+A_2x\}e^{-ax}]\frac{dx}{x^2}=-\Sigma A_2\log a-\Sigma A_1 a,$$

where $\Sigma A_1=0$, $\Sigma A_2=0$, and the real parts of a, b, c, ... are positive or zero.

Ex. 5. By differentiating $J=\int_0^\infty \frac{\sin(xy)}{x(1+x^2)}dx$ twice, we find that

$$\frac{d^2J}{dy^2}-J=\int_0^\infty \sin(xy)\frac{dx}{x}=\frac{\pi}{2}, \quad \text{if } y>0;$$

the last result following from (1) above.

Hence, since $\lim_{y\to 0} J=0$, and J remains finite as y tends to ∞ (see Ex. 1, Art. 172), we find $J=\frac{1}{2}\pi(1-e^{-y})$.

Thus, on differentiating, we find, if y is positive,

$$\int_0^\infty \frac{\cos(xy)}{1+x^2}dx=\tfrac{1}{2}\pi e^{-y}=\int_0^\infty \frac{x\sin(xy)}{1+x^2}.$$

When y is negative we find $J=\frac{1}{2}\pi(e^y-1)$, and so the other integrals become $\frac{1}{2}\pi e^y$, $-\frac{1}{2}\pi e^y$, respectively.

Thus J and the cosine integral are continuous at $y=0$. But the third integral is discontinuous there (see Exs. 1, 3, Art. 172).

In like manner, the integral $\int_0^\infty \sin(xy)\frac{dx}{x}$ has the value $\pm\frac{1}{2}\pi$, according to the sign of y, and vanishes for $y=0$.

Ex. 6. As an example of Tannery's theorem, we take the integral

$$I_n=\frac{1}{n}\int_0^b f(x)\left(\frac{\sin nx}{x}\right)^2 dx, \quad (b>0)$$

in which $|f(x)|$ is supposed less than the constant H in the interval $(0, b)$.

Here
$$I_n=\int_0^{nb} f\left(\frac{x}{n}\right)\frac{\sin^2 x}{x^2}dx,$$

so that
$$\lim_{n\to\infty} I_n=f(0)\int_0^\infty \frac{\sin^2 x}{x^2}dx=\frac{\pi}{2}f(0).$$

For
$$\int_0^\infty \frac{\sin^2 x}{x^2}dx=\left[-\frac{\sin^2 x}{x}\right]_0^\infty+\int_0^\infty \frac{\sin 2x}{x}dx=\frac{\pi}{2},$$

this result following from (1) above. In applying Tannery's theorem we can take $M(x)=H(\sin^2 x)/x^2$; and $f(x/n)$ tends to the limit $f(0)$ uniformly in any *fixed* interval for x. It is understood here that $f(0)$ denotes the limit of $f(x)$ as x tends to 0 through *positive* values.

Ex. 7. It follows at once from (6) that if

$$J_n=\frac{1}{n}\int_0^b f(x)\left(\frac{\sin nx}{\sin x}\right)^2 dx, \quad (0<b<\pi)$$

then
$$\lim_{n\to\infty} J_n=\tfrac{1}{2}\pi f(0).$$

The reader should prove that if $b=\pi$, this result must be replaced by
$$\lim J_n=\tfrac{1}{2}\pi\{f(0)+f(\pi)\}.$$

The integral J_n is interesting on account of an application to Fourier Series given by Fejér (see Ex. 5, Ch. XI.).

174. Some further theorems on integrals containing another variable.

Just as Tannery's theorem (Art. 172) resembles Weierstrass's test for uniform convergence, so there is a theorem related in a similar way to Abel's test (Art. 171).

If $f(x, n)$ is positive and steadily decreases (as x increases, n being kept constant) and if $\int_a^\infty \phi(x)dx$ is convergent, then

$$\lim_{n\to\infty} \int_a^{\lambda_n} f(x, n)\phi(x)dx = \int_a^\infty g(x)\phi(x)dx,$$

provided that $\lim \lambda_n = \infty$, that $f(x, n)$ tends to the limit $g(x)$ uniformly in any fixed interval, and that $f(a, n)$ is less than a constant A for all values of n.

For then we can write, by Art. 168,

$$\left|\int_\xi^{\lambda_n} f(x, n)\phi(x)dx\right| < f(\xi, n)H < f(a, n)H < AH,$$

where H is the upper limit to $\left|\int_\xi^{\xi'} \phi(x)dx\right|$ as ξ' ranges from ξ to ∞. Since the last integral converges when extended to infinity, we can find ξ so as to make $AH < \epsilon$; and then also

$$\left|\int_\xi^\infty g(x)\phi(x)dx\right| < g(\xi)H < AH < \epsilon,$$

because, as x increases, $g(x)$ decreases and does not exceed A.

Consequently we find

$$\left|\int_a^{\lambda_n} f(x, n)\phi(x)dx - \int_a^\infty g(x)\phi(x)dx\right|$$
$$< 2\epsilon + \left|\int_a^\xi \{f(x, n) - g(x)\}\phi(x)dx\right|.$$

Since ξ is fixed the limit of the last integral as n tends to ∞, is zero by Art. 45 (2); and so we have

$$\overline{\lim_{n\to\infty}} \left|\int_a^{\lambda_n} f(x, n)\phi(x)dx - \int_a^\infty g(x)\phi(x)dx\right| \leqq 2\epsilon.$$

It follows that this maximum limit is zero, or

$$\lim_{n\to\infty} \int_a^{\lambda_n} f(x, n)\phi(x)dx = \int_a^\infty g(x)\phi(x)dx.$$

Dirichlet's first integral.

As an application of this theorem, consider the integral

$$J_n = \int_0^b \frac{\sin nx}{x} f(x)\,dx, \qquad (b > 0).$$

If we change the variable of integration to x/n, we have

$$J_n = \int_0^{nb} f\left(\frac{x}{n}\right) \frac{\sin x}{x}\,dx.$$

Hence, if $f(x)$ is positive and never increases, our theorem can be applied, because*

$$f(0) \geqq f(x/n) > 0,$$

and $f(x/n)$ tends to the limit $f(0)$ uniformly in any fixed interval.

Hence $$\lim_{n \to \infty} J_n = f(0) \int_0^\infty \frac{\sin x}{x}\,dx = \frac{\pi}{2} f(0),$$

in virtue of Art. 173, Ex. 1, above.

It is, however, easy to remove the conditions from the function $f(x)$ of being positive and never increasing. Suppose, for example, that $f(x)$ first decreases in the interval $(0, c)$, and afterwards increases in the interval (c, b). Now consider the functions $F(x)$, $G(x)$ defined by

$$\left.\begin{aligned} F(x) &= f(x) + A, \\ G(x) &= \phantom{f(x) + {}} A \end{aligned}\right\} \quad (0 \leqq x \leqq c)$$

and

$$\left.\begin{aligned} F(x) &= f(c) + A, \\ G(x) &= f(c) + A - f(x), \end{aligned}\right\} \quad (c \leqq x \leqq b)$$

where A is a constant such that $f(c) + A$ and $f(c) + A - f(b)$ are both positive. Then the conditions of being positive and never increasing are satisfied by both $F(x)$ and $G(x)$, so that

$$\lim_{n \to \infty} \int_0^b F(x) \frac{\sin nx}{x}\,dx = \frac{\pi}{2} F(0),$$

with a similar equation for $G(x)$. But $F(x) - G(x) = f(x)$, so that

$$\lim_{n \to \infty} \int_0^b f(x) \frac{\sin nx}{x}\,dx = \frac{\pi}{2} f(0).$$

It is easy to see that *this result can be at once extended to any case where $f(x)$ has a limited number of maxima and minima and no infinities between* 0 *and* b.

---

* Here we use $f(0)$ to denote the limit of $f(x)$ as x approaches 0 through *positive* values; this limit exists in virtue of the monotonic property of $f(x)$.

Dirichlet's second integral.

Consider now

$$K_n = \int_0^b \frac{\sin(2n+1)x}{\sin x} f(x)dx, \quad (0 < b < \pi)$$

where n is an integer. We can write

$$K_n = \int_0^b \frac{\sin \nu x}{x} \phi(x)dx,$$

where $$\nu = 2n+1, \quad \phi(x) = \frac{x}{\sin x} f(x).$$

Since $x/\sin x$ steadily increases and has no infinity in the interval $(0, b)$, it follows that $\phi(x)$ will satisfy the conditions set forth in dealing with Dirichlet's first integral, provided that $f(x)$ satisfies them.*

Hence $$\lim_{n\to\infty} K_n = \tfrac{1}{2}\pi\phi(0) = \tfrac{1}{2}\pi f(0).$$

If, however, the range of integration extends up to π, we may write the integral in the form

$$\left(\int_0^{\frac{1}{2}\pi} + \int_{\frac{1}{2}\pi}^{\pi}\right) \frac{\sin \nu x}{\sin x} f(x)dx,$$

and then change the variable in the second part to $\pi - x$; this gives

$$\int_0^{\frac{1}{2}\pi} \frac{\sin \nu x}{\sin x} [f(x) + f(\pi - x)]\, dx.$$

Hence $$\lim_{n\to\infty} \int_0^{\pi} \frac{\sin(2n+1)x}{\sin x} f(x)dx = \frac{\pi}{2}[f(0) + f(\pi)].$$

Ex. 1. As a verification we note that

$$\frac{\sin(2n+1)x}{\sin x} dx = 1 + 2\cos 2x + 2\cos 4x + \dots + 2\cos 2nx,$$

so that $$\int_0^{\frac{1}{2}\pi} \frac{\sin(2n+1)x}{\sin x} dx = \tfrac{1}{2}\pi$$

and $$\int_0^{\pi} \frac{\sin(2n+1)x}{\sin x} dx = \pi,$$

which agree with the general theorems on writing $f(x) = 1$.

---

* For then we can write

$$x/\sin x = B - \chi(x), \quad f(x) = F(x) - G(x),$$

where $\chi(x)$, $F(x)$, $G(x)$ are positive and never increase in the interval $(0, b)$, while B is a positive constant. Then

$$\phi(x) = \{BF(x) + \chi(x)G(x)\} - \{BG(x) + \chi(x)F(x)\}.$$

Ex. 2. It is instructive to investigate the value of the integral

$$K_n = \int_0^b \frac{\sin(2n+1)x}{\sin x} dx, \quad (0 < b \leqq \tfrac{1}{2}\pi),$$

by means of the curve $y = \{\sin(2n+1)x\}/\sin x$. This curve is of the same general type as the one given in Fig. 46, Art. 169; except that the initial ordinate is $y = 2n+1$ and that the points of crossing the axis are

$$\pi/(2n+1), \quad 2\pi/(2n+1), \quad \ldots, \quad n\pi/(2n+1).$$

Then, using the argument given there, we see that the value of the integral K_n is expressed by a finite series of the form

$$v_0 - v_1 + v_2 - \ldots + (-)^k v_k,$$

where k is an integer such that $(2n+1)b$ lies between $(k-1)\pi$ and $k\pi$, and

$$v_0 > v_1 > v_2 > \ldots > v_k > 0.$$

Hence (if r is any integer less than k), the value of the integral K_n differs from

$$v_0 - v_1 + v_2 - \ldots + (-1)^{r-1} v_{r-1}$$

by less than v_r. Thus, changing the variable to x/ν, K_n lies between

$$\int_0^{r\pi} \frac{\sin x}{\nu \sin(x/\nu)} dx \quad \text{and} \quad \int_0^{(r+1)\pi} \frac{\sin x}{\nu \sin(x/\nu)} dx,$$

where $\nu = 2n+1$. If we make n tend to ∞, we find that the limit of the integral K_n lies between*

$$\int_0^{r\pi} \frac{\sin x}{x} dx \quad \text{and} \quad \int_0^{(r+1)\pi} \frac{\sin x}{x} dx,$$

where r is any positive integer. Thus

$$\lim_{n\to\infty} K_n = \int_0^\infty \frac{\sin x}{x} dx = \frac{\pi}{2}. \qquad \text{[DIRICHLET.]}$$

Ex. 3. It is easy to see (as in Ex. 2 or otherwise) that

$$\lim_{n\to\infty} \int_0^{b_n} \frac{\sin(2n+1)x}{\sin x} dx = \int_0^\lambda \frac{\sin x}{x} dx,$$

where $\lambda = \lim(2nb_n)$. The maxima of this integral are given by $\lambda = \pi, 3\pi, 5\pi, \ldots$ and the minima by $\lambda = 2\pi, 4\pi, 6\pi, \ldots$.

Glaisher (*Phil. Trans.*, vol. 160, 1870, p. 387) has given the following numerical values for the maxima and minima, where

$$I_r = \int_{r\pi}^\infty \frac{\sin x}{x} dx = \frac{\pi}{2} - \int_0^{r\pi} \frac{\sin x}{x} dx,$$

$$I_1 = -0{\cdot}28114, \quad I_3 = -0{\cdot}10397, \quad I_5 = -0{\cdot}06317,$$
$$I_2 = +0{\cdot}15264, \quad I_4 = +0{\cdot}07864, \quad I_6 = +0{\cdot}05276.$$

Thus the greatest value of the integral is $\frac{1}{2}\pi - I_1 = 1{\cdot}85194$, and the least (if $\lambda > \pi$) is $\frac{1}{2}\pi - I_2 = 1{\cdot}41816$. (See also Ex. 12, p. 351.)

---

* From the inequalities proved in Art. 59 and in the footnote p. 184, we see that (since $x/\nu < \frac{1}{2}\pi$),

$$\left|\frac{x}{\nu} - \sin\frac{x}{\nu}\right| < \frac{1}{6}\frac{x^3}{\nu^3}, \quad \operatorname{cosec}\frac{x}{\nu} < \frac{2\nu}{x}, \quad \left|\frac{\sin x}{x}\right| < 1.$$

Thus $\left|\frac{\sin x}{x} - \frac{\sin x}{\nu\sin(x/\nu)}\right| = \left|\frac{\sin x}{x}\right| \cdot \frac{1}{\sin(x/\nu)} \cdot \left|\frac{x}{\nu} - \sin\frac{x}{\nu}\right| < \frac{1}{3}\frac{x^2}{\nu^2} \leqq \frac{\pi^2}{3}\frac{(r+1)^2}{\nu^2}$,

and so this difference tends to zero uniformly in the interval $0 \leqq x \leqq (r+1)\pi$

Ex. 4. If $f(x)$ is positive and steadily decreases in such a way that $\Sigma f(n\pi)$ is convergent, we can prove that

$$\lim_{n\to\infty}\int_0^\infty \frac{\sin(2n+1)x}{\sin x} f(x)dx = \frac{\pi}{2}[f(0)+2f(\pi)+2f(2\pi)+\ldots];$$

and again, if $f(x)$ tends steadily to zero,

$$\lim_{n\to\infty}\int_0^\infty \frac{\sin(2n+1)x}{\sin x}\cos x f(x)dx = \frac{\pi}{2}[f(0)-2f(\pi)+2f(2\pi)-\ldots].$$

In particular, if $f(x)=e^{-cx}(c>0)$, the first limit is $\frac{1}{2}\pi\coth(\frac{1}{2}c\pi)$ and the second is $\frac{1}{2}\pi\tanh(\frac{1}{2}c\pi)$.

Ex. 5. Shew that if $f(x)$ satisfies the conditions of Dirichlet's integral, and $0<s<1$, then

$$\lim_{n\to\infty} n^s\int_0^b f(x)\frac{\sin nx}{x^{1-s}}dx = f(0)\int_0^\infty x^{s-1}\sin x\,dx = f(0)\,\Gamma(s)\sin(\tfrac{1}{2}s\pi),$$

$$\lim_{n\to\infty} n^s\int_0 f(x)\frac{\cos nx}{x^{1-s}}dx = f(0)\int_0^\infty x^{s-1}\cos x\,dx = f(0)\,\Gamma(s)\cos(\tfrac{1}{2}s\pi).$$

For the values of the integrals, see Ex. 36, p. 474.

Jordan's extension of Dirichlet's integral.

Suppose that $I_n=\int_0^b \phi(x,n)f(x)dx$, where $f(x)$ is positive and never decreases in the interval $(0, b)$, while $\phi(x, n)$ has the properties

$$\text{(i)}\quad \left|\int_0^\xi \phi(x,n)dx\right| < A, \quad \text{if } 0\leqq\xi\leqq b;$$

$$\text{(ii)}\quad \lim_{n\to\infty}\int_0^\xi \phi(x,n)dx = l, \quad \text{if } 0<c\leqq\xi\leqq b,$$

where A is a constant and c is arbitrary, but must be regarded as fixed in taking the limit (ii). Under these circumstances

$$\lim I_n = lf(0),$$

where $f(0)$ denotes the limit of $f(x)$, as on p. 444.

For we have from Art. 168, if $0<c<b$,

$$(h-h')[f(0)-f(c)]+h'f(0)<I_n<(H-H')[f(0)-f(c)]+H'f(0),$$

where H, h are the upper and lower limits of $\int_0^\xi \phi(x,n)dx$ as ξ varies from 0 to c and H', h' as ξ varies from c to b. Now, from (i) $\quad |h-h'|<2A, \quad |H-H'|<2A,$

so that $\quad |I_n-lf(0)|<2A[f(0)-f(c)]+\eta f(0),$

if η is the greater of $H'-l$ and $l-h'$.

Now choose c so that $2A[f(0)-f(c)]<\epsilon$; and having fixed c, make n tend to infinity. Since $\lim_{n\to\infty}\eta=0$ by condition (ii), it follows that

$$\overline{\lim_{n\to\infty}}\,|I_n-lf(0)|\leqq\epsilon.$$

Hence $$\lim I_n=lf(0).$$

This result can be at once extended to any function $f(x)$ of the type considered in dealing with Dirichlet's integral (see p. 444).

175. Integration of series, when infinities of the integrand occur in the range.

It is obvious that infinite integrals are excluded from the discussion of Art. 45: one case of practical importance presents itself when the terms of the series are of the form $\phi(x)f_n(x)$, where $\phi(x)\to\infty$ at, say, the upper limit b. Then we can easily establish the following result:

A. *If $\Sigma f_n(x)$ converges uniformly in the interval (a, b) and $\int_a^b|\phi(x)|\,dx$ is convergent (and has the value J), then*

$$\int_a^b\phi(x)[\Sigma f_n(x)]\,dx=\Sigma\int_a^b\phi(x)f_n(x)dx.$$

For then we can find m, independently of x, so that

$$\left|\sum_m^p f_n(x)\right|<\epsilon,\quad \text{if } p>m.$$

Thus $$\left|\sum_m^p\int_a^b\phi(x)f_n(x)dx\right|<\epsilon\int_a^b|\phi(x)|\,dx=\epsilon J.$$

It follows that $\sum_0^\infty\int_a^b\phi(x)f_n(x)dx$ converges, and that

$$\left|\sum_m^\infty\int_a^b\phi(x)f_n(x)dx\right|\leqq\epsilon J.$$

At the same time we have

$$\left|\sum_0^\infty f_n(x)-\sum_0^{m-1}f_n(x)\right|\leqq\epsilon,$$

so that $$\left|\int_a^b\phi(x)\left[\sum_0^\infty f_n(x)\right]dx-\sum_0^{m-1}\int_a^b\phi(x)f_n(x)dx\right|\leqq\epsilon J.$$

Thus we find

$$\left|\int_a^b\phi(x)\left[\sum_0^\infty f_n(x)\right]dx-\sum_0^\infty\int_a^b\phi(x)f_n(x)dx\right|\leqq 2\epsilon J,$$

and, since J is fixed, we can determine m to make $2\epsilon J$ as small as we please; but in the last inequality the left-hand side is *independent of m*, and therefore must be zero.

Thus $$\int_a^b \phi(x)\left[\sum_0^\infty f_n(x)\right]dx = \sum_0^\infty \int_a^b \phi(x) f_n(x)\,dx.$$

Ex. 1. This case is illustrated by

$$\int_0^1 \log x \log(1+x)\,dx = \Sigma(-1)^{n-1}\int_0^1 \frac{x^n}{n}\log x\,dx = \Sigma\frac{(-1)^n}{n(n+1)^2}$$

$$= \Sigma(-1)^n\left[\frac{1}{n} - \frac{1}{n+1} - \frac{1}{(n+1)^2}\right] = 2 - 2\log 2 - \frac{1}{12}\pi^2.$$

Here the series for $\log(1+x)$ converges uniformly from 0 to 1; but $\log x \to \infty$ as $x \to 0$.

On the other hand, the series may also tend to ∞ (or it may cease to be uniformly convergent) as $x \to b$; when this happens, we can often justify term-by-term integration by means of the theorem:

B. *Suppose that $\phi(x)$ is positive in the interval (a, b), and that the terms $f_n(x)$ are all positive, then the convergence of either the integral*

$$\int_a^b \phi(x)[\Sigma f_n(x)]\,dx,$$

or the series $$\Sigma\int_a^b \phi(x) f_n(x)\,dx,$$

is necessary and sufficient to allow of term-by-term integration.

It is obvious that both conditions are necessary: the only point to be proved is that *either* of them is sufficient.

Write $$F(\delta, m) = \int_a^{b-\delta} \phi(x)\left\{\sum_0^m f_n(x)\right\}dx, \quad (\delta > 0);$$

then, since $\phi(x)$ and $f_n(x)$ are never negative, the function $F(\delta, m)$ *never decreases* as δ tends to zero and m to infinity. Thus, as in Art. 31(5), we see that if either of the repeated limits

$$\lim_{\delta\to 0}\left\{\lim_{m\to\infty} F(\delta, m)\right\}, \quad \lim_{m\to\infty}\left\{\lim_{\delta\to 0} F(\delta, m)\right\},$$

is convergent, so also is the other, and their values are equal.

Now Art. 45 applies to the interval $(a, b-\delta)$, so that

$$\lim_{m\to\infty} F(\delta, m) = \int_a^{b-\delta} \phi(x)\left\{\sum_0^\infty f_n(x)\right\}dx,$$

and so $$\lim_{\delta\to 0}\left\{\lim_{m\to\infty} F(\delta, m)\right\} = \int_a^b \phi(x)\left\{\sum_0^\infty f_n(x)\right\}dx. \quad \text{..................(1)}$$

Similarly the other repeated limit is seen to be equal to the series

$$\sum_0^\infty \int_a^b \phi(x) f_n(x) dx. \quad \ldots\ldots\ldots\ldots\ldots\ldots\ldots\ldots(2)$$

Hence the theorem is established; for if either the integral (1) or the series (2) is convergent, so also is the other, and their values are equal.

Ex. 2. An application of Theorem B is given by the equation

$$\int_0^1 \frac{\log(1-x)}{\sqrt{(1-x)}} dx = -\sum_1^\infty \int_0^1 \frac{x^n dx}{n\sqrt{(1-x)}} = -\frac{4}{3}\left(1+\frac{2}{5}+\frac{2.4}{5.7}+\frac{2.4.6}{5.7.9}+\ldots\right).$$

This result is easily verified directly; by integration by parts, the integral is found to be -4, and the sum of the series in brackets is 3 (see Ex. 2, p. 42).

We get another illustration by expanding $1/\sqrt{(1-x)}$ instead of $\log(1-x)$.

Ex. 3. Another example is given by

$$\int_0^1 \frac{x^p}{1-x} \log x \, dx = \sum_1^\infty \int_0^1 x^{n+p-1} \log x \, dx = -\sum_1^\infty \frac{1}{(n+p)^2},$$

where $p+1$ is *positive*. Here we use Theorem A to include $x=0$ in the interval and Theorem B to include $x=1$.

The special case $p=0$ gives

$$\int_0^1 \frac{\log x}{1-x} dx = -\sum_1^\infty \frac{1}{n^2} = -\frac{\pi^2}{6};$$

and if $p=-\frac{1}{2}$ the integral can also be evaluated in finite terms.

C. *When the terms $f_n(x)$ are not all positive, we can apply a similar argument in case either the integral*

$$\int_a^b |\phi(x)| \{\Sigma |f_n(x)|\} dx \text{ or the series } \Sigma \int_a^b |\phi| \cdot |f_n| dx$$

*converges.** Here we write

$$\phi f_n = \{\phi + |\phi|\}\{f + |f_n|\} - |\phi|\{f + |f_n|\} - |f_n|\{\phi + |\phi|\} + |\phi| \cdot |f_n|,$$

and then, under either of the given conditions, Theorem B can be applied to each term on the right-hand side. Of course these conditions are easily seen to be *sufficient* only and not *necessary* here. Thus, for example, if

$$f_n(x) = (-1)^{n-1} x^n, \quad a=0, \quad b=1,$$

*Hardy, *Messenger of Mathematics*, vol. 35, 1905, p. 126; Bromwich, *ibid.*, vol. 36, 1906, p. 1. It should be noticed that the argument fails if we only know that $\int_a^b \phi(x)[\Sigma f_n(x)] dx$ is *absolutely* convergent.

we have

$$1-\tfrac{1}{2}+\tfrac{1}{3}-\tfrac{1}{4}+\dots=\int_0^1 \frac{dx}{1+x}, \quad \text{(see Arts. 47, 52)}$$

although here $\Sigma|f_n(x)|=1/(1-x)$ and $\int_0^1 dx/(1-x)$ diverges.

Ex. 4. To illustrate Theorem C, consider

$$\int_0^1 \frac{x^p}{1+x}\log x\,dx=\sum_1^\infty(-1)^{n-1}\int_0^1 x^{n+p-1}\log x\,dx=\sum_1^\infty\frac{(-1)^n}{(n+p)^2}, \quad p+1>0.$$

Here we take $\phi(x)=x^p\log x$ and $\Sigma|f_n(x)|=1/(1-x)$, and then the conditions of Theorem C are satisfied (compare Ex. 3). In particular, $p=0$ gives

$$\int^1 \frac{\log x}{1+x}dx=\sum_1^\infty\frac{(-1)^n}{n^2}=-\frac{\pi^2}{12}.$$

But Theorems B, C do not suffice in a number of comparatively simple cases which present themselves in practice and do not come under any really general theorem. In dealing with power-series, the remark made at the top of p. 134 is often useful; and in some cases we can apply Theorem C to $\Sigma|a_{n-1}-\lambda a_n|x^n$, taking λ to be $\lim(a_n/a_{n+1})$, and then proceed as in the following example:

Ex. 5. Consider the integral $\int_0^1 \frac{x^p\log x}{(1+x)^2}dx, \quad p+1>0.$

If $1/(1+x)^2=\Sigma a_n x^n=\Sigma f_n(x)$, $\Sigma|f_n(x)|=1/(1-x)^2$, and Theorem C fails because the integral diverges if $1/(1-x)^2$ is put in place of $1/(1+x)^2$.

Now $a_n=(-1)^n(n+1)$ and $\lambda=-1$. Also $1/(1+x)=\Sigma(a_{n-1}+a_n)x^n$; and by Theorem C,

$$\int_0^1 \frac{x^p\log x}{1+x}\Sigma(a_{n-1}+a_n)x^n=\Sigma(a_{n-1}+a_n)\int_0^1\frac{x^{n+p}\log x}{1+x}dx.$$

The coefficient of a_n on the right is

$$\int_0^1 x^{n+p}\log x\,dx=-1/(n+p+1)^2;$$

and so we find

$$\int_0^1 \frac{x^p\log x}{(1+x)^2}dx=\Sigma(-1)^{n-1}\frac{n+1}{(n+p+1)^2}.$$

In particular, if $p=0$, the series reduces to $-\log 2$, and it is easily verified that this is then the value of the integral; thus our work is confirmed.

Ex. 6. To illustrate the result given on p. 134, consider the equation

$$\int_0^1 \frac{x^p}{1+x}dx=\frac{1}{p+1}-\frac{1}{p+2}+\frac{1}{p+3}-\dots, \quad (p+1>0);$$

this is valid, because the resulting series converges.

Ex. 7. Further applications of Theorems B, C are given by the equations

$$\left.\begin{aligned}\int_0^1 \frac{(\log x)^{2r-1}}{1-x}\,dx &= -(2r-1)!\left(1+\frac{1}{2^{2r}}+\frac{1}{3^{2r}}+\dots\right) = -\frac{2^{2r-2}}{r}B_r\pi^{2r},\\ \int_0^1 \frac{(\log x)^{2r-1}}{1+x}\,dx &= -(2r-1)!\left(1-\frac{1}{2^{2r}}+\frac{1}{3^{2r}}-\dots\right) = -\frac{2^{2r-1}-1}{2r}B_r\pi^{2r},\end{aligned}\right\}\text{Art. 93,}$$

$$\int_0^1 \frac{dx}{x}\log\frac{1+x}{1-x} = 2\left(1+\frac{1}{3^2}+\frac{1}{5^2}+\dots\right) = \frac{\pi^2}{4}.$$

From the last integral we can obtain the results

$$\int_0^\infty \frac{dx}{x}\log\left(\frac{x+y}{x-y}\right)^2 = \pi^2, \qquad \text{if } y>0,$$

$$\text{or} = 0, \qquad \text{if } y=0,$$

$$\text{or} = -\pi^2, \quad \text{if } y<0.$$

Ex. 8. By changing the variable in Ex. 6 from x to t, where $x=e^{-2at}$, we find

$$\int_0^\infty \frac{\cosh(bt)}{\cosh(at)}\,dt = \frac{\pi}{2a}\sec\left(\frac{\pi b}{2a}\right), \qquad 0<b<a.$$

Similarly, starting from the equation

$$\int_0^1 \frac{1-x^{p-1}}{1-x}\,dx = 1-\frac{1}{p}+\frac{1}{2}-\frac{1}{p+1}+\dots, \qquad (p>0)$$

which can be established by Theorem B, we find

$$\int_0^\infty \frac{\sinh(bt)}{\sinh(at)}\,dt = \frac{\pi}{2a}\tan\left(\frac{\pi b}{2a}\right), \qquad 0<b<a.$$

176. Integration of an infinite series over an infinite interval.

The method of proof employed in Art. 45 does not justify the deduction of the equation

$$\int_a^\infty \phi(x)\left[\sum_0^\infty f_n(x)\right]dx = \sum_0^\infty \int_a^\infty \phi(x)f_n(x)\,dx,$$

from the knowledge that $\Sigma f_n(x)$ is uniformly convergent for all values of x greater than a.

A. However, if in addition we know that $\int_a^\infty |\phi(x)|\,dx$ is convergent, the method used to prove Theorem A of Art. 175 can be at once modified to establish the desired result.

But it is often necessary to justify the equation when either $\int_a^\infty |\phi(x)|\,dx$ is divergent or $\Sigma f_n(x)$ can only be proved to converge uniformly over a fixed interval;* and then some new test must be introduced.

Thus, for example, if

$$f_0(x)+f_1(x)+\ldots+f_n(x)=S_n(x)=(2x/n^2)e^{-x^2/n^2},\ \lim_{n\to\infty} S_n(x)=0;$$

and the maximum of $S_n(x)$ is $\sqrt{2}/n\sqrt{e}$, so that $S_n(x)$ converges uniformly to its limit in any interval for x. But yet we find, taking $\phi(x)=1$,

$$\int_0^\infty S_n(x)\,dx=1\,;\quad \text{so that } \lim_{n\to\infty}\int_0^\infty S_n(x)\,dx=1,$$

and this is not the same as $\int_0^\infty [\lim_{n\to\infty} S_n(x)]\,dx.$

This illustrates the case when $\int_a^\infty |\phi(x)|\,dx$ diverges (because $\phi(x)=1$); the other difficulty arises in the integration of series such as the exponential series $\Sigma x^n/n!$, which converges uniformly in any fixed interval (which may be arbitrarily great) but does not converge uniformly in an infinite interval.

B. Many cases of practical importance are covered by the following test:

If $\Sigma f_n(x)$ converges uniformly in any fixed interval $a \leqq x \leqq b$, where b is arbitrary, and if $\phi(x)$ is continuous for all finite values of x, then

$$\int_a^\infty \phi(x)[\Sigma f_n(x)]\,dx = \Sigma\int_a^\infty \phi(x) f_n(x)\,dx\,;$$

provided that either the integral $\int_a^\infty |\phi(x)|\{\Sigma|f_n(x)|\}\,dx$ or the series $\Sigma\int_a^\infty |\phi(x)|.|f_n(x)|\,dx$ is convergent.

For, by means of the identity

$$\phi f_n=\{\phi+|\phi|\}\{f_n+|f_n|\}-|\phi|.\{f_n+|f_n|\}-|f_n|.\{\phi+|\phi|\}+|f_n|.|\phi|,$$

we can at once reduce this theorem to the case in which ϕ and f_n are never negative.

In this case the function

$$F(\lambda,\mu)=\int_a^\lambda \phi(x)[\sum_0^\mu f_n(x)]\,dx$$

never decreases as λ, μ increase; and consequently we can repeat the

---

* The distinction between uniform convergence over a *fixed* and over an *infinite* interval may be illustrated by the two examples $S_n(x)=x/n$ and $S_n(x)=1/(x+n)$. The former converges uniformly to zero in any interval $(0, b)$, where b is *fixed*, but may be taken arbitrarily great; the latter converges uniformly to zero for *all* positive values of x.

arguments given in Art. 31 (5) to prove that if $\lim_{\lambda\to\infty}(\lim_{\mu\to\infty} F(\lambda\ \mu))$ exists, so also does the other repeated limit, and the two limits are equal.

But, in virtue of the uniform convergence of $\Sigma f_n(x)$, we have

$$\lim_{\mu\to\infty} F(\lambda, \mu) = \int_a^\lambda \phi(x)[\sum_0^\infty f_n(x)]dx,$$

so that

$$\lim_{\lambda\to\infty}\{\lim_{\mu\to\infty} F(\lambda, \mu)\} = \int_a^\infty \phi(x)[\sum_0^\infty f_n(x)]\,dx.$$

The other repeated limit is seen in the same way to be

$$\sum_0^\infty \int_a^\infty \phi(x) f_n(x)\,dx,$$

and so the test is established.

Ex. 1. Consider

$$\int_0^\infty \frac{\sin(bx)}{e^{ax}-1}\,dx,$$

where a is positive, and $b=p+iq$, where $|q|=s<a$; since

$$|\sin(bx)| = [\sinh^2(qx)+\sin^2(px)]^{\frac{1}{2}} < \cosh sx < e^{sx}$$

and the integral

$$\int_0^\infty [e^{sx}/(e^{ax}-1)]\,dx$$

is convergent, it follows from Theorem B that term-by-term integration is permissible,* because the terms in the series

$$1/(e^{ax}-1) = e^{-ax}+e^{-2ax}+e^{-3ax}+\ldots$$

are all *positive*. Thus we have

$$\int_0^\infty \frac{\sin(bx)}{e^{ax}-1}\,dx = \frac{b}{a^2+b^2}+\frac{b}{(2a)^2+b^2}+\frac{b}{(3a)^2+b^2}+\ldots.$$

In the case when $a=2\pi$, this expression is equal (by Art. 93) to

$$\frac{1}{2}\left(\frac{1}{e^b-1}-\frac{1}{b}+\frac{1}{2}\right),$$

and so in general it is equal to

$$\frac{\pi}{a}\left(\frac{1}{e^{2\pi b/a}-1}-\frac{a}{2\pi b}+\frac{1}{2}\right).$$

Ex. 2. In like manner we prove that

$$\int_0^\infty \frac{\sin(bx)}{e^{ax}+1}\,dx = \frac{b}{a^2+b^2}-\frac{b}{(2a)^2+b^2}+\frac{b}{(3a)^2+b^2}-\ldots$$

$$=\frac{1}{2}\left[\frac{1}{b}-\frac{\pi}{a\sinh(\pi b/a)}\right],$$

by Art. 92.

---

* Note that exactly the same argument enables us to include 0 in the range of integration, although the series diverges there.

Ex. 3. Taking the case $a=2\pi$, expand both sides of Ex. 1 in powers of b. In this case the application of the theorem depends upon the integral

$$\int_0^\infty \frac{\sinh\{|b|x\}}{e^{2\pi x}-1}dx,$$

which converges if $|b|<2\pi$. Thus we find

$$\int_0^\infty \frac{x^{2r-1}dx}{e^{2\pi x}-1}=\frac{B_r}{4r};$$

see Art. 93 and compare Art. 175, Ex. 7.

Ex. 4. Similarly, by expanding $\sin(bx)$ in powers of x, we find that if $0<b<a$,

$$\int_0^\infty e^{-ax}\sin(bx)dx=\frac{b}{a^2}\left(1-\frac{b^2}{a^2}+\frac{b^4}{a^4}+\dots\right)=\frac{b}{a^2+b^2}.$$

And without restriction on b, we have from the values of $\Gamma(\frac{1}{2})$, $\Gamma(\frac{3}{2})$, ...,

$$\int_0^\infty e^{-ax^2}\cos(2bx)dx=\frac{\sqrt{\pi}}{2\sqrt{a}}\left(1-\frac{b^2}{a}+\frac{1}{2!}\frac{b^4}{a^2}-\dots\right)=\frac{\sqrt{\pi}}{2\sqrt{a}}\exp\left(-\frac{b^2}{a}\right).$$

C. However, Theorem B does not cover all cases which are required. For example, it is not hard to see that the series

$$\sum_1^\infty \frac{\sin(nx)}{n^p x}, \quad (p>1)$$

can be integrated term-by-term between the limits 1 and ∞, although the test given above fails.* This case and others are covered by the following test:

Write $\int_a^x f_n(x)dx=g_n(x)$ *and suppose that the series* $\Sigma f_n(x)$ *converges uniformly in any fixed interval* (a, b), *while the series* $\Sigma g_n(x)$ *converges uniformly in an infinite interval* $(x\geqq a)$; *then*

(1) $\Sigma\left[\int_a^\infty f_n(x)\,dx\right]$ *converges,*

(2) $\int_a^\infty [\Sigma f_n(x)]\,dx$ *converges,*

(3) *the values of* (1) *and* (2) *are equal.*

[DINI.]

For, by Art. 45, we have

$$\int_a^x [\Sigma f_n(x)]\,dx=\Sigma g_n(x).$$

And, since $\Sigma g_n(x)$ is uniformly convergent, we have (Art. 45)

$$\int_a^\infty [\Sigma f_n(x)]\,dx=\lim_{x\to\infty}[\Sigma g_n(x)]=\Sigma\lim_{x\to\infty} g_n(x)=\Sigma\left[\int_a^\infty f_n(x)dx\right].$$

* See the second paper quoted on p. 450.

177. The inversion of a repeated infinite integral.

It is by no means easy to determine fairly general conditions under which the equation*

$$(1) \qquad \int_a^\infty dx \int_b^\infty f(x, y)dy = \int_b^\infty dy \int_a^\infty f(x, y)dx$$

is correct.

Here we shall simply consider the easiest case, when either $f(x, y)$ is *positive* or else the integrals still converge when $f(x, y)$ is replaced by $|f(x, y)|$.

Let us write

$$F(\lambda, \mu) = \int_a^\lambda dx \int_b^\mu f(x, y)dy = \int_b^\mu dy \int_a^\lambda f(x, y)dx,$$

this equation being valid (see p. 410) if, as we suppose, $f(x, y)$ is continuous for all finite values of x, y (or at least for all such as come under consideration). Further, write

$$\phi(x, \mu) = \int_b^\mu f(x, y)dy, \quad \psi(x) = \lim_{\mu\to\infty} \phi(x, \mu) = \int_b^\infty f(x, y)dy,$$

assuming the convergence of the last integral. Let the interval (a, λ) be subdivided by continued bisection into n sub-intervals, each of length l, and let $h_r(\mu)$ denote the minimum of $\phi(x, \mu)$ in the rth interval; then, as in Art. 163, we have

$$F(\lambda, \mu) = \lim_{n\to\infty} \sum_{r=1}^n lh_r(\mu).$$

Now this sum cannot decrease as n and μ tend to infinity;† and so we may use theorem (5), Art. 31, which gives

$$(2) \qquad \lim_{\mu\to\infty} \left\{ \lim_{n\to\infty} \sum_1^n lh_r(\mu)\right\} = \lim_{n\to\infty} \left\{ \lim_{\mu\to\infty} \sum_1^n lh_r(\mu)\right\},$$

provided that *one* of these limits converges. Thus

$$(3) \qquad \lim_{\mu\to\infty} F(\lambda, \mu) = \lim_{n\to\infty} \sum_1^n lk_r, \quad \text{if } k_r = \lim_{\mu\to\infty} h_r(\mu).$$

Now we shall prove below (see the small type, p. 457) that

---

* For wider conditions, see a paper in the *Proc. Lond. Math. Soc.* (2), vol. 1, 1903, p. 187, and other papers quoted there. Reference may also be made to Gibson's *Calculus*, Ch. XXI. (2nd ed.), and Jordan's *Cours d'Analyse*, t. 2, §§ 71, 72.

† As regards n, see the argument of Art. 163; and $\phi(x, \mu)$ increases with μ (because $f(x, y)$ is not negative), so that the same is true of $h_r(\mu)$.

k_r is the minimum of $\psi(x)=\lim \phi(x, \mu)$ in the rth interval; and so, using Art. 163 again, we have

(4) $$\lim_{n\to\infty} \sum_1^n lk_r = \int_a^\lambda \psi(x)dx.$$

Since the integral in (4) is supposed convergent (otherwise equation (1) would be obviously meaningless), the equation (4) shews that the right-hand limit in (2) exists; and so the assumption made above is justified. From the equations (3) and (4) we see that*

$$\int_b^\infty dy \int_a^\lambda f(x, y)dx = \int_a^\lambda dx \int_b^\infty f(x, y)dy.$$

From (3) and (4) it is also clear that

$$\int_a^\infty dx \int_b^\infty f(x, y)dy = \lim_{\lambda\to\infty} \{\lim_{\mu\to\infty} F(\lambda, \mu)\};$$

and similarly, we find that the second integral in (1) is equal to the repeated limit of $F(\lambda, \mu)$ taken in the reverse order.

Now $F(\lambda, \mu)$ cannot decrease, as λ and μ increase, so that we can again apply theorem (5) of Art. 31; and we obtain de la Vallée Poussin's theorem:

Equation (1) *above is correct, provided that* **both** *the integrals*

$$\int_a^\infty f(x, y)dx, \quad \int_b^\infty f(x, y)dy$$

are convergent, and that **either** *of the repeated integrals converges.*

It will be seen that (by using $f+|f|$ in place of f) we can extend the theorem to cases when f changes sign, provided that the integrals all remain convergent when $|f|$ is put in place of f.

We have still to prove that *if $h(\mu)$ is the minimum of $\phi(x, \mu)$ in any interval ($p \leqq x \leqq q$), then $h(\mu)$ tends to a limit k, which is the minimum of $\psi(x)$ in the same interval.*

From the definition of $h(\mu)$, we have

$$\phi(x, \mu) \geqq h(\mu),$$

and so, on making μ tend to infinity, we find

(5) $$\psi(x) \geqq k.$$

If it happens that $\phi(p, \mu) \leqq k$, it is evident that $\psi(p) \leqq k$, also; and so we see from (5) that $\psi(p)=k$, and consequently k is the minimum of $\psi(x)$ in the interval (p, q).

* Note that we do *not* use any condition of uniform convergence, as in Art. 172 (2); instead, we have the condition that f is nowhere negative.

But if $\phi(p, \mu) > k \geqq h(\mu)$, it follows from Ex. 21, p. 395, that the equation $\phi(x, \mu) = k$ has at least one root in the interval (p, q); let ξ_μ denote the *least* root.* Then, if $\nu > \mu$, we have †

$$\phi(\xi_\mu, \nu) \geqq \phi(\xi_\mu, \mu) = k,$$

so that $\xi_\nu \geqq \xi_\mu$, and ξ_ν therefore tends to a limit ξ as ν tends to infinity. Again

$$\phi(\xi_\nu, \mu) \leqq \phi(\xi_\nu, \nu) = k, \quad (\nu > \mu),$$

so that, on making ν tend to infinity, we have

$$\phi(\xi, \mu) \leqq k.$$

Thus $\psi(\xi) \leqq k$, and so from (5) we find that $\psi(\xi) = k$, which is therefore again the minimum of $\psi(x)$.

Ex. As an application we shall establish the equations

$$\int_0^\infty e^{-yt}\left(\frac{1}{e^t - 1} - \frac{1}{t} + \frac{1}{2}\right) dt = 2\int_0^\infty \frac{x\,dx}{(x^2+y^2)(e^{2\pi x}-1)},$$

$$\int_0^\infty \frac{e^{-yt}}{t}\left(\frac{1}{e^t - 1} - \frac{1}{t} + \frac{1}{2}\right) dt = 2\int_0^\infty \frac{\operatorname{arc\,tan}(x/y)}{e^{2\pi x}-1}\, dx,$$

where the real part of y is positive and the arc tan function is determined so as to vanish with x.

We have seen (Ex. 1, Art. 176) that

$$\frac{1}{e^t - 1} - \frac{1}{t} + \frac{1}{2} = 2\int_0^\infty \frac{\sin(xt)}{e^{2\pi x}-1}\, dx,$$

and therefore

$$\int_0^\infty e^{-yt}\left(\frac{1}{e^t - 1} - \frac{1}{t} + \frac{1}{2}\right) dt = 2\int_0^\infty e^{-yt}\, dt \int_0^\infty \frac{\sin(xt)}{e^{2\pi x}-1}\, dx.$$

Now the last integral is absolutely convergent, since

$$\int_0^\infty \frac{|\sin(xt)|}{e^{2\pi x}-1}\, dx < \int_0^\infty \frac{xt\,dx}{e^{2\pi x}-1} = \frac{t}{24}, \quad \text{(Ex. 3, Art. 176)},$$

and

$$|e^{-yt}| = e^{-\xi t}, \quad \text{if } y = \xi + i\eta.$$

Thus

$$2\int_0^\infty |e^{-yt}|\, dt \int_0^\infty \frac{|\sin(xt)|}{e^{2\pi x}-1}\, dx < \frac{1}{12\xi^2},$$

which proves the absolute convergence; we can therefore invert the order of integration without altering the value of the integral, and we then find

$$\int_0^\infty e^{-yt}\left(\frac{1}{e^t - 1} - \frac{1}{t} + \frac{1}{2}\right) dt = 2\int_0^\infty \frac{x\,dx}{(x^2+y^2)(e^{2\pi x}-1)}.$$

Now, if we write $y = \xi + i\eta$ in the last equation, we can integrate with respect to ξ under the integral sign, between ξ_0 and ∞; for

$$\left|e^{-yt}\left(\frac{1}{e^t - 1} - \frac{1}{t} + \frac{1}{2}\right)\right| < \frac{1}{12} t e^{-\xi t},$$

and so

$$\int_{\xi_0}^\infty d\xi \int_0^\infty \left|e^{-yt}\left(\frac{1}{e^t - 1} - \frac{1}{t} + \frac{1}{2}\right)\right| dt < \frac{1}{12\xi_0}, \quad (\xi_0 > 0)$$

---

* If the equation has an infinite set of roots, the limiting values of the set are also roots (because ϕ is continuous); and so the set *attains* its lower limit, which is therefore the least root.

† The reader is advised to use a figure in following the argument here.

so that this double integral is absolutely convergent. Similarly we find that the right-hand side is absolutely convergent, since $|x^2+y^2| \geqq \xi^2$, so that

$$\int_{\xi_0}^{\infty} d\xi \int_0^{\infty} \frac{x\,dx}{|x^2+y^2|.(e^{2\pi x}-1)} \leqq \int_{\xi_0}^{\infty} \frac{d\xi}{\xi^2} \int_0^{\infty} \frac{x\,dx}{e^{2\pi x}-1} = \frac{1}{24\xi_0}.$$

Thus we find the further equation

$$\int_0^{\infty} \left(\frac{1}{e^t-1} - \frac{1}{t} + \frac{1}{2}\right) dt \int_{\xi_0}^{\infty} e^{-yt}\, d\xi = 2\int_0^{\infty} \frac{dx}{e^{2\pi x}-1} \int_{\xi_0}^{\infty} \frac{x\,d\xi}{(x^2+y^2)},$$

which gives $\displaystyle \int_0^{\infty} \frac{e^{-y_0 t}}{t}\left(\frac{1}{e^t-1} - \frac{1}{t} + \frac{1}{2}\right) dt = 2\int_0^{\infty} \frac{\text{arc tan}\,(x/y_0)}{e^{2\pi x}-1}\, dx,$

where $y_0 = \xi_0 + i\eta$.

178. The Gamma-integral.

In Art. 42 we have seen that

$$\Gamma(1+x) = \lim_{n\to\infty} \frac{n^x . n!}{(1+x)(2+x)\dots(n+x)}.$$

We shall now express this function by means of an infinite integral when a, the real part of $1+x$, is positive.

Write $$I_s = \int_0^1 y^{x+s}(1-y)^{n-s}\, dy,$$

then, using the method of integration by parts, we find that

$$I_s/I_{s+1} = (n-s)/(1+s+x),$$

and so $$I_0/I_n = n!/\{(1+x)(2+x)\dots(n+x)\}.$$

But $$I_n = 1/(n+1+x),$$

so that $$\Gamma(1+x) = \lim n^{x+1} I_0,$$

or, changing the variable by writing $t = ny$, we have

$$\Gamma(1+x) = \lim_{n\to\infty} \int_0^n \left(1-\frac{t}{n}\right)^n t^x\, dt.$$

We can apply Tannery's theorem (Art. 172 (4)) to the last integral; for we have*

$$|e^{-t}t^x - (1-t/n)^n t^x| < t^{1+a}/(2n),$$

and so (since a is positive) the integrand converges to the limit $e^{-t}t^x$ uniformly in any *fixed* interval for t. Further, we have

$$|(1-t/n)^n t^x| < e^{-t}t^{a-1},$$

and the integral $\displaystyle\int_0^{\infty} e^{-t}t^{a-1}\, dt$ is convergent, because a is positive.

---

* Actually $1 - e^t\left(1-\frac{t}{n}\right)^n = \int_0^t e^v\left(1-\frac{v}{n}\right)^{n-1} \frac{v}{n}\, dv$, so that, when t is positive, $e^{-t} - (1-t/n)^n$ is positive and less than $t^2/(2n)$.

Thus all the conditions are satisfied, and so we find

$$\Gamma(1+x)=\int_0^\infty e^{-t}t^x dt.$$

A somewhat similar integral can be found for Euler's constant; we have seen (Art. 11) that

$$C=\lim_{n\to\infty}\left(1+\frac{1}{2}+\ldots+\frac{1}{n}-\log n\right).$$

But $$1+\frac{1}{2}+\ldots+\frac{1}{n}=\int_0^1(1+x+x^2+\ldots+x^{n-1})\,dx=\int_0^1\frac{1-x^n}{1-x}dx.$$

Thus we find, on writing $x=1-t/n$,

$$1+\frac{1}{2}+\ldots+\frac{1}{n}=\int_0^n\left[1-\left(1-\frac{t}{n}\right)^n\right]\frac{dt}{t}.$$

And $$\log n=\int_1^n\frac{dt}{t};$$

hence $$C=\lim_{n\to\infty}\left[\int_0^1\left\{1-\left(1-\frac{t}{n}\right)^n\right\}\frac{dt}{t}-\int_1^n\left(1-\frac{t}{n}\right)^n\frac{dt}{t}\right],$$

and, by the same method as before, we obtain as the limit

$$C=\int_0^1(1-e^{-t})\frac{dt}{t}-\int_1^\infty e^{-t}\frac{dt}{t}=-\lim_{\delta\to 0}\left[\log\delta+\int_\delta^\infty e^{-t}\frac{dt}{t}\right].$$

Since $$\int_\delta^\infty\frac{dt}{t(1+t)}=\log\frac{1+\delta}{\delta},$$

we see that $$C=\lim_{\delta\to 0}\left[\int_\delta^\infty\left(\frac{1}{1+t}-e^{-t}\right)\frac{dt}{t}-\log(1+\delta)\right]$$
$$=\int_0^\infty\left(\frac{1}{1+t}-e^{-t}\right)\frac{dt}{t}.$$

Another form is easily obtained by changing the variable from t to $1/t$ in the integral $\int_1^\infty e^{-t}dt/t$; this gives

$$C=\int_0^1(1-e^{-t}-e^{-1/t})\frac{dt}{t}.$$

A number of definite integrals for C can be obtained from the expression

$$-\lim_{\delta\to 0}\left[\log\delta+\int_\delta^\infty e^{-t}\frac{dt}{t}\right].$$

Amongst them are the following:

$$C=\int_0^\infty\left(\frac{1}{1+t^2}-e^{-t}\right)\frac{dt}{t}=\int_0^\infty\left(\frac{1}{e^t-1}-\frac{e^{-t}}{t}\right)dt=\int_0^\infty e^{-t}\log\frac{1}{t}\,dt.$$

It is easy to see from Art. 180 below that

$$\Gamma'(1)=\int_0^\infty\frac{dt}{t}\left(e^{-t}-\frac{t}{e^t-1}\right)=-C.$$

Useful properties of the Gamma-function.

1. When x is a positive integer, we can write

$$\Gamma(1+x)=\lim_{n\to\infty}\frac{n!\,.\,n^x}{(1+x)(2+x)\ldots(n+x)}=\lim_{n\to\infty}\frac{x!\,.\,n^x}{(1+n)(2+n)\ldots(x+n)}=x!;$$

a result which is also easily obtained from the definite integral, using the method of integration by parts.

2. $$\Gamma(x)\Gamma(1-x)=\lim \frac{n!\,n^{x-1}}{x(1+x)\dots(n-1+x)}\cdot\frac{n!\,n^{-x}}{(1-x)(2-x)\dots(n-x)}$$

$$=\lim \frac{n+x}{n}\,\frac{(n!)^2}{x(1-x^2)(2^2-x^2)\dots(n^2-x^2)}$$

$$=\lim\left[x\left(1-\frac{x^2}{1^2}\right)\left(1-\frac{x^2}{2^2}\right)\dots\left(1-\frac{x^2}{n^2}\right)\right]^{-1}$$

or $\Gamma(x)\Gamma(1-x)=\pi/\sin(\pi x)$. (Art. 91.)

3. Writing $x=\frac{1}{2}$ in the last result, we find, since $\Gamma(\frac{1}{2})$ is positive,

$$\Gamma(\tfrac{1}{2})=\sqrt{\pi}.$$

4. $$\Gamma(x+\tfrac{1}{2})\Gamma(x+1)=\lim\frac{(n!)^2\,n^{2x-\frac{1}{2}}}{(\frac{1}{2}+x)(1+x)(\frac{3}{2}+x)\dots(n+x)}$$

$$=\lim\frac{(n!)^2\,n^{2x-\frac{1}{2}}\,2^{2n}}{(1+2x)(2+2x)\dots(2n+2x)}.$$

Also $$\Gamma(2x+1)=\lim\frac{(2n)!\,(2n)^{2x}}{(1+2x)(2+2x)\dots(2n+2x)}.$$

Thus $$2^{2x}\frac{\Gamma(x+\frac{1}{2})\Gamma(x+1)}{\Gamma(2x+1)}=\lim\frac{(n!)^2\,2^{2n}}{(2n)!\,n^{\frac{1}{2}}}.$$

Since this last expression does not contain x, we can find its value by putting $x=0$; this gives $\Gamma(\frac{1}{2})$ or $\sqrt{\pi}$, a result which can also be obtained by appealing directly to the definition.

179. Stirling's asymptotic formula for the Gamma-function when x is real, large and positive.

In the integral

$$\Gamma(1+x)=\int_0^\infty e^{-t}t^x\,dt,$$

the maximum of the integrand is $e^{-x}x^x$ and occurs for $t=x$, so if we write

$$e^{-t}t^x=(e^{-x}x^x)\,e^{-y^2},$$

the range of values $(-\infty, 0, +\infty)$ for y will correspond precisely to the range $(0, x, \infty)$ for t.

Thus $$\Gamma(1+x)=e^{-x}x^x\int_{-\infty}^{\infty}e^{-y^2}\frac{dt}{dy}\,dy.$$

Now, taking logarithms, we have

$$y^2=(t-x)-x\log(t/x)$$

so that $$2y\frac{dy}{dt}=1-\frac{x}{t}.$$

But the properties of the logarithmic function shew that y^2 lies between*

$$\frac{x}{2}\left(\frac{t-x}{x}\right)^2 \quad \text{and} \quad \frac{x}{2}\left(\frac{t-x}{t}\right)^2.$$

Thus, since y has the same sign as $t-x$, we see that y lies between

$$\left(\frac{x}{2}\right)^{\frac{1}{2}}\frac{t-x}{x} \quad \text{and} \quad \left(\frac{x}{2}\right)^{\frac{1}{2}}\frac{t-x}{t}.$$

Thus, $(x/2)^{\frac{1}{2}}(1/y)$ lies between

$$x/(t-x) \quad \text{and} \quad t/(t-x).$$

And therefore, since $t/(t-x)=1+x/(t-x)$, we see that $t/(t-x)$ must lie between

$$\frac{1}{y}\left(\frac{x}{2}\right)^{\frac{1}{2}} \quad \text{and} \quad 1+\frac{1}{y}\left(\frac{x}{2}\right)^{\frac{1}{2}}.$$

Hence $\dfrac{dt}{dy}=\dfrac{2ty}{t-x}$ lies between

$$(2x)^{\frac{1}{2}} \quad \text{and} \quad 2y+(2x)^{\frac{1}{2}}.$$

Accordingly, we have

$$\Gamma(1+x)=e^{-x}x^x\int_{-\infty}^{\infty} e^{-y^2}[(2x)^{\frac{1}{2}}+\xi]\,dy,$$

where $|\xi|<2|y|$.

Now† $\displaystyle\int_{-\infty}^{+\infty} e^{-y^2}dy=\pi^{\frac{1}{2}}, \quad \int_{-\infty}^{\infty}|y|e^{-y^2}dy=1,$

and accordingly

$$\left|\frac{\Gamma(1+x)}{e^{-x}x^x(2\pi x)^{\frac{1}{2}}}-1\right|<\frac{2}{(2\pi x)^{\frac{1}{2}}}.$$

Hence $$\lim_{x\to\infty}\frac{\Gamma(1+x)}{e^{-x}x^x(2\pi x)^{\frac{1}{2}}}=1,$$

or, as we may write it,

$$\Gamma(1+x) \sim e^{-x}x^x(2\pi x)^{\frac{1}{2}}.$$

---

* We have, if $(t-x)/x=\tau, \quad y^2=x\int_0^\tau \theta\,d\theta/(1+\theta),$
which obviously lies between $\frac{1}{2}x\tau^2$ and $\frac{1}{2}x\tau^2/(1+\tau)^2$.

† $\displaystyle\int_{-\infty}^{\infty} e^{-y^2}dy=2\int_0^\infty e^{-y^2}dy=\int_0^\infty e^{-x}x^{-\frac{1}{2}}dx=\Gamma(\tfrac{1}{2})=\pi^{\frac{1}{2}}.$ (Art. 178 (3).)

Again, we see that

$$|\log\Gamma(1+x)-\log\{e^{-x}x^x(2\pi x)^{\frac{1}{2}}\}|<2/[(2\pi x)^{\frac{1}{2}}-2],$$

so that $\log\Gamma(1+x)\sim(x+\frac{1}{2})\log x-x+\frac{1}{2}\log(2\pi)$,

using the symbol $\sim$ in the extended sense explained in Art. 133.

If we subtract $\log x$, we obtain

$$\log\Gamma(x)\sim(x-\tfrac{1}{2})\log x-x+\tfrac{1}{2}\log(2\pi).$$

The foregoing method is due to Liouville, who gave it in his *Journal de Mathématiques* (t. 11, 1846, p. 464).

Ex. Consider the value of

$$\phi(x)=n^{nx}\Gamma(x)\Gamma\left(x+\frac{1}{n}\right)\Gamma\left(x+\frac{2}{n}\right)\ldots\Gamma\left(x+\frac{n-1}{n}\right)\Big/\Gamma(nx),$$

where n is a positive integer.

If we change x to $x+1$, we see that

$$\phi(x+1)/\phi(x)=n^n\left[x\left(x+\frac{1}{n}\right)\ldots\left(x+\frac{n-1}{n}\right)\right]\Big/nx(nx+1)\ldots(nx+n-1)=1.$$

Hence $\phi(x)=\phi(x+1)=\phi(x+2)=\ldots=\phi(x+s)$.

But when y is large, $\Gamma(y+a)\sim\Gamma(y).y^a$ (Art. 42), so that

$$\phi(y)\sim n^{ny}y^{\frac{1}{2}(n-1)}[\Gamma(y)]^n/\Gamma(ny),$$

or, using the asymptotic formula above,

$$\phi(y)\sim n^{ny}y^{\frac{1}{2}(n-1)}[e^{-ny}y^{ny}(2\pi/y)^{\frac{1}{2}n}][e^{ny}(ny)^{-ny}(2\pi/ny)^{-\frac{1}{2}}].$$

Hence, as y tends to infinity, $\phi(y)$ tends to the limit $(2\pi)^{\frac{1}{2}(n-1)}n^{\frac{1}{2}}$, and we have already proved that $\phi(x)=\phi(x+s)$, where s is an arbitrarily great positive integer, so that we must have $\phi(x)=(2\pi)^{\frac{1}{2}(n-1)}n^{\frac{1}{2}}$.

The special case $n=2$ has been discussed in Art. 178 (4).

180. Integrals for $\log\Gamma(1+x)$.

We have proved (Ex. 1, Art. 173) that if the real parts of a, b are positive,

$$\log\frac{b}{a}=\int_0^\infty\frac{dt}{t}(e^{-at}-e^{-bt}).$$

Hence, *if the real part of* $1+x$ *is positive,* we have

$$\log\left(\frac{r}{r+x}\right)=\int_0^\infty\frac{dt}{t}(1-e^{-xt})e^{-rt}.\quad(r=1,2,3,\ldots)$$

Now $$\log\Gamma(1+x)=\lim_{n\to\infty}\log\frac{n^x n!}{(1+x)(2+x)\ldots(n+x)}$$

$$=\lim_{n\to\infty}\left[x\log n+\sum_1^n\log\left(\frac{r}{r+x}\right)\right].$$

Thus we are led to consider the function

$$S(x, n) = x \log n + \sum_1^n \log\left(\frac{r}{r+x}\right)$$

$$= x\int_0^\infty (e^{-t} - e^{-nt})\frac{dt}{t} + \sum_1^n \int_0^\infty (1-e^{-xt})e^{-rt}\frac{dt}{t}.$$

Now $$\sum_1^n e^{-rt} = e^{-t}(1-e^{-nt})/(1-e^{-t}) = (1-e^{-nt})/(e^t-1),$$

so that $$S(x, n) = \int_0^\infty \left(xe^{-t} - \frac{1-e^{-xt}}{e^t-1}\right)\frac{dt}{t} - \int_0^\infty e^{-nt}\left(x - \frac{1-e^{-xt}}{e^t-1}\right)\frac{dt}{t}$$

$$= F(x) + G(x, n), \text{ say.}$$

It is to be observed that both in $F(x)$ and in $G(x, n)$ the integrands are finite at $t=0$.

For if $t < 2\pi$, we can write (Art. 93)

$$\frac{1-e^{-xt}}{e^t-1} = x - \tfrac{1}{2}(x+x^2)t + \dots,$$

so that $$\frac{1}{t}\left(x - \frac{1-e^{-xt}}{e^t-1}\right) = \tfrac{1}{2}(x+x^2) + X_1 t + X_2 t^2 + \dots,$$

and similarly for the other integrand.

Thus, when $t<1$, $\frac{1}{t}\left|x - \frac{1-e^{-xt}}{e^t-1}\right|$ cannot exceed some fixed value, independent of t; but if $t>1$, this expression is less than $|x| + \frac{e+1}{e-1}$, because $|e^{-xt}| < e^t$ (since the real part of $1+x$ is positive). Thus we can determine a value X, independent of t, such that

$$\left|\frac{1}{t}\left(x - \frac{1-e^{-xt}}{e^t-1}\right)\right| < X.$$

Then $$|G(x, n)| < \int_0^\infty Xe^{-nt}\,dt$$

or $$< X/n,$$

so that $$\lim_{n\to\infty} G(x, n) = 0.$$

Hence $$\log\Gamma(1+x) = \lim_{n\to\infty} S(x, n) = F(x)$$

$$= \int_0^\infty \frac{dt}{t}\left(xe^{-t} - \frac{1-e^{-xt}}{e^t-1}\right).$$

This integral can be divided into two parts, and we find

$$\log\Gamma(1+x)=\phi(x)+\psi(x),$$

where $$\phi(x)=\int_0^\infty\left[xe^{-t}-\frac{1}{e^t-1}+\left(\frac{1}{t}-\frac{1}{2}\right)e^{-xt}\right]\frac{dt}{t}$$

and $$\psi(x)=\int_0^\infty\left(\frac{1}{e^t-1}-\frac{1}{t}+\frac{1}{2}\right)e^{-xt}\frac{dt}{t}=2\int_0^\infty\frac{\operatorname{arc\,tan}(y/x)}{e^{2\pi y}-1}dy,$$

the last expression following from the example of Art. 177. The advantage of this transformation is due to two facts, first that the value of $\phi(x)$ can be found in terms of elementary functions; and secondly that $\psi(x)$ tends to zero if $|x|$ tends to ∞ in such a way that the real part of x also tends to ∞.

For, in the course of the example of Art. 177, we proved that

$$|\psi(x)|<1/12\xi,$$

where ξ is the real part of x. Thus when ξ tends to ∞, we have

$$\lim\psi(x)=0.$$

The limit is also 0, when η tends to ∞, ξ being kept positive (see Ex. 56, p. 478).

As regards $\phi(x)$, we have

$$\phi(x)-\phi(1)=\int_0^\infty\left[(x-1)e^{-t}+\left(\frac{1}{t}-\frac{1}{2}\right)(e^{-xt}-e^{-t})\right]\frac{dt}{t}$$
$$=(x+\tfrac{1}{2})\log x-(x-1)$$

by Ex. 3, Art. 173. Thus we see that, if $A=1+\phi(1)$,

$$\phi(x)=(x+\tfrac{1}{2})\log x-x+A.$$

To determine A, we make use of (4), Art. 178, which gives

$$\log\Gamma(x+\tfrac{1}{2})+\log\Gamma(x+1)+2x\log 2-\log\Gamma(2x+1)=\tfrac{1}{2}\log\pi.$$

Thus we have, since $\lim\psi(x)=0$,

$$\lim\left[\phi(x-\tfrac{1}{2})+\phi(x)+2x\log 2-\phi(2x)\right]=\tfrac{1}{2}\log\pi,$$

which gives, on inserting the value of $\phi(x)$,

$$\lim\left[A+x\log\left(1-\frac{1}{2x}\right)+\frac{1}{2}-\frac{1}{2}\log 2\right]=\frac{1}{2}\log\pi$$

or* $$A=\tfrac{1}{2}\log(2\pi)\qquad\text{(compare Ex. 55, p. 478).}$$

* The value of A can also be found from Stirling's asymptotic formula (Art. 179); or by a device due to Pringsheim (*Math. Annalen*, Bd. 31, p. 473).

Thus we can write

$$\log\Gamma(1+x)=(x+\tfrac{1}{2})\log x-x+\tfrac{1}{2}\log 2\pi+\psi(x).$$

where
$$\psi(x)=2\int_0^\infty \frac{\arctan(y/x)}{e^{2\pi y}-1}\,dy$$

and
$$|\psi(x)|<1/12\xi.$$

It is often convenient to have a formula for $\Gamma(1+x)$, when x is real and varies in the neighbourhood of a fixed large value. Thus, if we write
$$x=\nu+a,$$
where ν is large and a may be large, but is of the order $\sqrt{\nu}$ at most, we obtain the asymptotic expression*

$$\log\Gamma(1+x)\backsim(\nu+a+\tfrac{1}{2})\log\nu-\nu+\tfrac{1}{2}\log 2\pi+\tfrac{1}{2}a^2/\nu,$$

where the error is of order a/ν.

Hence
$$\Gamma(1+\nu+a)\backsim(2\pi\nu)^{\frac{1}{2}}\nu^{\nu+a}e^{-\nu+\frac{1}{2}a^2/\nu}.$$

In several books on analysis, the integrals for $\log\Gamma(1+x)$ are found by a somewhat different method due to Dirichlet.

In outline, this proof is as follows:

(1) Differentiate the Gamma-integral, and we find

$$\Gamma'(1+x)=\int_0^\infty e^{-t}t^x\log t\,dt=\int_0^\infty e^{-t}t^x\,dt\int_0^\infty(e^{-v}-e^{-vt})\frac{dv}{v}.$$

(2) Invert the order of integration, and we obtain

$$\frac{\Gamma'(1+x)}{\Gamma(1+x)}=\int_0^\infty[e^{-v}-(1+v)^{-(1+x)}]\frac{dv}{v}=\lim_{\delta\to 0}\left[\int_\delta^\infty e^{-v}\frac{dv}{v}-\int_\delta^\infty(1+v)^{-(1+x)}\frac{dv}{v}\right].$$

Also
$$\int_\delta^\infty\frac{dv}{v}(1+v)^{-(1+x)}=\int_{\log(1+\delta)}^\infty\frac{e^{-xy}}{e^y-1}\,dy,\quad \text{if } 1+v=e^y.$$

(3) We must next prove that

$$\lim_{\delta\to 0}\int_{\log(1+\delta)}^{\delta}\frac{e^{-xy}}{e^y-1}\,dy=0,$$

and then we have
$$\frac{\Gamma'(1+x)}{\Gamma(1+x)}=\int_0^\infty\left(\frac{e^{-y}}{y}-\frac{e^{-xy}}{e^y-1}\right)dy.$$

(4) Finally, if we integrate the last equation, we arrive at the same integral as before for $\log\Gamma(1+x)$.

The reader will find it a good exercise in the use of Arts. 166, 172, 177, to shew that the steps (1)–(4) are legitimate. Proofs will be found in Jordan's *Cours d'Analyse* (t. 2, 2me éd., pp. 176–182).

---

* We have

$$\phi(\nu+a)=\phi(\nu)+a\phi'(\nu)+\tfrac{1}{2}a^2\phi''(\nu+\theta a),\qquad(0<\theta<1),$$

so that here we get

$$\left(\nu+\frac{1}{2}\right)\log\nu-\nu+\frac{1}{2}\log(2\pi)+a\left(\log\nu+\frac{1}{2\nu}\right)+\frac{1}{2}a^2\left[\frac{1}{\nu+\theta a}-\frac{1}{2}\frac{1}{(\nu+\theta a)^2}\right].$$

EXAMPLES.

Tests of Convergence.

1. Determine the values of a, b for which the integrals

$$(1)\ \int_0^\infty x^{a-1}\cos x\,dx,\quad (2)\ \int_0^\infty x^{a-1}\sin x\,dx,\quad (3)\ \int_0^\infty \frac{x^{a-1}dx}{1+x},\quad (4)\ \int_0^\infty \frac{x^{a-1}-x^{b-1}}{1-x}dx$$

are convergent.

2. Discuss the continuity of the integral

$$\int_0^\infty \frac{\sin y\,dx}{1-2x\cos y+x^2}$$

regarded as a function of y. Sketch its graph. [*Math. Trip.* 1904.]

3. Discuss the convergence of the integrals

$$\int_0^\infty \tanh x\left(\frac{x+1}{x^3+x+1}\right)^{\frac{1}{2}}dx,\quad \int_0^\infty \frac{\sin(xy)dx}{\sqrt{(x^2-x+1)}}.$$ [*Math. Trip.* 1893.]

4. If $0<\kappa\leqq\frac{1}{2}$, both the series and the integral

$$\sum^\infty \frac{\sin n\theta}{n^\kappa+a\sin n\theta},\quad \int^\infty \frac{\sin x\,dx}{x^\kappa+a\sin x}$$

are divergent if $a>0$, although both converge if $a=0$. When $\kappa>\frac{1}{2}$, the series and integral are both convergent.

Reconcile these results with Dirichlet's tests (Arts. 20 and 171). [Hardy.]

5. If $f(x)$ tends *steadily* to zero as $x\to\infty$, prove from Art. 166 that we can infer the convergence of $\int^\infty xf'(x)dx$ from that of $\int^\infty f(x)dx$, provided that $f(x)$ is monotonic.

Similarly, shew that if (a_n) is a monotonic sequence, the convergence of $\Sigma n(a_n-a_{n+1})$ can be deduced from that of Σa_n.

6. Apply the method of Art. 166 to prove that, if α, β, γ are positive, the integral

$$\int^\infty \frac{e^{\alpha x}dx}{e^{\beta x}\sin^2 x+e^{\gamma x}\cos^2 x}$$

converges if $\beta+\gamma>2\alpha$, and diverges if $\beta+\gamma\leqq 2\alpha$.

Deduce that, if $\beta>1>\gamma>0$ and $\beta+\gamma>2$, the integral

$$\int^\infty \frac{dt}{t[(l_1t)^\beta\sin^2(l_2t)+(l_1t)^\gamma\cos^2(l_2t)]}$$

is convergent, where $l_1t=\log t$, $l_2t=\log(\log t)$.

Shew that in the last integral the integrand tends *steadily* to zero, but that no test of the logarithmic scale suffices to establish the convergence of the integral.

State and prove corresponding results for series. [Hardy.]

7. Shew that the series $\sum \dfrac{n^{a-1}}{a+n^\beta \sin^2(n\pi\lambda)}$

diverges if a is positive and λ is *rational* (in contrast to the corresponding integral in Art. 166). But if λ is the root of an algebraic equation of degree $m>1$, the series converges if $\beta > a+2m$. However, irrational values of λ can be constructed for which the series will diverge, whatever β may be.

[Compare a paper by Hardy, *Proc. Lond. Math. Soc.* (2), vol. 3, pp. 444–9.]

8. Shew that the integrals

$$\int^\infty \cos\{f(x)\}dx, \quad \int^\infty \sin\{f(x)\}dx$$

are convergent, provided that $f'(x)$ tends steadily to infinity with x.

Prove also that $\int^\infty f'(x)\sin\{e^{f(x)}\}dx$ is convergent no matter how rapidly $f'(x)$ tends to infinity.

[In the first case it is *not* sufficient that $f(x)$ tends steadily to infinity, as we may see by taking $f(x)=x$.]

9. Although (see Ex. 8) $\int^\infty \cos(x^2)dx$ and $\int^\infty \sin(x^2)dx$ are convergent, prove that $\Sigma\cos(n^2\theta)$ and $\Sigma\sin(n^2\theta)$ cannot converge if θ/π is rational (see Exs. A, 13-17, Ch. XI., and Ex. 10, p. 485).

Change of Variables.

10. If $g(\xi)$ is an odd function of ξ, prove by dividing the range into intervals $(0, \frac{1}{2}\pi)$, $(\frac{1}{2}\pi, \pi)$, $(\pi, \frac{3}{2}\pi)$, ... and introducing the new variables x, $\pi-x$, $x-\pi$, $2\pi-x$, ... respectively, that

$$\int_0^\infty g(\sin x)\frac{dx}{x} = \int_0^{\frac{1}{2}\pi} g(\sin x)\frac{dx}{\sin x},$$

provided that both integrals converge.

Deduce that

$$\int_0^\infty \sin^{2n+1}x\frac{dx}{x} = \tfrac{1}{2}\pi\frac{1.3\ldots(2n-1)}{2.4\ldots 2n},$$

$$\int_0^\infty \tan^{-1}(a\sin x)\frac{dx}{x} = \tfrac{1}{2}\pi\sinh^{-1}a, \qquad \text{(Ex. 15.)}$$

$$\int_0^\infty (\log\cos^2 x)\frac{\sin x}{x}dx = -\pi\log 2. \qquad \text{[Wolstenholme.]}$$

11. Apply the same method to prove that, if $f(\xi)$ is an even function of ξ, $a>0$ and $0 \leqq \kappa \leqq 2$

(1) $\displaystyle\int_0^\infty f(\sin x)\frac{dx}{x^2} = \int_0^{\frac{1}{2}\pi} f(\sin x)\frac{dx}{\sin^2 x}$;

(2) $\displaystyle\int_0^\infty f(\sin x)\frac{2a\,dx}{a^2+x^2} = \sinh 2a\int_0^{\frac{1}{2}\pi}\frac{f(\sin x)dx}{\sinh^2 a+\sin^2 x}$

$$= 2\int_0^{\frac{1}{2}\pi} f(\sin x)\left(1+2\sum_1^\infty e^{-2na}\cos 2nx\right)dx;$$

(3) $$\int_0^\infty f(\sin x)\frac{2a\cos(\kappa x)}{a^2+x^2}\,dx$$

$$=\int_0^{\frac{1}{2}\pi} f(\sin x)\left[\cosh(\kappa a)\frac{\sinh 2a}{\sinh^2 a+\sin^2 x}-\sinh(\kappa a)\right]dx$$

$$=2\int_0^{\frac{1}{2}\pi} f(\sin x)\left[e^{-\kappa a}+2\cosh(\kappa a)\sum_1^\infty e^{-2na}\cos 2nx\right]dx;$$

(4) $$\int_0^\infty f(\sin x)\cos(\kappa x)\frac{dx}{x^2}=\int_0^{\frac{1}{2}\pi} f(\sin x)(\operatorname{cosec}^2 x-\kappa)\,dx.$$

[It is understood that all the integrals converge; the series used are given in Ex. 17, p. 190, Ex. 31, p. 259, and Ex. 5, p. 290.]

12. Illustrations of the last example are:

(1) $$\int_0^\infty (\log\cos^2 x)\frac{dx}{x^2}=-\pi,\quad \int_0^\infty (\log\cos^2 x)^2\frac{dx}{x^2}=4\pi\log 2,$$

$$\int_0^\infty (\log\cos^2 x)(\log\sin^2 x)\frac{dx}{x^2}=2\pi(2\log 2-1).$$

(2) $$\int_0^\infty \cos^2 x\,\frac{dx}{a^2+x^2}=\frac{\pi}{4a}(1+e^{-2a}),\quad \int_0^\infty (\log\cos^2 x)\frac{dx}{a^2+x^2}=\frac{\pi}{a}\log\tfrac{1}{2}(1+e^{-2a}),$$

and a similar formula containing $\log\sin^2 x$ and $\log\frac{1}{2}(1-e^{-2a})$.

(3) $$\int_0^\infty \cos(\kappa x)(\log\cos^2 x)\frac{dx}{a^2+x^2}=\frac{\pi}{a}[\cosh(\kappa a)\log(1+e^{-2a})-e^{-\kappa a}\log 2],$$

and a similar formula containing $\log\sin^2 x$ and $\log(1-e^{-2a})$.

(4) $$\int_0^\infty \cos(\kappa x)(\log\cos^2 x)\frac{dx}{x^2}=\pi(\kappa\log 2-1);$$

but in this case there is no corresponding formula with $\log\sin^2 x$.

[DE LA VALLÉE POUSSIN and HARDY.]

Differentiation and Integration.

13. Calculate the integrals

$$u=\int_0^1 \log(x^2+y^2)\,dx,\quad u_0=\int_0^1 \log x^2\,dx,$$

and prove that $$\lim_{y\to 0}(u-u_0)/y=\pm\pi,$$

the ambiguous sign being the same as the sign of y. Explain why this limit is not the same as the integral

$$\int_0^1 \lim_{y\to 0}[\{\log(x^2+y^2)-\log x^2\}/y]\,dx.$$ [STOLZ.]

14. Prove by differentiation, or by expanding in powers of a, that

$$\int_0^\infty \log(1+a\operatorname{sech} x)\,dx=\tfrac{1}{2}[\pi\sin^{-1}a-(\sin^{-1}a)^2].$$

Obtain two other integrals by writing $a=i\beta$, where β is real; and verify these results by differentiation and expansion. [HARDY.]

15. Prove similarly that

$$\int_0^{\frac{1}{2}\pi} \log(1+a\sin x)\frac{dx}{\sin x} = \tfrac{1}{2}[\pi\sin^{-1}a - (\sin^{-1}a)^2]$$

and

$$\int_0^{\frac{1}{2}\pi} \log(1+a\sin^2 x)\frac{dx}{\sin^2 x} = \pi[\sqrt{(1+a)}-1].$$

Obtain four other integrals by writing $a=i\beta$. [WOLSTENHOLME.]

16. By integrating the equation

$$\int_0^\infty \frac{\cos(xy)}{a^2+x^2}\,dx = \frac{\pi}{2a}e^{-ay}, \text{ where } a>0,\ y>0,$$

with respect to y, prove that

$$\int_0^\infty \frac{\sin(xy)}{x(a^2+x^2)}\,dx = \frac{\pi}{2a^2}(1-e^{-ay}),\quad \int_0^\infty \frac{1-\cos(xy)}{x^2(a^2+x^2)}\,dx = \frac{\pi}{2a^3}(e^{-ay}+ay-1),$$

$$\int_0^\infty \frac{xy-\sin(xy)}{x^3(a^2+x^2)}\,dx = \frac{\pi}{2a^4}[1-ay+\tfrac{1}{2}(ay)^2-e^{-ay}],$$

and so on, the terms introduced on the left being those of the sine and cosine power-series and the terms on the right being those of the exponential series. [*Math. Trip.* 1902.]

17. Justify differentiating the integral

$$\int_0^{\frac{1}{2}\pi} \tan^{-1}(a^2\tan^2 x)\,dx, \qquad (a>0)$$

under the integral sign, and so prove that its value is $\pi\tan^{-1}\{a/(a+\sqrt{2})\}$.

Change the variable to θ, where $a^2\tan^2 x = \cot\theta$, and deduce that if we put $\sqrt{\kappa} = 2a/(a^2-1)$,

$$\int_0^{\frac{1}{2}\pi} \tan^{-1}\left\{\frac{2\sqrt{(\tan\theta)}}{\sqrt{\kappa}(1+\tan\theta)}\right\}d\theta = \pi\tan^{-1}\left\{\frac{1}{\sqrt{(2\kappa)}+\sqrt{(1+\kappa)}}\right\}.$$

Examine the special cases $\kappa=2$ (Wolstenholme) and $\kappa=8$ (*Oxford Senior Scholarship*).

18. By differentiation or otherwise, prove that if a, b are positive,

$$\int_0^\infty \log\left(1+\frac{a^2}{x^2}\right)dx = \pi a,\quad \int_0^\infty \frac{\log(a^2+x^2)}{b^2+x^2}\,dx = \frac{\pi}{b}\log(a+b),$$

$$\int_0^\infty \tan^{-1}(ax)\tan^{-1}(bx)\frac{dx}{x^2} = \tfrac{1}{2}\pi\left[a\log\left(1+\frac{b}{a}\right)+b\log\left(1+\frac{a}{b}\right)\right],$$

$$\int_0^\infty \log\left(1+\frac{a^2}{x^2}\right)\log\left(1+\frac{b^2}{x^2}\right)dx = 2\pi\left[a\log\left(1+\frac{b}{a}\right)+b\log\left(1+\frac{a}{b}\right)\right].$$

19. By differentiation or otherwise, prove that, if a is positive,

$$\int_0^{\frac{1}{2}\pi} \tan^{-1}(\sinh a\sin x)\,dx = \int_0^a \frac{t\,dt}{\sinh t}$$

$$= \frac{1}{4}\pi^2 - 2\left[(1+a)e^{-a} + \frac{1}{3^2}(1+3a)e^{-3a} + \frac{1}{5^2}(1+5a)e^{-5a} + \dots\right].$$

[*Math. Trip.* 1892.]

20. Shew that if a, b, c are positive, a being the greatest of the three,

$$\int_0^\infty \sin ax \cos bx \cos cx \frac{dx}{x} = \tfrac{1}{2}\pi, \qquad \text{if } a > b+c,$$

$$\text{or } \tfrac{1}{4}\pi, \qquad \text{if } a < b+c.$$

Deduce by integration that

$$\int_0^\infty \sin ax \sin bx \cos cx \frac{dx}{x^2} = \tfrac{1}{2}\pi b, \qquad \text{if } a > b+c,$$

$$\text{or } \tfrac{1}{4}\pi(a+b-c), \qquad \text{if } a < b+c,$$

and
$$\int_0^\infty \sin ax \sin bx \sin cx \frac{dx}{x^3} = \tfrac{1}{2}\pi bc, \qquad \text{if } a > b+c,$$

$$\text{or } \tfrac{1}{8}\pi(2bc+2ca+2ab-a^2-b^2-c^2), \qquad \text{if } a < b+c.$$

In particular, $\displaystyle\int_0^\infty \sin tx \sin^2 x \frac{dx}{x^3} = \tfrac{1}{2}\pi t(1-\tfrac{1}{4}t), \qquad \text{if } 0 < t < 2,$

$$\text{or } \tfrac{1}{2}\pi, \qquad \text{if } t > 2.$$

21. Prove that, if $t > |a_1| + |a_2| + \ldots + |a_n|$,

$$\int_0^\infty \sin tx \,.\, \prod_1^n \sin a_r x \,.\, \prod_1^p \cos b_r x \,.\, \frac{dx}{x^{n+1}} = \tfrac{1}{2}\pi(a_1 a_2 \ldots a_n). \qquad \text{[Störmer.]}$$

22. The results of Exs. 20, 21 can also be found by integration by parts; this method gives at once

$$\int_0^\infty (\Sigma A \cos ax) \frac{dx}{x^{2n+2}} = (-1)^{n+1} \tfrac{1}{2}\pi\, \Sigma A \frac{a^{2n+1}}{(2n+1)!},$$

$$\int_0^\infty (\Sigma A \cos ax) \frac{dx}{x^{2n+1}} = (-1)^n\, \Sigma A \log a \frac{a^{2n}}{(2n)!},$$

where
$$\Sigma A = 0, \quad \Sigma A a^2 = 0, \quad \Sigma A a^4 = 0, \quad \ldots, \quad \Sigma A a^{2n} = 0.$$

Establish similar formulae for integrals which contain sums of sines; and prove that

$$\int_0^\infty \left(\frac{\sin x}{x}\right)^n dx = \frac{1}{(n-1)!} \frac{\pi}{2^n} \left[n^{n-1} - n(n-2)^{n-1} + \frac{n(n-1)}{2!}(n-4)^{n-1} - \ldots\right],$$

the number of terms in the bracket being $\tfrac{1}{2}n$ or $\tfrac{1}{2}(n+1)$.

[Wolstenholme.]

Dirichlet's Integrals.

23. Apply the theorem of Art. 172 (4) to justify the equation

$$\lim_{c\to 0} \int_0^\infty f(x+ct) e^{-t^2} dt = \tfrac{1}{2}\sqrt{\pi}\, f(x),$$

and deduce that

$$\lim_{c\to 0} \frac{1}{c} \int_{-\infty}^\infty f(y) e^{-(y-x)^2/c^2} dy = \tfrac{1}{2}\sqrt{\pi} \lim_{\delta\to 0} [f(x+\delta) + f(x-\delta)].$$

[Weierstrass.]

24. Apply Abel's test of uniform convergence to prove that if $f(t)$ is monotonic (at least after a certain stage) and continuous, then

$$\lim_{n\to\infty} \int_a^\infty f(t) \sin(nt) \frac{dt}{t} = \pi f(0), \quad \tfrac{1}{2}\pi f(0), \quad \text{or } 0,$$

according as a is negative, zero, or positive.

Deduce that if x is positive and $\int^{\infty} f(t)dt$ is convergent, then

$$\int_0^{\infty} \cos(xv)dv \int_0^{\infty} f(t)\cos(vt)dt = \frac{\pi}{2} f(x),$$

and the same result is true if the cosines are *both* replaced by sines. [FOURIER.]

25. By taking $f(x)=e^{-ax}$ $(a>0)$, deduce from the last example that

$$\int_0^{\infty} \frac{v\sin(xv)}{a^2+v^2}dv = \frac{\pi}{2}e^{-ax} = \int_0^{\infty} \frac{a\cos(xv)}{a^2+v^2}dv.$$

Consider similarly the integrals given by taking $f(x)=1$ from $x=0$ to 1, and $f(x)=0$ from 1 to ∞. [FOURIER.]

26. From the integrals

$$\operatorname{sech} x = 2\int_0^{\infty} \frac{\cos 2xt}{\cosh \pi t}dt, \quad \operatorname{sech}^2 x = 4\int_0^{\infty} \frac{t\cos 2xt}{\sinh \pi t}dt,$$

$$e^{-x^2} = \frac{2}{\sqrt{\pi}}\int_0^{\infty} e^{-t^2}\cos 2xt\,dt,$$

prove by the method of Ex. 4, Art. 174, that

$$\sum_{-\infty}^{\infty} \operatorname{sech}(x+n\omega) = \frac{\pi}{\omega}\sum_{-\infty}^{\infty} \frac{\cos(2n\pi x/\omega)}{\cosh(n\pi^2/\omega)},$$

$$\sum_{-\infty}^{\infty} \operatorname{sech}^2(x+n\omega) = \frac{2\pi^2}{\omega^2}\sum_{-\infty}^{\infty} \frac{n\cos(2n\pi x/\omega)}{\sinh(n\pi^2/\omega)},$$

$$\sum_{-\infty}^{\infty} e^{-(x+n\omega)^2} = \frac{\sqrt{\pi}}{\omega}\sum_{-\infty}^{\infty} e^{-n^2\pi^2/\omega^2}\cos(2n\pi x/\omega).$$ [SCHLÖMILCH.]

Integration of Series.

27. Prove that (see Ex. 42, p. 167), if a, b are positive,

$$\int_0^{\frac{1}{2}\pi} \log(a^2\cos^2 x + b^2\sin^2 x)dx = \pi\log\left(\frac{a+b}{2}\right),$$

$$\int_0^{\frac{1}{2}\pi} \log(a^2\cos^2 x + b^2\sin^2 x)\cos 2nx\,dx = -\frac{\pi}{n}\left(\frac{b-a}{b+a}\right)^n,$$

and verify that these results remain correct when $b=0$ and when $a=0$.

Deduce that, if $r^2<1$ and p, q are positive,

$$(1-r^2)\int_0^{\pi} \frac{\log(a^2\cos^2 x + b^2\sin^2 x)}{1-2r\cos x+r^2}dx = 2\pi\log[\tfrac{1}{2}\{a(1+r^2)+b(1-r^2)\}],$$

$$\int_0^{\pi} \frac{\log(a^2\cos^2 x + b^2\sin^2 x)}{p^2\cos^2 x + q^2\sin^2 x}dx = \frac{2\pi}{pq}\log\left(\frac{aq+bp}{p+q}\right).$$

Compare Ex. 2, p. 347.

28. Using the series of Ex. 7, Ch. IX., prove that

$$\int_0^{\pi} \frac{\cos\frac{1}{2}\phi\,d\phi}{1+2t^2\cos\phi+t^4} = \frac{2\tanh^{-1}t}{t(1+t^2)}, \quad \int_0^{\pi} \frac{\cos\frac{1}{2}\phi\,d\phi}{1-2t^2\cos\phi+t^4} = \frac{2\tan^{-1}t}{t(1-t^2)}.$$

Deduce that $$\int_0^{\pi} \tan^{-1}\left(\frac{2t^2\sin\phi}{1-t^4}\right)\frac{d\phi}{\sin\frac{1}{2}\phi} = 8\tan^{-1}t\,.\,\tanh^{-1}t,$$

and verify this result by expanding in powers of t. [HARDY.]

29. From Art. 65, shew that (if $r^2<1$)

$$\frac{(1-r^2)\sin x}{(1-2r\cos x+r^2)^2}=\sin x+2r\sin 2x+3r^2\sin 3x+\dots.$$

Shew also that

$$\int_0^\pi \frac{\sin^2 x\cos x\,dx}{(1-2r\cos x+r^2)^2}=\frac{\pi r}{1-r^2},\quad \int_0^\pi \frac{x\sin x\,dx}{1-2r\cos x+r^2}=\frac{\pi}{r}\log(1+r).$$

30. (1) Prove that $\displaystyle\int_0^\infty \frac{t^{2n-1}dt}{\sinh(\pi t)}=\frac{2^{2n}-1}{2n}B_n$ (Ex. 7, Art. 175)

and $\displaystyle\int_0^\infty \frac{t^{2n}dt}{\cosh(\pi t)}=\frac{E_n}{2^{2n+1}},$ (Ex. 8, Art. 175)

where E_n is Euler's number (Ex. 38, p. 260).

(2) By expanding in powers of a, shew that

$$\int_0^\infty e^{-x}(1-e^{-ax})\frac{dx}{x}=\log(1+a).$$

31. From the series for $\log(4\sin^2 x)$, $\log(4\cos^2 x)$ (see Art. 65), prove that

$$\int_0^{\frac{1}{2}\pi}\cos 2nx\log(4\sin^2 x)dx=-\frac{\pi}{2n},\quad \int_0^{\frac{1}{2}\pi}\cos 2nx\log(4\cos^2 x)dx=(-1)^{n-1}\frac{\pi}{2n},$$

$$\int_0^{\frac{1}{4}\pi}\log(\cot^2 x)dx=2\left(1-\frac{1}{3^2}+\frac{1}{5^2}-\frac{1}{7^2}+\dots\right),$$

$$\int_0^{\frac{1}{12}\pi}\log(\cot^2 3x)dx=\frac{4}{3}\left(1-\frac{1}{3^2}+\frac{1}{5^2}-\frac{1}{7^2}+\dots\right).$$

Deduce that

$$\int_0^{\frac{1}{2}\pi}\{\log(4\sin^2 x)\}^2dx=\frac{1}{6}\pi^3=\int_0^{\frac{1}{2}\pi}\{\log(4\cos^2 x)\}^2dx,$$

$$\int_0^{\frac{1}{2}\pi}\log(4\sin^2 x).\log(4\cos^2 x)dx=-\frac{1}{12}\pi^3.$$ [WOLSTENHOLME.]

[Compare Exs. 46, 47; and note that the only difficulties arise in extending the rule for term-by-term integration up to the limits.]

32. Use Art. 175 to justify the following transformations;

$$\sum_1^\infty \frac{1}{n^2}\frac{1}{2^n}=\int_0^1\log\left(\frac{1}{x}\right)\frac{dx}{2-x}$$

$$=\int_0^1\log\left(\frac{1}{1-x}\right)(1-x+x^2-\dots)dx$$

$$=1-\tfrac{1}{2}(1+\tfrac{1}{2})+\tfrac{1}{3}(1+\tfrac{1}{2}+\tfrac{1}{3})-\dots$$

$$=\tfrac{1}{12}\pi^2-\tfrac{1}{2}(\log 2)^2.$$ [LEGENDRE.]

33. Shew that

$$\int_0^\infty e^{-t^2}\sin(2tx)\frac{dt}{t}=\sqrt{\pi}\int_0^x e^{-v^2}dv$$ (Ex. 4, Art. 176),

and verify the equation by making x tend to ∞.

34. If $\sinh x.\sinh y=1$, prove that $\displaystyle\int_0^\infty y\,dx=\tfrac{1}{4}\pi^2$.

[Write $t=e^{-x}$ and use Ex. 7, Art. 175.] [*Math. Trip.* 1902.]

Gamma Functions, etc.

35. Use the method of Art. 106 to shew that if the real parts of r, s are positive,

$$\Gamma(r)\Gamma(s)=\int_0^\infty e^{-x}x^{r-1}dx\int_0^\infty e^{-y}y^{s-1}dy=\int_0^\infty e^{-\xi}\xi^{r+s-1}d\xi\int_0^1 \eta^{r-1}(1-\eta)^{s-1}d\eta,$$

and deduce that

$$\Gamma(r)\Gamma(s)/\Gamma(r+s)=\int_0^1 \eta^{r-1}(1-\eta)^{s-1}d\eta.$$

36. If $U=\int_0^\infty e^{-xt}t^{n-1}dt$, where $x=\xi+i\eta$ and $\xi>0$, shew that

$$\frac{\partial U}{\partial \xi}=-\frac{n}{x}U,\quad \frac{\partial U}{\partial \eta}=-\frac{in}{x}U,$$

and hence prove that $\qquad U=\Gamma(n)/x^n, \quad$ if $n>0$.

By using Ex. 2, Art. 172, deduce that if $0<n<1$,

$$\int_0^\infty \cos t\,.\,t^{n-1}dt=\Gamma(n)\cos(\tfrac{1}{2}n\pi),\quad \int_0^\infty \sin t\,.\,t^{n-1}dt=\Gamma(n)\sin(\tfrac{1}{2}n\pi),$$

and verify that the last result is correct if $-1<n<1$.

Obtain the corresponding formulae for

$$\int_0^\infty \cos(x^p)dx \quad\text{and}\quad \int_0^\infty \sin(x^p)dx, \quad p>1.$$ [CAUCHY.]

37. If the real part of x lies between $-k$ and $-(k+1)$, where k is a positive integer, prove that

$$\Gamma(x)=\int_0^\infty t^{x-1}\left[e^{-t}-1+t-\ldots+(-1)^{k+1}\frac{t^k}{k!}\right]dt.$$ [CAUCHY.]

[Apply the process of integration by parts to the integral for $\Gamma(x+k)$.]

38. Shew that if α, β are real,

$$\left\{\frac{\Gamma(\alpha)}{|\Gamma(\alpha+i\beta)|}\right\}^2=\prod_0^\infty\left[1+\frac{\beta^2}{(\alpha+n)^2}\right].$$ [MELLIN.]

If $x=i\eta$, shew that

$$|\Gamma(1+x)|=\sqrt{\{(\pi\eta)/\sinh(\pi\eta)\}}.$$

39. If

$$A=\int_0^1\frac{dx}{\sqrt{(1-x^4)}},\quad B=\int_0^1\frac{x^2dx}{\sqrt{(1-x^4)}},$$

express A, B in terms of Gamma-functions, and prove that

$$\Gamma(\tfrac{1}{4})=(\tfrac{1}{8}\pi)^{\frac{1}{4}}A^{\frac{1}{2}}.$$

Assuming the value of $\Gamma(\frac{1}{4})$ given in Ex. 41, deduce that

$$A=1{\cdot}311029,\quad B=0{\cdot}599070.$$ [GAUSS.]

40. Similarly express the integrals

$$\int_0^1\frac{dx}{\sqrt{(1-x^3)}},\quad \int_0^1\frac{x\,dx}{\sqrt{(1-x^3)}}$$

in terms of $\Gamma(\frac{1}{3})$, and so obtain numerical values for them. [GAUSS.]

41. Deduce from the product formula for $\Gamma(1+x)$ that if $|x|<2$,

$$\log\Gamma(1+x)=\tfrac{1}{2}\log\left\{\frac{\pi x}{\sin(\pi x)}\right\}-\tfrac{1}{2}\log\left(\frac{1+x}{1-x}\right)+C_1x-C_3x^3-C_5x^5-\dots,$$

where

$$C_1=1-C=0{\cdot}422783,\qquad C_7=\tfrac{1}{7}\sum_2^\infty n^{-7}=0{\cdot}0011928,$$

$$C_3=\tfrac{1}{3}\sum_2^\infty n^{-3}=\quad 673530,\qquad C_9=\tfrac{1}{9}\sum_2^\infty n^{-9}=\qquad 2232,$$

$$C_5=\tfrac{1}{5}\sum_2^\infty n^{-5}=\quad 73856,\qquad C_{11}=\tfrac{1}{11}\sum_2^\infty n^{-11}=\qquad 449.$$

As a numerical exercise, prove that

$$\log_{10}\Gamma(\tfrac{5}{4})=\bar{1}{\cdot}957321,\quad \log_{10}\Gamma(\tfrac{4}{3})=\bar{1}{\cdot}950841.$$

It will also be found from this series that if $\Gamma(1+i)=re^{i\theta}$, then

$$\log_{10}r=\bar{1}{\cdot}71731\quad\text{and}\quad\theta=-{\cdot}30163.$$

These give $\qquad \Gamma(1+i)=0{\cdot}49802-(0{\cdot}15495)i$;

a result calculated to 7 decimals by Gauss, from Stirling's series (Art. 132), writing $x=10+i$.

42. If

$$\psi(x)=\Gamma'(x)/\Gamma(x)=\frac{d}{dx}\{\log\Gamma(x)\},$$

prove that

$$\psi(x)=\lim_{n\to\infty}\left[\log n-\left(\frac{1}{x}+\frac{1}{1+x}+\dots+\frac{1}{n+x}\right)\right]$$

$$=-C-\frac{1}{x}+\sum_1^\infty\left(\frac{1}{n}-\frac{1}{n+x}\right),$$

where C is Euler's constant.

Shew that

$$\psi(x)-\psi(y)=\sum_0^\infty\left(\frac{1}{n+y}-\frac{1}{n+x}\right).$$

Deduce from Arts. 178, 179 that

$$\psi(1+x)-\psi(x)=1/x,\quad \psi(1-x)-\psi(x)=\pi\cot(\pi x),$$

$$\psi(2x)=\tfrac{1}{2}[\psi(x)+\psi(x+\tfrac{1}{2})]+\log 2,$$

$$\psi(rx)=\frac{1}{r}\left[\psi(x)+\psi\left(x+\frac{1}{r}\right)+\dots+\psi\left(x+\frac{r-1}{r}\right)\right]+\log r,$$

$$\psi(x)+C=\int_0^1\frac{1-t^{x-1}}{1-t}\,dt\quad\text{(if the real part of } x \text{ is positive).}$$

Obtain the particular results,

$$\psi(1)=-C,\quad \psi(2)=1-C,\quad \psi(3)=\tfrac{3}{2}-C,\ \dots,$$

$$\psi(\tfrac{1}{2})=-C-2\log 2,\quad \psi(\tfrac{3}{2})=2-C-2\log 2.$$

Shew also that

$$\psi'(1)=\sum_1^\infty(1/n^2)=\tfrac{1}{6}\pi^2,\quad \psi'(\tfrac{1}{2})=\sum_0^\infty 4/(2n+1)^2=\tfrac{1}{2}\pi^2.$$

43. Shew that, if p, q are positive integers,

$$\psi(p/q)+C=\lim_{t\to 1}f(t),$$

where

$$f(t)=-t^p\log(1-t^q)-q\sum_0^\infty\frac{t^{p+nq}}{p+nq}$$

$$=-t^p\log\{(1-t^q)/(1-t)\}+(1-t^p)\log(1-t)+\sum_{r=1}^{q-1}\omega^{-pr}\log(1-\omega^r t),$$

if $\qquad \omega=\cos(2\pi/q)+i\sin(2\pi/q)$.

Deduce from this and the corresponding formula with $q-p$ in place of p, that

$$\psi\left(\frac{p}{q}\right)+C=-\log q-\tfrac{1}{2}\pi\cot\left(\frac{p\pi}{q}\right)+\sum_{r=1}^{q-1}\cos\left(\frac{2\pi rp}{q}\right)\log\left\{2\sin\left(\frac{r\pi}{q}\right)\right\}.$$

Obtain the particular results,

$$\psi(\tfrac{1}{3})+C=-\tfrac{3}{2}\log 3-\tfrac{1}{6}\pi\sqrt{3},\quad \psi(\tfrac{1}{4})+C=-3\log 2-\tfrac{1}{2}\pi,$$
$$\psi(\tfrac{2}{3})+C=-\tfrac{3}{2}\log 3+\tfrac{1}{6}\pi\sqrt{3},\quad \psi(\tfrac{3}{4})+C=-3\log 2+\tfrac{1}{2}\pi.\quad \text{[GAUSS.]}$$

44. Similar results can be obtained for the function $\beta(x)=\sum_{0}^{\infty}\frac{(-1)^n}{x+n}$; thus, shew that

$$\beta(x)+\beta(1+x)=1/x,\quad \beta(x)+\beta(1-x)=\pi\operatorname{cosec}(\pi x),$$
$$\beta(x)=\frac{1}{2}\left[\psi\left(\frac{x+1}{2}\right)-\psi\left(\frac{x}{2}\right)\right]=\psi(x)-\psi\left(\frac{x}{2}\right)-\log 2,$$
$$\lim_{x\to\infty}[\psi(x)-\log x]=0,\quad \lim_{x\to\infty}\beta(x)=0,$$
$$\beta(x)=\int_0^1\frac{t^{x-1}}{1+t}dt \quad \text{(if the real part of } x \text{ is positive).}$$

In particular, prove that

$$\beta(1)=\log 2,\qquad \beta(\tfrac{1}{2})=\tfrac{1}{2}\pi,$$
$$\beta(\tfrac{1}{3})=\log 2+\tfrac{1}{3}\pi\sqrt{3},\quad \beta(\tfrac{2}{3})=-\log 2+\tfrac{2}{3}\pi\sqrt{3}.$$

45. If

$$f(a)=\int_0^{\frac{1}{2}\pi}\sin^{2a-1}x\,dx=\frac{\sqrt{\pi}}{2}\frac{\Gamma(a)}{\Gamma(a+\frac{1}{2})},$$

prove from Art. 172 (3) that we may differentiate under the integral sign, provided that a is positive.

Hence

$$f'(a)=2\int_0^{\frac{1}{2}\pi}\sin^{2a-1}x\,.\log\sin x\,.\,dx=\frac{\sqrt{\pi}}{2}\frac{\Gamma(a)}{\Gamma(a+\frac{1}{2})}[\psi(a)-\psi(a+\tfrac{1}{2})]$$

and

$$f''(a)=4\int_0^{\frac{1}{2}\pi}\sin^{2a-1}x\,.(\log\sin x)^2.\,dx$$
$$=\frac{\sqrt{\pi}}{2}\frac{\Gamma(a)}{\Gamma(a+\frac{1}{2})}[\{\psi(a)-\psi(a+\tfrac{1}{2})\}^2+\psi'(a)-\psi'(a+\tfrac{1}{2})].$$

46. Shew from Ex. 45 that

$$\int_0^{\frac{1}{2}\pi}\sin x\,.\log\sin x\,.\,dx=\log 2-1,$$
$$\int_0^{\frac{1}{2}\pi}\sin x\,.(\log\sin x)^2.\,dx=(\log 2-1)^2+1-\tfrac{1}{12}\pi^2,$$
$$\int_0^{\frac{1}{2}\pi}\frac{\log\sin x}{\sqrt{(\sin x)}}dx=-\frac{\sqrt{\pi}}{4\sqrt{2}}[\Gamma(\tfrac{1}{4})]^2,$$
$$\int_0^{\frac{1}{2}\pi}\sqrt{(\sin x)}\,.\log\sin x\,.\,dx=\sqrt{2}\pi^{\frac{3}{2}}(\pi-4)/\{\Gamma(\tfrac{1}{4})\}^2,$$
$$\int_0^{\frac{1}{2}\pi}\log\sin x\,.\,dx=-\tfrac{1}{2}\pi\log 2,$$
$$\int^{\frac{1}{2}\pi}(\log\sin x)^2.\,dx=\tfrac{1}{2}\pi[(\log 2)^2+\tfrac{1}{12}\pi^2].$$

47. Justify the differentiation of the equation (Ex. 35)

$$\int_0^{\frac{1}{2}\pi} \sin^{2a-1} x \cos^{2\beta-1} x\, dx = \frac{1}{2}\frac{\Gamma(a)\Gamma(\beta)}{\Gamma(a+\beta)}.$$

Deduce that

$$\int_0^{\frac{1}{2}\pi} \log\sin x \,.\, \log\cos x \,.\, dx = \tfrac{1}{2}\pi[(\log 2)^2 - \tfrac{1}{24}\pi^2],$$

$$\int_0^{\frac{1}{2}\pi} \sin x \,.\, \log\sin x \,.\, \log\cos x \,.\, dx = 2 - \log 2 - \tfrac{1}{8}\pi^2.$$

Miscellaneous.

48. From the series

$$\operatorname{sech} x = 2(e^{-x} - e^{-3x} + e^{-5x} - \ldots),$$

prove that if the real part of c is greater than -1,

$$\int_0^\infty \frac{e^{-cx}}{\cosh x}\, dx = \tfrac{1}{2}\left[\psi\left(\frac{c+3}{4}\right) - \psi\left(\frac{c+1}{4}\right)\right].$$

[See Ex. 42 and use Art. 52 (3).]

49. From the last example deduce that, if the real part of a is positive and not greater than 1,

$$\int_0^\infty \frac{\sinh ax}{\cosh x}\frac{dx}{x} = \log\cot\tfrac{1}{4}(1-a)\pi,$$

and hence, if λ is real, prove that

$$\int_0^\infty \cos\lambda x \tanh x \frac{dx}{x} = \log\coth\tfrac{1}{4}\lambda\pi, \qquad [\textit{Math. Trip.}\ 1889.]$$

$$\int_0^\infty \frac{\sin\lambda x}{\cosh x}\frac{dx}{x} = 2\tan^{-1}(\tanh\tfrac{1}{4}\lambda\pi). \qquad [\text{Hardy.}]$$

50. From Ex. 8, Art. 175, prove that if the real part of a is positive and not greater than $\frac{1}{2}$,

$$\int_0^\infty \frac{\sinh^2 ax}{\sinh x}\frac{dx}{x} = \tfrac{1}{2}\log\sec a\pi. \qquad [\textit{Math. Trip.}\ 1895.]$$

51. Deduce from Ex. 35 and Art. 175 B, that if x and a are positive

$$\frac{\Gamma(x)\Gamma(a)}{\Gamma(x+a)} = \frac{1}{x} - \frac{a-1}{x+1} + \frac{(a-1)(a-2)}{2!(x+2)} - \frac{(a-1)(a-2)(a-3)}{3!(x+3)} + \ldots.$$

Shew also that

$$\left[\frac{\Gamma(x)}{\Gamma(x+\frac{1}{2})}\right]^2 = \frac{1}{x} + \frac{1^2}{4}\cdot\frac{1}{x(x+1)} + \frac{1^2 . 3^2}{4 . 8}\frac{1}{x(x+1)(x+2)} + \ldots$$

[To obtain the latter series, expand $\eta^{-\frac{1}{2}}(1-\eta)^{x-1}$ in the form

$$\eta^{-\frac{1}{2}}(1-\eta)^{x-\frac{1}{2}}\left(1 + \tfrac{1}{2}\eta + \frac{1 . 3}{2 . 4}\eta^2 + \ldots\right).]$$

52. Obtain the first integral of Ex. 49 from the series

$$\operatorname{sech} x = 2(e^{-x} - e^{-3x} + e^{-5x} - \ldots)$$

by applying Frullani's integral to the separate terms.

Obtain similarly the following integrals:

$$\int_0^\infty e^{-ax} \tanh x \frac{dx}{x} = \log \frac{a}{4} + 2\log\left\{\Gamma\left(\frac{a}{4}\right)\Big/\Gamma\left(\frac{a+2}{4}\right)\right\},$$

$$\int_0^\infty e^{-ax}(1 - \operatorname{sech} x)\frac{dx}{x} = -\log\frac{a}{4} + 2\log\left\{\Gamma\left(\frac{a+3}{4}\right)\Big/\Gamma\left(\frac{a+1}{4}\right)\right\},$$

where the real part of a is positive. [HARDY.]

53. Write down the form of Frullani's integral when $\phi(x) = 1/(1+e^{-x})$; and deduce that when p is positive,

$$\int_0^\infty \left(\frac{\sinh px}{\cosh px + \cos qx} - \frac{\sinh px}{\cosh px + \cos rx}\right)\frac{dx}{x} = \tfrac{1}{2}\log\left(\frac{p^2+q^2}{p^2+r^2}\right).$$

[*Math. Trip.* 1890.]

54. The following integrals are allied to Frullani's integral

$$\left.\begin{aligned}&\int_0^\infty (\sin mx - \sin nx)^2 \frac{dx}{x^2} = \tfrac{1}{2}\pi\,|m-n|,\\ &\int_0^\infty (e^{-mx} - e^{-nx})^2\frac{dx}{x^2} = 2m\log\frac{m+n}{2m} + 2n\log\frac{m+n}{2n}\end{aligned}\right\}\quad m, n \geqq 0.$$

$$\int_{-\infty}^\infty [\phi(x-a) - \phi(x-b)]\,dx = (b-a)[\phi(\infty) - \phi(-\infty)].$$

Evaluate the first of these integrals when m, n have opposite signs.

55. By changing the variable from t to $2t$ in Art. 180, prove that

$$\phi(1) = A - 1 = \int_0^\infty \frac{dt}{t}\left[\left(\frac{1}{t}+1\right)e^{-2t} - \frac{1}{t}e^{-t}\right] + \frac{1}{2}\int_0^\infty \frac{1-e^{-t}}{e^t+1}\frac{dt}{t}.$$

Shew from Ex. 3, Art. 173, that of these integrals the first is equal to $\log 2 - 1$ and the second to $\frac{1}{2}\log(\frac{1}{2}\pi)$. (See Ex. 52.)

56. Prove that if ξ is positive, the function

$$\frac{e^{-\xi t}}{t}\left(\frac{1}{e^t-1} - \frac{1}{t} + \frac{1}{2}\right)$$

steadily decreases as t increases from 0 to ∞ (see p. 234). By applying Art. 168, deduce that, if $x = \xi + i\eta$ in the formulae of p. 465,

$$|\psi(x)| < \tfrac{1}{8}/|\eta|.$$

57. Deduce from Ex. 56 that if $x = \xi + i\eta$, where ξ is positive and fixed, but η tends to infinity,

$$|\Gamma(1+x)| \sim \sqrt{(2\pi)}\, r^{\xi+\frac{1}{2}} e^{-\frac{1}{2}\pi\eta},$$

where $r = |x|$. (Compare Ex. 38, p. 474.) [PINCHERLE.]

EASY MISCELLANEOUS EXAMPLES.

1. Shew that $\quad \sum_1^\infty \left[n \log\left(\frac{2n+1}{2n-1}\right) - 1\right] = \frac{1}{2}(1 - \log 2).$

[The series can be summed to n terms; or we may express the general term in the form $\int_0^1 \{x^2/(4n^2 - x^2)\}\, dx$.]

2. Discuss the convergence of the series

$$\Sigma n^k[\sqrt{(n+1)} - 2\sqrt{n} + \sqrt{(n-1)}]. \qquad [\textit{Math. Trip.}\ 1890.]$$

3. If
$$C_n = 1 + \frac{1}{2} + \dots + \frac{1}{n} - \log n,$$
prove that
$$C_n - C_{2n} = \int_0^1 \frac{t^{2n}}{1+t}\, dt,$$
and deduce that Euler's constant is equal to
$$1 - \int_0^1 \frac{dt}{1+t}(t^2 + t^4 + t^8 + t^{16} + \dots). \qquad [\text{Catalan.}]$$

4. Prove that, if $\mu > \nu$, the sum
$$\sum_{n=0}^{\mu}{}' \frac{2\nu}{\nu^2 - n^2}, \quad (n = \nu \text{ excluded}),$$
tends to the limit $\log\{(1+k)/(1-k)\}$, when μ, ν tend to infinity in such a way that ν/μ tends to k. [*Math. Trip.* 1894.]

5. Apply Euler's method (Art. 24) to shew that
$$1 - \frac{1}{3^2} + \frac{1}{5^2} - \frac{1}{7^2} + \dots = \cdot 915965\dots, \quad 1 - \frac{1}{3^4} + \frac{1}{5^4} - \frac{1}{7^4} + \dots = \cdot 988944\dots.$$

[For other methods of transforming and evaluating these series see Glaisher, *Messenger of Maths.*, vol. 33, 1903, pp. 1, 20.]

6. If s_n denotes $2^{-n} + 3^{-n} + 4^{-n} + \dots$ to ∞, prove by conversion into double series that
$$s_2 + s_3 + s_4 + \dots = 1, \qquad s_2 + s_4 + s_6 + \dots = \tfrac{3}{4},$$
$$\tfrac{1}{2}s_2 + \tfrac{1}{3}s_3 + \tfrac{1}{4}s_4 + \dots = 1 - C, \quad s_2 + \tfrac{1}{2}s_4 + \tfrac{1}{3}s_6 + \dots = \log 2.$$

[See Woolsey Johnson, *Bull. Am. Math. Soc.*, vol. 12, 1906, p. 477.]

7. If θ is positive, prove that

$$\sum_1^\infty (3\coth^4 n\theta - 4\coth^2 n\theta + 1) = 4\sum_1^\infty n^3(\coth n\theta - 1).$$

[Write $x = e^{-2\theta}$ and convert into a double series.] [*Math. Trip.* 1894.]

8. Shew that the series

$$\sum \frac{x^n(x^{2n+1}-1)}{(x^{2n}+1)(x^{2n+2}+1)}$$

does not converge uniformly in any interval including $x=1$.

[*Math. Trip.* 1901.]

9. (1) If $f_n(x) = nx^{n-1} - (n+1)x^n$, prove that

$$\sum_1^\infty f_n(x) = 1, \quad 0 \leqq |x| < 1,$$

and deduce that $\int_0^1 \sum f_n(x)\,dx = 1$, while $\sum \int_0^1 f_n(x)\,dx = 0$.

(2) Prove that the series obtained by differentiating

$$\sum \frac{1}{n^2}\log(1+nx^2)$$

is uniformly convergent for all real values of x, including $x=0$. Is the same true of the given series?

10. Prove that

$$\left(\frac{1}{a} + \frac{1}{2}\frac{x}{a+2} + \frac{1.3}{2.4}\frac{x^2}{a+4} + \dots\right)\left(1 + \frac{1}{2}x + \frac{1.3}{2.4}x^2 + \dots\right)$$

$$= \frac{1}{a}\left[1 + \frac{a+1}{a+2}x + \frac{(a+1)(a+3)}{(a+2)(a+4)}x^2 + \dots\right],$$

$$\log(1+x^2).\tan^{-1}x = 2[\tfrac{1}{3}(1+\tfrac{1}{2})x^3 - \tfrac{1}{5}(1+\tfrac{1}{2}+\tfrac{1}{3}+\tfrac{1}{4})x^5 + \dots],$$

$$[F(\tfrac{1}{2}, 1, 2, x)]^2 = F(1, \tfrac{3}{2}, 3, x), \quad [F(1, \tfrac{3}{2}, 3, x)]^2 = F(2, \tfrac{5}{2}, 5, x).$$

[All these can be obtained by direct multiplication; but the law of the coefficients is more quickly determined by differentiation or some other special device.]

11. Shew how to calculate log 2, log 3, log 5, log 7 from the five series a, b, c, d, e given by writing

$$x = \tfrac{1}{19}, \tfrac{1}{49}, \tfrac{1}{161}, \tfrac{1}{99}, \tfrac{1}{251}$$

respectively in the series for $\log\{(1+x)/(1-x)\}$; and prove that

$$a - 2b + c = d + 2e.$$

[For results to 260 decimals, see ADAMS, *Math. Papers*, vol. 1, p. 459.]

12. Prove that as x tends to 1,

$$x + \tfrac{1}{2}x^4 + \tfrac{1}{3}x^9 + \tfrac{1}{4}x^{16} + \dots \sim -\tfrac{1}{2}\log(1-x).$$ [CESÀRO.]

13. Sketch the graphs from 0 to 2π of the functions

$$\sin 5x + \frac{1}{2}\sin 10x + \frac{1}{3}\sin 15x + \dots,$$

$$\cos 2x + \frac{1}{3^2}\cos 6x + \frac{1}{5^2}\cos 10x + \dots,$$

$$\sin 2x + \frac{1}{3}\sin 6x + \frac{1}{5}\sin 10x + \dots.$$

14. If $$f(x)=a_0+\sum_{n=1}^{\infty}(a_n\cos nx+b_n\sin nx),$$

prove that $$\frac{1}{s}[f(x)+f(x+a)+f(x+2a)+\ldots+f\{x+(s-1)a\}],$$

where $a=2\pi/s$, contains only those terms of the original series in which n is a multiple of s.

[This result is of some importance in the numerical applications of Fourier's series; see WEDMORE, *Journal Instit. Elect. Engineers*, vol. 25, 1896, p. 224, and LYLE, *Phil. Mag.* (6), vol. 11, 1906, p. 25.]

15. If $\nu=4n^2-1$, so that ν takes the values (1.3), (3.5), (5.7), ... for $n=1, 2, 3, \ldots$, prove that

$$\Sigma\frac{1}{\nu}=\frac{1}{2},\quad \Sigma\frac{1}{\nu^2}=\frac{1}{16}(\pi^2-8),\quad \Sigma\frac{1}{\nu^3}=\frac{1}{64}(32-3\pi^2),\quad \Sigma\frac{1}{\nu^4}=\frac{1}{768}(\pi^4+30\pi^2-384).$$

[Take $x=\frac{1}{2}$ in the series of Ex. 17, p. 190.]

16. Shew that $$1-\frac{1}{5}+\frac{1}{7}-\frac{1}{11}+\frac{1}{13}-\frac{1}{17}+\ldots=\frac{\pi}{2\sqrt{3}},$$

$$\frac{1}{5^2}+\frac{1}{7^2}+\frac{1}{17^2}+\frac{1}{19^2}+\frac{1}{29^2}+\frac{1}{31^2}+\ldots=\frac{\pi^2}{36}(2-\sqrt{3}),$$

by giving x special values in the series of Ex. 17, p. 190.

17. Prove that if n is even

$$\sum_{r=1}^{n}\tan^{-1}\left(\sec\frac{2r\pi}{2n+1}\sinh x\right)=\tan^{-1}\left\{\frac{\sinh nx}{\cosh(n+1)x}\right\}.$$

[*Math. Trip.* 1907.]

18. Shew that the remainder after n terms in the first series of Ex. 37, p. 260, is

$$(-1)^{n-1}\frac{x^{2n+1}}{(2n)!\sin x}\int_0^1\cos(xt)\,\phi_{2n}(t)\,dt,$$

where $\phi_n(t)$ denotes the Bernoullian function defined in Art. 94.

[*Math. Trip.* 1905.]

19. Shew that

$$\lim_{n\to\infty}\left[\left(\frac{1}{n}\right)^n+\left(\frac{2}{n}\right)^n+\ldots+\left(\frac{n-1}{n}\right)^n\right]=\frac{1}{e-1}.$$

[WOLSTENHOLME.]

[Use Arts. 95, 49 and the series for $1/(e^x-1)$ in Art. 93.]

20. Discuss the convergence of the series

$$\sum_{-\infty}^{\infty}\frac{\cos(x+n\alpha)}{\cos(y+n\beta)},$$

where x, y, α, β are complex. [*Math. Trip.* 1892.]

Discuss the convergence of the products

$$\prod_1^{\infty}\left(1+\frac{1}{n^2x}\right),\quad \prod_1^{\infty}\left[1+\frac{1}{n^2(x^n-1)}\right],\quad \prod_1^{\infty}\frac{1+e^{2nx}}{1+e^{(2n-1)x}},$$

for all values of the complex variable x. [*Math. Trip.* 1893.]

21. Prove that if

$$S_0 = 1! - 2! + 3! - 4! + \dots, \qquad \text{(Arts. 98, 132)}$$

then

$$1(1!) - 2(2!) + 3(3!) - 4(4!) + \dots = 1 - 2S_0,$$

$$1^2(1!) - 2^2(2!) + 3^2(3!) - 4^2(4!) + \dots = 5S_0 - 2,$$

and generally $\Sigma(-1)^{n-1}n^k(n!)$ is of the form $aS_0 + \beta$, where a, β are integers (positive or negative).

22. Shew that

$$\int_0^x \frac{e^t}{\sqrt{t}}\,dt \sim \frac{e^x}{\sqrt{x}}\left(1 + \frac{1}{2x} + \frac{1.3}{2^2x^2} + \frac{1.3.5}{2^3x^3} + \frac{1.3.5.7}{2^4x^4} + \dots\right).$$

[For an application to an important physical problem, see LOVE, *Phil. Trans.*, A, vol. 207, 1907, pp. 195–197.]

23. If

$$X_n = \int_0^x e^{-x}\frac{x^n}{n!}\,dx,$$

prove that

$$\lim_{x\to\infty}\left(\lim_{n\to\infty} X_n\right) = 0, \quad \lim_{n\to\infty}\left(\lim_{x\to\infty} X_n\right) = 1.$$

24. From the power-series for $\frac{1}{2}[\log(1+x)]^2$, shew that if $-\pi < \theta < \pi$,

$$[\log(4\cos^2\tfrac{1}{2}\theta)]^2 - \theta^2 = 8[\tfrac{1}{2}\cos 2\theta - \tfrac{1}{3}(1+\tfrac{1}{2})\cos 3\theta + \tfrac{1}{4}(1+\tfrac{1}{2}+\tfrac{1}{3})\cos 4\theta - \dots].$$

and obtain a corresponding formula for $[\log(4\sin^2\frac{1}{2}\theta)]^2$.

Shew also that

$$\theta^2 = \tfrac{1}{3}\pi^2 - 4\left(\cos\theta - \frac{1}{2^2}\cos 2\theta + \frac{1}{3^2}\cos 3\theta - \dots\right).$$

Deduce the integrals of Ex. 31, p. 473.

25. If

$$u_n = -2n\int_0^{\frac{1}{2}\pi}\cos 2nx\log(2\cos\tfrac{1}{2}x)\,dx,$$

prove that

$$u_n - u_{n+1} = (-1)^n/(2n+1),$$

and that

$$u_n = \frac{\pi}{4} - \left(1 - \frac{1}{3} + \frac{1}{5} - \dots \pm \frac{1}{2n-1}\right). \qquad \text{[CATALAN.]}$$

26. Prove that if k is an integer and $|r| < 1$,

$$\int_0^\pi \frac{(1-r^2)\cos nx}{1 - 2r\cos kx + r^2}\,dx = 0, \text{ if } n \text{ is not divisible by } k,$$

$$= \pi r^\nu, \text{ if } n = \nu k.$$

Deduce that if $k = a\kappa$ and $l = a\lambda$, where κ, λ are co-prime integers,

$$\int_0^\pi \frac{(1-r^2)(1-s^2)\,dx}{(1-2r\cos kx + r^2)(1-2s\cos lx + s^2)} = \pi\frac{1 + r^\kappa s^\lambda}{1 - r^\kappa s^\lambda}. \qquad \text{[HARDY.]}$$

27. Prove that if a is positive and $y^2 = ax^2 + 2bx + c$,

$$\int_{x_0}^{x_1}\frac{dx}{y} = \frac{2}{\sqrt{a}}\tanh^{-1}\left\{\frac{(x_1 - x_0)\sqrt{a}}{y_1 + y_0}\right\}.$$

If also $ac - b^2$ is positive and equal to p^2, prove that

$$\int_{x_0}^{x_1}\frac{dx}{y^3} = \frac{1}{p}\tan^{-1}\left\{\frac{p(x_1 - x_0)}{ax_1x_0 + b(x_1 + x_0) + c}\right\},$$

where the angle lies between 0 and π.

[For a discussion of these and other similar cases, see BROMWICH, *Messenger of Maths.*, vol. 35, 1906, p. 131.]

28. Prove that if λ, μ are real but not necessarily positive,

$$\int_0^{\frac{1}{2}\pi} \log(\lambda\cos^2\theta + \mu\sin^2\theta)^2 d\theta = 2\pi\log\{\tfrac{1}{2}|(\sqrt{\lambda}+\sqrt{\mu})|\}.$$

(Ex. 2, p. 347.)

Deduce that if $u = ax^2 + 2bx + c$, $u' = a'x^2 + 2b'x + c'$, where u is positive for all real values of x (so that $p = ac - b^2 > 0$),

$$\int_{-\infty}^{\infty} \log u^2 \frac{dx}{u} = \frac{2\pi}{\sqrt{p}} \log\frac{4p}{a}, \quad \int_{-\infty}^{\infty} \log\left(\frac{u'}{u}\right)^2 \frac{dx}{u} = \frac{2\pi}{\sqrt{p}}\log\left\{\frac{|q+\sqrt{pp'}|}{2p}\right\},$$

where $q = \frac{1}{2}(ac' + a'c) - bb'$, $p' = a'c' - b'^2$. Express the second result in a real form, when p' is negative.

29. Prove that if the integral $\int_0^{\frac{1}{2}\pi} f(\sin^2 x)dx$ is convergent and equal to $\frac{1}{2}A\pi$, then the integral

$$\int^{\infty} [A - f(\sin^2 x)]\phi(x)dx$$

converges, provided that $\phi(x)$ tends steadily to zero.

Deduce the convergence of

$$\int^{\infty} \log(4\cos^2 x)\phi(x)dx, \quad \int^{\infty} \log(4\sin^2 x)\phi(x)dx. \qquad \text{[Hardy.]}$$

30. Prove that in the sense defined by Pringsheim (Ch. V.),

$$\lim_{\lambda,\,\mu\to\infty} \int_0^{\lambda}\int_0^{\mu} \sin(ax+by)x^{r-1}y^{s-1}dx\,dy \qquad (r, s > 0)$$

$$= a^{-r}b^{-s}\Gamma(r)\Gamma(s)\sin\tfrac{1}{2}(r+s)\pi.$$

Prove also that if $0 < a < \pi$,

$$\lim_{\lambda,\,\mu\to\infty} \int_0^{\lambda}\int_0^{\mu} e^{i(x^2+2xy\cos a+y^2)}dx\,dy = \frac{ia}{2\sin a}.$$

[See Hardy, *Messenger of Maths.*, vol. 32, 1903, pp. 92, 159.]

31. Prove that if the real part of x is positive,

$$(1) \quad \int_0^1 e^{-t}t^{x-1}dt = \frac{1}{e}\sum_0^{\infty} \frac{1}{x(x+1)\dots(x+n)},$$

$$(2) \quad \int_0^1 \frac{t^{x-1}dt}{1+t} = \frac{1}{2}\sum_0^{\infty} \frac{n!}{x(x+1)\dots(x+n)}\,\frac{1}{2^n}.$$

32. Prove that if in the interval (a, b) the function $|f(x, n)|$ is less than 1 for all values of n, and if the function $\phi(x)$ is positive and has a convergent integral from a to b, then

$$\lim_{n\to\infty} \int_a^b f(x, n)\phi(x)dx = \int_a^b \{\lim_{n\to\infty} f(x, n)\}\phi(x)dx,$$

provided that $f(x, n)$ tends to its limit uniformly in any interval which does not contain $x = a$. [See also Ex. 51, p. 495.]

Deduce that

$$\lim_{n\to\infty} \int_0^{\frac{1}{2}\pi} e^{-n\sin x}\phi(x)dx = 0,$$

$$\lim_{n\to\infty} \int_0^{\frac{1}{2}\pi} [p + (1-p)\sin^2 x]^{\frac{1}{2}n}\phi(x)dx = 0, \qquad (0 < p < 1).$$

HARDER MISCELLANEOUS EXAMPLES,

WITH REFERENCES.

1. Prove that the function which is equal to 1 when x is rational, and to 0 when x is irrational, can be defined by either of the following repeated limits:

$$(1)\ \lim_{m\to\infty}\left[\lim_{n\to\infty}\{\cos(m!\,\pi x)\}^{2n}\right].$$

[PRINGSHEIM and LEBESGUE.]

$$(2)\ \lim_{m\to\infty}\left[\operatorname{sgn}\{\sin^2(m!\,\pi x)\}\right],$$

where $$\operatorname{sgn} X=\lim_{n\to\infty}\frac{2}{\pi}\arctan(nX).$$ [PIERPOINT.]

[This function *cannot* be represented as a *single* limit of a *continuous* function of x, n: BAIRE, *Leçons sur les Fonctions Discontinues*, pp. 75, 83.]

2. If $f(x)$ is equal to q, when x is represented by the rational fraction p/q, and to 0, when x is irrational, prove that $f(x)$ *is not finite in any interval, however small*, although $f(x)$ is finite for every value of x.

3. Discuss the solution of an equation

$$f(x)\phi(x)=k,$$

where f, ϕ are real monotonic functions, by forming the sequence (a_n) in which $$f(a_{n+1})=k/\phi(a_n).$$

[See SOMMERFELD, *Göttingen Nachrichten*, 1898, p. 360; and compare Exs. 10, 11, p. 18. An extension to complex variables has been made by FATOU, *Comptes Rendus*, t. 143, 1906, p. 546.]

4. Prove that if $\quad v_n=\left(\dfrac{\sin n\alpha}{n\alpha}\right)^2, \quad 0<\alpha<\dfrac{\pi}{2},$

then $$\sum_1^\infty v_n<\frac{2}{\alpha},\quad \sum_1^\infty |v_n-v_{n+1}|<1+\frac{4}{\pi}+\frac{4}{\pi^2}.$$

[To prove the first result take an integer r so that $r\alpha\leqq 1<(r+1)\alpha$; then

$$\sum_1^r v_n<\frac{1}{\alpha},\quad \sum_{r+1}^\infty v_n<\frac{1}{\alpha}.$$

To prove the second, take the integer s so that $s\alpha\leqq\pi<(s+1)\alpha$; then

$$\sum_1^{s-1}|v_n-v_{n+1}|=v_1-v_s<1,\quad v_{n+1}-v_n=\int_{n\alpha}^{(n+1)\alpha}\frac{d}{dx}\left(\frac{\sin^2 x}{x^2}\right)dx,$$

so that $$\sum_s^\infty |v_n-v_{n+1}|<\int_{s\alpha}^\infty\left(\frac{2}{x^2}+\frac{2}{x^3}\right)dx<\frac{4}{\pi}+\frac{4}{\pi^2}.]$$

5. If $f(x)$ is a function which tends steadily to infinity with x (so that $f'(x)$ is positive), prove that, when the sequence $a_n/f'(n)$ is *monotonic*, the condition

$$\lim_{n\to\infty} \{a_n f(n)/f'(n)\}=0$$

is *necessary* for convergence.

For example, with $f(n)=\log n$, $na_n \log n$ must tend to zero when (na_n) is monotonic. [PRINGSHEIM.]

[Compare Arts. 9 and 166.]

6. If $f(n)$ is a function of n which tends steadily to zero as n tends to infinity, then the two series $\Sigma f(n)$, $\Sigma f(\kappa n+\lambda)$ converge and diverge together if κ is positive.

7. (1) If both (b_n) and (a_n/b_n) tend steadily to infinity with n, prove that $(a_{n+1}-a_n)/(b_{n+1}-b_n)$ tends also to infinity with a rapidity not less than that of (a_n/b_n).

(2) If (a_n), (b_n) both tend steadily to infinity with n, and (a_n/b_n) tends steadily to zero, prove that $(a_{n+1}-a_n)/(b_{n+1}-b_n)$ tends also to zero with a rapidity not less than that of (a_n/b_n).

[For proofs of these and a number of similar theorems, see BORTOLOTTI, *Annali di Matematica* (3), t. 11, 1905, p. 29.]

8. Du Bois Reymond has constructed a series defying all the logarithmic criteria as follows:

$$\sum_1^{n_1}\frac{1}{n}+\sum_{n_1}^{n_2}\frac{1}{n\log n}+\sum_{n_2}^{n_3}\frac{1}{n\log n\log(\log n)}+\dots,$$

where $n_1, n_2, n_3, \dots$ are the least integers, such that

$$\log n_1>1,\ \log\left(\frac{\log n_2}{\log n_1}\right)>1,\ \log\left\{\frac{\log(\log n_3)}{\log(\log n_2)}\right\}>1,\ \dots.$$

The numbers $n_1, n_2, n_3, \dots$ increase very rapidly, thus $n_1=3$, $n_2=20$, but n_3 lies between 10^7 and 10^8.

Prove that this series is divergent.

[Compare BOREL, *Séries à termes positifs*, 1902, p. 12.]

9. Shew (see Ex. 6, p. 391) that if θ is irrational we can find a sequence $n_1, n_2, n_3, \dots$ which tends to ∞ and is such that

$$\lim_{r=\infty}\sum_1^{n_r}\sin(n\theta\pi)=A,$$

where A is any number between the extreme limits of $\Sigma\sin(n\theta\pi)$.

10. Prove that if $|\sin(n^2\theta)|<\epsilon$ and $|\sin(n+1)^2\theta|<\epsilon$,

then $$|\sin(2n+1)\theta|<2\epsilon.$$

Deduce that $\Sigma\sin(n^2\theta)$ cannot converge unless θ is a multiple of π.

By means of Exs. 13-17, pp. 243, 244, shew that $\Sigma\sin(n^2\theta)$ is divergent if $\theta/\pi=(2\lambda+1)/2\mu$ or $2\lambda/(4\mu+3)$, where λ and μ are integers; and that it oscillates finitely if $\theta/\pi=(2\lambda+1)/(2\mu+1)$ or $2\lambda/(4\mu+1)$.

Discuss similarly the series $\Sigma\cos(n^2\theta)$.

[Little appears to be known as to the nature of these series when θ/π is irrational.]

11. Prove in the same way that $\Sigma \sin(2^n\theta)$ can only converge if $\theta/\pi = \lambda/2^\mu$, where λ and μ are integers.

12. Shew that $\Sigma \sin(n!\pi x)$ converges for all *rational* values of x; and also for

$$x = (2m+1)e,\ 2m/e,\ \sin 1,\ \cos 1. \qquad \text{[Riemann.]}$$

For instance, if $x=e$, we have $\sin(n!\pi e) = (-1)^{n-1}\sin(R_n\pi)$, where

$$R_n = \frac{1}{n+1} + \frac{1}{(n+1)(n+2)} + \dots.$$

We can now apply Art. 21.

We note for comparison with Exs. 10-12 that the integrals

$$\int_0^\infty \sin(x^2)dx, \quad \int_0^\infty \sin(2^x)dx, \quad \int_0^\infty \sin\{\Gamma(x)\}dx$$

are all convergent (see Ex. 8, p. 468).

13. Prove that if the series $f(x) = \sum_0^\infty a_n x^n$ converges absolutely for $|x|<1$, the two series

$$\phi(x) = \sum_0^\infty \frac{a_n x^n}{1-yc^n}, \quad \psi(x) = \sum_0^\infty y^n f(xc^n)$$

both converge and are equal if $|x|<1$, $|c|<1$, $|y|<1$.

[Compare Ex. 6, p. 126.]

[The series obtained by writing $a_n = (-1)^n/n!$, $y=c$ in Ex. 13 have been considered in Dedekind's edition of Dirichlet's *Zahlentheorie*; and those found by writing $x=c$, $a_n=(-1)^n/n!$ or $1/n!$, have been discussed by Pringsheim, *Chicago Math. Congress Papers*, p. 288.]

14. If the two series of the last example are not equal, but are convergent, prove that

$$\phi(x) - y\phi(xc) = f(x) = \psi(x) - y\psi(xc);$$

and deduce that if $\chi(x) = \psi(x) - \phi(x)$, then

$$\chi(x) = y\chi(xc).$$

[That the series need not be equal when $c>1$ is clear from Ex. 28.]

15. If we put $y=1$ in Ex. 13, we must omit the first term in $\phi(x)$ and write

$$\phi_1(x) = \sum_1^\infty \frac{a_n x^n}{1-c^n};$$

then prove that

$$\phi_1(x) - \phi_1(xc) = f(x) - a_0.$$

Deduce that if

$$\chi_1(x) = \psi(x) - \phi_1(x),$$

then

$$\chi_1(x) = \chi_1(xc) + a_0.$$

16. Deduce from Exs. 20, 21, p. 251, that if $0 \leqq y < x \leqq \frac{1}{2}$,

$$\sum_{n=1}^\infty \left\{ \sum_{m=1}^\infty \frac{\cos(2m\pi x)\cos(2n\pi y)}{m^2-n^2} \right\} = \pi^2[\tfrac{3}{4}\{\phi_2(y) - \phi_2(x)\} + \tfrac{1}{4}(y-\tfrac{1}{2})],$$

where $m=n$ is omitted from the series; but that the sum is zero if $x=y$.

Shew further that the order of summation is immaterial except when $x=0$, $y=0$ (see Ex. 11, p. 92). [Hardy.]

17. Obtain the sum of the repeated series

$$\sum_{(m)}\sum_{(n)}\frac{\exp\{2\pi i(mx+ny)\}}{a+m\omega_2+n\omega_1},\quad 0<x,\ y<1,$$

from Ex. 3, Art. 90, and Ex. 73 below. [KRONECKER.]

[For a proof that the *double* series is convergent, see HARDY, *Messenger of Maths.*, vol. 34, p. 146.]

18. Let the symbol $\{\xi\}$ denote $\xi-\nu$, where ν is the integer *nearest* to ξ, so that $\{\xi\}$ lies between $-\frac{1}{2}$ and $+\frac{1}{2}$; this definition, however, does not determine $\{n+\frac{1}{2}\}$ to which we attach the value 0. Thus, as ξ tends to $n+\frac{1}{2}$ through smaller values, $\{\xi\}$ tends to $+\frac{1}{2}$; as ξ tends to $n+\frac{1}{2}$ through larger values, $\{\xi\}$ tends to $-\frac{1}{2}$; but $\{\xi\}$ has no other discontinuities.

Prove that the series

$$\{x\}+\frac{\{2x\}}{2^2}+\frac{\{3x\}}{3^2}+\ldots+\frac{\{nx\}}{n^2}+\ldots$$

is uniformly convergent for all values of x; but that the series has a discontinuity $\frac{1}{8}\pi^2/m^2$ at every point $x=p/2m$, where p is odd and prime to m. Thus the series is discontinuous infinitely often in every interval. [RIEMANN.]

19. Prove on the lines of Art. 55 that the equation

$$y=x\psi(x,\ y),$$

where $\psi(x,\ y)$ is a double power-series, absolutely convergent for $|x|=a$, $|y|=b$ can be satisfied by a power-series

$$y=\sum_1^\infty c_n x^n,$$

which certainly converges if $|x|<4aM/(4M+ab)$, where M is a certain constant determined by the terms in the series for $\psi(a,\ b)$.

20. Shew that if x and y are related as in Art. 56, we can obtain an expansion in powers of y for a function $h(x)=g(x)/x^k$, where $g(x)$ is a power-series in x, and k is a positive integer. Prove also that if we write

$$h(x)=\sum_{-k}^\infty B_n y^n,$$

then nB_n is the coefficient of $1/x$ in the expansion of $h'(x)/y^n$ in ascending powers of x; this determines all the coefficients except B_0, which is the coefficient of $1/x$ in the expansion of

$$\frac{h(x)}{y}\frac{dy}{dx}=h(x)\frac{d}{dx}(\log y).$$

Apply the method to the examples of Art. 56, taking $h(x)=1/x^k$. [TEIXEIRA.]

21. If $\phi(x)=x+\frac{x^2}{2^2}+\frac{x^3}{3^2}+\frac{x^4}{4^2}+\ldots$, prove that

(1) $\phi(-x)+\phi(x)=\frac{1}{2}\phi(x^2)$.

(2) $\phi(x)+\phi(1-x)=\frac{1}{6}\pi^2-\log x\,.\log(1-x)$. [ABEL.]

(3) $\phi(-x)+\phi\left(\frac{x}{1+x}\right)=-\frac{1}{2}\{\log(1+x)\}^2$. [LEGENDRE.]

Deduce the value of $\phi(\frac{1}{2})$ (see also Ex. 32, p. 473), and that if $x=\frac{1}{2}(\sqrt{5}-1)$, so that $x^2+x=1$, then

$$\phi(x)=\tfrac{1}{10}\pi^2-(\log x)^2, \quad \phi(x^2)=\tfrac{1}{15}\pi^2-(\log x)^2.$$

[BERTRAND, *Calc. Int.*, § 270; ROGERS, *Proc. Lond. Math. Soc.* (2), vol. 4, 1906, p. 169.]

22. (1) Prove that the differential equation

$$\frac{d^2y}{dx^2}+(1+\lambda x)y=0$$

can be solved by a series

$$u_0+\lambda u_1+\lambda^2 u_2+\lambda^3 u_3+\dots,$$

where $\quad u_0=A\cos x+B\sin x, \quad u_n(x)=-\int_0^x t u_{n-1}(t)\sin(x-t)\,dt.$

Shew that $\quad |u_n|<|x|^{3n+1}/n!.$

(2) Prove that if $v(x)$ is continuous in an interval (a, b), the equation

$$\frac{d^2y}{dx^2}-v(x)y=0$$

has a solution $y=\Sigma u_n$ in the same interval, where

$$u_0=A+Bx, \quad u_n(x)=\int_0^x (x-t)v(t)u_{n-1}(t)\,dt.$$

Shew that if β, γ are certain constants depending on the nature of $v(x)$,

$$|u_n|<\beta\gamma^n/(2n)!.$$

[A. C. DIXON.]

Results allied to Abel's theorem.

23. If $\phi(x)$ tends steadily to zero as x increases, but the series $\Sigma\phi(n)$ diverges, shew that as a tends to zero (from the positive side),

$$\Sigma\phi(n)e^{-an} \sim \int_c^\infty \phi(x)e^{-ax}\,dx, \quad c \geqq 0.$$

An illustration has already occurred in Ex. 5, p. 419; another is given by

$$\sum_2^\infty \frac{x^n}{(\log n)^p} \sim \frac{1}{(1-x)[\log\{1/(1-x)\}]^p}, \quad \text{as } x\to 1.$$

[LE ROY, *Bulletin des sci. math.*, t. 24, 1900, p. 245.]

24. Extend the results of the last example to double series, and prove in particular that if $0<s<1$,

$$\Sigma\Sigma \frac{x^m y^n}{(m^2+n^2)^s} \sim \frac{\Gamma(k)}{r^k}\int_0^{\frac{1}{2}\pi} \frac{d\theta}{\cos^k(\theta-\omega)}, \quad (\text{as } x, y\to 1),$$

where $r\cos\omega=1-x$, $r\sin\omega=1-y$, $k=2-2s$. [HARDY.]

25. Abel's theorem (Arts. 51, 83) can be extended as follows:

Suppose that Σa_n is r-ply indeterminate (in Cesàro's sense) and has the sum l, and that v_n is a function of x which satisfies the following conditions:

$$\lim_{n\to\infty} n^r v_n=0, \quad \Sigma n^r|\Delta^{r+1}v_n|<K, \qquad (x>c),$$

while $$\lim_{x \to c} v_n = 1,$$

then $$\lim_{x \to c} \Sigma a_n v_n = l.$$

[BROMWICH, *Math. Annalen*, Bd. 65; other results included under this theorem have been found by HARDY, *Math. Annalen*, Bd. 64, 1907, and C. N. MOORE, *Trans. Amer. Math. Soc.*, vol. 8, 1907. See also HARDY, *Proc. Lond. Math. Soc.* (2), vol. 4.]

26. Apply Euler's formula (Art. 131) to obtain the asymptotic formulae (as x tends to 0)

$$\frac{1}{1+x}+\frac{1}{2(1+2x)}+\frac{1}{3(1+3x)}+\ldots \sim C+\log\frac{1}{x}+\frac{x}{2}-B_1\frac{x^2}{2}+B_2\frac{x^4}{4}-\ldots.$$

Use Ex. 26, p. 472, to prove that

$$\operatorname{sech} x+\operatorname{sech} 2x+\operatorname{sech} 3x+\ldots \sim \tfrac{1}{2}[(\pi/x)-1].$$

If we attempt to continue the last asymptotic formula as a power-series in x, all the coefficients will be found to be zero: and as a matter of fact, the next term in the approximation is $(2\pi/x)e^{-\pi^2/x}$.

27. Apply the method of Art. 135 to the series

$$F(q)=\Sigma x^n q^{n^2} \quad (x>1,\ q<1),$$

and prove that as $q \to 1$,

$$F(q) \sim \exp\left\{\frac{1}{4}\frac{(\log x)^2}{\log(1/q)}\right\}.$$

Discuss in the same way the series $\Sigma x^n q^{n^3}$.

28. Deduce from Ex. 5, p. 348, and the power-series for $\sin x$ and $\cos x$, that if $f(x)$ is any continuous function in the interval $(0, 2\pi)$, a polynomial

$$P(x)=A_0+A_1x+A_2x^2+\ldots+A_nx^n$$

can be determined so that

$$|f(x)-P(x)|<\epsilon.$$ [WEIERSTRASS.]

[Compare PICARD, *Traité d'Analyse*, t. 1, Ch. IX., § V.]

29. Shew that by proper choice of n the maximum value of $t-t^n$, and the corresponding value of t, can both be brought as near to 1 as we please.

Deduce that, as x tends to 1, the series

$$x-x^{2!}+x^{3!}-x^{4!}+\ldots$$

oscillates between the limits 0 and 1. [HARDY.]

30. If we write $a_n=(-1)^n/n!$, $y=-1$, and $-\log x$ for x in Ex. 13, we get the series

$$f(x)=\psi(-\log x)=x-x^c+x^{c^2}-x^{c^3}+\ldots,$$

$$g(x)=\phi(-\log x)=\frac{1}{2}+\frac{\log x}{1+c}+\frac{1}{2!}\frac{(\log x)^2}{1+c^2}+\frac{1}{3!}\frac{(\log x)^3}{1+c^3}+\ldots.$$

Prove that if $F(x)=f(x)-g(x)$, then

$$F(x)+F(x^c)=0.$$

Verify also that, to five places of decimals,

$$F(1/\sqrt{2}) = \cdot 00275, \quad \text{if } c=2,$$

and that

$$F(1/6^{\frac{1}{4}}) = \cdot 13185, \quad \text{if } c=6.$$

Shew further that, as x tends to 1, $g(x)$ tends to $\frac{1}{2}$; and deduce that $f(x)$ oscillates between limits at least as wide as

$$\cdot 50275 \text{ and } \cdot 49725, \quad \text{if } c=2,$$

or

$$\cdot 63185 \text{ and } \cdot 36815, \quad \text{if } c=6.$$

31. If we write $a_n = (-1)^n/n!$ and $-\log x$ for x in Ex. 15, we get the series

$$f(x) = \psi(-\log x) = x + x^c + x^{c^2} + x^{c^3} + \dots,$$

$$g(x) = \phi_1(-\log x) = -\frac{\log x}{c-1} - \frac{1}{2!}\frac{(\log x)^2}{c^2-1} - \frac{1}{3!}\frac{(\log x)^3}{c^3-1} - \dots.$$

Prove that if $F(x) = f(x) - g(x)$, then

$$F(x) = F(x^c) + 1,$$

and verify that

$$F(1/\sqrt{2}) = 1\cdot 1960, \quad \text{if } c=2.$$

Shew also that $g(x)$ tends to 0 as x tends to 1, and deduce Cesàro's result that $f(x)$ has the asymptotic representation

$$-\log\left(\log\frac{1}{x}\right)\Big/\log c. \quad \text{[See Ex. 4, p. 133.]}$$

[For a treatment of the series in Exs. 30, 31 by means of contour integrals, see §§ 8–13 of a paper by HARDY, *Quarterly Journal*, vol. 38, 1907, p. 269; some similar results have been obtained for the series

$$\sum a_n e^{-\lambda_n s},$$

where λ_n tends steadily to infinity, by CAHEN, *Annales de l'École Normale Sup.* (3), t. 11, 1894, p. 75.]

Functions without a derivate.

32. Shew that if a continuous function $F(x)$ has a finite derivate b at $x=a$, an interval $(a-\delta, a+\delta)$ can be found such that the quotient $\{F(x_2) - F(x_1)\}/(x_2 - x_1)$ lies between $b-\epsilon$ and $b+\epsilon$ when x_1 is *any* point such that $a-\delta < x_1 < a$, and x_2 is any point such that $a < x_2 < a+\delta$.

In like manner if $F(x)$ has an infinite derivate (positive) at $x=a$, prove that the interval can be chosen so that the quotient is greater than N, however large N may be.

33. Write $F(x) = \sum r^n \cos(s^n \pi x)$, where $0 < r < 1$ and s is an odd integer; and let $S_m(x)$ denote the sum of the first m terms in $F(x)$, then shew that

$$|S_m'(x)| < \pi (rs)^m/(rs-1);$$

while if $a = p/s^m$, where p is an integer,

$$F(a) = S_m(a) + (-1)^p r^m/(1-r).$$

Hence prove that, if q is $p+1$ or $p+3$, and $\beta = q/s^m$, then

$$\frac{F(\beta) - F(a)}{\beta - a} = \frac{(rs)^m}{q-p}\left[(-1)^q \frac{2}{1-r} + M\right],$$

where

$$|M| < 3\pi/(rs-1).$$

Deduce that if $rs > 1 + \frac{3}{2}\pi(1-r)$, the quotient has the same sign as $(-1)^q$, and can be made arbitrarily great by increasing m sufficiently.

[The conditions on r, s are easily satisfied; for example, take $r = \frac{1}{2}$, $s \geqq 7$; or $r = \frac{2}{3}$, $s \geqq 5$.]

34. If any point a is given, and an arbitrarily small interval $(a-\delta, a+\delta)$ containing a, shew that points α_1, α_2, β_1, β_2 can be found of the types considered in the last example, and such that α_1, α_2 fall in the interval $(a-\delta, a)$ and β_1, β_2 fall in the interval $(a, a+\delta)$, while

$$\{F(\beta_1) - F(\alpha_1)\}/(\beta_1 - \alpha_1) > N,$$

and

$$\{F(\beta_2) - F(\alpha_2)\}/(\beta_2 - \alpha_2) < -N,$$

however large N may be.

Deduce that the function $F(x)$, although continuous for all values of x, has nowhere a definite derivate. [WEIERSTRASS.]

[The investigation usually given for Weierstrass's function leads to the unnecessarily narrow condition $rs > 1 + \frac{3}{2}\pi$. Thus $r = \frac{1}{2}$ would require $s \geqq 13$, instead of $s \geqq 7$; and $r = \frac{2}{3}$, $s \geqq 9$ instead of $s \geqq 5$.]

35. Shew from Ex. 33 that if

$$\alpha = (q-1)/t^m, \quad \beta = q/t^m, \quad \gamma = (q+1)/t^m,$$

where q is an odd integer, $F(\beta)$ is less than both $F(\alpha)$ and $F(\gamma)$; and deduce that $F(x)$ has at least one minimum in the interval (α, γ). Hence prove that $F(x)$ has an infinity of maxima and minima in any interval, however small.

[It has sometimes been stated that α is a maximum, β a minimum of $F(x)$; but this cannot be proved at any rate by the foregoing line of argument.]

36. (1) If s is even instead of odd in Weierstrass's function, shew that a similar discussion leads to the condition $rs > 1 + \frac{3}{2}\pi$. Examine also the corresponding cases when sines take the place of cosines in the series.

(2) Prove that with the notation of Exs. 33, 34,

$$\left|\frac{S_m(\beta_1) - S_m(\alpha_1)}{\beta_1 - \alpha_1} - \frac{S_m(\beta_2) - S_m(\alpha_2)}{\beta_2 - \alpha_2}\right| < \frac{\pi^2}{2}(\beta_1 - \alpha_1 + \beta_2 - \alpha_2)\frac{r^m s^{2m}}{rs^2 - 1};$$

and deduce that Weierstrass's function has nowhere a definite *finite* derivate if $rs^2 > 1 + \frac{3}{4}\pi^2(1-r)$.

[The last condition is satisfied by $r = \frac{2}{3}$, $s = 3$; and by $r = 1/s$, if $s \geqq 9$. For other similar functions, see DINI, *Grundlagen*, Kap. X.]

Complex Series.

37. Deduce from Ex. 16, p. 244, that if p is a prime number of the form $4k+3$,

$$\sum_{n=1}^{\infty}\left(\frac{n}{p}\right)\frac{1}{n} = \frac{\pi}{p\sqrt{p}}\sum_{m=1}^{p-1}\left(\frac{m}{p}\right)m,$$

where $\left(\frac{n}{p}\right)$ is $+1$ if n is a quadratic residue* of p, -1 if n is a non-residue, and 0 if n is a multiple of p.

* That is, if we can find an integer x such that $x^2 - n$ is a multiple of p.

In particular, with $p=7$, the residues less than 7 are 1, 2, 4, and so

$$1+\frac{1}{2}-\frac{1}{3}+\frac{1}{4}-\frac{1}{5}-\frac{1}{6}+\frac{1}{8}+\frac{1}{9}-\frac{1}{10}+\frac{1}{11}-\frac{1}{12}-\frac{1}{13}+\ldots=\frac{\pi}{\sqrt{7}}.$$

[DIRICHLET: see also GLAISHER, *Messenger of Maths.*, vol. 31, p. 98.]

38. If the real part of p is positive, prove that (see Art. 89),

$$\mathfrak{M}\, x^{-n}(1+x)^p=p(p-1)\ldots(p-n+1)/n!,$$

the mean value being taken along the circle $|x|=1$.

Deduce that if the real part of q is also positive,

$$\mathfrak{M}(1+x)^p(1+1/x)^q=F(-p,\ -q,\ 1,\ 1)=\frac{\Gamma(1+p+q)}{\Gamma(1+p)\Gamma(1+q)},$$

and hence that

$$\int_0^{\frac{1}{2}\pi}\cos 2m\theta\cos^{2n}\theta\, d\theta=\frac{\pi}{2^{2n+1}}\frac{\Gamma(2n+1)}{\Gamma(n+m+1)\Gamma(n-m+1)},$$

provided that the real parts of m, $n-m$ are positive. [CAUCHY.]

39. (1) If (a_n) is a sequence such that *every* subsequence selected from (a_n) has zero as *one* limiting value, prove that (a_n) tends to zero as a limit.

(2) If for *every* value of x from 0 to a $(a>0)$,

$$\lim c_n \sin nx=0,$$

prove that (c_n) tends to zero.

(3) If for *all* values of x from α to β (both real),

$$\lim(a_n\cos nx+b_n\sin nx)=0,$$

prove that (a_n), (b_n) both tend to zero.

[G. CANTOR, *Math. Ann.*, Bd. 4, p. 141.]

40. If the imaginary part of $\Sigma a_n x^n$ $(x=\xi+i\eta)$ converges at *all* points of an arc of a regular curve (which is not a radius through the origin), prove (from Ex. 39) that $\Sigma a_n x^n$ has a radius of convergence not less than p, if p is the least distance from the origin to the given arc.

It is not enough to suppose the convergence given at all points of a *dense* set along the arc; thus the imaginary part of $\Sigma x^{n!}$ converges for *any* value of r, if $x=r(\cos\theta+i\sin\theta)$ and θ/π is rational.

[KALMÁN, *Math. Annalen*, Bd. 63, 1907, p. 322.]

41. For any series of the type

$$\Sigma\frac{a_n\,.\,n!}{x(x+1)\ldots(x+n)}$$

there are two constants λ, μ such that the series diverges if $R(x)<\lambda$, converges if $\lambda<R(x)<\mu$, and converges absolutely if $R(x)>\mu$. These constants are connected by the conditions

$$\lambda\leqq\mu\leqq\lambda+1.$$

Also $\qquad\lambda=\overline{\lim}\,\{\log|A_n|\}/\log n,\quad \mu=\overline{\lim}\,(\log B_n)/\log n,$

where $\qquad A_n=a_1+a_2+\ldots+a_n,\qquad B_n=|a_1|+|a_2|+\ldots+|a_n|.$

It is assumed that the coefficients a_n are such that $\Sigma a_n x^n$ has a radius of convergence equal to unity (for other cases see Exs. 1, 2, p. 254).

Corresponding theorems can be established for the series

$$\Sigma a_n \frac{x(x-1)\dots(x-n+1)}{1.2\dots n}.$$

[LANDAU, *Münchener Sitzungsber.*, Bd. 36, 1906, pp. 192-208.]

42. Shew that if the sequence (M_n) of real numbers tends *steadily* to infinity, the series $\Sigma a_n M_n^{-x}$ converges if, and only if,

$$R(x) > \overline{\lim}\, \{\log |A_n|\}/\log M_n, \qquad \text{[CAHEN.]}$$

where $A_n = a_1 + a_2 + \dots + a_n$. It is assumed that the series Σa_n does not converge; if it does, this condition gives $R(x) > 0$, which is sufficient, but need not be necessary for convergence.

[Compare Exs. 9, 10, p. 255.]

43. Consider the series

$$\Sigma \frac{c_n x^n}{\sin(n\pi\lambda)},$$

where λ is not real and rational. Prove that if $c_n = 1/n!$, different values of λ can be chosen so that the series

(1) converges for every value of x,
(2) has a finite radius of convergence,
(3) diverges for every non-zero value of x.

[See HARDY, *Proc. Lond. Math. Soc.* (2), vol. 3, 1905, p. 441.]

Riemann's ζ-function.

44. Prove that the series

$$1 + 2^{-s} + 3^{-s} + 4^{-s} + \dots = \zeta(s)$$

converges if, and only if, $R(s)$ (the real part of s) is greater than 1.

When this condition is not satisfied, $\zeta(s)$ is defined as the constant term in the asymptotic expansion (Arts. 130, 131)

$$1 + 2^{-s} + 3^{-s} + \dots + n^{-s} \backsim \frac{n^{1-s}}{1-s} + \frac{1}{2}n^{-s} - \frac{B_1}{2!}sn^{-s-1}$$
$$+ \frac{B_2}{4!}s(s+1)(s+2)n^{-s-3} - \dots + \zeta(s) + \dots.$$

Thus if $0 < R(s) \leqq 1$, we have

$$\zeta(s) = \lim_{n\to\infty}\left[(1 + 2^{-s} + 3^{-s} + \dots + n^{-s}) - \frac{n^{1-s}}{1-s}\right],$$

and so on. It can be proved by more elaborate methods that the function $\zeta(s)$ is analytic and has only one singularity, a pole at $s=1$; the presence of such a singularity follows from Ex. 11, Ch. II., which gives

$$\lim_{s\to 1}\,[\zeta(s) - 1/(s-1)] = C.$$

[RIEMANN, *Ges. Werke*, p. 136.]

45. The following are easy examples on the ζ-function—assuming $R(s)>1$:

(1) $[\zeta(s)]^{-1}=\Pi(1-1/p^s)$, where p is any *prime* number. [EULER.]

(2) $[\zeta(s)]^2=\Sigma\lambda(n)/n^s$, where $\lambda(n)$ is the number of divisors of n.

(3) From (1) deduce that

$$[\zeta(s)]^q=\Sigma(\alpha_q\beta_q\ldots\kappa_q)/n^s,$$

where n is of the form $a^\alpha b^\beta\ldots k^\kappa$, when expressed in terms of its prime factors, and $\alpha_q=q(q+1)\ldots(q+\alpha-1)/\alpha!$.

(4) From (2), (3) we obtain the familiar result

$$\lambda(n)=(\alpha+1)(\beta+1)\ldots(\kappa+1).$$

(5) The coefficient of $1/n^s$ in the product $\zeta(s)\zeta(s-1)$ is the *sum* of the divisors of n.

(6) $[\zeta(s)]^{-1}=\Sigma\mu(n)/n^s$, where $\mu(n)$ is 0 if n has any repeated factor, $+1$ if n has an even number of factors (all different), and -1 if n has an odd number of factors (all different).

(7) Let $\phi(n)$ denote the number of numbers less than and prime to n; and let d denote any divisor of n, and write $\psi(n)=\Sigma\phi(d)$. We have then

$$\Sigma\psi(n)/n^s=\zeta(s)\Sigma\phi(n)/n.$$

(8) $$\zeta(s-1)/\zeta(s)=\Sigma\phi(n)/n^s.$$

(9) $$\zeta(2s)/\zeta(s)=\Sigma(-1)^{\alpha+\beta+\ldots+\kappa}/n^s=\Pi(1+1/p^s)^{-1}.$$ [CAHEN.]

46. Shew that the series

$$\xi(s)=1-2^{-s}+3^{-s}-4^{-s}+\ldots$$

converges if, and only if, $R(s)$ is positive, and prove that then

$$\xi(s)=(1-2^{1-s})\zeta(s).$$

Shew also that, if $0<R(s)<1$,

$$\Gamma(s)\zeta(s)=\lim_{n\to\infty}\int_0^\infty x^{s-1}(1-e^{-nx})\left(\frac{1}{e^x-1}-\frac{1}{x}\right)dx=\int_0^\infty x^{s-1}\left(\frac{1}{e^x-1}-\frac{1}{x}\right)dx.$$

Deduce that

$$\left(1-\frac{1}{2^s}\right)\Gamma(s)\zeta(s)=\int_0^\infty\left(\sum_1^\infty\frac{(-1)^n x^s}{x^2+n^2\pi^2}\right)dx=-\frac{\pi^s}{2\cos\left(\frac{1}{2}s\pi\right)}\xi(1-s),$$

and so prove that $\quad 2^{1-s}\Gamma(s)\zeta(s)\cos\left(\frac{1}{2}s\pi\right)=\pi^s\zeta(1-s).$

47. If $0<R(s)<1$, shew that the series

$$\eta(s)=1-3^{-s}+5^{-s}-7^{-s}+\ldots$$

is convergent, and that

$$2^s\Gamma(s)\eta(s)\sin\left(\frac{1}{2}s\pi\right)=\pi^s\eta(1-s).$$

48. Prove from the definition of the ζ-function that

$$\zeta(0)=-\tfrac{1}{2},\quad \zeta(2m)=2^{2m-1}\pi^{2m}B_m/(2m)!,$$
$$\zeta(-2m)=0,\quad \zeta(-2m+1)=\tfrac{1}{2}(-1)^m B_m/m,$$

where m is any positive integer.

Deduce that the relation between $\zeta(s)$ and $\zeta(1-s)$ given in Ex. 46 is true when s is any integer; that it is true for all values of s is proved by Riemann (*l.c.*).

49. The series, in which p takes all odd prime values,

$$\Sigma(-1)^{\frac{1}{2}(p+1)}\frac{1}{p^s}=\frac{1}{3^s}-\frac{1}{5^s}+\frac{1}{7^s}+\frac{1}{11^s}-\frac{1}{13^s}-\frac{1}{17^s}+\ldots$$

converges if $s=1$ or if $R(s)\geqq 1$.

It has been conjectured that the series converges if $\frac{1}{2}<s<1$, but the proof is not yet complete.

[See LANDAU, *Math. Annalen*, Bd. 61, 1906, p. 527, and earlier papers quoted there.]

Infinite Integrals.

50. Prove that if $\phi(x)$ is *positive* and tends to ∞ at one or more places in (a, b), but in such a way that $\int_a^b \phi(x)dx$ converges, then the value of this integral is equal to

$$\lim_{n\to\infty}\sum_1^n \delta h_r,$$

where the interval (a, b) is subdivided by continued bisection into n equal parts δ, and h_r is the *lower limit* of $\phi(x)$ in the rth subdivision.

51. Apply the last example and the theorem proved in the small type of Art. 177 (p. 457) to prove that the result of Ex. 32 (p. 483) is true *without supposing that $f(x, n)$ converges uniformly* in any part of the interval (a, b).

52. By writing $x=e^u\sqrt{b/a}$, $m=2\sqrt{ab}$, $t=m\sinh u$, prove that

$$\int_0^\infty f(ax+b/x)\frac{dx}{x}=2\int_0^\infty f(m\cosh u)du,$$

$$\int_0^\infty f(ax+b/x)\log x\frac{dx}{x}=\log(b/a)\int_0^\infty f(m\cosh u)du,$$

$$\int_0^\infty f(a^2x^2+b^2/x^2)dx=\frac{1}{a}\int_0^\infty f(t^2+2ab)dt,$$

$$\int_0^\infty (a^2x^2-b^2/x^2)f(a^2x^2+b^2/x^2)dx=\frac{1}{a}\int_0^\infty t^2f(t^2+2ab)dt$$

As special examples of the last two, we find

$$\int_0^\infty (a^2x^2+b^2/x^2)e^{-(a^2x^2+b^2/x^2)}dx=\sqrt{\pi}(b+1/4a)e^{-2ab},$$

$$\int_0^\infty (a^2x^2-b^2/x^2)e^{-(a^2x^2+b^2/x^2)}dx=(\sqrt{\pi}/4a)e^{-2ab}.$$

53. By the same transformation as in the last example, we find

$$\int_0^\infty f(ax-b/x)\frac{dx}{x}=2\int_0^\infty f(m\sinh u)du,$$

$$\int_0^\infty f(ax-b/x)\frac{dx}{1\pm x^2}=\int_0^\infty f\{(a\pm b)t\}\frac{dt}{1\pm t^2},$$

where $f(\xi)$ is an *even* function; on both sides of the *second* equation principal values are to be used if necessary.

Similarly, if $g(\xi)$ is an *odd* function,

$$\int_0^\infty g(ax-b/x)\frac{x\,dx}{1\pm x^2}=\int_0^\infty g\{(a\pm b)t\}\frac{t\,dt}{1\pm t^2}.$$

Special examples of these are given by taking $f(\xi)=\cos\xi$, $g(\xi)=\sin\xi$. [Compare HARDY, *Quarterly Journal*, vol. 32, p. 374.]

54. If an integral $\int_a^\infty f(x)dx$ oscillates, we may call it *summable*, if the limit

$$\lim_{t\to 0}\int_a^\infty e^{-tx}f(x)dx$$

exists. We may, when convenient, indicate this limit by writing G before the sign of integration. Deduce from Ex. 2, Art. 172, that the "condition of consistency" (Art. 100) is satisfied. Prove that (compare Ex. 36, p. 74),

$$G\int_0^\infty x^{n-1}\begin{pmatrix}\cos ax\\ \sin ax\end{pmatrix}dx=\frac{\Gamma(n)}{a^n}\begin{Bmatrix}\cos(\frac{1}{2}n\pi)\\ \sin(\frac{1}{2}n\pi)\end{Bmatrix},\quad a>0,\ n\geqq 1.$$

Thus $$G\int_0^\infty \cos x\,dx=0,\quad G\int_0^\infty \sin x\,dx=1.$$

[HARDY, *Quarterly Journal*, vol. 35, 1903, p. 22.]

55. Use the result of Ex. 54 to prove that if the series $F(x)=\Sigma a_n x^n$ converges for all values of x, then

$$\int_0^\infty \cos mx\,F(x)dx=-\frac{a_1}{m^2}+\frac{3!\,a_3}{m^4}-\frac{5!\,a_5}{m^6}+\dots,$$

$$\int_0^\infty \sin mx\,F(x)dx=\ \frac{a_0}{m}-\frac{2!\,a_2}{m^3}+\frac{4!\,a_4}{m^5}-\dots,$$

provided that *both* series on the right are convergent.

Deduce that

$$\int_0^\infty \cos mx\frac{\cos\sqrt{x}}{\sqrt{x}}dx=\sqrt{\frac{\pi}{2m}}\left(\cos\frac{1}{4m}+\sin\frac{1}{4m}\right),$$

$$\int_0^\infty \sin mx\,J_0(x)dx\ =1/\sqrt{(m^2-1)},\quad (m>1),$$

where $$J_0(x)=1-\frac{x^2}{2^2}+\frac{x^4}{2^2.4^2}-\frac{x^6}{2^2.4^2.6^2}+\dots.$$

56. Following the lines of Art. 109, frame a definition of *uniformly summable* integrals; and prove that if $G\int_a^\infty f(x,\alpha)dx$ is uniformly summable in an interval, its value is a continuous function of α; and that integration is permissible under the sign of integration.

57. Discuss also the question of differentiation with respect to a parameter. In particular if

$$u=\int_0^\infty \frac{\sin ax}{1+x^2}dx,$$

shew that $$\frac{d^2u}{da^2}-u=-G\int_0^\infty \sin ax\,dx=-\frac{1}{a}.$$

Deduce that $u=\frac{1}{2}[-e^{a}\mathrm{li}(e^{-a})+e^{-a}\mathrm{li}(e^{a})]$.

Evaluate in the same way

$$\int_0^\infty \frac{\cos ax}{1+x^2}\,dx. \qquad \text{[See also Ex. 5, Art. 173.]}$$

58. Using the expansion

$$\tanh \tfrac{1}{2}\pi x = 1-2e^{-\pi x}+2e^{-2\pi x}-\dots,$$

shew that $$G\int_0^\infty \tanh \tfrac{1}{2}\pi x \,.\, \sin ax \,.\, dx = \operatorname{cosech} a,$$

and deduce that

$$\int_0^\infty \tanh \tfrac{1}{2}\pi x \frac{\sin ax}{1+x^2}\,dx = \tfrac{1}{2}[e^{-a}\log(e^{2a}-1)-e^{a}\log(1-e^{-2a})].$$

59. Prove that if $f(x)$ tends steadily to infinity and $\int^\infty \sin\{f(x)\}\,dx$ converges or oscillates between finite limits, then

$$G\int_a^\infty \cos\{f(x)\}f'(x)\,dx = -\sin\{f(a)\}.$$

Establish similarly the corresponding results with sines and cosines interchanged.

Deduce Stokes's result that if

$$u=\int_0^\infty \cos(x^3-xy)\,dx, \quad v=\int_0^\infty \sin(x^3-xy)\,dx,$$

then $$\frac{d^2u}{dy^2}+\tfrac{1}{3}yu=0, \quad \frac{d^2v}{dy^2}+\tfrac{1}{3}yv=-\tfrac{1}{3}.$$

60. Obtain the asymptotic solution of the differential equation

$$\frac{d^2u}{dy^2}-\tfrac{1}{3}yu=0, \qquad \text{(compare Ex. 58)}$$

by writing $y=3z^{\frac{2}{3}}$, $u=vz^{-\frac{1}{6}}$; and prove that the equation reduces to

$$\frac{d^2v}{dz^2}-\left(4-\frac{5}{36z^2}\right)v=0,$$

which gives the solution

$$v\sim e^{\pm 2z}\left(1+\frac{1.5}{1}\zeta+\frac{1.5.7.11}{1.2}\zeta^2+\frac{1.5.7.11.13.17}{1.2.3}\zeta^3+\dots\right),$$

where $\zeta=\pm 1/144z$.

[Stokes, *Math. and Phys. Papers*, vol. 2, p. 329; vol. 4, pp. 77, 283.]

Trigonometrical Series.

61. Shew that, if p, q are positive integers ($q<p$),

$$\Sigma' \frac{1}{n^2}\frac{\sin^2(na)}{\sin^2(nqa)}=\frac{\pi^2}{6}\frac{p^2-1}{p^2},$$

where $a=\pi/p$, and in the summation n takes all positive integral values except multiples of p. [H. N. Davis.]

[Write $n=mp\pm r$, and sum with respect to m by Ex. 17, p. 190.]

62. Shew that

$$\Sigma \frac{1}{n} \sin 2n\theta \sin^2 n\phi = \frac{1}{4}\pi, \quad \left\{ \begin{matrix} 0 < 2\theta < \pi \\ \theta < \phi < \pi - \theta \end{matrix} \right\}.$$

Deduce that

$$\Sigma \frac{1}{n^2} \sin^2 n\theta \sin^2 n\phi = \frac{1}{4}\pi\theta, \quad \Sigma \frac{1}{n^2} \sin^4 n\theta \sin^2 n\phi = \frac{1}{8}\pi\theta,$$

$$\Sigma \frac{1}{n^4} \sin^4 n\theta \sin^2 n\phi = \frac{1}{6}\pi\theta^3,$$

with certain restrictions on θ and ϕ. [H. N. Davis.]

63. If p is any integer and q is an integer not divisible by p, prove that

$$E\left(\frac{q}{p}\right) = \frac{q}{p} - \frac{1}{2} + \frac{1}{2p} \sum_{n=1}^{p-1} \sin(2qn\theta) \cot(n\theta),$$

where $E(x)$ denotes the integral part of x and $\theta = \pi r/p$, r being any integer prime to p.

Writing $r=2$, deduce that if p, q are odd,

$$\sum_{n=1}^{\frac{1}{2}(p-1)} E\left(\frac{nq}{p}\right) = \frac{q(p^2-1)}{8p} - \frac{p-1}{4} - \frac{1}{2p} \sum_{n=1}^{\frac{1}{2}(p-1)} \tan(qn\alpha) \cot(2n\alpha),$$

where $\alpha = \pi/p$. [Eisenstein.]

[See Ex. 21, p. 245, for the first part.]

64. Shew that if (a_n) and (b_n) tend to 0, the series

$$F(x) = \tfrac{1}{2} a_0 x^2 - \sum_1^\infty \frac{1}{n^2} (a_n \cos nx + b_n \sin nx)$$

is uniformly convergent for all values of x.

Prove that

$$\frac{1}{4\alpha^2} \{ F(x+2\alpha) + F(x-2\alpha) - 2F(x) \} = a_0 + \sum_1^\infty \left(\frac{\sin n\alpha}{n\alpha} \right)^2 (a_n \cos nx + b_n \sin nx).$$

Deduce from Ex. 4 and Art. 80 that

$$\lim_{\alpha \to 0} \frac{1}{4\alpha^2} \{ F(x+2\alpha) + F(x-2\alpha) - 2F(x) \} = a_0 + \sum_1^\infty (a_n \cos nx + b_n \sin nx)$$

for all values of x for which the right-hand side can be proved to converge. [Riemann, *Ges. Werke*, p. 232.]

65. Prove by means of Ex. 4 and Weierstrass's M-test that if (A_n) tends to zero, the series

$$\alpha \Sigma A_n \left(\frac{\sin n\alpha}{n\alpha} \right)^2$$

converges uniformly for all values of α.

Hence shew that with the notation of the last example,

$$\lim_{\alpha \to 0} \frac{1}{2\alpha} [F(x+2\alpha) + F(x-2\alpha) - 2F(x)] = 0,$$

provided that (a_n) and (b_n) tend to zero, *no matter whether the series* $\Sigma(a_n \cos nx + b_n \sin nx)$ *converge or not.* [Riemann, *l.c.*]

66. Deduce from the last two examples that if the series

$$a_0+\Sigma(a_n \cos nx + b_n \sin nx)$$

is equal to zero at all but a finite number of points in the interval $(0, 2\pi)$, the coefficients a_n, b_n are all zero.

[G. CANTOR, *Crelle's Journal*, Bd. 72; see PICARD, *Traité d'Analyse*, t. 1. Ch. IX.; and HOBSON, *Theory of Functions of a Real Variable*, §§ 480-5.]

Mittag-Leffler's method of representing functions.

67. If $F(x)$ tends to infinity at each point of an infinite sequence (a_n) in such a way that

$$\lim_{x\to a_n} (x-a_n)F(x)=A_n,$$

then we can write

$$F(x)=\Sigma\frac{A_n}{x-a_n}\left(\frac{x}{a_n}\right)^{m_n}+G(x),$$

where m_n is a positive integer chosen so as to make the series convergent, and $G(x)$ is a function devoid of singularity in the finite part of the plane. It is assumed that the sequence of moduli $|a_n|$ never decreases, and tends to infinity with n.

Shew that if $|A_n/a_n|^{\frac{1}{n}}$ remains less than a fixed value, m_n need never exceed n. [See Exs. 6-8, p. 255.]

68. (1) Indicate the connexion between Mittag-Leffler's representation and the series of partial fractions for $\pi \cot \pi x$, $\pi \operatorname{cosec} \pi x$ (see Art. 92).

(2) Prove that $\qquad \lim\,(x+n)\Gamma(x)=(-1)^n/n!, \qquad$ as x tends to $-n$,

and deduce that

$$\Gamma(x)=\frac{1}{x}+\sum_1^\infty\frac{(-1)^n}{n!(x+n)}+G(x).$$

Shew also that

$$G(x)=\int_1^\infty e^{-t}t^{x-1}dt.$$

(3) Indicate the relation of Mittag-Leffler's theorem to the equations (a being positive)

$$\psi(x)=\frac{\Gamma'(x)}{\Gamma(x)}=-\frac{1}{x}+\sum_1^\infty\frac{x}{n(x+n)}-C, \quad \text{(Ex. 42, p. 475.)}$$

$$\frac{\Gamma(x)\Gamma(a)}{\Gamma(x+a)}=\frac{1}{x}+\frac{1-a}{1}\,\frac{1}{x+1}+\frac{(1-a)(2-a)}{1.2}\,\frac{1}{x+2}+\ldots,$$

(Ex. 51, p. 477.)

69. (1) Shew, as in Ex. 3, Art. 90, that

$$\frac{2\pi i e^{ixy}}{e^{2\pi i x}-1}=\frac{1}{x}+\sum_{-\infty}^{\infty}{}'\frac{e^{iny}}{x-n}+G(x),$$

where $G(x)=0$, if $0<y<2\pi$, but $G(x)=-2\pi i e^{ix(y-2\pi)}$ if $2\pi<y<4\pi$. Indicate the relation of this result to Mittag-Leffler's representation.

(2) Prove also that if θ is positive,

$$-e^{\theta x}\frac{\Gamma'(x)}{\Gamma(x)}=\sum_0^\infty\frac{e^{-n\theta}}{x+n}+G(x),$$

where

$$G(x)=e^{\theta x}\left[C+\log(e^\theta-1)-\int_0^\theta\frac{1-e^{-\theta x}}{1-e^{-\theta}}d\theta\right].$$

Jacobi's Theta-functions.

70. If we put $x = \pm e^{2i\phi}$ in the functions of Ex. 16, Chap. V., the functions obtained are called theta-functions; thus

$$\vartheta_1(\phi) = g(-e^{2i\phi}, q) = 2q^{\frac{1}{4}} \sin\phi - 2q^{\frac{9}{4}} \sin 3\phi + 2q^{\frac{25}{4}} \sin 5\phi - \dots,$$
$$\vartheta_2(\phi) = g(\ e^{2i\phi}, q) = 2q^{\frac{1}{4}} \cos\phi + 2q^{\frac{9}{4}} \cos 3\phi + 2q^{\frac{25}{4}} \cos 5\phi + \dots,$$
$$\vartheta_3(\phi) = f(\ e^{2i\phi}, q) = 1 + 2q\cos 2\phi + 2q^4 \cos 4\phi + 2q^9 \cos 6\phi + \dots,$$
$$\vartheta_4(\phi) = f(-e^{2i\phi}, q) = 1 - 2q\cos 2\phi + 2q^4 \cos 4\phi - 2q^9 \cos 6\phi + \dots.$$

[There is some divergence as to notation; we adopt that of Tannery and Molk so far as the suffixes are concerned.]

71. From Ex. 16, Ch. V., prove that

$$\vartheta_1(\alpha, \sqrt{q})\vartheta_1(\beta, \sqrt{q}) = \vartheta_3(\phi)\vartheta_2(\psi) - \vartheta_2(\phi)\vartheta_3(\psi),$$

where $\alpha = \frac{1}{2}(\phi+\psi)$, $\beta = \frac{1}{2}(\phi - \psi)$; and obtain corresponding formulae for the products $\vartheta_2(\alpha)\vartheta_2(\beta)$, etc.

Deduce that

$$\vartheta_1(\alpha, \sqrt{q})\vartheta_2(0, \sqrt{q}) = 2\vartheta_1(\alpha)\vartheta_4(\alpha),$$
$$\vartheta_2(\alpha, \sqrt{q})\vartheta_2(0, \sqrt{q}) = 2\vartheta_2(\alpha)\vartheta_3(\alpha),$$

and obtain other similar formulae.

72. From Exs. 18, 20, Ch. VI., prove that

$$q_3^4\vartheta_4^2 = q_2^4\vartheta_3^2 - 4q^{\frac{1}{4}}q_1^4\vartheta_2^2, \quad q_3^4\vartheta_3^2 = q_2^4\vartheta_4^2 - 4q^{\frac{1}{4}}q_1^4\vartheta_1^2,$$
$$\vartheta_1'(0) = 2q^{\frac{1}{4}}q_0^3, \quad \vartheta_2(0) = 2q^{\frac{1}{4}}q_0q_1^2, \quad \vartheta_3(0) = q_0q_2^2, \quad \vartheta_4(0) = q_0q_3^2,$$
$$\vartheta_1'(0) = \vartheta_2(0)\vartheta_3(0)\vartheta_4(0), \quad \vartheta_3^4(0) = \vartheta_2^4(0) + \vartheta_4^4(0).$$

73. If $\quad F(x) = iq^{\frac{1}{4}}\frac{1-x^2}{x}\prod_1^\infty \left[(1-q^{2n}x^2)\left(1-\frac{q^{2n}}{x^2}\right)(1-q^{2n})\right],$

prove that

$$F(e^{i\phi}) = \vartheta_1(\phi),$$
$$F(q^n x) = (-1)^n F(x)/(q^{n^2}x^{2n}),$$

where n may be any integer, positive or negative.

Shew also that, as x tends to $1/q^n$,

$$\lim F(x)/(1-x^2q^{2n}) = (-1)^n iq^{\frac{1}{4}}q_0^3/q^{n^2},$$

where (as in Ex. 14, p. 105), q_0 denotes $\Pi(1-q^{2n})$.

Deduce by Mittag-Leffler's method that, if $|p|$ lies between 1 and $|1/q$

$$iq^{\frac{1}{4}}q_0^3\frac{F(px)}{F(x)F(p)} = \sum_{-\infty}^{\infty}\frac{x^2q^{2n}p^2}{1-x^2q^{2n}} + G(x).$$

[It may be proved by more advanced methods (see Tannery and Molk, *Fonctions Elliptiques*, t. 3, Art. 479) that $G(x)$ is zero here.]

74. Assuming that in the formula of the last example $G(x)$ is zero, shew that

$$\left(\frac{q_0q_2}{q_1q_3}\right)^2 = -2iq^{\frac{1}{4}}q_0^3\frac{F(i/\sqrt{q})}{F(i)F(1/\sqrt{q})} = 2\sum_{-\infty}^{\infty}\frac{q^n}{1+q^{2n}}.$$

Deduce that (see Ex. 18, p. 106),

$$1 + 4\sum_1^\infty \frac{q^n}{1+q^{2n}} = (1 + 2q + 2q^4 + 2q^9 + \dots)^2.$$

75. Deduce from Ex. 73 that if $|p|$ lies between 1 and $|1/q|$,

$$iq^{\frac{1}{4}}q_0{}^3\frac{F(px)}{F(p)F(x)}=\frac{x^2}{1-x^2}+\frac{1}{1-p^2}+\sum_1^\infty\frac{x^2(pq)^{2n}}{1-x^2q^{2n}}-\sum_1^\infty\frac{(q/p)^{2n}}{x^2-q^{2n}}.$$

Write $x=i$ and let p tend to i (after division by $1+p^2$), and deduce that

$$\left(\frac{q_0}{q_1}\right)^4=1+8\sum_1^\infty\frac{(-1)^n nq^{2n}}{1+q^{2n}},$$

or
$$(1-2q^2+2q^8-2q^{18}+\ldots)^4=1-8\left(\frac{q^2}{1+q^2}-\frac{2q^4}{1+q^4}+\frac{3q^6}{1+q^6}-\ldots\right),$$

where the law of the indices on the left is $2n^2$.

76. From Ex. 74 prove that

$$(1+2q+2q^4+2q^9+\ldots)^4=1+8\left(\frac{q}{1-q}+\frac{2q^2}{1+q^2}+\frac{3q^3}{1-q^3}+\ldots\right).$$

Deduce that every positive integer can be expressed as the sum of four squares. [JACOBI, *Ges. Werke*, Bd. 1, pp. 239, 247, 423.]

77. Writing $x=\sqrt{q}$, $p=i$ in Ex. 74, prove that

$$(1+2q+2q^4+2q^9+\ldots)^2=1+4\left(\frac{q}{1-q}-\frac{q^3}{1-q^3}+\frac{q^5}{1-q^5}-\ldots\right).$$

Deduce that any prime of the form $4k+1$ can be expressed (in one way only) as the sum of two squares; but no prime of the form $4k+3$ can be so expressed. [JACOBI, *l.c.*]

Series defining functions which have no analytical continuation.

78. Prove (as in Ex. 4, Art. 51) that if a is positive and $0<r<1$, then

$$\lim_{r\to1}\sqrt{(1-r)}\,\Sigma r^{(c+na)^2}=\tfrac{1}{2}\sqrt{\pi/a}.$$

Deduce that when a and b are positive integers and x is allowed to approach the point $\exp(2\pi ib/a)$ of the unit circle along the radius,

$$\Sigma x^{n^2}\sim P\sqrt{\{\pi/a(1-r)\}},\quad r=|x|,$$

where (see Exs. 13-17, p. 243)

$$P=\pm1,\text{ if } a=4k+1;\quad P=\pm i,\text{ if } a=4k+3;\quad P=\pm1\pm i,\text{ if } a=4k.$$

[The method fails when a is of the form $4k+2$; thus with $b=1$, $a=2$, we get
$$\Sigma x^{n^2}=\Sigma(-1)^n r^{n^2},$$
and so the series tends to the limit $\frac{1}{2}$ (Ex. 4, Art. 51).]

79. From the last example, prove that the function Σx^{n^2} has a singularity in every arc of the unit-circle, however small; and that *the function cannot be continued beyond the circle.*

Deduce from Ex. 74 a corresponding property for the function

$$\Sigma x^n/(1+x^{2n}).$$

[WEIERSTRASS, *Ges. Werke*, Bd. 2, p. 228.]

80. Shew that the functions

$$\Sigma x^{a^n}, \quad \Sigma x^{n!}$$

tend to infinity as x approaches the points

$$\exp(2\pi ib/a^m), \quad \exp(2\pi ib/m!),$$

respectively along the radii.

Deduce that these functions cannot be continued beyond the unit-circle.

81. Apply Ex. 31, p. 253, to prove that the function

$$f(x) = \Sigma n^{-n} x^{a^n}$$

has a singularity at every point of the type $\exp(2\pi ib/a^m)$. The function therefore cannot be continued beyond the unit-circle, although $f(x)$, $f_1(x)$, $f_2(x)$, ... (Art. 84) all converge absolutely at every point of the circumference. [PRINGSHEIM.]

82. Assuming Borel's theorem (see Ex. 83), prove that the function

$$\phi(x) = \Sigma p^n x^{n^2}, \quad (0 < p < 1),$$

cannot be continued beyond the unit-circle, although $\phi(x)$, $\phi_1(x)$, $\phi_2(x)$, ... all converge absolutely at every point of the circumference. [FREDHOLM.]

83. Although it is beyond the range of the methods given in this book, we state the following theorem, which includes Exs. 78-82 as special cases:

The function $\Sigma c_n x^{a_n}$ cannot be continued beyond the unit-circle, provided that the integers a_n increase fast enough to satisfy the condition

$$\varliminf (a_{n+1} - a_n)/\sqrt{a_n} > 0,$$

and that $|x| = 1$ is the circle of convergence.

A narrower form of the second condition is often convenient in practice, namely

$$\lim \frac{\log\{|c_n|/|c_{n+1}|\}}{a_{n+1} - a_n} = 0.$$

[BOREL, *Liouville's Journal de Math.* (5), t. 2, 1896, p. 441.]

84. Prove (by means of Ex. 79 or Ex. 83) that the function

$$f(x) = \sum_1^\infty n^{-4} x^{n^2}$$

has the unit-circle as a natural boundary.

Prove also that if $\quad |x| < 1$ and $|x_1| < 1$,

then

$$\left|\frac{f(x_1) - f(x)}{x_1 - x}\right| \geqq 1 - \sum_2^\infty \frac{1}{n^2} > \frac{1}{4};$$

and so shew that *the function inverse to $f(x)$ is single-valued.*

Extend the argument to the function

$$\phi(x) = \Sigma c_n x^{a_n}/a_n, \quad a_0 = 1,$$

where Σc_n is a convergent series of positive terms whose sum is less than $2c_0$ and the indices satisfy Borel's condition (Ex. 83).

85. It is evident that the series given in Exs. 78-84 have all large gaps in the sequence of indices; but this is not essential to secure that the function cannot be continued beyond the unit-circle.

Thus, no continuation exists for the function

$$\Sigma x^n + \Sigma x^{n^2},$$

which has no gaps in the sequence of indices.

[Compare PRINGSHEIM, *Math. Annalen*, Bd. 44, pp. 49-51.]

Kummer's Series for $\log \Gamma(1+x)$.

86. We find from Art. 180 that, if the real part of x lies between 0 and 1,

$$\log \frac{\Gamma(x)}{\Gamma(1-x)} = \int_0^\infty \left[(2x-1)e^{-t} + \frac{\sinh(\frac{1}{2}-x)t}{\sinh \frac{1}{2}t}\right]\frac{dt}{t}.$$

Also we have $\quad \dfrac{\sinh(\frac{1}{2}-x)t}{\sinh\frac{1}{2}t} = \sum_1^\infty \dfrac{8n\pi}{t^2+4n^2\pi^2}\sin(2n\pi x) \qquad$ (Ex. 22, p. 257)

and $\quad 1-2x = \sum_1^\infty \dfrac{2}{n\pi}\sin(2n\pi x). \qquad$ (Art. 65)

Thus, *assuming that the order of integration and summation can be inverted,** we find

$$\log\frac{\Gamma(x)}{\Gamma(1-x)} = 2\sum_1^\infty a_n \sin(2n\pi)x, \quad \text{if } a_n = \int_0^\infty\left(\frac{4n\pi}{t^2+4n^2\pi^2} - \frac{e^{-t}}{n\pi}\right)\frac{dt}{t}.$$

Now it is easy to verify that

$$\int_\delta^\infty\left(\frac{4p}{t^2+4p^2} - \frac{e^{-t}}{p}\right)\frac{dt}{t} = \frac{1}{2p}\log(4p^2+\delta^2) + \frac{1}{p}\left(\log\frac{1}{\delta} - \int_\delta^\infty e^{-t}\frac{dt}{t}\right),$$

so that $\quad \displaystyle\int_0^\infty\left(\frac{4p}{t^2+4p^2} - \frac{e^{-t}}{p}\right)\frac{dt}{t} = \frac{1}{p}[\log(2p)+C], \qquad$ (see p. 460)

where C is Euler's constant.

Thus $\quad \displaystyle\log\frac{\Gamma(x)}{\Gamma(1-x)} = 2\sum_1^\infty \frac{1}{n\pi}[C+\log(2n\pi)]\sin(2n\pi x),$

and $\quad \Gamma(x)\Gamma(1-x) = \pi/\sin(\pi x), \qquad$ (Art. 177)

so that we obtain Kummer's series for $\log\Gamma(x)$, namely,

$$(\tfrac{1}{2}-x)(C+\log 2) + (1-x)\log\pi - \tfrac{1}{2}\log\sin(\pi x) + \sum_1^\infty \frac{\log n}{n\pi}\sin(2n\pi x).$$

Power Series in Two Variables.

87. If the series $\Sigma a_{m,n}x^m y^n$ is absolutely convergent for $x=x_0$, $y=y_0$, prove that it is absolutely convergent for all values of x and y, such that

$$|x|<|x_0|, \quad |y|<|y_0|. \qquad \text{[ABEL.]}$$

Prove that it is also uniformly convergent for all values of x and y, such that

$$|x| \leqq |x_0|-\delta, \quad |y| \leqq |y_0|-\delta,$$

where δ is any positive number. [WEIERSTRASS.]

* The theorems A, B in Art. 176 do not appear to be sufficiently delicate to justify this step; but a justification can be obtained from theorem C, although the analysis appears to be rather tedious.

88. Suppose that r is given; then there will be a maximum value of r', such that the series is absolutely convergent within the circles $|x|=r$, $|y|=r'$. This maximum value of r' may be regarded as a function of r, given by a relation

$$\phi(r, r')=0;$$

and r, r' are then called a *pair of associated radii of convergence*.

Let
$$\lambda(k)=\overline{\lim_{n\to\infty}} \sqrt[n]{a_{p,q}k^q},$$
where $p+q=n$. Then prove that

$$r=\frac{1}{\lambda(k)}, \quad r'=\frac{k}{\lambda(k)}$$

are a pair of associated radii. Eliminating k, shew that the relation between r and r' is

$$r\lambda\left(\frac{r'}{r}\right)=1.$$

[LEMAIRE, *Bull. des sci. math.*, t. 20, 1896, p. 286.]

89. Prove that the series $\quad \Sigma 2^{-n-1} x^m y^n$
is convergent if
$$|x|<1, \quad |y|<2,$$
and deduce that r' is defined as a function of r by the equations

$$r'=2, \quad 0 \leqq r<1; \quad r'=0, \quad r \geqq 1.$$

For the series
$$\Sigma \frac{m+n!}{m!\,n!} x^m y^n,$$
shew that the relation is $r+r'=1$.

90. Evidently $r'=\phi(r)$ is a non-increasing function of r Let R denote the upper limit of the values of r for which $\phi(r)>0$, so that

$$\phi(r)>0, \quad 0 \leqq r<R; \quad \phi(r)=0, \quad r>R.$$

Shew that for the series which is the expansion of

$$\frac{1}{(3-x)(1-xy)(2-y)},$$

we have $R=3$, and

$$r'=2, \ (r \leqq \tfrac{1}{2}); \quad r'=1/r, \ (\tfrac{1}{2}<r<3); \quad r'=0, \ (r \geqq 3).$$

[For an application of this theory to the differential equation

$$\frac{dy}{dx}=\Sigma a_{m,n} x^m y^n,$$

see GOURSAT, *Bull. de la Soc. math. de France*, t. 35, 1907, p. 81.]

91. The function $r'=\phi(r)$ satisfies the inequality

$$\begin{vmatrix} 1, & \log r_1, & \log \phi(r_1) \\ 1, & \log r_2, & \log \phi(r_2) \\ 1, & \log r_3, & \log \phi(r_3) \end{vmatrix} \leqq 0,$$

where $0<r_1<r_2<r_3<R$.

[FABRY, *Comptes Rendus*, t. 137; HARTOGS, *Math. Annalen*, Bd. 62.]

92. From Ex. 91 it may be deduced that $\phi(r)$ is continuous for any value of r in $(0, R)$, the limits excluded. For an example of discontinuity for $r=R$, see Ex. 90 above. For one of discontinuity for $r=0$, consider

$$x+xy+xy^2+xy^3+\dots,$$

for which $\qquad \phi(r)=\infty, \ (r=0), \quad \phi(r)=1, \ (r>0);$

here $R=\infty$.

93. Shew also that $\phi(r)$ can only be constant (and not zero) in an interval beginning with $r=0$ (though, as Ex. 92 shews, not necessarily including $r=0$).

[For solutions of Exs. 91-93, see HARTOGS, *l.c.*]

94. The series $\Sigma a_{m,n}x^m y^n$ may be convergent, though not absolutely, for special sets of values of x and y outside the region of absolute convergence. Thus, if $a_{m,n}$ is given by the scheme of Ex. 2, p. 90, with $a_n=b_n=2^n$, the power series, which is the expansion of

$$(1-y)/(1-2x)+(1-x)/(1-2y),$$

converges for $x=1$, $y=1$; but for absolute convergence we must have

$$|x|<\tfrac{1}{2}, \quad |y|<\tfrac{1}{2}.$$

[BROMWICH AND HARDY, *Proc. Lond. Math. Soc.*, vol. 2, 1904, p. 161.]

95. This kind of convergence can only present itself for particular, isolated pairs of values of x or y.

[HARTOGS, *Inauguraldissertation*, Leipzig, 1904, p. 21.]

96. The series may converge in a wider region than that of absolute convergence, when summed by rows, or columns, or diagonals. Thus prove that the series for $1/(1-x-y)$ converges absolutely if $|x|+|y|<1$; it converges by diagonals if $|x+y|<1$; and it converges by rows if

$$|x|<1, \quad |y|<|1-x|;$$

or by columns if $\qquad |y|<1, \quad |x|<|1-y|.$

Represent the regions of convergence (x, y being real) in a diagram.

[CAUCHY.]

97. Let $f(x)=\sum_0^\infty c_n x^n$ be a series convergent for $|x|<1$. Then the double power series

$$\begin{aligned} f(x+y)= \ & c_0 + c_1 x + c_2 x^2 + \dots, \\ & c_1 y + 2c_2 xy + 3c_3 x^2 y + \dots, \\ & + c_2 y^2 + 3c_3 xy + 6c_4 x^2 y^2 + \dots, \\ & + \dots \end{aligned}$$

converges absolutely if $|x|+|y|<1$, by diagonals if $|x+y|<1$, by rows if $|x|<1$, $|y<1-x|$, and by columns if $|y|<1$, $|x|<|1-y|$.

[HARTOGS.]

98. If the series $\Sigma\Sigma a_{m,n}$ satisfies the condition

$$\left|\sum_{m=0}^{\mu}\sum_{n=0}^{\nu} a_{m,n}\right| < K$$

for all values of μ, ν, and if any one of the three series

$$\sum_{(m,n)} a_{m,n}, \quad \sum_{(m)}\sum_{(n)} a_{m,n}, \quad \sum_{(n)}\sum_{(m)} a_{m,n}$$

is convergent, the corresponding one of the three limits

$$\lim_{(x,y)} f(x, y), \quad \lim_{(x)(y)} f(x, y), \quad \lim_{(y)(x)} f(x, y)$$

is determinate and equal to the sum of the series. In these limits x, y are supposed to approach the common limit 1, simultaneously or successively.

[Bromwich and Hardy, *l.c.*, p. 168.]

99. Extend the last example to cases in which the series $\Sigma a_{m,n}$ can only be summed by a mean-value process similar to Cesàro's.

[Bromwich and Hardy, *l.c.*, p. 173.]

100. An example of the theorems of Ex. 98 is given by the coefficients of Ex. 13, p. 92, for which

$$\Sigma a_{m,n} x^m y^n = (x-y)/(2-x-y).$$

An example of the mean-value method is given by writing

$$a_{m,n} = (-1)^{m+n} mn/(m+n)^2.$$

[Bromwich and Hardy, *l.c.*, pp. 169, 175.]

[For various extensions of the results given in Exs. 98-100, see Bromwich, *Proc. Lond. Math. Soc.*, vol. 6, 1907, pp. 67, 74.]

INDEX OF SPECIAL INTEGRALS, PRODUCTS AND SERIES.

(The numbers refer to pages, not to articles.)

Asymptotic series.

Euler's constant, 322; Stirling's series, 324, 329, 335, 355; logarithmic integral, 325; Fresnel's integrals, 327; Bessel's functions, 345.

Elliptic function series.

Theta-functions, 93, 106, 107, 350, 500; various series of fractions, 94, 114, 172, 353, 355, 472, 480.

Integrals.

Beta, 474.
Borel's, 268 *et seq.*; 301 (extension).
Dirichlet's, 444-447, 471.
Elliptic, 17, 162, 474.
Error function, 352, 353, 473, 482.
Euler's constant, 460.
Exponential or Logarithmic, 325, 424, 440.
Fejér's, 442.
Fourier's, 472.
Fresnel's, 327.
Frullani's, 432, 478.
Gamma, 459, 463.
Jordan's, 447.
Poisson's, 210, 212.
Sine and cosine, 351, 446; complete sine integral, 418, 419, 441, 468, 471.

Non-convergent series.

$\Sigma(-x)^n$, 274, 298, 303, 318; $\Sigma(-1)^n \log n$, 310, 351; $\Sigma n^p \cos n\theta$, etc., 275, 288, 317; $\Sigma(-x)^n n!$, 267, 327, 482.

See also under Fourier series in general Index.

Numerical series.

$\Sigma 1/n!$, 22; $\Sigma 1/n$, 23, 30, 44, 92, 322, 353, 413; $\Sigma 1/n^p$, $(p>1)$, 30, 71, 93, 187, 235, 324, 353; $\Sigma(\log n)^p/n^q$, 31, 413; $\Sigma(-1)^n/n$, 51, 63, 70, 85, 121, 153; $\Sigma(-1)^n/n^p$, 51, 70, 71, 85, 479.

Double series, $\Sigma m^{-\alpha} n^{-\beta}$, $\Sigma(m+n)^{-\alpha}$, $\Sigma f(am^2+2bmn+cn^2)$, 79.

Series for π, 156, 157, 168, 324 (Ex. 7).

Numerical values.

$\Sigma 1/n!$, 22; $\Sigma 1/n$ and Euler's constant, 323; $\Sigma(-1)^n/n$, 58, 155; $\Sigma(-1)^n/(2n+1)$, 156; $\Sigma(-1)^n/(an+b)$, 161, 481; $\Sigma(-1)^n/\sqrt{n}$, 57; $\Sigma 1/n^2$, 59, 71, 93, 187, 324; $\Sigma(-1)^n/(2n+1)^2$, 479; $\Sigma 1/n^3$, 59, 324; $\Sigma 1/n^{2r}$, 234, 324, 481.

$\log 2$, $\log 3$, ... 154, 480; $\sqrt{2}$, $\sqrt[3]{2}$, 171; $\sin(\frac{1}{2}\pi)$, $\cos(\frac{1}{2}\pi)$, $e^{\frac{1}{2}\pi}$, e^{π}, 146, 147; $\Gamma(\frac{4}{3})$, $\Gamma(\frac{5}{4})$, $\Gamma(1+i)$, 475.

Power-series.

Binomial, 89, 150, 225; sum of squares of coefficients in, 166 (Ex. 33), 249 (Ex. 21).

Exponential, 143, 217, 406.

Geometric, 15.

Hypergeometric, 35, 52; limit as $x \to 1$, 42 (Ex. 2), 105 (Ex. 13), 161 (Ex. 5), 171 (Ex. 13).

Inverse sine and tangent, 155, 169, 170, 224.

Lagrange's, 140, 174 (Exs. 30, 31), 487 (Exs. 19, 20).

Logarithmic, 152, 162, 223.

Sine and cosine, 146, 221.

Theta, 93, 106, 107, 350, 500.

$\Sigma n^{p-1}x^n$, Σx^{n^2}, $\Sigma(-1)^n x^{n^2}$, Σx^{a^n}, $\Sigma x^{n!}$, 133, 389, 489, 501.

$x/(e^x-1)$, 234, $(e^{xt}-1)/(e^x-1)$, 235.

Products.

$\Pi(1 \pm q^n)$, $\Pi\{(1+q^{2n-1}x)(1+q^{2n-1}/x)\}$, 105, 106.

Gamma-product, 102, 461, 463; sine and cosine, 184, 231.

Wallis's product, 184.

Trigonometrical series.

Convergence of $\Sigma v_n \cos n\theta$, $\Sigma v_n \sin n\theta$, 50; $\Sigma \sin(n^2\theta)$, $\Sigma \sin(2^n\theta)$, $\Sigma \sin(n!\theta)$, 485, 486.

$\Sigma r^n \begin{pmatrix}\cos\\ \sin\end{pmatrix}(n\theta)$, $\Sigma \frac{r^n}{n}\begin{pmatrix}\cos\\ \sin\end{pmatrix} n\theta$, 157.

$\Sigma \frac{\sin n\theta}{n}$, 159, 160, 189, 230, 290, 480; $\Sigma \frac{\cos n\theta}{n}$, 159, 230, 290.

$\Sigma \frac{\cos n\theta}{n^p}$, 114, 115, 167, 168, 256, 257, 480, 498; $\Sigma\left(\frac{\sin na}{na}\right)^2$, 484.

$\pi\cot(\pi x) = \frac{1}{x} + \Sigma \frac{2x}{x^2-n^2}$, $\pi\operatorname{cosec}(\pi x) = \frac{1}{x} + \Sigma \frac{(-1)^n 2x}{x^2-n^2}$, 187-190, 231-233.

$\Sigma 1/(x-n)^r$, 190, 475, 476.

$\Sigma \frac{\cos nx}{n^2+a^2}$, $\Sigma \frac{n \sin nx}{n^2+a^2}$, 257; $\Sigma \frac{\cos(n+a)x}{(n+a)^2+b^2}$, 290.

See also under Fourier series in general Index.

Zeta series, 493-495.

GENERAL INDEX.*

(*The numbers refer to pages, not to articles.*)

Abel's Lemma, 54, 205 (for series); 381 (extension); 426 (for integrals).

Abel's test for convergence, 48, 205 (of series); 113, 207 (uniform); 429, 434 (of integrals).

Abel's theorem on continuity of power-series, 130, 210, 251; 172 (Exs. 16, 17), 488 (Ex. 25), (extensions of); 291 (for summable series); 506 (Exs. 98-100) (for double series).

Absolute convergence, 47 (of real series); 81 (of double series); 198 (of complex series); 429 (of integrals).

Absolute summability, 276-284.

Area of a curve, arithmetic treatment, 407, 409.

Asymptotic series, 322-329 (Euler's method); 330-337 (Poincaré's theory); 338-346 (Borel's theory).

Bendixson's test for uniform convergence, 127.

Bernoulli's numbers, 234; functions, 235.

Bessel's functions, 345.

Borel's theory of non-convergent series, 267-307, 338-346.

Cauchy's double series theorem, 67, 87.

Cauchy's inequalities for coefficients of a power-series, 209, 247.

Cauchy's tests for convergence, 24-27, 80; see also under Maclaurin.

Cesàro's mean-value, 310-322.

Cesàro's theorem, 132 (on comparison of divergent series), 171 (Ex. 15); 314 (on multiplication of series).

Complex series, convergence of, 197 (general principle); 200 (Pringsheim's tests); 204-206 (Abel's and Dirichlet's tests).

Continuity, 115 (of series); 286 (of summable series); 395 (of functions); 436 (of integrals).

Convergence, absolute, see Absolute convergence.

Convergence, circle of, 208.

Convergence, general principle of, 8 (for real sequences); 196 (for complex sequences); 376 (proof).

Convergence, interval of, 128.

Convergence, uniform, see Uniform convergence.

Decimals, infinite, 357-366.

Dedekind's definition of irrational numbers, 366-370.

Definite integral, as limit of a sum, 407, 409, 417, 495 (Ex. 50).

Derangement of series, 63-67; 68 (Riemann's and Pringsheim's theorems).

Differential equations, existence theorems, 174-176 (simple cases of results due to Cauchy, Fuchs and Picard).

Differentiation, 118 (of series); 228 (of trigonometrical series); 287 (of summable series); 437-439 (of infinite integrals).

* No attempt has been made to index references to authors; but where a theorem is usually quoted in the text under an author's name, it appears under that name in the index.

Dirichlet's integrals, 444, 445, 471.

Dirichlet's test for convergence, 49, 206 (of series); 114, 207 (uniform); 430, 435 (of integrals).

Double integrals, 410 (arithmetic definition); 456 (inversion of repeated).

Double series, 73-89 (convergence); 76-79 (of positive terms); 503-506 (power-series).

Ermakoff's tests of convergence, 37.

Euler's constant, 30, 323, 460.
  summation formula, 238, 322.
  transformation, 55, 169, 302.

Exponential function, 145, 402 (real variable); 217 (complex variable).

Exponential series, 143, 217, 406.

Féjer's theorem on Fourier series, 348.

Fourier integrals, 447, 471, 472.

Fourier series, 167, 168, 189, 230, 256, 257, 263, 472, 480-482, 498, 503; non-convergent, 275, 276, 284, 287-290, 347-350, 354.

Fresnel's integrals, 327.

Frobenius's theorem, 132, 313 (extension).

Functions, 395 (continuous); 490 (without derivates); 501 (without continuation). For other special functions, see under Exponential, Gamma, Logarithmic, Zeta, etc.

Gamma function, 102 (product); 459 (integral); 461 (Stirling's formula); 463 (formulae for logarithm); 461, 474, 478 (miscellaneous properties); 503 (Kummer's series).

Goursat's Lemma, 394.

Huygens' zones in Optics, 58.

Infinite integral, 414 (limit of an integral); 417, 495 (limit of a sum); 420, 429 (tests for convergence).

Integration, 116, 448, 452 (of series); 286 (of summable series); 437 (of integrals).

Inversion of repeated integrals, 410, 456.

Irrational numbers, 358 (as decimals); 366-370 (Dedekind's definition).

Jacobi's theta-functions, 500.

Jordan's extension of Dirichlet's integrals, 447.

Kummer's series, 503.

Kummer's tests for convergence, 32.

Lagrange's series, 140, 172, 249, 487.

Le Roy's methods for summation of non-convergent series, 299.

Limits, 2 (definition); 9 (rules of combination); 246 (of point-sets); 377 (of quotients); 383-389 (miscellaneous theorems).

Limits, maximum and minimum (or extreme), 12, 375 (of sequences); 394 (of infinite sets).

Limits, upper and lower, 11 (of sequences); 394 (of infinite sets).

Logarithmic function, 221 (complex variable); 396-402 (real variable).

Logarithmic scale of infinity, 405.

Maclaurin's theorem* connecting the convergence of series and integrals, 29; for double series, 80.

Mertens' theorem, 85; 284 (for non-convergent series).

Mittag Leffler's theorem, 499.

Monotonic sequences, 5; 373 (proof of convergence); 393 (quasi-monotonic sequences).

Multiplication of series, 66, 82-86; 280, 284, 314, 331, 341 (non-convergent series).

Non-convergent series, 261-267 (general remarks); see also under Summable Series, Asymptotic Series.

Numbers, irrational, 358 (as decimals); 366-370 (Dedekind's definition).

Poincaré's theory of asymptotic series, 330-337.

Poisson's integral, 210, 212.

Power-series, 128-142 (real); 202-216 (complex); 293-296 (summable).

Pringsheim's tests of convergence, 200.

Pringsheim's theorem on multiplication of series, 86.

Products, infinite, 96-102 (real); 197-199 (complex).

Repeated integrals, inversion of order of integration, 410, 456.

Riemann's theorem on derangement of series, 68.

Riemann's theorems on Fourier series, 498 (Exs. 64-66).

Riemann's Zeta function, 493-495.

*Commonly attributed to Cauchy; it occurs in Maclaurin's *Fluxions*, 1742, Art. 350.

Second theorem of mean value, 426.

Stirling's series, 324, 329, 335, 355, 461.

Stirling's test for convergence of products, 104 (Ex. 3).

Summable series, 269 (Borel's integral); 271 (addition of terms); 297 (Borel's limit); 299 (Le Roy's methods); 301 (extension of Borel's integral); 302 (Euler's method); 310 (Cesàro's method).

Symbols, ∞, 4; ϵ, 5; $\rightarrow$, 2; $\underline{\lim}$, $\overline{\lim}$, 13; $\sim$, 11, 330.

Tannery's theorem, 123 (series); 124 (products); 438 (integrals); 443 (extension).

Taylor's theorem, 214, 252.

Tests for convergence, see under Abel, Bendixson, Cauchy, Dirichlet, Ermakoff, Kummer, Maclaurin, Pringsheim, Stirling, Weierstrass. General remarks on the tests, 35, 40, 422, 467, 485.

of series, 26, 27, 29, 32-34, 37, 39, 47-49.

Tests for convergence, of double series, 79, 80, 89.

of integrals, 421, 422, 429, 430, 434, 435.

of products, 96, 99, 101, 199.

Trigonometrical formulae, 177, 178 (for $\cos n\theta$ and $\sin n\theta$); 184, 231 (product for $\sin\theta$, $\cos\theta$); 186, 232 (series for $\cot\theta$, $\operatorname{cosec}\theta$); 195 (de Moivre's theorem); 221 (definition of $\sin x$, $\cos x$, when x is complex); $|\sin x| < \frac{6}{5}|x|$, $|\cos x| < 2$, when $|x| < 1$, 221.

Uniform convergence, 108 (sequence); 112, 206 (series); 113, 114, 127, 207 (tests); 121 (products); 433-435 (integrals).

Weierstrass's tests for convergence, 113 (uniform); 204 (power-series); 434 (integrals).

Weierstrass's theorem on double series, 253.

Zeta function, (Riemann's), 493-495.

Lightning Source UK Ltd.
Milton Keynes UK
UKHW030607211019
351998UK00006B/701/P